Springer Series on Naval Architecture, Marine Engineering, Shipbuilding and Shipping

Volume 19

Series Editor

Nikolas I. Xiros, University of New Orleans, New Orleans, LA, USA

The Naval Architecture, Marine Engineering, Shipbuilding and Shipping (NAMESS) series publishes state-of-art research and applications in the fields of design, construction, maintenance and operation of marine vessels and structures. The series publishes monographs, edited books, as well as selected PhD theses and conference proceedings focusing on all theoretical and technical aspects of naval architecture (including naval hydrodynamics, ship design, shipbuilding, shipyards, traditional and non-motorized vessels), marine engineering (including ship propulsion, electric power shipboard, ancillary machinery, marine engines and gas turbines, control systems, unmanned surface and underwater marine vehicles) and shipping (including transport logistics, route-planning as well as legislative and economical aspects).

The books of the series are submitted for indexing to Web of Science.

All books published in the series are submitted for consideration in Web of Science.

Alexander Olsen

Ship Operations in Extreme Low Temperature Environments

Springer

Alexander Olsen
Southampton, UK

ISSN 2194-8445 ISSN 2194-8453 (electronic)
Springer Series on Naval Architecture, Marine Engineering, Shipbuilding and Shipping
ISBN 978-3-031-52512-4 ISBN 978-3-031-52513-1 (eBook)
https://doi.org/10.1007/978-3-031-52513-1

This Springer imprint is published by the registered company Springer Nature Switzerland AG
The registered company address is: Gewerbestrasse 11, 6330 Cham, Switzerland

Paper in this product is recyclable.

Preface

The operation of merchant and naval vessels in low temperature environments presents many challenges for designers, builders, owners, and operators. These challenges include both hardware issues related directly to the construction, outfitting, and operation of vessels and those issues pertaining to the ability of the crew to function in a difficult environment. Moreover, operations in high latitudes provide unique challenges to planning, seamanship, ingenuity, endurance, and foresight. The elements, always dangerous, become hostile. Mountainous seas, storm force winds, and near-zero visibility for days on end put tremendous strain on mariners and their vessels. The Arctic has been defined in a variety of ways; for ease, this book has adopted the NATO definition of the Arctic polar region which includes the areas surrounding the geographic North Pole consisting of a deep central basin; the peripheral shallow seas (Bering, Chukchi, East Siberian, Laptev, Kara, Barents, and Norwegian); ice-covered portions of the Greenland and Norwegian Seas; Baffin Bay, Canadian Archipelago, Seas of Japan and Okhotsk; the continental margins of Canada and Alaska; and the Beaufort Sea.

To assist the marine industry, this book has been written to provide guidance to those employed onboard vessels operating in extreme low temperature environments and for those who are shore based but have a tangible interest in ensuring the vessels under their responsibility are operated in accordance with Class mandated requirements. To that end, this book consists of four main parts, with each part relating to a specific element of extreme low temperature maritime operations. Part I considers the geography and oceanography of the Arctic, and the kind of conditions mariners should expect when working in these northern latitudes. Part II examines the Class specific requirements pertaining to vessel design, hull structure materials, welding and coatings, equipment, systems and machinery, safety systems and specific vessel type requirements. Part III examines the Class rules for Polar Class vessels including the structural and machinery requirements; requirements needed to meet the enhanced Polar Class notation, and requirements for vessels who are intended to navigate in first-year ice. This part also provides important guidance for vessel owners and operators seeking a Baltic Ice Class notation. Part IV has been written to provide vessel crew members with essential guidance and support to their health, wellbeing, and safety whilst living and working in extreme low temperature environments.

It is anticipated this book will be of interest, and use, to mariners around the world and especially those who are tasked with the unenviable and difficult task of living and working in the most extraordinarily difficult climates.

Southampton, UK
2023

Alexander Olsen

The original version of the book has been revised. A correction to this book can be found at https://doi.org/10.1007/978-3-031-52513-1_18

Temperature Conversion Table

°C	°F	°C	°F	°C	°F	°C	°F
−50	−58.0	−10	14	26	78.8	66	150.8
−49	−56.2	−9	15.8	27	80.6	67	152.6
−48	−54.4	−8	17.6	28	82.4	68	154.4
−47	−52.6	−7	19.4	29	84.2	69	156.2
−46	−50.8	−6	21.2	30	86.0	70	158.0
−45	−49.0	−5	23	31	87.8	71	159.8
−44	−47.2	−4	24.8	32	89.6	72	161.6
−43	−45.4	−3	26.6	33	91.4	73	163.4
−42	−43.6	−2	28.4	34	93.2	74	165.2
−41	−41.8	−1	30.2	35	95.0	75	167.0
−40	−40	Freezing Point of Water		36	96.8	76	168.8
−39	−32.2			37	98.6	77	170.6
−38	−36.4			38	100.4	78	172.4
−37	−34.6	0	32	39	102.2	79	174.2
−36	−32.8			40	104.0	80	176.0
−35	−31.0	+1	33.8	41	105.8	81	177.8
−34	−29.2	2	35.6	42	107.6	82	179.6
−33	−27.4	3	37.4	43	109.4	83	181.4

(continued)

(continued)

°C	°F	°C	°F	°C	°F	°C	°F
−32	−25.6	4	39.2	44	111.2	84	183.2
−31	−23.8	5	41.0	45	113.0	85	185.0
−30	−22.0	6	42.8	46	114.8	86	186.8
−29	−20.2	7	44.6	47	116.6	87	188.6
−28	−18.4	8	46.4	48	118.4	88	190.4
−27	−16.6	9	48.2	49	120.2	89	192.2
−26	−14.8	10	50.0	50	122.0	90	194.0
−25	−13.0	11	51.8	51	123.8	91	195.8
−24	−11.2	12	53.6	52	125.6	92	197.6
−23	−9.4	13	55.4	53	127.4	93	199.4
−22	−7.6	14	57.2	54	129.2	94	201.2
−21	−5.8	15	59.0	55	131.0	95	203.0
−20	−4.0	16	60.8	56	132.8	96	204.8
−19	−2.2	17	62.6	57	134.6	97	206.6
−18	−0.4	18	64.4	58	136.4	98	208.4
−17	1.4	19	66.2	59	138.2	99	210.2
−16	3.2	20	68.0	60	140.0		
−15	5	21	69.8	61	141.8	Boiling Point of Water	
−14	6.8	22	71.6	62	143.6		
−13	8.6	23	73.4	63	145.4		
−12	10.4	24	75.2	64	147.2	100	212
−11	12.2	25	77.0	65	149.0		

Knots to Kilometres/Miles per Hour Conversion Table

5	Knots	=	5.8	Mi/h
10	Knots	=	11.5	Mi/h
15	Knots	=	17.3	Mi/h
20	Knots	=	23.0	Mi/h
25	Knots	=	28.8	Mi/h
30	Knots	=	34.6	Mi/h
35	Knots	=	40.3	Mi/h
40	Knots	=	46.1	Mi/h
45	Knots	=	51.8	Mi/h
50	Knots	=	57.6	Mi/h
55	Knots	=	63.4	Mi/h
60	Knots	=	69.1	Mi/h
65	Knots	=	74.9	Mi/h
70	Knots	=	80.6	Mi/h
75	Knots	=	86.4	Mi/h
80	Knots	=	92.2	Mi/h
85	Knots	=	97.9	Mi/h
90	Knots	=	103.7	Mi/h
95	Knots	=	109.4	Mi/h
100	Knots	=	115.2	Mi/h
105	Knots	=	121.0	Mi/h
110	Knots	=	126.7	Mi/h
115	Knots	=	132.5	Mi/h
120	Knots	=	138.2	Mi/h

(continued)

(continued)

125	Knots	=	144.0	Mi/h
130	Knots	=	149.8	Mi/h
135	Knots	=	155.5	Mi/h
140	Knots	=	161.3	Mi/h
145	Knots	=	167.0	Mi/h
150	Knots	=	172.0	Mi/h
200	Knots	=	230.15	Mi/h
300	Knots	=	345.23	Mi/h
400	Knots	=	460.31	Mi/h
500	Knots	=	575.38	Mi/h

Nautical Miles/Kilometres/Miles Conversion Table

Nautical miles	=	Kilometres	=	Miles
1	=	1.85	=	1.15
2	=	3.70	=	2.30
3	=	5.55	=	3.45
4	=	7.40	=	4.60
5	=	9.26	=	5.75
6	=	11.11	=	6.90
7	=	12.96	=	8.05
8	=	14.81	=	9.20
9	=	16.66	=	10.35
10	=	18.52	=	11.50
15	=	27.78	=	17.26
20	=	37.04	=	23.01
25	=	46.30	=	28.76
30	=	55.56	=	34.52
35	=	64.82	=	40.27
40	=	74.08	=	46.03
45	=	83.34	=	51.78
50	=	92.60	=	57.53
60	=	111.12	=	69.04
70	=	129.60	=	80.55
80	=	148.16	=	92.06
90	=	166.68	=	103.57
100	=	185.20	=	115.07

(continued)

(continued)

Nautical miles	=	Kilometres	=	Miles
200	=	370.40	=	230.15
300	=	555.60	=	345.23
400	=	740.80	=	460.31
500	=	926.00	=	575.39
1000	=	1852.00	=	1150.78

Contents

Part III Polar Class Notations

List of Figures

List of Tables

Part I
The Arctic Environment

Chapter 1
Introduction

1.1 Introduction

The guidance provided in this book is designed to apply to all vessels that are designed, equipped and intended to operate in low temperature environments. Special attention is given to those vessels operating for extended periods in the Arctic regions, as this presents specific and unique challenges for vessels and crew members. The application of the guidance in this book is optional. When a vessel is designed, equipped, built and surveyed in accordance with the relevant Class rules, and when found satisfactory during Class notation survey, a classification notation may be granted which demonstrates the vessel's compliance with the appropriate Class requirements for vessels operating in low temperature environments. Those vessels that are designed to meet the requirements of an ice class are typically required to meet specific Class rules around "strengthening for navigation in ice" or other equivalent and recognised Ice Class Rules. Noting that regional climatic conditions usually vary, this book provides guidance on notations related to the vessel's operating area. For example, for vessels intended to operate in Polar regions (Arctic or Antarctic) on a year-round basis, the requirements set by Class will be different from the requirements for a vessel operating in the Polar region only during the local summer period. Similarly, other vessels may be non-ice class yet operate in low temperatures. Subsequently, this book recognises the fact that the vessels' intended operational profile may vary as some vessels are intended to operate with the assistance of an ice breaker and others are intended to operate independently. Accordingly, this book also provides guidance related to the requirements which address the duration of emergency electrical power. This extended emergency power duration is expressed in hours and may be appended to the base optional class notations. To provide as much context as possible, this book refers to the most relevant international regulations and standards that are considered to be applicable. It is recommended that readers of this book refer to the most recent text of those regulations and standards when seeking to

A. Olsen, *Ship Operations in Extreme Low Temperature Environments*, Springer Series on Naval Architecture, Marine Engineering, Shipbuilding and Shipping 19,
https://doi.org/10.1007/978-3-031-52513-1_1

apply the guidance set out herein, as it is the intent of the book to remain consistent with the pertinent international regulations and standards developed by the global maritime industry.

1.2 Objective of the Book

The objective of this book is to provide supplemental guidance that is not addressed by ice class Rules. Low temperature environments present numerous challenges related to the operation of maritime equipment, systems, and structures, as well as vessel maintenance and safety equipment. Additionally, the vessel's crew performance is often adversely affected by low ambient temperatures and weather conditions prevalent in the most northern and southern maritime regions. Vessels designed and constructed without addressing the effects of low temperatures may experience increased structural and equipment failures and non-functioning systems. Personnel performance will typically be reduced by the effects of low temperatures. Vessel systems and exposed structures designed and equipped in accordance with the guidance in this book may obtain a Class notation indicating the vessel is designed to operate in a low temperature environment which may include the presence of ice and/or at a design service temperature specified by the vessel owner/operator. This book is therefore intended to provide guidance and information to naval architects, vessel designers, vessel owners/operators and ship's personnel where there is a design service temperature of −10 °C (14 °F) or less, excluding ice class requirements, if specified. It is recognised that vessels operating in low temperatures may have unique operating characteristics for which the requirements in this book may not be applicable. For such cases, Class will usually consider alternative arrangements, provided substantiating information and/or a risk analysis is submitted to Class for their review.

1.3 Classification Notations

Where requested by the vessel owner/operator, vessels designed, equipped, built and surveyed in accordance with the requirements stipulated by Class (and discussed in this book) may be assigned Class notations which are based on the following indicative examples:

- **CCO (TDST, TMAT)** is a basic notation for vessels operating in a low temperature and is applicable to vessels designed, constructed, and surveyed in accordance with the guidance contained in Chaps. 2 through 7. The design service temperature and the minimum anticipated temperature in °C for which the vessel's structure and exposed machinery are designed respectively are annotated in place of "**TDST**, **TMAT**" in the parentheses. Included in this notation are vessels which typically operate in the following low temperature environments: the Baltic region,

Okhotsk Sea, Gulf of St. Lawrence, etc.). Vessels which operate in a Polar region during the local summer season exclusively may also be eligible for this optional notation.

- **CCO-POLAR (TDST, TMAT)** is a notation available for those vessels intended to operate in Polar regions on a continuous basis. The requirements contained in Chaps. 2 through 7 are applicable to vessels seeking this optional notation. The design service temperature and the minimum anticipated temperature in °C for which the vessel's structure and exposed machinery are designed respectively are annotated in place of "**TDST**, **TMAT**" in the parentheses.
- **(HR HOURS)** notation can be appended to the **CCO-POLAR (TDST, TMAT)** notation to indicate that a vessel is equipped and provided with arrangements for emergency power in excess of the minimum 18 h specified within Regulation II-1/43 of the 1974 Safety of Life at Sea Convention (SOLAS). This notation would be expressed with the designation **HR** in parentheses with the total number of hours (i.e., **CCO- POLAR (–30, –50)(HR24)**) or similar.

Vessels receiving either the **CCO** or the **CCO-POLAR** notation may also be required to obtain the Class notation **POT**, which is discussed in Chap. 3.

- **DE-ICE** is a notation available for vessels occasionally operating in low temperatures subject to ice accretion. The requirements in Chap. 10 are applicable to vessels seeking this optional notation.

As an option, vessels designed, built, and surveyed in accordance with the requirements for any of the previous notations may have the + appended to the notation to indicate the placement of additional equipment onboard for the crew and specific low temperature environment training for the crew as per Chaps. 8 and 9 of this book.

1.4 Definitions

The following definitions are used throughout this book.

Ambient temperature: The air temperature in the vicinity of the vessel.

Anti-icing: Use of a system and/or a procedure that when in operation will keep ice from forming on the protected surfaces.

Deck machinery: Any machinery or equipment onboard the vessel that is directly exposed to the low temperature operating environment. Some examples include winches, windlasses, cranes or other lifting appliances, hatch covers, vehicle ramps, boat davits and mooring fittings.

De-icing: Use of a system and/or a procedure that will reduce ice levels on the surfaces being protected. De-icing is commenced after ice accretion has occurred, resulting in either complete or partial removal of the ice.

Design internal temperature: The design internal temperature, also denoted as TDIT, is applicable for machinery located in closed, unheated spaces. It is determined from the Design Service Temperature plus 20 °C (36 °F) (TDIT = TDST + 20). In no case is the design internal temperature to be greater than 0 °C (32 °F).

Design service temperature: The design service temperature, also denoted by TDST, is to be taken as the lowest mean daily average air temperature in the area of operation where:

- *Mean*: Statistical mean over observation period (at least 20 years).
- *Average*: Average for one day and night.
- *Lowest*: Lowest during year.

For seasonally restricted service the lowest value within the period of operation applies. Figure 1.1 illustrates this definition graphically.

EPIRB: Emergency Position Indicating Radio Beacon.

Emergency: A serious, unexpected, often dangerous event requiring immediate action.

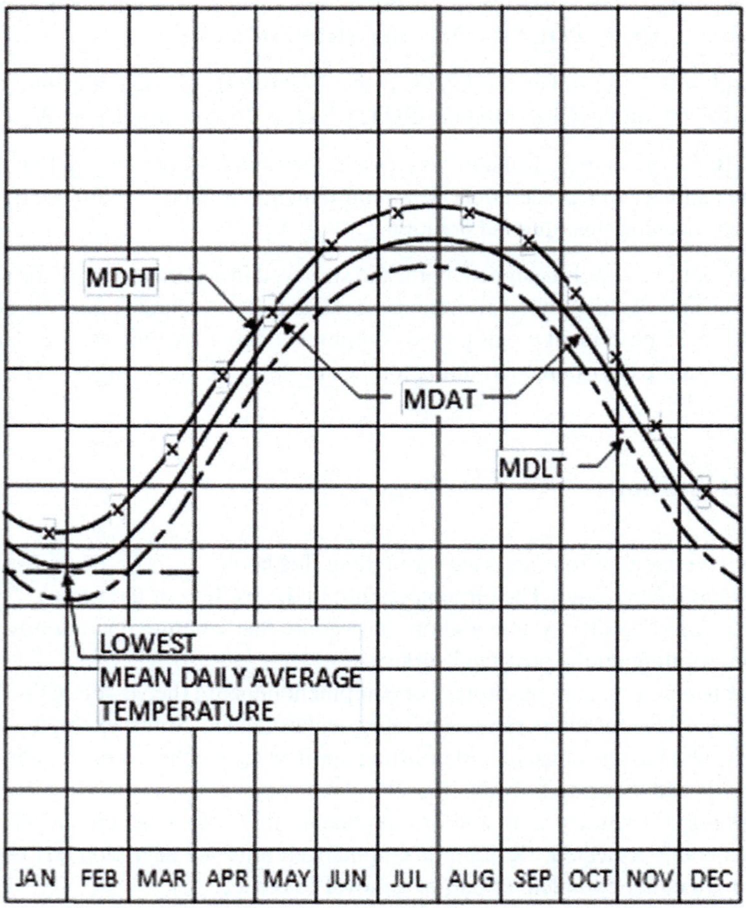

Fig. 1.1 Graphical depiction of design service temperature

Emergency service hours: The number of hours which the emergency source of electrical power can provide. SOLAS Regulation II-1/43 require the minimum time to be 18 h.

Essential: Systems or equipment necessary to maintain vessel propulsion, manoeuvring, electrical services, firefighting (along with bilge system), ballasting or personnel safety.

Guide for Certification of Lifting Appliances: The latest edition of Class guide for certifying lifting appliances.

In the notch: An operational mode in which a trailing vessel is moored in a specially constructed notch in the stern of an ice breaker.

Low temperature environment: Areas of vessel operations with ambient temperatures less than or equal to −10 °C (14 °F).

Machinery: Equipment and systems subject to the design requirements set by Class.

Manoeuvring mode: The lowest vessel speed necessary to maintain manoeuvrability.

MCR: The maximum continuous rating of the prime mover.

Minimum anticipated temperature: The minimum anticipated temperature, also denoted as TMAT, can be specified by the Designer, Owner, or Builder. TMAT may be a probabilistic temperature, taken as 2 standard deviations below the mean daily low temperature, or defined as a value below the TDST such as TMAT = TDST − 20 °C, or TMAT = LMDLT −10 °C (where LMDLT is Lowest Mean Daily Low Temperature).

1.5 Certification Procedure

1.5.1 Engineering Review

Vessels operating in low temperature environments need to be provided with certain design characteristics. These design characteristics are addressed in Chaps. 2 through 5 of this publication. Certain vessel types have unique features or operating characteristics unlike other vessels. Requirements for four specific vessel types are listed in Chap. 6. The design characteristics and/or equipment discussed in these Chapters are assumed to be permanently installed (e.g., welded or bolted) in/on the vessel at the completion of construction. This book also lists the requirements for additional equipment that is not permanently installed (e.g., clothing for personnel protection, special safety equipment) and additional personnel training described in Chaps. 8 and 9. The vessel owner/operator may wish to indefinitely delay the implementation of the requirements of Chaps. 8 and 9 for those instances where vessel operations in the low temperature environment are not anticipated. However, where these requirements are delayed, prior to the vessel operating in a low temperature environment, the vessel owner/operator must provide this additional equipment onboard and suitable personnel training together with a vessel operations and training manual. Typically, in these situations, arrangements are to be made for a Class approved surveyor to verify the equipment is onboard and the personnel have received proper training via onboard supporting documentation.

1.5.2 Plans and Particulars to Be Submitted (The Winterisation Plan)

Where stipulated by Class, the vessel owner/operator may be required to submit a winterisation plan to Class for review. Where required, the plan is to indicate the measures to be taken in different conditions, and should include, at a minimum, the following items:

1. Proposed methodology for anti-icing or de-icing on:
   - External superstructure horizontal surfaces (decks)
   - External superstructure vertical surfaces (bulkheads)
   - Radar antenna
   - Navigational lights
   - Search lights
   - Communication antenna
   - Lifesaving appliances
   - Vents for tanks
   - Escape routes deck surfaces, rails, doors, and stairs (must be ice accretion prevention)
   - Fuelling stations
   - Mooring equipment and controls
   - Calculations to support the above arrangements.
2. Means to prevent the following systems from freezing or becoming uncontrollably viscous:
   - Firefighting systems in tanks, cargo spaces, or on deck
   - Fresh water lines in tanks, cargo spaces, or on deck
   - Sanitary drains, black and grey water lines
   - Ballast lines, in tanks, cargo spaces, or on deck
   - Fuel and oil lines
   - Ballast tanks
   - Fresh water tanks
   - Fuel and oil tanks

- Any tank and associated piping containing a fluid that is susceptible to low temperatures.
- Combustion air intake
- Ventilation air intake
- Calculations to support the above arrangements.

3. Means of heating and ventilation, including calculations:
   - Crew and passenger cabins
   - Public areas in accommodation
   - Enclosed working spaces
   - Combustion air (preheating)
   - Engine room(s)
   - Steering gear compartment
   - Pump room
4. Test plan(s) to demonstrate the functionality of the above systems
5. Electrical single-line diagram of any electrical heat tracing systems (if-fitted)
6. Piping diagrams for any thermal fluids used for heat tracing (if-fitted)
7. Instructions for crew on how and when to exercise the various aspects of the plan.

1.5.3 Initial Survey

Vessels complying with the guidance contained in Chaps. 8 through 10 and confirmed by satisfactory Class survey may receive the **CCO** notation, or similar, which will be listed in the vessel's official Record. The **CCO** notation may be maintained indefinitely provided the hull and machinery are to the satisfaction of the attending Class surveyor at the subsequent Annual Surveys. Whenever Class is notified that the implementation of the requirements in Chaps. 8 and 9 have also been complied with and confirmed by satisfactory survey, the **CCO** notation with + will usually be appended to the vessel's official Record.

1.5.4 Annual Surveys

Simultaneously with each Annual Survey—Hull and Annual Survey—Machinery, the hull and machinery subject to Chaps. 2 through 6 are usually examined by the attending Class surveyor. The additional equipment and personnel training requirements in Chaps. 8 and 9 are also usually verified by the attending Class surveyor, where applicable.

1.5.5 Notation Changes

The vessel owner/operator may advise Class, via written notice prior to or during the Annual Survey that the additional equipment and personnel training requirements described in Chaps. 8 through 9 will not be complied with. In this instance, Class will amend the notation to **CCO** without the +.

1.5.6 National Administration Requirements

National administrations may have additional requirements for vessels operating in their territorial waters. These requirements may address additional vessel features, equipment, personnel training and instruction manuals. These requirements are not discussed in this book; however, a list of administrations is provided in the appendices.

1.5.7 Design Service Temperature

The design service temperature is to be selected based on the months and the regions the vessel is designed to operate in. Temperature data for northern and southern areas for the first day and the fifteenth day of each month are provided for illustrative purposes in the appendices under "Air Temperature". A listing of various meteorological organisations is also provided for guidance in the appendices.

1.5.8 Ergonomics

The maritime industry is aware of the significant role of the human element for effective operation, safety standards and practices. For example, effective heating of accommodation spaces to an acceptable temperature, provision of proper clothing, or heated shelters on deck. Additional information is available in the book *Human Factors in Ship Design and Marine Operations*[1] which promotes the application and understanding of ergonomic data and principles to vessel and offshore installation design. Although the ergonomic guidance provided in this book are not currently required by Class, it is strongly urged that vessel designers, owners, and operators to adopt these principles to the greatest extent feasible.

---

[1] Olsen, A. and Karkori, F. *Human Factors in Ship Design and Marine Operations*. Southampton: Magellan Maritime Press, 2022.

In this introductory chapter we have discussed the role and function of classification notations for ships employed in low temperature environments. It should be noted that whilst any vessel may be called upon to support operations in extreme temperature environments, only those which are designed and constructed to Class standards for that purpose should be employed therein for any length of time. In Chap. 2, we will examine the Arctic environment in terms of its physiology and climate in more detail.

Chapter 2
The Arctic Environment

2.1 Introduction

For the purposes of this book, the term "Arctic" is used to indicate the terrestrial basin which is filled by the Arctic or Polar Ocean. "Arctic" will also include the seas contiguous to the Arctic Ocean, the shores surrounding these seas, and in some cases, the navigable rivers which flow into these seas and the Arctic Ocean. The criteria commonly used for defining the Arctic regions are discussed here. The word Arctic is a derivation of the Greek word "Bear," which connotes that area lying under the "Big Dipper." Astronomically, it is the whole area lying north of 66° 33′ N (the Arctic Circle). If this definition is used, some areas of temperate climate are included and some areas of very intemperate climate excluded. Some scientists have suggested that the boundary of the Arctic be defined northward of the isotherm in which the average temperature of the warmest month is below 10 °C (50 °F). This closely coincides with the tree-line. Because it is peculiarly a polar phenomenon, the auroral zone has also been advanced as a criterion for a geophysical definition of the Arctic. The Aurora Borealis, observed in high latitudes as a luminous circumpolar feature, is centred about a point where the geomagnetic pole intersects the earth surface, and its zone of maximum frequency of occurrence can extend south as far as Fort Churchill.

The US Army's definition, accepted by the quadripartite countries, is …

> *… that portion of the northern hemisphere characterised by having an average temperature of less than 0 °C [32 °F] and an average temperature of the warmest month of less than 10 °C [50 °F]. This coincides with the southern boundary of the zone of discontinuous permafrost.*

Oceanographers consider the Arctic to be that region in which only pure "Arctic water" is found at the surface, at or near 0 °C (32 °F), and with a salinity of approximately 30 parts per thousand. This water is formed by a combination of:

(1) Water from the Atlantic and Pacific Oceans
(2) Water drained from surrounding land areas, and
(3) Water resulting from the melting of sea ice.

A. Olsen, *Ship Operations in Extreme Low Temperature Environments*, Springer Series on Naval Architecture, Marine Engineering, Shipbuilding and Shipping 19,
https://doi.org/10.1007/978-3-031-52513-1_2

The most important feature is that this water flows out of the Arctic Ocean on the surface, along the east coast of Greenland, through the Arctic Islands, spreading over Foxe Basin and Baffin Bay, and is carried as far south as Newfoundland by the Labrador current. Using this as a criterion, a line can be drawn north of which all waters may be considered as Arctic waters and which encompasses much of the Subarctic and Arctic, but which excludes the northern coasts of Norway and the Barents Sea. For the mariner, the most significant factors to be considered are latitude and the extent of the so-called "Arctic waters." For land operations, the line of discontinuous permafrost is the principal boundary to be considered (refer to Fig. 2.1). Over most of the Arctic, and in particular the Arctic Ocean, minimum winter air temperatures range between −45 °C (−49 °F) and −48 °C (−54.4 °F), which is comparable to Winnipeg in Canada. The coldest temperatures in the northern hemisphere are found not in the Arctic but in the subarctic latitudes. Generally speaking, throughout most of the Arctic (except in the coastal areas which come under the influence of maritime climatic systems), in Labrador, northern Norway and Kamchatka, precipitation is so low that "desert" conditions prevail. Precipitation falls during the summer months as a fine rain and is prevented from sinking into the ground by permafrost, the surface in summer can usually be found to be wet. The main body of the Arctic Ocean is covered year-round with pack-ice which is continually on the move. It is due to this constant movement of the ice that, even in winter, coverage of the Arctic Ocean is not complete. Ice coverage in winter is about 90%, decreasing in the summer to 70–80%, and in some coastal areas even less. The fringing seas become in summer either ice-free or reduced in ice concentration, with the exception of the northwest part of the Canadian Archipelago where the ice usually remains fast.

2.2 The Arctic Environment

2.2.1 Vegetation in the Arctic

The vegetation of the Arctic falls into two main divisions in the Arctic and Subarctic: tundra and forest. Although the individual plant species involved may vary from area to area, the type and form of vegetation is remarkably uniform, and it is this consistent association of vegetation with climate that makes it possible to speak of the Arctic as a single region in spite of all other variations. The chief characteristic of the tundra is the absence of trees. Tundra comprises creeping shrubs, grass-like plants, and mosses. Several zones are recognised, grading from north to south, from almost barren ground to fairly dense growth with small shrubs several feet high. The forest is the coniferous boreal or taiga, with larch (tamarack) and spruce as the dominant species in the Canadian Arctic, and a maximum of birch in Russia. Generally speaking, the whole forest area is interspersed with low wetland, known in Canada as "muskeg." By definition, muskeg consists of "more than 15 cm (5.9 in)

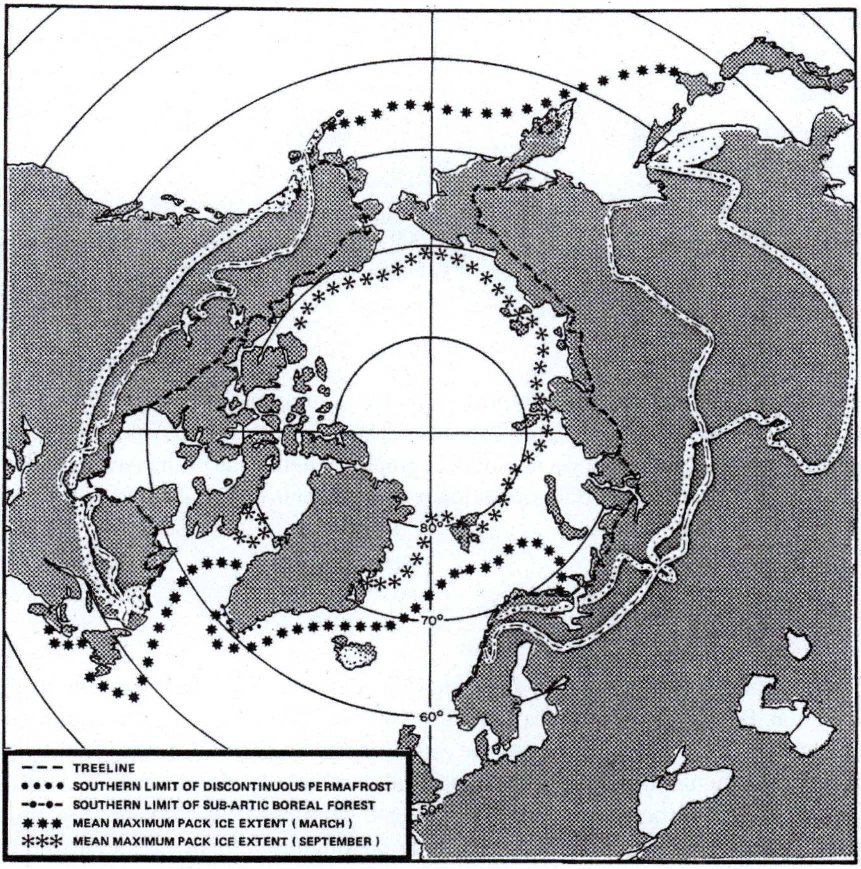

Fig. 2.1 Boundaries of the Arctic and sub-Arctic environments

of organic material overlying mineral soils". It may assume several forms such as "perched," "confined," or "continuous," and may include woody material (trees) up to 15 cm (5.9 in) in diameter. These trees always develop a tuft at the top and are seldom more than 4 m (13.1 ft) in height; therefore, muskegs are readily recognisable in the wooded areas. Where muskegs have developed on permafrost, the trees may lean in all directions due to frost heaving. When in this condition, they are often referred to as "drunken forests," particularly in Russia.

Generally speaking, in the Arctic, the tropopause (the level at which the fall of temperature with height ceases or reverses sign) is lower than in the tropics. The height of the tropopause varies seasonally in the Arctic from 7 km (4.3 mi) in February to 10 km (6.2 mi) in August. There is a sharp separation from the tropical tropopause at about 45°N and this separation is usually the region of the jet stream. The characteristics of the stratosphere above the Arctic tropopause are such that in winter a giant westerly vortex forms around the geographic pole, reaching maximum

Table 2.1 Arctic land environment (by political units)

	km^2	mi^2
Canada	3,100,000	1,926,250
Greenland	1,175,000	730,111
Iceland	105,000	65,243
Other North Atlantic islands	62,000	38,525
Continental Scandinavia	18,000	11,184
USA (Alaska)	545,000	338,647
Russian Federation	2,055,000	1,276,917

intensity in February and disappearing rapidly, forming in turn an equally vast but slow-moving easterly vortex in mid-summer. These features of the Arctic tropopause and stratosphere constitute yet another geographical definition of the Arctic, and one which is significant for certain operational requirements (Table 2.1).

2.2.2 *Limits of Arctic Lands*

The core of the polar regions is conveniently defined as the lands and seas around the Polar Basin, itself centred on the North Pole. When considering Arctic land areas alone, the southern limits of the Arctic are most usefully demarcated by the tree-line. Although this boundary is not ideal, particularly in mountainous areas, it marks a real break between the high winds (and wind-chill) and driving snow of the barren grounds that are usually underlain by continuous permafrost, and the milder conditions prevailing in the forests to the south. When this definition is in effect, for physiographic purposes three northern countries—Canada, Greenland and the Russia—contain 90% of the so-called Arctic Lands.

2.2.3 *Permafrost and Terrain*

The mean annual air temperatures below −2 °C (28.4 °F) leads to subsoil temperatures that are permanently below freezing point. If the ground beneath the annual surface thawing zone, known as the active layer, remains below freezing point for more than two years, permafrost conditions are said to prevail. Permafrost has an important influence on Arctic soils, landforms and vegetation, and affects the whole pattern of life in northern areas. Where mean annual air temperatures below −2 °C (28.4 °F) prevail, permanently frozen ground is found everywhere except beneath the largest lakes and rivers. In general, above the treeline, permafrost is continuous except in certain coastal areas. Continuous permafrost is therefore a phenomenon primarily of continental Arctic areas. Where winter temperatures in the Arctic average about

−2 °C (28.4 °F), as they do in the Aleutians, Iceland, coastal south Greenland and northern Scandinavia, and in the sub-arctic forests, areas of permafrost are separated by unfrozen ground and the permafrost is said then to be discontinuous. This zone extends farthest south in the eastern interior of continents and approaches the Gulf of St. Lawrence in eastern Quebec, and northern Korea in East Asia. In Arctic continental areas, permafrost is commonly from 300 to 500 m (984–1640 ft) deep. However, greater depths of up to 600 m (1968 ft) have been found in the northern Russia and there is some evidence that permafrost may be up to 1500 m (4921 ft) deep in a few Siberian areas.

From an engineering point of view, the critical part of the permafrost is the active layer close to the surface where freezing and thawing occurs. It varies in thickness and is dependent on site, exposure, soil drainage and vegetation. In the high Arctic at poorly drained sites, the active layer may be less than 10 cm (3.9 in), whereas near the southern margins of permafrost, in favourable localities, it may be 2 to 3 m (6.5–9.8 in) thick at the end of the summer. The maximum depth of the active layer under most favourable conditions may be in excess of 4 m (13.1 ft). The chief significance of this layer is that moisture content is often abnormally high, which creates serious problems for construction engineers and, in some cases, completely inhibits summer ground mobility. Many other landforms are found only in northern areas. One striking feature in these areas is the vast quantity of shattered bedrock lying on the surface and covering the valley sides. This broken bedrock results from frost shattering. The presence of ground ice leads also to unusual landforms. Ground ice exists in most areas where there is permafrost, occurring as ice in the soil pores, and as veins, wedges and sheets. The most spectacular ground ice occurs on the coasts of central and eastern Arctic Siberia where massive occurrences of ice, up to 70 m (229 ft) in thickness, have been reported. Elsewhere in the Arctic, particularly in the presence of silts, and especially in deltaic areas, ice mounds may develop covered either by organic or by mineral soils. These mounds vary in height from a few centimetres to more than 50 m (164 ft) and are known as "pingos". They are found in the Siberian river deltas, the Mackenzie delta, and the coastal plains of northern Alaska and Siberia.

If a change in the environment takes place resulting in the melting of the ground ice, hollows with uneven topography known as "thermokarst" form. Tundra ponds also develop when ground ice melts. A conspicuous feature of the terrain in Arctic areas is the tendency for soils to develop regular textural patterns. The term "patterned ground" includes all such features that occur in vegetation and soils, on horizontal and on sloping ground. It is caused primarily by a complex process of sorting of the materials comprising the soil, due to the freeze–thaw cycle, or by the development of regular hexagonal patterns due to the thermal contraction of the active layer. The surface layers of the soil often become supersaturated with water during the spring and summer, and movement (which may be perceptible through a period of a few days) may occur on gentle slopes. This phenomenon, known as solifluction, develops throughout the Arctic and results in patterns which are usually called "soil glaciers." In mountainous regions, features known as "rock glaciers" or "rock streams" sometimes occur, in which, instead of soil, boulders are in motion "lubricated" by ground ice or a mixture of soil and ground ice. Heavy snowfall is not common in the Arctic and, indeed, exceptionally heavy snowfalls are restricted to mountainous and maritime areas. However, snow lies for extended periods and has an important influence on northern terrains. The killing effect of the wind, on vegetation during the winter months, is reduced by a snow cover and in the southern Arctic the height of bushes is commonly controlled by the depth of the snowfall.

2.2.4 Permanent Land Ice

Less than 5% of the Arctic lands are covered with permanent ice (refer to Fig. 2.3). Greenland is the only area where glaciers dominate the land, and an ice sheet comparable to the great ice-caps of the glacial periods, 6000 to 12,000 years ago, survives. The main Greenland ice-cap is 2400 km (1491 mi) from north to south and is 1100 km (683 mi) across at its widest. Beyond the confines of Greenland, the ice sheet has a considerable influence on the climate of some of the northern circumpolar lands and is the main source of icebergs in the northern seas. It has been estimated that if all the ice in the Greenland ice sheet melted, the level of the world's oceans would rise by 6.5 m (21.3 ft). Greenland has the shape of an elongated basin surrounded by a mountain rim, which is highest and widest along the eastern side. The interior of the rock basin is close to sea level on the average, while a third of its total area (500,000 km^2 (310,685 mi^2) is below sea level. Inland from the coast the first hundred kilometres of ice is often crevassed and, particularly along the east coast, there are "nunataks" where the underlying mountains break through the ice sheet. Further inland the crevasses disappear, and the smooth, gentle slopes of the icecap prevail. In the interior, the ice sheet reaches over 3000 m (9842 ft). Great ice streams flow from this source through gaps in the mountain walls to the sea (refer to Fig. 2.2).

The distribution of the remainder of the glaciers in Arctic North America is closely related to topography and moisture supply. In Arctic Canada, glaciers occur plentiful north of Hudson Strait, and the most southerly icecaps occur geographically in three groups. The most southerly group (which are semi-permanent snowfields) is in the Torngat Mountains of Labrador. Permanent ice is more plentiful north of Hudson Strait, and the most southerly icecaps occur on the uplands on both sides of Frobisher. Larger ice-caps, often completely submerging the underlying mountains, are found on Baffin Island north of Cumberland Sound, on Bylot lsland, East Devon Island, and in three southern sectors of Ellesmere Island. A second group of Canadian glaciers are adjacent to the Arctic Ocean in northern Ellesmere Island and on Axel Heiberg Island. The third group consists of the four small icecaps on Melville Island. The remaining glaciers of Canada are found in the western mountain ranges and are not strictly in character. In Alaska there is a significant difference between the small number of Arctic glaciers of the Brooks Range and the Romanozov Mountains, and the vast glacier systems on the mountains adjacent to the Pacific Ocean, including the border ranges between Alaska and the Yukon, the Aleutian Range, and smaller

Fig. 2.2 Chamberlain Glacier entering Wolstenholme Fjord (Northwest Greenland)

glaciers of the Alaska Peninsula. Snowfall is not as heavy on the northwestern side of the Pacific Ocean, and glaciers are restricted to the Kamchatka Peninsula.

The North Atlantic Ocean and the Norwegian Sea constitute a major source of moisture for Arctic glaciers in Scandinavia. This moisture is carried by the winds as far as north-central Siberia (the Urals), and wherever there is sufficient relief there are small icecaps. However, the mainland of the Russian Arctic outside the Urals has no permanent ice and neither do the eastern islands of northeastern Siberia, excepting Bennett Island. In Scandinavia, permanent ice is restricted to the highest land, normally in excess of 2000 m (6561 ft). The altitude of glaciation becomes progressively lower in a north eastward direction, until it is below 500 m (1640 ft) on the Severnaya Zemlya Islands. A large expanse of ice covers Iceland, where the largest glacier is Vatnajokull. The other large glaciers are all in the southeastern or central parts of the island. North of Iceland, Jan Mayen Island supports an icecap,

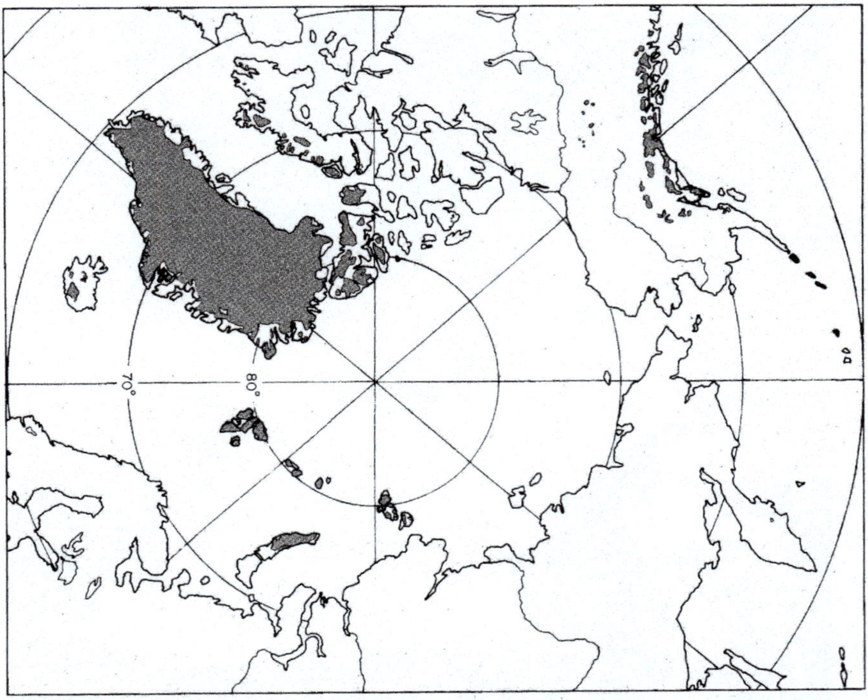

Fig. 2.3 Contemporary Arctic glaciers (shaded areas are glaciated)

but Bear Island farther northeast is much lower and has no permanent ice. There are numerous glaciers in Svalbard, and Northeast Land is covered with thin plateau ice. Other islands around the Barents Sea, including the northern third of Novaya Zemlya, Franz Josef Land and Severnaya Zemlya all carry large ice- caps (Fig. 2.3).

2.2.5 *Beaches and Cliffs*

Arctic littoral processes in ice-free summer months do not differ from those in temperate latitudes. However, several distinctive processes are operative at break-up and freeze-up of sea ice. Sea ice has, in general, a secondary role to waves as an agent in erosion, transportation, and deposition. With the onset of winter, when air temperatures drop below freezing, an ice shelf (storm-ice foot or ice rampart) commonly forms along Arctic beaches where there is a low tidal range. Water from storm waves and spray freeze on the beach, leaving layers of ice-cemented gravel and ice. Sand and gravel are often washed onto the developing ice shelf and become incorporated. This ice shelf "armour" protects the beach against subsequent wave

Fig. 2.4 Ice-shove ridges, Nansen Sound, Northern Canada

action and ice-shove. If the pack-ice should move onshore early in the winter, shore-fast ice with adhering debris may be shoved inland onto the beach (refer to Fig. 2.4). During the winter, shore-fast ice and offshore ice stop all wave action. Pack-ice at break-up, and later in the summer, may push onto the beach and even impinge against the sea cliff, if present. Ice moving onto a beach tends to act either as a plow or a raft. Where plowing occurs, the ice planes and pushes debris into mounds and ridges, usually below the upper limit of storm waves. Where rafting occurs, debris frozen and incorporated in the ice may be transported onto the beach, locally above the reach of storm waves. Usually, storm waves soon rework most of the transported material and, by freeze-up, little evidence of it is visible; however, in the high Arctic bays and channels that are permanently choked with ice, ice-pushed beach ramparts are a normal feature at the rear of beaches. Ice moving onshore can cause damage to infrastructure such as roads and buildings.

On coasts where the tidal range is considerable, boulder barricades are the most conspicuous sign of the action of sea ice. Typically, there is a narrow string of boulders parallel to the shore and several hundred feet out. At low tide they are clearly visible above the water surface whereas at high tide they are submerged. They represent a navigational danger on the approach to many open beaches. The depth of water increases rapidly beyond the boulder limit. The pushing and rafting of boulders by the moving sea ice that forms the boulder barricades is also responsible for the development of boulder-covered flats which are exposed at low tide near the heads of Arctic bays. Cliffs of unconsolidated materials, which may rise to 30 m (98.4 ft) or more, front many miles of coast in the Pacific area. The frozen sea cliffs retreat by undercutting, thermal erosion and slumping. The greatest erosion occurs during storm surges. When there is a combination of a high-water level and strong waves, the slumped debris at the foot of the sea cliff may be removed and a notch 1.5–4.5 m (4.9–14.7 ft) deep eroded into the cliff to produce an overhang in the frozen sediments, and this is followed by sediment failure and retreat of the cliff face. Thermal erosion may also produce glacier-like mud streams. Differential thermal erosion is particularly marked along coasts where large ice-wedges of tundra polygons are present. Gullies selectively etch out the ice-wedges to produce indentations or hanging valleys along the cliffs. In winter, the sea cliff and beach zone are an accumulation site for large snowdrifts which persist and protect the cliff foot in early summer. Rapid slumping from above may bury the snow-banks completely and some survive over the summer, thus adding additional protection to the cliff foot for a second summer season. The type of sediment in the sea cliff tends to be reflected in the number and size of adjacent beaches and bars. As a generalisation, beach and bar formation diminishes directly with reduction of grain size; icy muds may have no associated beach and bar deposits.

Offshore bars, bay-mouth bars, spits, tombolos, and other bar deposits are found along much of the Arctic coasts. This is to be expected where offshore depths are modest and the supply of debris from coastal recession and rivers is large. Offshore bars may parallel hundreds of miles of coast, being separated from the coast by a lagoon several miles wide. Spits build out where there are directional changes in the coast, in response to winds and currents. Driftwood and large logs may become stranded on these bars and spits. The main characteristics of the shores of Baffin Bay, Greenland, Spitzbergen and Norway are the spectacular cliffs, which can be found in many sectors (refer to Fig. 2.5.). In the Precambrian Shield areas, the majority of cliffs are preglacial, or, in the case of fiords, of glacial origin. These cliffs show little evidence of strong contemporary erosion subaerially or by waves. In contrast, many of the cliffs in younger sedimentary and igneous rocks, particularly in Svalbard and Iceland, have developed massive postglacial scree slopes that are active today. In some parts of the Atlantic Arctic, notably Scandinavia, a rock platform that varies in width up to several kilometres separates the cliffs from the sea. This feature, known as the strandflat, may be drowned, thus forming a "skerry coast."

2.2.6 *Details of Coastal Topography*

Detailed descriptions of the Arctic coastline of the areas covered in this manual are contained in the appropriate Sailing directions.

2.3 The Arctic Climate

Knowledge of the Arctic climate and weather conditions has reached a fairly satisfactory state. Coastal and island stations have been in operation for a period long enough to provide statistical information sufficient for a description of the regional

Fig. 2.5 The Labrador Coast

climates. In addition, observing stations on ice islands or on the pack-ice itself have, since 1950, contributed to an increased understanding of the weather and climate over the Arctic Ocean. Refer to Fig. 2.6 and Table 2.2 for the locations of the main observing stations. Since much of the northern land surface is of low elevation, there is little obstruction to free atmospheric flow between the middle latitudes and the Arctic. The areas of highest relief are Greenland, Ellesmere and Baffin Islands in Canada, the Brooks and Alaska ranges in Alaska, and the mountains of Scandinavia. The special and typical conditions met with in the Arctic are those resulting from:

(1) The distinctive regime of daylight and darkness together with low solar elevation which result in a prolonged period of radiational loss from the surface.
(2) The high reflectivity of the surface as long as snow remains on the ground; and
(3) The low moisture content of the air in winter in many areas of the Arctic, because of the ice cover on lakes and ocean.

The wetness, warmth, roughness, relief and albedo (reflectivity) of the earth's surface are among the major determinants of weather and climate. The stability of the air changes with the characteristics of the underlying surface. Cooling produces a stable stratification and a shallow layer of modified air, while heating from below steepens the temperature lapse rate and causes instability. The roughness, texture, wetness and colour of the Arctic landscape change little within each of the two seasons but vary markedly from summer to winter. The ice-covered Arctic Ocean is the most

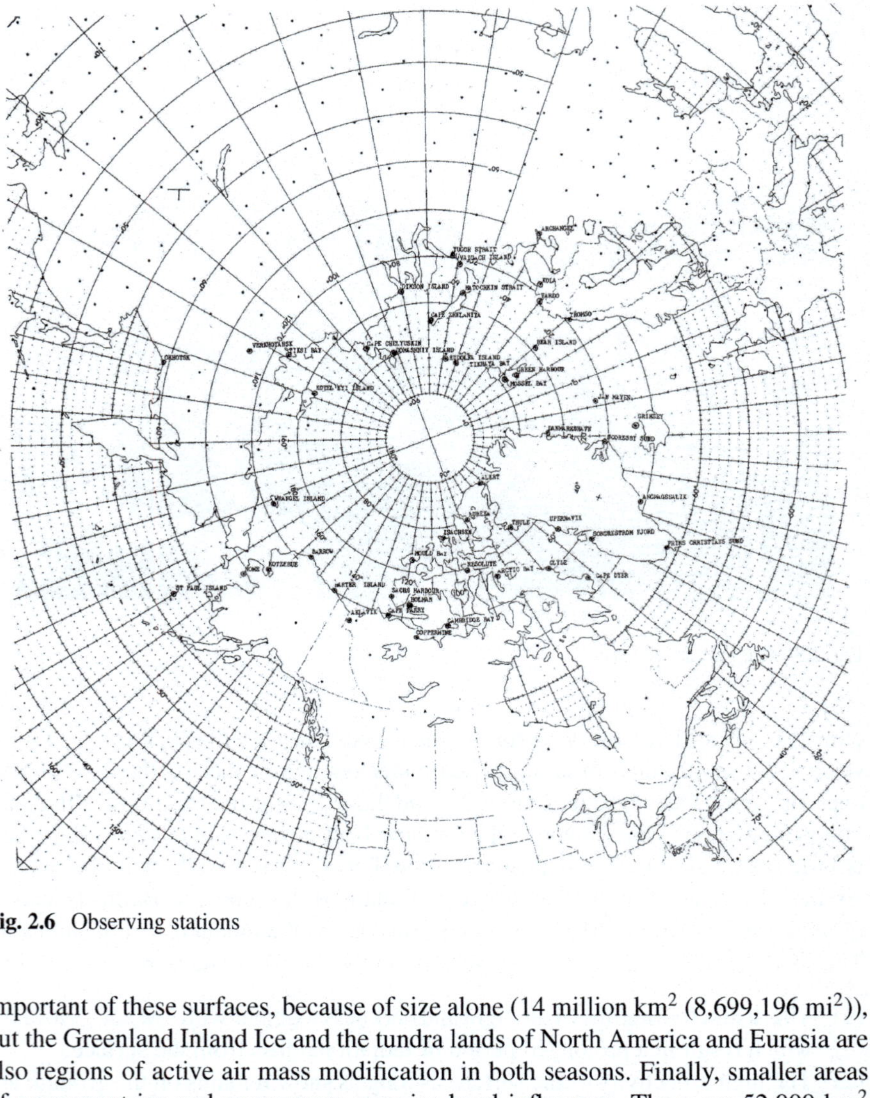

Fig. 2.6 Observing stations

important of these surfaces, because of size alone (14 million km^2 (8,699,196 mi^2)), but the Greenland Inland Ice and the tundra lands of North America and Eurasia are also regions of active air mass modification in both seasons. Finally, smaller areas of permanent ice and snow cover exercise local influences. There are 52,000 km^2 (32,311mi^2) of glaciers and icecaps in Russia, 130,000 km^2 (80,778 mi^2) in northern Canada, and 57,000 km^2 (35,418 mi^2) in the Svalbard Archipelago. The extension of polar sea ice is smallest in September, when it covers approximately 6.52 million km^2 (4,051,340 mi^2), and greatest in March (approximately 15.49 million km^2 (9,625,039 mi^2) or 11% of the sea area of the northern hemisphere). The neighbouring seas are also frozen in March. The Hudson Bay and Hudson Strait add 1.3 million km^2 (807,782 mi^2) to the ice-covered surface in winter, and the Bering Sea, Sea of Okhotsk, White Sea and the Gulfs of Bothnia, Finland and St. Lawrence form southern fringe areas of a snow- and ice-covered polar cap. Therefore, the polar

Table 2.2 Observing station locations (refer to Fig. 2.6)

Station	Latitude	Longitude	Elevation in m (ft)	Station	Latitude	Longitude	Elevation in m (ft)
Green Harbour	7802°N	1415°E	11 (36)	Barter Island	7008°N	14,338°W	12 (39)
Mossel Bay	7954°N	1627°E	–	Aklavik	6814°N	13,500°W	9 (30)
Tromsø	6939°N	1857°E	102 (335)	Sachs Harbour	7157°N	12,444°W	84 (277)
Bear Island	7428°N	1917°E	40 (133)	Cape Parry	7010°N	12,441°W	16 (53)
Vardo	7022°N	3106°E	12 (39)	Mould Bay	7614°N	11,920°W	15 (50)
Kola	6853°N	3301°E	7 (23)	Holman	7030°N	11,738°W	9 (30)
Archangel	6428°N	4031°E	6 (20)	Coppermine	6749°N	11,505°W	8.5 (28)
Tikhaya Bay	8019°N	5248°E	6 (20)	Cambridge Bay	6907°N	10,501°W	14 (47)
Matochkin Strait	7316°N	5624°E	18 (59)	Isachsen	7847°N	10,332°W	25 (83)
Rudolfa Island	8148°N	5757°E	48 (157)	Resolute	7443°N	9459°W	63 (209)
Vaigach Island	7024°N	5848°E	11 (36)	Eureka	8000°N	8556°W	2.4 (8)
Yugor Strait	6949°N	6046°E	13 (43)	Arctic Bay	7300°N	8518°W	11 (36)
Cape Zhelaniya	7657°N	6834°E	8 (26)	Thule	7634°N	6848°W	39 (129)
Dikson Island	7330°N	8024°E	20 (66)	Clyde	7027°N	6833°W	3 (10)
Domashniy Island	7930°N	9108°E	3 (10)	Alert	8230°N	6220°W	62 (205)
Cape Chelyuskin	7743°N	10,417°E	5 (16)	Cape Dyer	6635°N	6137°W	368 (1,207)
Tiksi Bay	7135°N	12,856°E	10 (33)	Upernavik	7247°N	5609°W	64 (210)
Yerhoyansk	6735°N	13,330°E	121 (400)	Sondrestrom Fjord	6700°N	5048°W	58 (190)
Kotel'nyi Island	7602°N	13,806°E	3.3 (11)	Prins Christian Sund	6003°N	4312°W	76 (250)
Cape Shalaurova	7311°N	14,314°E	8 (26)	Angmagssalik	6530°N	3733°W	29 (95)
Okhotsk	5921°N	14,317°E	5 (16)	Scoresby Sound	7028°N	2158°W	17 (56)
Wrangel Island	7058°N	17,833°W	3 (10)	Danmarkshavn	7646°N	1845°W	6 (20)
St Paul Island	5709°N	17,013°W	7 (22)	Grimsey	6630°N	1801°W	22 (72)

(continued)

Table 2.2 (continued)

Station	Latitude	Longitude	Elevation in m (ft)	Station	Latitude	Longitude	Elevation in m (ft)
Nome	6430°N	16,526°W	14 (46)	Jan Mayen	7059°N	0820°W	23 (76)
Kotzebue	6652°N	16,238°W	5 (16)				
Barrow	7118°N	15,647°N	4 (13)				

cap extends south of 60°N everywhere except in the North Atlantic region, and in the Norwegian and Barents Seas where Atlantic water enters the Polar Basin and submerges to form the intermediate layer of Atlantic water in the Arctic Ocean.

2.3.1 Air Pressure

See Table 2.2.

2.3.2 Winter Months

The distribution of mean sea-level pressure for January is illustrated in Fig. 2.7. The circulation is dominated by four large cells two continental highs and two oceanic lows. The large, semi-permanent Siberian high is joined to a smaller high over the Mackenzie Valley (1023 mbar) by an elongated ridge across the Arctic Ocean. The Iceland low (994 mbar) controls the mean circulation along and over northern Scandinavia and the sea areas on the Eurasian side of the Pole. An elongated trough follows the open water and runs from Iceland north of Norway to Novaya Zemlya and the Chelyuskin Peninsula. Another trough covers Baffin Bay. Both troughs are actually the scene of frequent cyclonic activity. The remaining cell is the smaller Aleutian low. The circulation over the Pole itself is dominated by the cyclonic flow around the Icelandic low. There is no "Arctic high" on the January mean map, as stipulated by older theories for the Arctic Ocean, although strong anticyclones are common over the Basin. The "Polar easterlies" exist, in winter, in the mean flow only over the Norwegian-Barents Sea area and along the north flank of the Aleutian low. Because of the extreme asymmetry and eccentricity of the sea -level pressure field, no single latitude circle shows any appreciable integrated easterly flow. Over the central Arctic Ocean, the main air stream in January is directed from the middle and western Siberian coast towards the Pole and thence, southward across the Greenland-Svalbard area. On the Bering Strait side of the Pole the Siberian air masses cross the Polar Ocean and invade the Canadian Archipelago, mixing with air from the Alaskan-Yukon area. Over the regions of elevated topography, the pressure gradients are fictitious, as for example over Greenland where the surface winds are not accurately depicted by the pressure map.

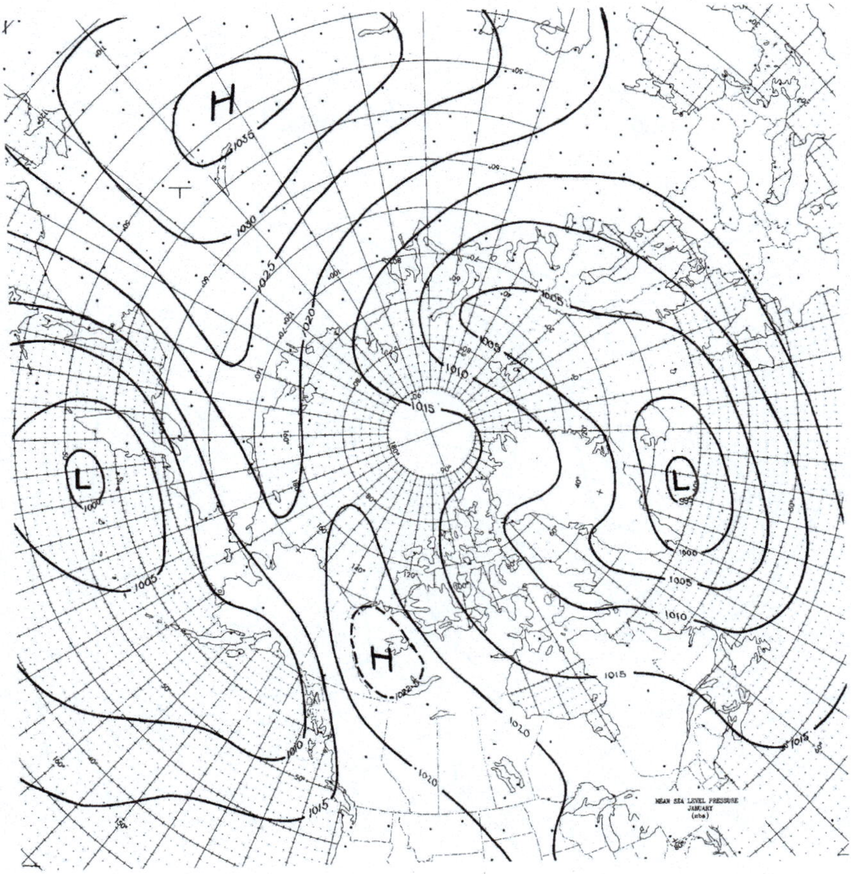

Fig. 2.7 Mean sea-level pressure—January

2.3.3 Spring and Summer Months

The January pattern persists into March. Thereafter the mean pressure field undergoes rapid change. There is pronounced weakening of the low cells and of the Siberian continental high and a relative intensification of high pressure over the Arctic Ocean. It is in the spring that strong anti-cyclonic activity is most likely to occur over the central Arctic, and it is at this time that the concept of a "Polar anticyclone" is most nearly fulfilled. In summer (Fig. 2.8), the mean pressure gradient over the central Arctic Ocean is weak. A closed anticyclone is found over the edge of the Arctic Ocean covering the Barents Sea (1013 mbar), while a weak ridge is located over the Beaufort Sea (1013 mbar). The Aleutian low diminishes to a feeble trough in summer, and the Icelandic low is also weak (1010 mbar). The Siberian high disappears completely and is replaced by a thermal low-pressure area over the heated Asiatic land mass.

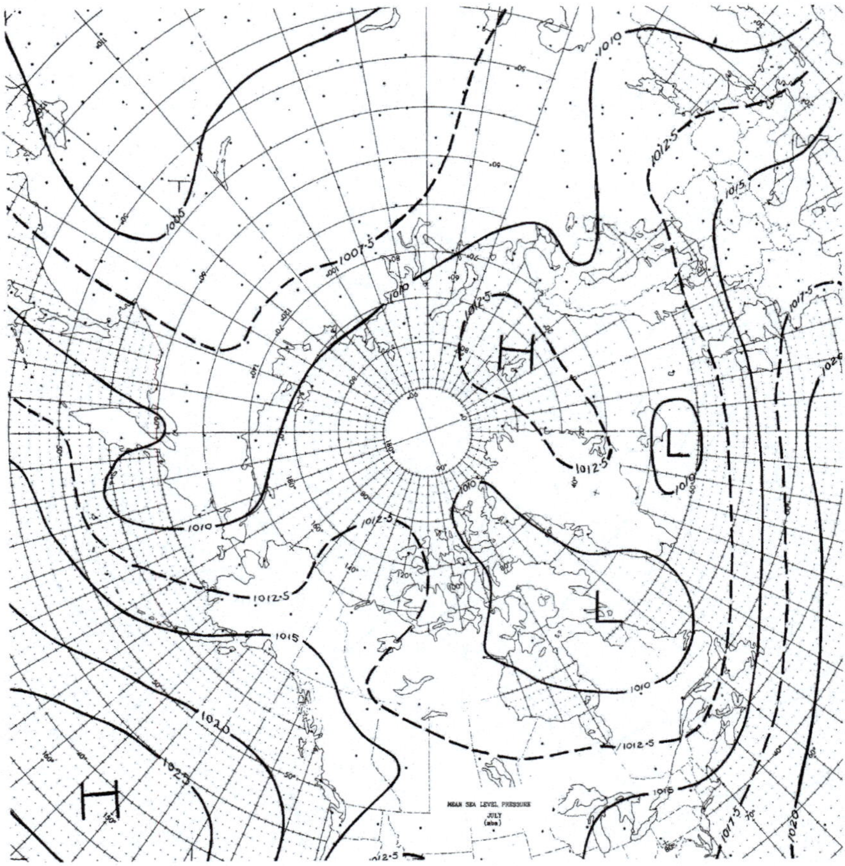

Fig. 2.8 Mean sea-level pressure—July

The July mean pressure map thus shows a feeble pattern with a sluggish resultant mean flow. The prevailing air currents over the Eurasian coast are directed from the Arctic Basin to the coast, with a marked easterly component, so that on the Siberian coast the winds are ENE, while on the coast westward of 90°E the prevailing wind is from the NE.

2.3.4 Autumn Months

By October, the pressure patterns show a superficial resemblance to the spring conditions. High pressure occurs over the central Arctic Ocean and the sub-polar lows are back in their customary positions. The mean pattern, however, represents a compromise between contrast of vigorous cyclonic circulation systems and periods of anti-cyclonic control. The Aleutian low regains its full intensity and the Icelandic low approaches wintertime strength, the extension over Novaya Zemlya being re-established. The Siberian high attains moderate intensity in October (1018 mbar).

2.3.5 Principal Tracks of Cyclones and Synoptic Systems

Figure 2.9 supplement the charts of mean pressure by indicating the most frequent areas of the northern hemisphere where cyclones form and dissipate, and the major tracks which they follow in winter (January) and summer (July).

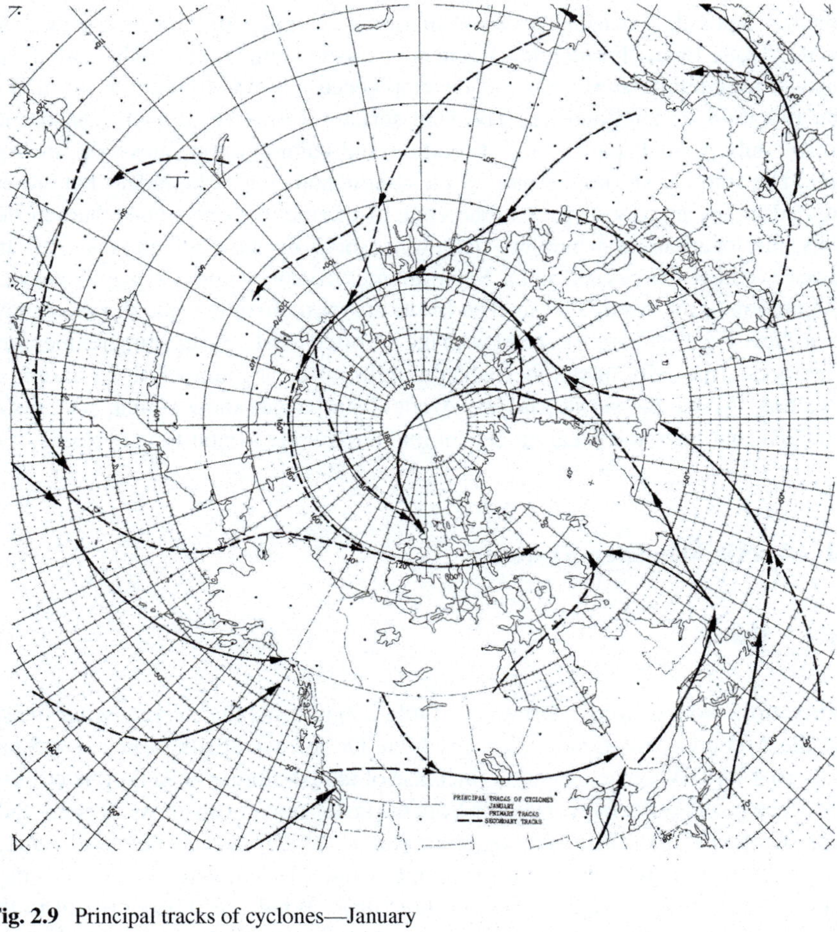

Fig. 2.9 Principal tracks of cyclones—January

Over the Polar Basin, many cyclones spiral in toward the Pole. Storms breaking off the Icelandic low usually move northeast to the Norwegian-Barents Sea and a primary track continues along 75°N. These frequent storms play an important part in the climate of the north European and Siberian coastlands. This primary track is joined by tracks from the Baltic, Black and Caspian Seas. Some storms stagnate near Novaya Zemlya, others regenerate and curve in towards the Pole. There is a difference in the climatic importance of synoptic weather systems when the middle latitudes are compared with the Arctic Ocean. In the former regions, synoptic systems control the weather and climate by an ever-changing sequence of different air masses. Over the Arctic Basin in winter, cyclonic patterns do not influence the surface climate except as far as they control mixing in the inversion layer. Warm and cold spells in temperate latitudes are most often air mass phenomena (advection); over the Arctic Ocean cold spells are caused by radiation, not by advection, while milder surface conditions require vertical transport downward of warmer air from above the inversion. Sporadic invasions of warm maritime air do occur, however, sometimes into the Arctic Basin. Frontal development occurs in summer, along the continental coasts. Strong temperature contrasts develop because of the different absorption of solar radiation by land and ice. The land surface warms appreciably, but the ice absorbs only a small part of the insolation and cannot warm above the melting point. The summer storms develop in the continental coastal boundary but are not confined to any narrow strip. A broad zone of thermal contrasts may sharpen into fronts, the locations of which are determined both by the circulation and by the presence of heat sources and sinks. Such summer frontal cyclones bring the heaviest precipitation to the Arctic Ocean. The fronts are more distinct some distance above the surface. The frigid air in contact with the ice often masks the existing temperature differences, especially in summer when a uniform surface temperature is found in all sectors. Above the surface layers, Arctic disturbances show similar features of cloud distribution to the usual models of middle-latitude frontal-type cyclones.

2.4 Arctic Weather Systems

2.4.1 Wind

Observations from drifting ice stations (Table 2.2) indicate that surface wind speeds over the Arctic Ocean are not extremely high. On the average the force is 4–5 m/s (13.1–16.4 ft/s). When an inversion (increase of temperature with height) is present, the surface layer is effectively isolated from the faster-moving air above. It is this fact, in combination with the lack of topographic effects, which results in the small number of occurrences of fierce winds. Although light winds predominate, fierce winds may occur and persist for periods of one to three days. Wind speeds are critical in the Arctic because in addition to the winds intensifying the feeling of cold, they are responsible for blowing snow-the major deterrent to travel in winter. The annual mean wind speeds are greatest at exposed coastal stations near cyclone tracks, and it is in these same locations that gales are most frequent (for example, at Jan Mayen, northern Norway to Dikson Island and Bering Strait) (Table 2.2). On the other hand, at sheltered locations or those remote from normal active storm tracks, the annual mean wind speeds are low (as at Eureka and Aklavik).

2.4.2 Temperature

The most common misconception of the Arctic is that the land areas are covered with eternal ice and snow and that there is everlasting winter with intense cold. A large part of Greenland is a striking example of an ice-covered land possessing these qualities and from it, the rest of the north has been pictured by analogy. The high elevation of Greenland and Ellesmere Island highly favours glaciation. In general, however other Arctic lands, possess neither of these characteristics and over a substantial portion of the Arctic the scanty snows melt rapidly with the approach of summer. Most of the little snow that does fall is soon swept by the wind into gullies and the lee of hills, so that over much of the Arctic lands the winter snow cover is quite thin. Next consider the intense cold: at the North Pole, the lowest temperatures do not fall below −55 °C (−67 °F), a figure that is occasionally reached in some parts of the Canadian Prairies. The Siberian winter is by no means as unpleasant as its extreme temperatures might suggest; the air is often calm and the skies clear. Danger to man and beast occurs only when the wild "buran" or "purga" blows. Such fearful blizzards also occur in the interior of Canada. The irregularities of the winter isotherms over the Arctic Ocean are produced by the influence of the Atlantic and Pacific Oceans. In the North Atlantic, the isotherms are pushed far northward, and it is here that one finds the greatest positive anomaly of temperature in the world (the greatest departure from latitudinal average). The winter temperatures in the vicinity of the Bering Strait are influenced by the open waters of the Bering Sea. North of the strait the influence is limited to a small area in which large local differences occur with different wind directions. Large and rapid temperature fluctuations are otherwise not a common feature over the Arctic Ocean. During the greater part of the year, the area is covered by a thin layer of frigid air which, to a great extent is isolated from the atmosphere above. The air temperature near the surface is primarily dependent on the temperature of the ice surface itself.

The winter temperatures over the pack-ice remain constant for a considerable time. This "flat" minimum represents the temperature at which the loss of heat, by radiation from the snow and ice surface, balances the amount of heat which, is conducted to the surface from the water under the ice. It also depends on the heat transported into the Arctic by warm air advection in advance of cyclones. During the extended periods of infrequent cyclone activity in winter, heat transfer to the surface from the atmosphere is small, since the eddy conductivity is exceedingly small in the inversion layer. The presence, or nearness, of open water in the winter half of the year, is reflected in the mean air temperature pattern. Broad tongues of warmth extend into the Barents Sea. The strongest influx of heat by cyclonic activity occurs in the Atlantic section of the Arctic Ocean in January and February. From April until June the temperature rises quite rapidly, until the ice begins to melt. The mean summer surface temperatures remain fairly constant, conforming to the nature of the underlying surface. All regions are warmest in July (Table 2.2). Temperatures close to the melting point prevail over the pack-ice and along the fringes of the Greenland ice-cap and outer islands of the Archipelago. Maximum temperatures in the Arctic Basin do not exceed 5 °C (41 °F). An examination of large departures of monthly mean temperatures from normal values for 20 coastal or island stations, with 20 or more years of observations, showed that the greatest departures are found in winter (from November to April) and small departures are most probable in July. In winter, the greatest departures were observed in the Kara Sea, the smallest in the East Siberian Sea. In summer, the greatest departures from normal temperatures were observed on the coasts of the peripheral seas and decreasing toward the north. The temperature inversions observed in all Arctic regions are especially prominent over the pack-ice. During the darkness of winter in the high Arctic, the diurnal variation in temperature is irregular and the maximum and minimum temperatures can occur at any time during the 24 h of the day. Variations are due to changes in cloud cover or wind speed. Temperatures raise when cloud cover increases or winds strengthen and fall when winds decrease or skies clear. Small temperature variability in the summer is a characteristic common to all stations in the inner Arctic Ocean. The temperatures over the pack-ice never deviate far from the freezing point. The number of days with maximum temperature slightly above the freezing point is very the same all along the latitude of 75°N-around 40 days. The diurnal ranges are small.

2.4.3 Fog and Visibility

The range of visibility varies in the Arctic. Arctic air masses are pure, and the visual range may be very great. Poor visibilities in the lower layers of the atmosphere result from the presence of falling precipitation, fog or blowing snow. Haze and smoke pollution of air masses as they occur in lower latitudes are rare in the Arctic. During the warmer months, the most common type of fog is caused by the advection of warm and moist air over the melting ice and freezing water. When the air moves in such a way that it is cooled by contact with the colder underlying surface at a temperature below its dew point, it can no longer hold the excess moisture which condenses to form mist or fog. When the air near the surface is very stable, the fog will be shallow. On the other hand, turbulence in the lower levels of the air may give conditions under which this excess moisture becomes visible as low stratus clouds. The areas, which are most favourable for the formation of this type of summer fog or low clouds, are the open waters of the Kara, Laptev, East Siberian and Chukchi Seas. Fifteen to 20 days with fog is a normal condition for these areas during the summer months (Table 2.2). This advection-type fog occurs most frequently over coastal waters adjacent to a local cold-air source or over open leads in the pack-ice.

Over the Arctic Basin it is extremely frequent and widespread during the period when the ice surface is melting. During the cold season, a type of fog known as “steam fog” or “Arctic Sea Smoke” occurs when very frigid air moves over open water. It occurs only when the contrast between air and water temperatures is very great. This type of fog occurs most frequently over rivers, unfrozen lakes, open leads or polynyas in the Arctic ice. Over the Arctic Basin steam fog serves the important purpose of indicating the presence of open water.

Another winter phenomenon is the occurrence of “ice fog.” This is defined as a type of fog composed of suspended particles of ice, which occurs at extremely low temperatures, below −30 °C (−22 °F), and usually during clear, calm weather at high latitudes. It occurs locally in the vicinity of human habitation, herds of animals, vehicular or aircraft operation, and open water areas of fast-running streams. During the winter months, blowing snow is the most common cause of reduced visibilities, particularly in the more unprotected continental or insular locations. The frequency and severity of blowing snow is of course related directly to the frequency of high winds and the presence of new or powdery snow cover. Much of the snow of the Arctic falls in the form of loose granular flakes, which are very readily whipped up by even relatively light winds. Observations show that with winds in excess of 24 km/h (14.9 mi/h), blowing snow is always present in some degree in the cold Arctic. From December to April the visibilities at most Canadian Arctic Stations are reduced to 10 km (6.21 mi/h) or less in blowing snow about one third of the time. In the case of winds over 48 km/h (29.8 mi/h), more than 80% are associated with visibilities lower than 5 km (3.1 mi). Provided no restrictions such as fog, precipitation or blowing snow are present in the Arctic atmosphere, visibilities are remarkably good. The records of many travellers are filled with accounts of the extreme range of visibility, and indeed it is common to see dark mountains 150 km (93.2 mi) distant. On the other hand, the lack of contrast, particularly where all surface objects are covered with new snow, results in the inability to distinguish objects nearby. The addition of a uniformly overcast sky to an unbroken snow landscape gives rise to the condition usually referred to as “Arctic white-out.” The frequent well-marked temperature inversions of the Arctic region explain the many accounts of mirages. Objects that are known beyond any doubt to be below the horizon are not infrequently visible as mirages and the periods of daytime and twilight are lengthened as the normal index of refraction is altered. The inversion may also interfere with the identification of landmarks through distortion, and estimation of vertical distances is made much more difficult.

2.4.4 Cloud

The general character of cloud cover over the Arctic differs from that considered typical foremost temperate regions. The uniform and contourless stratus clouds, which are by far the most frequent type observed, give to the Arctic its reputation of a dull and monotonous appearance. Over the Arctic Ocean during the summer, the low stratus-type of cloud constitutes from 70 to 80% of all cloud observed. The reason is the continuous cooling of the air from below over the pack-ice fields and the presence of a source of moisture in the surface melt-water. These summer clouds are extremely uniform, extending as large sheets over a much wider areas than other cloud types. Along the coast in more southerly locations, the stratus clouds are proportionately less frequent. In these regions, the cloud decks tend to be more often broken up by convective currents, with a resulting increase in the proportion of stratocumulus cloud reported. In the Norwegian Sea, mean cloudiness is high throughout the year, with a slight maximum in the summer even though this is the season with the lowest cyclone frequency. For example, the mean cloudiness at 74°N, 0°W is 80% in January, but 90% in July (Table 2.3). Over the Arctic Ocean the cloud amount is lowest in winter and spring when the water content of the very frigid air is too low for cloud formation. A further characteristic of this area is that the seasonal change in cloudiness takes place over a noticeably short transitional period. Similar abrupt changes are found only in monsoon areas when a complete change overtakes place in the circulation pattern. Areas showing high cloud amounts in winter are the Norwegian and Barents Seas, where the January, mean cloudiness exceeds 80%. These percentages decrease to below 70% in the Kara Sea, 50% in the Laptev Sea, to about 40% in the East Siberian and Beaufort Seas (Table 2.3). With few cyclones penetrating eastward beyond Novaya Zemlya, and little or no surface moisture, only thin middle clouds and cirrus clouds prevail. The main feature of the winter cloud pattern over the North American continent is the gradual decrease in cloudiness towards the north.

2.4.5 *Precipitation*

Precipitation over most of the Arctic is noticeably light and the annual amounts are so small that the region is classified as a desert based on annual precipitation. Over the Polar Basin the annual precipitation amounts average less than 250 mm (9.8 in), while over the Siberian- American portion of the Arctic Ocean and the northern part of the Canadian Archipelago the amounts are of the order of 100 mm (3.9 in) or less (Table 2.3). The greatest portion of the annual total precipitation falls in the summer as rain, although the rate of fall is usually light, and in most cases, would be classed as drizzle rather than rain. Heavier showers may occasionally occur. Over much of the Polar Basin snow may be expected during all months of the year, and in the far north even in summer the number of days with snow may exceed the number of days with rain. Because of the low moisture concentration possible at extremely low temperatures, however, the winter snow is exceptionally fine and when it occurs on windy days its measurement becomes exceedingly difficult. Freezing rain and freezing drizzle can also occur in the summer months. Available data indicate that on the average freezing precipitation occurs less than 10 h per year. The rate of

Table 2.3 Arctic region weather trends (sheet 1 and 2)

1 PLACE	2 PERCENT CLOUD COVER		3 MEAN AIR TEMPERATURE °C		4 MEAN PRECIPITATION (INCHES)		5 MEAN MONTHLY PRECIPITATION x FREQUENCY		6 MEAN VISIBILITY x<2NM//x<5NM		7 MEAN NUMBER OF DAYS VISIBILITY <1/2 MI		8 SEA AND SWELL ≥5'//≥12'		9 MEAN VECTOR WINDS DIR/SPEED(KT)		10 PERCENT OCCURRENCE FOR PARTICULAR WIND SPEEDS ≤10/>17/>28/>34(KTS)	
	JAN	JUL	JAN	JUL	JAN	JUL	JAN	JUL	JAN	JUL	JAN	JUL	JAN	JUL	JAN	JUL	JAN	JUL
CANADA					RAIN/SNOW	RAIN/SNOW	RAIN/SNOW	RAIN/SNOW										
ARCTIC BAY	40	65	-28	2			20/20	15/10	10/75	10/85			ICE	<10	NW/10	NE/5	45/ / /<5	50/ / /0
BAFFIN ISLAND	45	70	-26	4			15/15	15/10	15/80	10/85			ICE	<10	NW/10	NW/<2.5	40/ / /<5	45/ / /0
BANKS ISLAND	43	70	-32	4			20/20	10/10	<10/70	10/70			ICE	–	N/10	NW/<2.5	40/ / /<5	50/ / /0
CAMBRIDGE BAY	42	65	-32	8			20/20	15/05	10/70	10/80					W/10	NE/05	40/ / /<5	50/ / /0
DAWSON	50	70	-28	16			25/15	20/0	10/80	10/70					NE/10	SE/05	30/ / /<10	50/ / /0
ELLESMERE ISLAND	40	70	-32	6			0/75	02/17	30/50	30/40					N/10	SW/10	-	50/ / /0
FROBISHER BAY	45	75	-24	4			15/15	20/05	15/70	10/85					NW/10	VAR/05	35/ / /05	50/ / /0
INUVIK	45	70	-28	8			20/20	20/05	10/70	20/70					SW/2.5	NW/05	45/ / /<10	50/ / /0
NUGVIK	45	70	-26	8			10/10	20/<5	20/70	10/80					SW/5	S/<2.5	30/ / /05	50/ / /0
QUEEN ELIZABETH ISLAND	40	75	-32	4			0/11	04/07	15/20	30/45					W/10	N/10	-	50/ / /0
RESOLUT	40	65	-32	4			0/11	04/07	15/20	30/45					W/10	N/05	45/ / /0	50/ / /0
REPULSE BAY	38	60	-28	8			20/15	20	20/70	10/90					NW/10	NW/05	40/ / /<5	50/ / /0
VICTORIA ISLAND	48	65	-32	8			20/20	15/05	10/70	10/80					NW/10	NE/05	40/ / /<5	50/ / /0
WHITEHORSE	55	70	-20	12			30/15	20/0	10/80	10/70					NE/10	SE/05	20/ / /<10	50/ / /0
DENMARK																		
FAROE ISLANDS	75	80	-12	4	8.0/2.7	5.1/0.0	30/10	20/0	10/75	10/70			50/15	3/1	SW/10	SW/<2.5	20/30/25/15	40/18/3/0
FINLAND																		
HELSINKI	80	55	-8	16	1.70/–	2.20/0.0		2/0			2.9	0.8		0/0		W/11	/3.5/0.1/	46/1.6/0.1/0
OULU	80	55	-8	16	2.90/–	2.36/0.0		2/0			3.2	0.5		1/0		S/10	/3.3/0.1/	45/1.4/0.0/0
KIRUNA	80	55	-8	16														
GREENLAND																		
AMMASSALIK	50	60	-4	4	2.90/–	1.50/–	5/13	5/3			N/A	N/A	6/0	0/0	NE/30		23/20/28/0	40/28/1.1/0
GOTHAB	60	60	-12	4	1.40/–	2.20/-					N/A	N/A					N/A	N/A
ITSEQQORTOOMIIT	60	60	-12	4	1.80/–	1.50/–					N/A	N/A					N/A	N/A
NORD	45	60	-28	4	0.84/–	1.00/–					N/A	N/A					/9.8/2.4/	/6.5/0.6/
QAQORTOQ	65	60	-6	4	2.78/–	1.20/–					N/A	N/A					N/A	N/A
QEQERTARSUAQ	60	60	-16	4	0.61/–	2.13/–					N/A	N/A					N/A	N/A
THULE	40	60	-28	4	0.36/3.3	0.67/0.0					5.0	7.4					/8.1/2.3/	/6.7/1.0/
JAPAN																		
KUSHIRO	60	88			1.80/–	4.40/0.0	3/9	9/0			6.8	12.7	7/.5	1/0	NW/15	E/8	/8.8/0.4/	/1.5/0.0/
MORIOKA					2.33/–	6.64/0.0					N/A	N/A					N/A	N/A
GIFU AB					2.76/–	9.48/0.0					N/A	N/A					/1.7/0.0/	/0.4/0.0/
NIIGATA	88	94			7.50/42.6	6.40/0.0	6/25	10/0			6.5	1.3	7/4	8/0	NW/18	SW/8	/19.9/2.0/	/0.2/0.0/
ICELAND																		
AKUHEYRI	70	75	0	9	1.70/–	1.30/–					2.8	0.4	3/10	7/0	NE/22	NE/8	32/9.3/0.9/0	42/1.7/0.1/0
REYKJAVIK	75	75	0	9	4.00/–	2.00/–					2.8	0.3	1/0	4/0	SW/23	N/13	22/35.9/10.5/0	39/9.4/0.6/0
IRELAND																		
MALIN HEAD	80	78			2.60/–	2.84/0.0	12/2	8/0			0.1	0.8	4/0	2/0	SW/17	NW/12	31/33.2/4.4/0	43/10.2/0.1/0
NORWAY																		
BERGEN	75	69	-2	12	7.90/–	5.20/0.0	10/25	15/0	05/20	25/40	N/A	N/A	30/15	10/10	S/10	SW/<2.5	20/N/A/N/A/<10	40/N/A/N/A/0
BJORNOYA ISLAND	75	69	-4	4	SEE CLIMATIC ATLAS		20/35	10/<5	25/70	25/40	SEE CLIMATIC ATLAS		30/<10	10/<10	N/10	NE/10	20/ / /<10	20/ / /<5
JAN MAYEN ISLAND	75	69	-4	6	SEE CLIMATIC ATLAS		05/30	25/<5	25/35	20/30	SEE CLIMATIC ATLAS		10/10	10/<10	N/15	CALM	20/ / /20	30/ / /<5
HAMMERFEST	75	69	-4	12	SEE CLIMATIC ATLAS		20/35	10/<5	25/70	25/40	SEE CLIMATIC ATLAS		30/15	10/<10	N/10	NW/<2.5	30/ / /10	40/ / /<5
NARVIK	75	69	-2	12	2.80/–	2.32/0.0					5.3	0.5					/14.0/2.4/	/2.1/0.1/
SVALBARD	75	69	-12	4	SEE CLIMATIC ATLAS		25/25	20/<5	10/15	05/10	SEE CLIMATIC ATLAS		–	10/<10	NE/15	S/15	05/ / /10	20/ / /<5
TRONDHEIM	75	69	-2	12	2.14/–	3.00/0.0					1.2	0.2					/7.2/1.2/	/6.8/0.1/

(continued)

Table 2.3 (continued)

PLACE 1	PERCENT CLOUD COVER 2 JAN	JUL	MEAN AIR TEMPERATURE °C 3 JAN	JUL	MEAN PRECIPITATION (INCHES) 4 JAN	JUL	MEAN MONTHLY PRECIPITATION x FREQUENCY 5 JAN	JUL	MEAN VISIBILITY x<2NM//x<5NM 6 JAN	JUL	MEAN NUMBER OF DAYS VISIBILITY <1/2 MI 7 JAN	JUL	SEA AND SWELL ≥5'//≥12' 8 JAN	JUL	MEAN VECTOR WINDS DIR/SPEED(KT) 9 JAN	JUL	10 PERCENT OCCURRENCE FOR PARTICULAR WIND SPEEDS ≤10/>17/>28/>34(KTS) JAN	≤10/>17/>28/>34(KTS) JUL
SCOTLAND					RAIN/SNOW	RAIN/SNOW	RAIN/SNOW	RAIN/SNOW										
PRESTWICK					3.48/–	3.02/0.0					1.2	0.4					/20.1/1.5/	/6.5/0.0/
SHETLAND ISLANDS	70	75	4	12			30/10	20/0	10/80	10/75			30/10	10/<10	SW/10	WSW/05	20/ / /15	40/ / /0
WICK					2.93/–	2/57/0.0					0.3	3.5					/32.2/11.8/	/6.2/11.6/
SOUTH KOREA																		
SEOUL					1.20/4.9	14.80/0.0					1.9	1.5					/1.2/0.0/	/0.7/0.0/
SACHON	74	70			0.78/1.1	10.34/0.0	3/4	6/0			0.2	0.9	14/3	2/0	NW/15	E/10	28/1.9/0.0/0	49/1.7/0.0/0
SWEDEN																		
GAVLE	70	67	–8	16	1.10/–	2.60/0.0					N/A	N/A				N/12	/N/A/N/A/	50/12.4/1/7/0
KIRUNA	70	67	–8	16	0.70/–	2.70/0.0					N/A	N/A					/N/A/N/A/	/N/A/N/A/
UMEA	70	67	–8	16	1.40/–	1.90/0.0					N/A	N/A					/N/A/N/A/	/N/A/N/A/
USA (ALASKA)																		
ALASKA PENINSULA	80	85	–4	12														
A. COLD BAY	80	90			2.37/8.0	2.20/0.0	7/6	6/0	/18	/38	5.1	5.3	1/1	4/1	E/20	W/13	23/43.0/9.0/0	36/31.4/2.7/0
B. KING COVE	80	90			1.08/6.10	2.13/0.0	7/6	6/0	/18	/38	2.6	4.8	1/1	4/1	E/20	W/13	23/19.1/2.4/0	36/5.6/0.3/0
ANCHORAGE	80	80	–16	13	1.08/11.8	2.37/0.0			/14	/2	7.7	0.6					/2.4/0.2/	/0.6/0.0/
BARROW	50	75	–24	6	0.19/2.4	0.89/0.7			/17	/39	N/A	N/A		0/2		E/13	/N/A/N/A/	42/N/A/N/A/0
BETHEL	70	85	–16	13	1.12/10.5	2.17/0.0			/16	/22	4.7	2.7					/17.9/0.8/	/0.8/0.2/
FAIRBANKS	55	75	–24	14	0.90/12.0	1.90/0.0			/20	/4	7.1	1.3					/0.5/0.0/	/0.4/0.0/
NOME	60	80	–16	12	1.10/10.5	2.70/0.0			/18	/27	5.4	3.0		1/0		SSE/16	/22.0/2.6/	39/5.7/0.1/0
PRUDHOE BAY	45	75	–28	10	N/A	N/A	25/25	20/05	<10/70	20/70	SEE CLIMATIC ATLAS				W/<2.5	NW/05	50/ / /0	50/ / /0
VALDEZ	70	80	–16	13	5.21/60.5	4.22/0.0					N/A	N/A					/N/A/N/A/	/N/A/N/A/
USA (ALEUTIANS)	80	85	0	8														
ADAK	89	95			6.74/15.7	2.99/0.0	7/17	10/0			3.3	3.3	2/.5	3/.5	SE/20	W/12	28/36.7/8.2/0	38/20.1/2.0/0
UMNAK	90	96			2.70/–	3.70/0.0	7/14	7/0			N/A	N/A	2/1	3/0	W/20	W/14	23/N/A/N/A/0	35/N/A/N/A/0
USSR																		
ANADYR	55	80	–16	8	2.45/–	1.35/0.0					3.0	1.2					/40.4/16.4/	/26.9/2.9/
EVENSK	70	80	–28	12	1.26/–	2.09/0.0		7/0			1.1	2.5		3/0		SW/12	/48.9/16.6/	43/2.7/0.0/0
FRANZ JOSEF LAND	70	90	–16	2														
A. NAGURSKOYE					1.27/–	1.09/–					1.9	3.8					/19.5/4.7/	/13.8/1.3/
B. OSTROV HEYSA					1.74/–	1.52/–					1.3	3.0					/27.5/6.5/	/13.2/0.9/
LENINGRAD	80	55	–8	16	1.12/–	2.90/0.0					0.3	0.2					/4.8/0.1/	/0.9/0.0/
TIKSI	50	75	–36	10	1.45/–	1.78/–					2.9	2.3					/26.0/8.7/	/8.6/0.3/
MURMANSK	75	70	–12	12	1.49/–	2.19/0.0	1/17	7/0			2.5	0.3	1/0	0/0	SW/21	E/10	18/27.6/3.8/0	49/5.0/0.0/0
NEW SIBERIAN ISLANDS	35	85	–32	2														
A. OSTROV KOTELNYY					0.41/–	1.32/–					0.3	3.0					/20.6/2.0/	/22.4/2.1/
B. MYS SHALAVROVA					0.59/–	1.5/–					1.3	3.9					/19.8/3.7/	/22.6/1.1/
NOVAYA ZEMLYA	60	80	–16	8														
A. MYS ZHELANIYA					1.98/–	1.11/–					3.2	5.1					/43.4/17.1/	/33.5/8.5/
B. MALYYE KARMAKULY					1.67/–	2.40/–					1.5	2.0					/48.3/21.3/	/23.6/8.0/
PROVIDENIYA	70	60	–16	6	3.77/–	2.93/0.0	0/21	4/0			1.5	2.7	0/0	2/0	NE/23	SW/14	20/25.7/4.4/0	28/1.0/0.0/0
SEVERNAYA ZEMLYA	55	85	–28	4	0.67/–	1.99/–					1.0	5.9					/18.9/2.4/	/11.1/0.5/
WRANGLE ISLAND	59	81	–24	4	1.14/–	1.39/–					1.5	6.0					/27.4/8.2/	/9.1/0.9/

accumulation is quite low compared to the rapid buildup of ice that can occur in more southerly latitudes. The characteristics of the snow cover, such as thickness and duration, have important climatic effects on the heat and moisture exchange at the surface. Over the central Arctic Ocean, the snow cover becomes established in late August. The average thickness of the scoured and drifted snow cover may be about 350–400 mm (13.7–15.7 in) by late spring. Steady snow melt usually begins by the middle of June-caused by solar radiation-and, although there are marked differences from year to year, the pack-ice is usually snow-free by the middle of July. South of latitude 75°, the annual precipitation amounts are higher. Some points in Greenland and the European Arctic for example present an annual precipitation in excess 1000 mm (39.3 in). As would be expected from the high frequency of cyclonic activity, the Norwegian Sea area has the greatest annual precipitation of all regions north of the Arctic Circle, although a secondary maximum of cyclonic activity in Baffin Bay accounts for high precipitation amounts along the west Greenland coast and the eastern coasts of the Canadian Arctic Archipelago.

2.5 Arctic Sea Ice

The single feature that makes the Arctic Ocean markedly different from most of the world's oceans is the presence of a perennial cover of sea ice. The extent and location of the sea ice varies from year to year. Figure 2.10 shows the extent to which icebergs can extend below the waterline. This presents a hazard to passing shipping.

2.5.1 Ages and Stages of Development of Sea Ice

There are four broad categories of sea ice age: new, young, first-year, and old. There are several stages of development that occur within these ages:

(1) *New ice.* New sea ice is very elastic sea ice that may be up to 10 cm (3.9 in) thick. Depending upon the conditions under which it was formed, new sea ice may have several distinct stages:

   (a) Slush, shuga, frazil and grease ice are all comprised of unconsolidated ice crystals or platelets.
   (b) Nilas is a consolidated form of new ice that remains very elastic. It is divided into dark nilas (0-5 cm; 0–1.9 in) and light nilas (5–10 cm; (1.9–3.9 in).

(2) *Young ice.* Young ice is harder, thicker (10–30 cm; (3.9–11.8 in), and more brittle than new ice. There are two subdivisions of young ice:

   (a) Grey ice (10–15 cm; 11.8–5.9 in); breaks on swell. Usually rafts under pressure, and

Fig. 2.10 *MS Columbus Caravelle* passing Jakobshaven, Gönland

(b) Grey-white ice under pressure more likely to ridge than raft (15–30 cm; 5.9–11.8 in).

(3) *First-year ice.* Once ice has grown past 30 cm (11.8 in) in thickness, it is considered to be first-year ice. Although there is no upper limit on the thickness of first-year ice, it does not exceed 2 m. First year ice is subdivided into three categories based on thickness:

(a) First-year thin (30–70 cm; 11.8–27.5 in).
(b) First-year medium (70–120 cm; 27.5–47.2 in); and
(c) First-year thick (greater than 120 cm; > 47.2 in).

(4) *Old ice.* Old ice is ice that has survived at least one summer's melt. This ice is less saline and harder than first-year ice. There are no thickness limits on multi-year ice as the distinction is based on physical properties and not thickness. There are two categories of old ice:

(a) Second-year ice—ice that has survived one summer's melt; and
(b) Multi-year ice—ice that has survived two or more summers' melt.

*Approximately

Fig. 2.11 Floe ice

2.5.2 Concentration

The amount of ice or the area coverage of sea ice is measured in tenths. The following descriptions apply to various concentrations:

(1) Very open ice—1 to 3 tenths.
(2) Open ice—4 to 6 tenths.
(3) Close ice—7 to 8 tenths.
(4) Very close ice—9 to less than 10 tenths; and
(5) Compact or consolidated ice—10 tenths.

2.5.3 Floe Size

Floe size varies with location and time of year. Waves can break the floes up into smaller pieces. A descriptive representation of floe size is found in Fig. 2.11.

2.5.4 Ice Topography

When driven by the forces of wind, seas and currents, sea ice may take many forms. As the ice is forced together, it will fracture to raft or ridge on top of itself. When the ice is forced apart, it will break to form fractures and leads. Similarly, at the ice edge, the forces of nature may create "ice tongues," "belts and strips," or a "diffuse ice edge." During the summer melt, melt ponds and thaw holes will appear on the ice (Figs. 2.12 and 2.13).

Fig. 2.12 Frazil ice

Fig. 2.13 Nilas ice

2.5.5 Ice Formation

When salt is added to fresh water, the temperature at which it freezes is lowered. The higher the salt content, the lower the freezing point of the salt solution. Seawater, with a salinity of approximately 32 parts per thousand in the Arctic, freezes at about −1.8 °C (28.7 °F). Unlike fresh water, the temperature of the maximum density of seawater is lower than the freezing point (for salinities greater than 24.7 parts per thousand). Consequently, as seawater cool, it becomes increasingly dense and sinks. Theoretically, for ice to form, the entire water column from the surface to the bottom must be cooled to the freezing point. In reality, only the upper layers of the water column must be cooled because deep water that is more saline provides the water column stability that is necessary for ice formation. As the ice forms, there is no room in the crystal structure for the dissolved salts. As a result, these salts are expelled from the ice. If the ice were formed very slowly it would be pure. However, the freezing process is never a slow one and the salts are trapped within the ice structure as it freezes. The salinity of sea ice is about 4–6 parts per thousand.

2.5.6 Structure and Properties of Sea Ice

When seawater is cooled to its freezing point and more heat is removed, ice forms initially as very thin disks or platelets known as frazil ice. These platelets average 2–3 mm (0.07–0.11 in) in width and about 0.5 mm (0.01 in) in thickness but vary in shape from hexagonal "snowflakes" to almost square plates. As further heat is removed, these pure ice crystals grow and multiply. They are less dense than water, so they float to the surface and give the water a slightly oily appearance (grease ice). If the water is turbulent; these ice crystals will be entrained in the water. Further cooling results in growth of the ice crystals and mechanical entrapment of small brine cells between them. These cells become separated from the water below the ice by selective downward growth of the ice crystals. The physical properties of sea ice are entirely dependent on its salt content. The detailed crystal structure, which results dominates these properties to the extent that one can relate any of them to "brine volume." The brine volume is defined as the fraction of the volume of sea ice occupied by fluid (liquid brine or air bubbles). The strength of the ice is dependent upon the brine volume. The larger the brine volume, the weaker the ice is. For this reason, old ice has a much lower salinity than first-year ice and is much stronger. For first year sea ice, the strength of the ice is also related to the temperature of the ice. The colder the ice, the stronger it will be. When the daily air temperature is near zero, the ice will have approximately 10% of its mid-winter strength.

2.5.7 *Annual Ice Cycle*

The process of ice formation has been briefly discussed above, but this is only the initial stage of ice growth, which is always dependent upon heat loss from the sea. This heat must flow upward through the ice layer and through any snow lying on the ice so that the insulating qualities of these two materials are key factors in determining the rate and amount of ice growth. Air temperature and the amount of radiant energy falling on the surface are also key factors. In the Arctic, radiation is completely dominant in determining the ice surface temperature and hence the duration of ice growth. This will continue until sometime in spring when the increasing solar radiation changes the heat budget of the ice from a loss to a gain. This usually happens before the air temperature rises above the melting point of the snow cover. Pure white snow reflects as much as 90% of the radiation falling on it, and although the quantity of heat absorbed by the ice is small at first, the 24 h of daylight soon result in a gradual increase in the temperature of the ice. When the air temperature reaches the melting point, the snow surface begins to melt rapidly; it absorbs about 60% of the radiant energy and puddles of melt-water become very extensive. In temperate latitudes, the air temperature can contribute to the change from heat loss to heat gain by the ice, for it can rise above the melting point for appreciable periods, if only in the daytime. The situation is thus more complicated, but the reversal of heat flux is still the controlling factor.

Puddling is the first apparent stage of deterioration of the ice cover. The water surface absorbs heat readily, permitting the puddle to widen and deepen and also to warm the ice. Later, flaws and cracks develop in the floe through which much of the surface water drains away, leaving dry hummocks of ice separated by ponds and streams of melt-water. When the ice has this appearance, operations on it must be conducted with caution for the bearing strength is uncertain and a wind can cause it to break up rapidly. In temperate latitudes the ice is reduced to a grey, water-saturated matrix (rotten ice), which finally melts, and the cycle is complete. Farther north, the summer is too brief for complete melting to take place, and by late August or September puddles begin to freeze and new ice forms between the floes. After a time, the floes themselves start growing again. A floe formed in one year, which survives through the following summer differs chemically and physically from ice that is less than one year old. During the summer most of the brine drains out of the ice so that the typical salinity of old ice (secondary multi-year ice) is about 0.5–1 part per thousand. Melt-water from this ice is quite potable. The crystal structure of the ice becomes less regular, the crystals themselves are smaller, and the ice is extraordinarily tough, even in summer. In summer, old floes may be distinguished from first-year ice by their colour. The melt- water puddles on an old floe have a very characteristic pale-blue colour, which persists after they freeze. On first-year ice, these puddles have a green-to-brownish appearance. The old ice itself has a pale-blue colour whereas first-year ice is much more a greenish-white colour. The surface of first year ice is comparatively smooth except for pressure ridges and hummocks. Old ice has a characteristic uneven surface as a result of the differential melting of puddles and old hummocks.

2.5.8 *Occurrence of Sea Ice*

In the Arctic there are no generalizations that can be made about the occurrence of sea ice in relation to latitude. This is evident when comparing the winter maximum extent of sea ice in the Sea of Okhotsk and the Gulf of St. Lawrence, which extend south to 45° North, and winter maximum extent in the Norwegian Sea, which never penetrates south of 70° North. The amount of seasonal variation in the sea ice cover is considerable and varies markedly from region to region. The Permanent Sea Ice Zone is the region that is perennially covered with sea ice. The seasonal sea ice zone is the region that is ice covered only part of the year and extends from the summer minimum sea ice extent to the winter maximum sea ice extent. The mean thickness of sea ice varies by region. The average thickness for undisturbed old sea ice in the central Arctic is about 3.5 m (11.4 ft). These thickness averages are misleading because the dynamic nature of sea ice creates ridges and the corresponding keels in the ice. Ridges of up to 15 m (49.2 ft) and keels of over 45 m (147 ft) have been found. This type of ridging is particularly evident where the ice pressure is impinging upon a coast. Submarine and, ice-breaker observations have shown that the normal ratio between ridge height and keel depth is between 1:4 and 1:6, with ridge/keel ratios being observed over a very wide range.

2.5.9 *Ice Forecasting*

The date of freeze-up of the sea depends on both the oceanographical and meteorological regimes encountered in the area. Only rarely is the knowledge of water currents, and of the actual temperature-salinity variations with depth, available for forecasts of this type to be made. The rate of ice growth on the other hand can be predicted with fair accuracy from meteorological data, as can the maximum thickness, which will be obtained. It is much more difficult to predict rates of decay, for the process is slow and long-range predictions of both wind and cloudiness are required for accurate results. One of the most important forecasting problems is ice motion. The general pattern of sea ice motion in the Arctic is indicated in Fig. 2.14. Recent studies have revealed that 70% of the ice motion is due to wind forcing. This pattern is of course affected by tides and currents as well as, land masses, and bathymetric features such as shoals. A general rule of thumb is that the ice will drift about 45° to the right of the wind direction at about two percent of the wind speed. In the eastern part of Greenland, the lack of information about the drifting ice makes navigations hazardous. Most of ice charts of this area figure the sea ice edge, but not the drifting

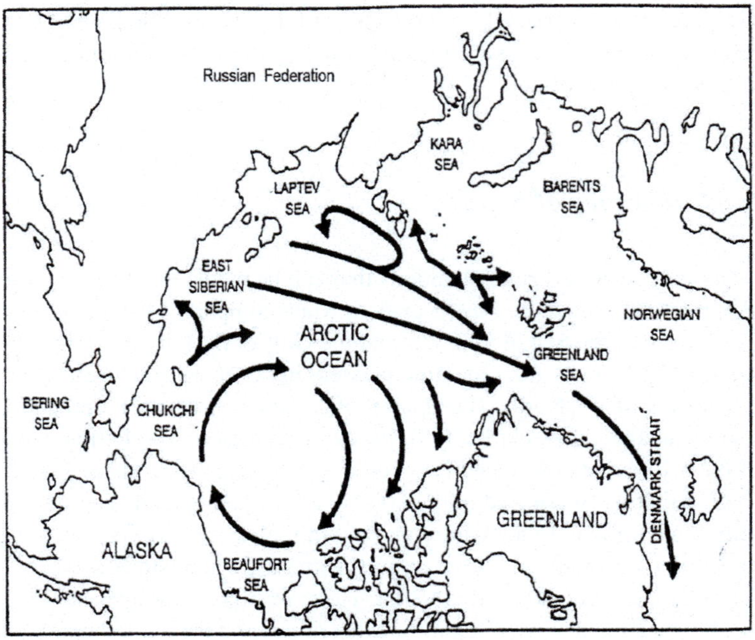

Fig. 2.14 General pattern of ice movement in the Arctic Ocean

ice. Further information about ice is available on the website: https://arcticweb.e-navigation.net/ but this website needs to be registered first. Surface temperature prediction charts are the most relevant tool to estimate ice position precisely, considering the following scale (Fig. 2.15):

- Less than 0 °C (32 °F) = ice edge.
- Between 0 °C (32 °F) and 4 °C (39.2 °F) = bergy waters.
- More than 4 °C (>39.2 °F) = open waters.

Ice routing information for the Arctic can be obtained from the following national agencies:

State	Agency	Services provided
Canada	Environment and Climate Change Canada, Canadian Ice Service (CIS), Ottawa	Information on sea ice and icebergs as charts, bulletins and warnings are available through weather and marine radio broadcasts and web sites: https://www.ec.gc.ca/glaces-ice/ Detailed ice analysis and satellite imagery are also available. Data is available in paper as well as electronic formats

(continued)

(continued)

State	Agency	Services provided
Canada	Canadian Coast Guard Ice Operations	Ice routing information for the Canadian Arctic can be obtained via any Coast Guard Radio Station. Further information on contacting Ice Operations may be obtained from the Ice Navigation in Canadian waters (TP 5064E)
Denmark	Danish Ice Information Service (DIIS), Narssarssuaq, Greenland	Ice charts of the areas East, West and South of Greenland are promulgated on a regular basis. Ice charts and ice reports can be obtained from all Greenland Coastal Radio Stations and on the internet at http://www.iserit.greenet.gl/isc/ice Information about icebergs in Greenland waters is available
Germany	Federal Maritime and Hydrographic Agency	Ice routing information can be obtained from this civil agency http://www.bsh.de/oceanography/ice/ice/htm
United States	National Ice Center (NIC), Washington DC	The NIC provides operational sea ice analyses and forecasts for the Arctic, Antarctic, Great Lakes and Chesapeake Bay. Their products and services can be accessed at http://www.natice.noaa.gov
United States	US Coast Guard International Ice Patrol (IIP), Groton, CT	The USCG IIP monitors iceberg conditions in the vicinity of the Grand Banks of Newfoundland, and pending severity, broadcasts the Southeastern, Southern and Southwestern limits of all known icebergs in two daily message bulletins. The IIP can also provide a daily fax chart containing ice information. For a comprehensive listing of products and services and instructions on accessing them use the IIP homepage at http://www.uscg.mil/lantarea/iip/home.html

2.6 Other Ice Formations Encountered at Sea

2.6.1 River Ice

There is appreciable difference in the strength of freshwater ice and sea ice, and consequently, in or near the estuaries of major rivers an additional hazard to shipping may be encountered in the spring when river ice is carried into the sea. This is particularly true of the rivers of the Russian Arctic.

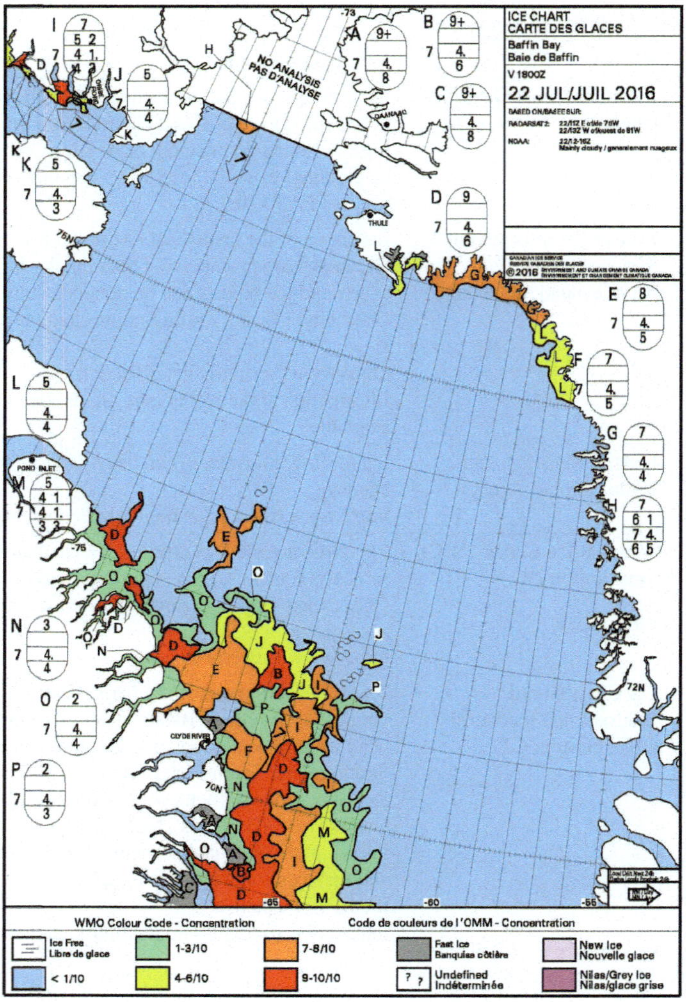

Fig. 2.15 Example of an ice chart using the WMO egg code

2.6.1.1 Ice Islands

These are masses of ice that have broken away from an ice shelf and have an undulating surface. They may have thicknesses of up to 60 m (196 ft) with areas of up to 400 km^2 (248.5 mi^2). In the final stages of melting, they usually break up into a group of tabular icebergs.

Fig. 2.16 Iceberg in the Arctic Sea with underside exposed

2.6.1.2 Icebergs

Icebergs (Fig. 2.16) are large masses of freshwater ice and compacted snow that have broken away or "calved" from a glacier. The Greenland Ice Cap is the single largest source of icebergs, with the largest concentration of icebergs found in Baffin Bay and the Davis Strait. In the East Greenland Sea icebergs are found imbedded in and sometimes outside the drift-ice, particularly south of Scores by Sound. Icebergs have been found in excess of 90 m (295 ft) in height and 500 m (1640 ft) in length. Their draft varies from two to more than 10 times their height due to their irregular shapes. Icebergs do not always travel in the direction of the wind. They have a small sail area relative to total size and travel in the direction of the current, which may be against the wind. Figure 2.17 shows the general drift pattern of Atlantic icebergs. Figure 2.18 shows typical areas where icebergs can be anticipated in the Pacific region.

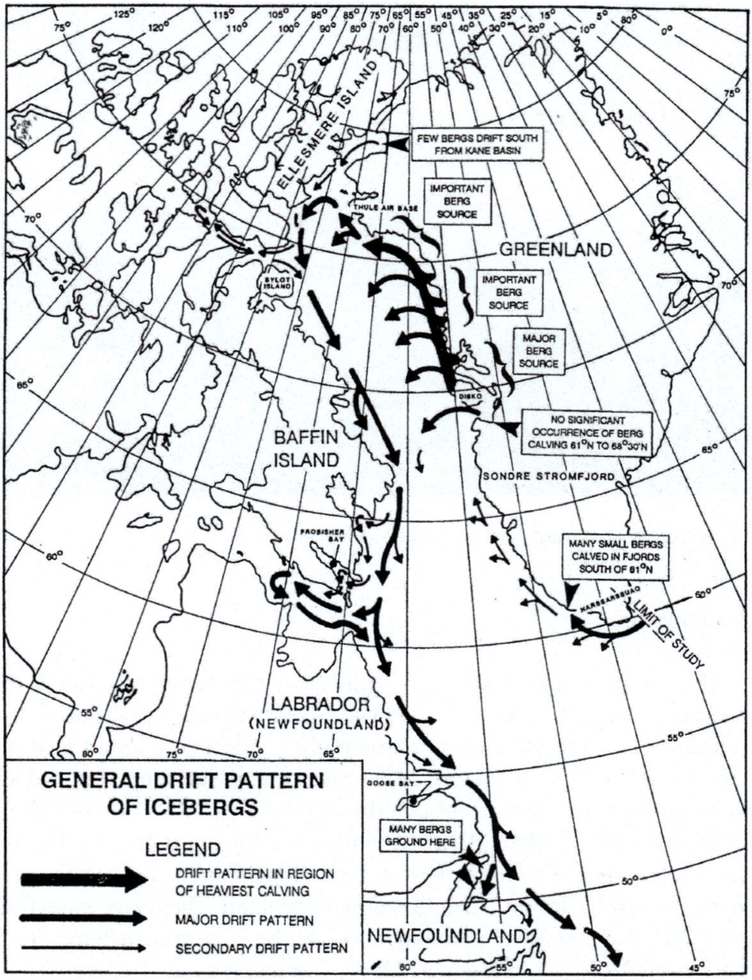

Fig. 2.17 General drift pattern of Atlantic icebergs

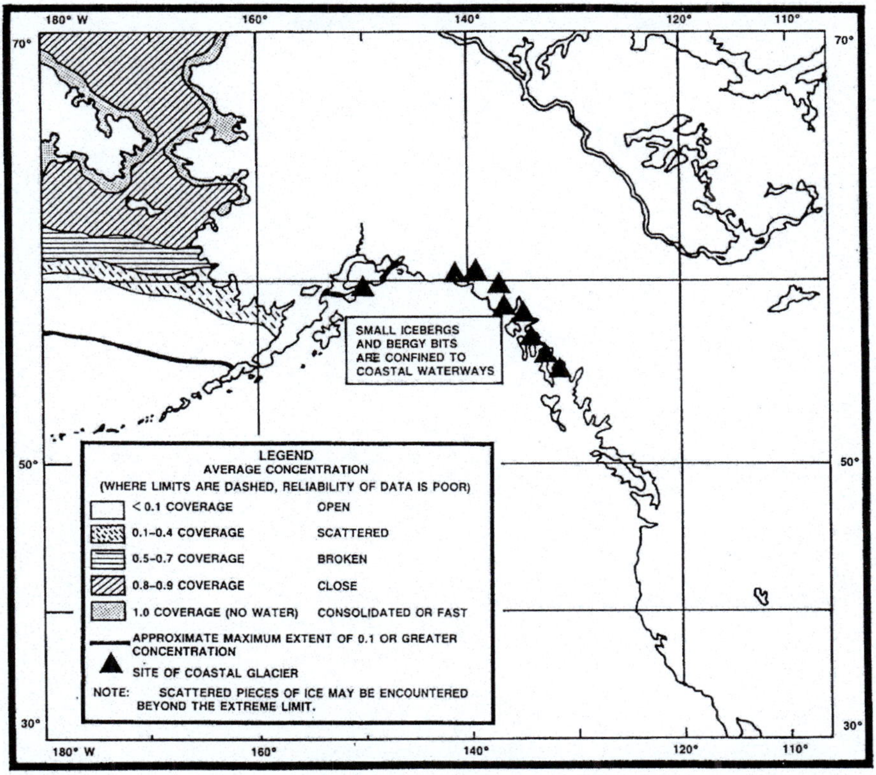

Fig. 2.18 Pacific Ocean icebergs

Chapter 3
Arctic Oceanography

3.1 Introduction

The Arctic Ocean surrounds the North Pole and is defined by the bordering land masses, of Siberia, Alaska, Canada and Greenland. In an oceanographic sense, it is a single large basin connected primarily with the Atlantic Ocean through two other major Arctic seas: Baffin Bay and the Greenland Sea. The water characteristics within these smaller basins can be traced in large part to North Atlantic water characteristics, which, have been modified through surface-acting processes associated with the unique climatic conditions of the Arctic. The characteristics of the peripheral continental Arctic seas are highly variable.

3.2 Bathymetry

The bathymetry of the Arctic basins (Fig. 3.1) is now generally known, though in certain areas, notably along the edge of the continental shelf within the Arctic Ocean, detailed information is lacking. The continental shelf on the North American side of the Arctic Ocean is narrow, 27–40 nm (50–75 km; 31–46.6 mi), whereas the sector bounded by Europe and Siberia is shallow and very broad, 220 nm (400 km; 248.5 mi), with peninsulas and islands dividing it into five marginal seas: the Barents, Kara, Laptev, East Siberian, and Chukchi. The Eurasian marginal seas occupy 36% of the area of the Arctic Ocean but contain only 2% of its volume of water. All the major continental rivers reaching the Arctic Ocean, excepting the Mackenzie, flow into these seas. The combination of the marginal seas, characterized by a high ratio of exposed surface to total volume, and a substantial input of fresh water in summer, influences surface water conditions in the Arctic Ocean. The margin of the continental shelf is indented by numerous submarine canyons in the northern Kara Sea and sea valleys in the Chukchi Sea. Oceanographically, the canyons are

A. Olsen, *Ship Operations in Extreme Low Temperature Environments*, Springer Series on Naval Architecture, Marine Engineering, Shipbuilding and Shipping 19,
https://doi.org/10.1007/978-3-031-52513-1_3

important as they function as preferential pathways for the egress of water from the warm Atlantic intermediate layer onto areas of the continental shelf, where it comes within the influence of strong mixing processes and can locally modify the surface waters and ice cover.

The Lomonosov Ridge with sill depth of about 1390 m (4560 ft; 760 fathoms) divides the Polar Basin into two deep basins: the Eurasian on the European side (3600 m; 11,811 ft; 1970 fathoms), and the Canadian on the North American side (3250 m; 10,662 ft; 1775 fathoms). The latter is further subdivided by the Alpha Rise, which runs parallel to the Lomonosov Ridge at 270 nm (500 km; 310 mi) distance. The continental shelf along east Greenland, south to 77°N, is broad (160 nm; 300 km; 186 mi) and contains a system of banks no deeper than 200 m (656 ft; 110 fathoms). South of 77°N, the shelf narrows until between 75°N and the Kap Farvel (Cape Farewell) it is less than 54 nm (100 km; 62.1 mi) wide. The shelf is marked by several deep indentations, particularly in the southern part. At about 79°N, the Greenland

Fig. 3.1 Arctic Ocean (bathymetry)

Sea is separated from the Polar Basin by a rise, with a sill depth of about 2800 m (9186 ft; 1530 fathoms). South of the rise, the Greenland Sea contains two deep basins. At 71°N, the Jan Mayen Ridge extends towards Greenland, with a maximum depth of about 1520 m (4968 ft; 830 fathoms), while in Denmark Strait the sill depth is about 550 m (1804 ft; 300 fathoms). Around Baffin Bay, the shelf is everywhere, narrow and nowhere exceeds a width of 54 nm (100 km; 62.1 mi). The shelf is widest off middle Baffin Island, Labrador, and west Greenland south of 78°N. The latter shelf contains numerous important fishing banks. Central Baffin Bay is a basin with depths exceeding 2000 m (6561 ft; 1,100 fathoms). The sill across Davis Strait appears to be about 600 m (1968 ft; 330 fathoms) deep. Baffin Bay connects with the Arctic Ocean through a network of channels ending in three sounds: Smith, Jones and Lancaster. The Jones Sound connection is quite restricted but the channels leading to Lancaster and Smith sounds are deep, the approximate sill depths being 150–200 m (492–656 ft; 82–110 fathoms) respectively.

3.2.1 Water Masses

In the Arctic, the most important processes conditioning and modifying the ocean water are:

(1) Addition of fresh water from the land, primarily from the large Siberian Rivers.
(2) Addition of fresh water locally through melting of ice.
(3) Heat gain through absorption of solar radiation in non-ice-covered areas during summer.
(4) Concentration of salt, and hence increase of density of surface water, through freezing and convective overturn.
(5) Heat loss to the atmosphere through any open water surface, including leads in the central Arctic pack-ice.

The addition of fresh water and heat gain (1, 2 and 3) lead to decreases in the density of the ocean water and occur only during the summer (June–September). The concentration of salt and heat loss (4 and 5) lead to increases in the density of the surface water. On the basis of temperature, the three water masses may be defined as follows:

(1) Surface water (Arctic water), from the surface down to about 200 m (656 ft; 110 fathoms), which has varying characteristics. In ice-covered areas the water temperature is close to that of freezing for the salt content. In the usually ice-free areas (eastern Greenland Sea; along west Greenland north through Davis Strait) temperatures may be a few degrees above freezing. Areas that are ice-free in summer (Chukchi Sea; near-shore areas of other peripheral seas; most of Baffin Bay) may seasonally exhibit surface temperatures of 1–2 °C (33.8–35.6 °F) or more. Temperatures below the surface are typically always cold, except in the Canadian Basin of the Arctic Ocean, where there may appear a

small temperature maximum (−1.0 °C (30.2 °F) in the 75–100 m (246–328 ft; 40–55 fathom) layer. The salinity of the surface layer may be uniform down to about 45 m (147 ft; 25 fathoms) and then increase until at 200 m (656 ft; 110 fathoms) it is 34.5 parts per 1000. The layer below the Arctic water, from about 200–900 m (656–2952 ft; 110–490 fathoms), known as the Atlantic layer, has temperatures above 0 °C (32 °F), with a maximum at 300–500 m (984–1640 ft; 165–275 fathoms). Salinities continue to increase over the surface values until by 400 m (1312 ft; 220 fathoms), and in many instances shallower depths, they attain in the Greenland Sea and Arctic Ocean uniform values in the range 34.9–35.0 parts per 1000, and in Baffin Bay, 34.5–34.6 parts per 1000.

(2) Beneath the Atlantic layer is bottom water with temperatures below 0 °C (32 °F) and the same uniform salinities attained in the Atlantic layer. Deep temperatures vary slightly from basin to basin: in the Canadian Basin they are about −0.5 °C (31.1 °F); in the Eurasian Basin −0.9 °C (30.38 °F); in the Greenland Sea −1.2 °C (29.84 °F); and in Baffin Bay −0.45 °C (31.19 °F).

The density of freezing water is much more strongly influenced by salinity than by temperature: thus, the vertical distribution of density closely parallels that of salinity. On the basis of density, the Arctic waters show a two-layer system, with a thin, less dense surface layer separated from the main body of water of quite uniform density. Vertical motion and the vertical transfer of heat and salt are therefore restricted, and hence the surface layer acts as a "lid" over the large masses of warmer water below. There is little spatial variation of surface temperature throughout the Arctic. Only those areas that are normally ice-free all year-round exhibit temperatures significantly above freezing. These areas are influenced by currents carrying warmer water into the Arctic (eastern Greenland Sea-West Spitsbergen Current, and the northward-flowing current along west Greenland) and remain ice-free for that reason. Seasonal temperature fluctuations occur in areas that are typically ice-free seasonally (July–September) and include the coastal sectors of the peripheral seas of the Arctic Ocean, north of Bering Strait, and around eastern and northern Baffin Bay. Areas in which major currents carry Arctic water towards the North Atlantic (western Baffin Bay-Canadian Current; western Greenland Sea-East Greenland Current) remain ice-covered for longer periods and have temperatures close to freezing at all times. In winter (October–April), a process of considerable importance in modifying the surface waters takes place. When ice grows from sea-water the salt is excluded from the ice and results in a local increase in salinity and hence density of the remaining water. This process takes place over the entire surface of the Arctic areas, and the average ice growth is about 1.5 m (4.92 ft) each winter. However, the ice cover is not continuous, and leads are continually opening and closing. The heat loss to the atmosphere from the open water of leads occurs 100 times faster than through ice (also producing fog in the atmosphere).

3.2.2 Currents

The circulation of the Arctic water is created by the rotation of the earth, and by density differences, and is also wind-induced. In the Arctic Ocean, the dynamic topography provides a quantitative estimate of the currents (refer to Fig. 3.2). The net effect of tides is unknown though there may be some asymmetry in their action, which would modify the circulations. The surface waters from the whole Eurasian side of the Arctic Ocean tend to move towards the North Pole. This flow is on the average slow, 2–3 cm/s (0.7–1.1 in/s; 0.04–0.06 kn), but after passing the region of the Pole it becomes more concentrated and then exits from the basin as the East Greenland Current. In the Beaufort Sea, the surface waters have a clockwise movement, a result of the general wind pattern; they tend to flow to the west along the shelf off the Canadian Arctic Islands, and to the north in the area north of the Bering Strait. Around the Greenland Sea there is a large cyclonic circulation, with average speeds in the range 10–20 cm/s (3.9–7.8 in/s; 0.2–0.4 kn). Inflow of North Atlantic water, both at the surface and at deep levels, occurs along the east side of the sea as the West Spitsbergen Current. The East Greenland Current is the major flow south on the west side. The surface water from the Arctic Ocean contributes to the upper layers, while the deeper waters are from the West Spitsbergen Current, completing the cyclonic gyre. The current closely follows the continental slope. Over the wide continental shelf of the northern area (77–80°N) the currents tend to be weak and variable. The East Greenland Current seems to accelerate towards the south, attaining speeds of 15-40 cm/s (5.9–15.7 in/s; 0.3–0.8 kn) near the Denmark Strait. The same general pattern of circulation is found in Baffin Bay. A cyclonic circulation dominates the bay; inflow of North Atlantic water occurs along western Greenland through the Davis Strait, and inflow from the Arctic Ocean through Smith, Jones and Lancaster sounds. The Canadian Current runs south along Baffin Island and, as it accumulates water from the various inflows, it in general shows higher speeds towards the south.

The circulation of the Atlantic layer (refer to Fig. 3.3) has been deduced from the distributions of temperature and salinity. Recent direct current measurements from Russian and US drifting stations in the Arctic Ocean and East Greenland Current in this layer (300–1000 m; 984–3280 ft; 165–550 fathoms), shown as vectors in Fig. 3.3, confirm the general pattern of motion. On entering the Eurasian basin from the Greenland Sea much of the water flows east along the edge of the Eurasian continental slope. The water enters the Canadian Basin on a broad front across the Lomonosov Ridge. There appears to be a general cyclonic circulation in the Eurasian Basin, and a smaller anticyclonic gyre in the Beaufort Sea. Speeds are everywhere low, less than 5 cm/s (1.9 in/s; 0.1 kn).

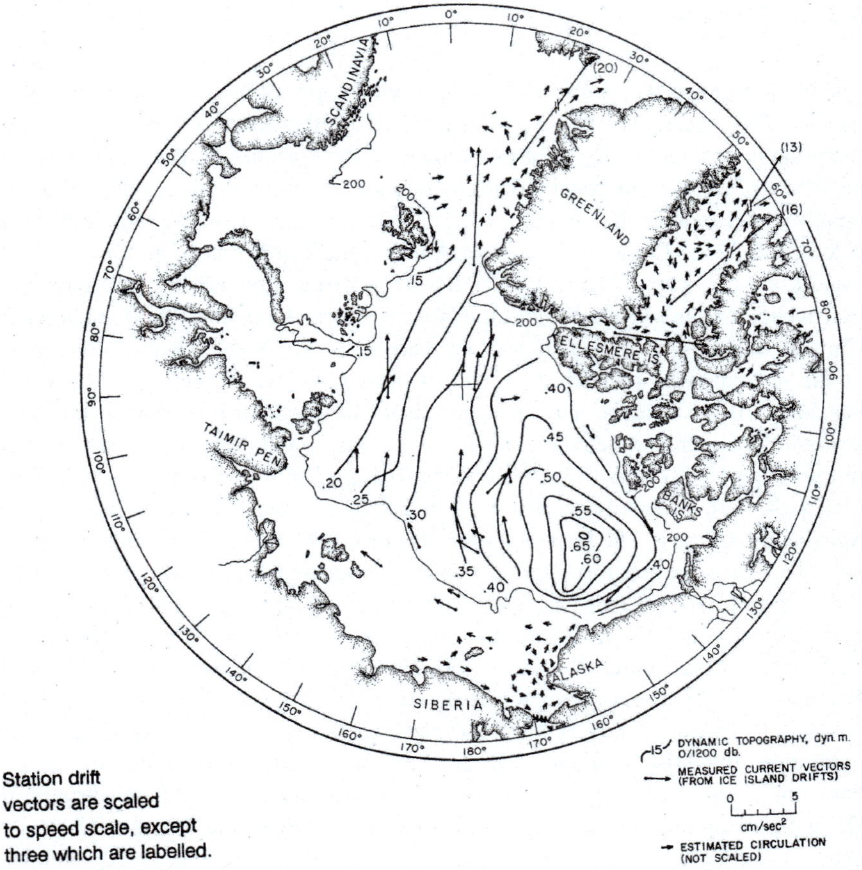

Fig. 3.2 Composite surface circulation

3.2.3 *Advection Boundaries*

Boundaries between the Arctic seas and the Atlantic and Pacific, and between the Arctic Ocean and Baffin Bay, are of prime importance because of the significant role of advection in determining the characteristics of Arctic waters. The best known oceanographically of these is Bering Strait, through which, in summer, Bering Sea surface water flows north into the Arctic Ocean through the Bering Strait. In the eastern channel of the strait, speeds normally range between 50 and 100 cm/s (19.6–39.3 in/s; 1–2 kn) though speeds over 150 cm/s (59 in/s; 2.9 kn) have been measured. Speeds are less in the western channel. The volume transport is about $1 \times 106\text{m}^3\ \text{s}^{-1}$. The situation in winter is unknown, though it has been suggested that the northward flow may be only one-fourth that of summer and may even reverse on occasion. The general flow through the Canadian Archipelago is from the Arctic Ocean towards Baffin Bay. Recent documentation of the drift of ice island WH-5 through Nares

Fig. 3.3 Composite Atlantic layer circulation

Strait (connecting to Smith Sound) confirms the general flow out of the Arctic Ocean through this strait. However, there was evidence of large pulsations in the southerly flow; the indicated periodicity of a few days to weeks suggests that major atmospheric disturbances may be important in significantly altering the flows through these channels. The strait between Greenland and Svalbard provides the primary connection between the waters of the North Atlantic and the Arctic Ocean, water flows into the Arctic Ocean on the eastern side of the strait and out of the Arctic Ocean on the western side as the East Greenland Current. A new concept of the circulation in the Greenland Sea and of the East Greenland Current has resulted from analysis of current measurements made from ice island Arlis-II during its drift along eastern Greenland during winter, 1965. The measurements showed the volume transport of the current to be about $40 \times 106\text{m}^3\ \text{s}^{-1}$, an order of magnitude larger than previously estimated. This large transport is a major circulation internal to the Greenland-Norwegian seas; the outflow and inflow from the Arctic Ocean represent only minor contributions and subtractions from a large cyclonic circulation. The Arctic water portion of the current to a considerable extent controls the ice distribution, and so its presence is significant out of all proportion to its small contribution to the total transport. The flow pattern in Denmark Strait is similar to that in the strait between Greenland and Svalbard; the East Greenland Current flowing south dominates the western side, and a flow from the North Atlantic runs north and east around Iceland. The East Greenland Current, transporting polar ice south, usually occupies three fourths or more of the surface of the strait. The flow regime varies seasonally, with summer usually the time of minimum ice. The Atlantic water does exercise considerable influence in the Norwegian Sea and along West Greenland, where it is at the surface and the surface is exposed. The Atlantic water also exerts some influences on ice cover and climate in the seas bordering Siberia. It enters these seas along submarine canyons, particularly in the Kara Sea, where it works its way up to shallower depths. Through a more pronounced influence of tidal action in the shallower water and the vigorous development of vertical convection in winter, the heat of the Atlantic water is carried to the surface where it is responsible for areas of much thinner ice cover and even for semi-permanent areas of open water (polynyas). A similar effect occurs in northern Baffin Bay, where an area of 20,000 km^2 (12,427 mi^2) of Smith Sound has typically very thin ice or even open water all winter ("North Water"). The cause of the "North Water" is unknown, though clearly some source of warmer water near the surface, as with the Siberian polynyas, is required for its existence. In general, within the deep-water part of the Arctic seas, the warm Atlantic layer has negligible influence on climate.

3.3 Western North American Waters

3.3.1 Bathymetry

The continental shelf between Point Barrow and Cape Prince Alfred varies in width from less than 43 nm (80 km; 128.7 mi) in the west to 86 nm (160 km; 99.4 mi) in the centre and northeast. Off Alaska the cross profile is characteristic of the continental oceanic margin in all parts of the world, consisting of the flat shelf and a steep upper continental slope that decreases as the Canada Basin is reached at about 3660 m (12,007 ft; 2000 fathoms). East of the border the slope diverges from the mainland in a swinging arc towards northwestern Banks Island, and the upper slope is gentler in this area than farther west. The continental shelf is crossed by several deep valleys. In the west is the Barrow Sea Valley, which originates off Cape Franklin and deepens north eastwards to become 19 nm (35 km; 21.7 mi) wide and U-shaped north of Point Barrow. A second valley is 50 nm (90 km; 55.9 mi) to the east. Northwest of

Mackenzie Bay is the Mackenzie Sea Valley, a broad, flat-bottomed valley leading from the outlet of the Mackenzie River; 120 nm (225 km; 139 mi) to the northeast a smaller valley has been recognised. The largest of the shelf valleys is a broad, asymmetrical feature, the Amundsen Trough, between Banks Island and the mainland. It has an average width of 22 nm (40 km; 24.8 mi) and the floor is at about 440 m (1443 ft; 240 fathoms).

3.3.2 *Currents*

Surface currents in the Beaufort Sea are light and irregular, depending on wind and pressure changes. The clockwise circulation of the Arctic basin sets southwest and west at 2.5 nm/day (5 cm/s; 1.9 in/s) over the continental slope, but there is little evidence that it is found inshore. Along the Alaskan coast there is thought to be a reverse (eastwards) current.

3.4 Eastern North American Waters (North of a Line, Resolution Island—Kap Farvel)

3.4.1 *General*

The eastern North American Arctic is centred around the waters of Baffin Bay which is connected to the Arctic Ocean through the channels leading west and north through the Queen Elizabeth Islands, and to the Atlantic Ocean through Davis Strait into the Labrador Sea. The body of water separating west Greenland and northeastern Canada forms a 1025 nm-long channel (1900 km; 1180 mi) from the Arctic Ocean to the Labrador Sea. The waters in eastern North America reflect the meeting of warmer waters from the Atlantic with the frigid waters of the Arctic Ocean and are in a large measure controlled by the bathymetry of the area. Baffin Basin, with a maximum depth of 2000 m (6561 ft; 1100 fathoms), is separated from the Labrador Sea by the Davis Strait ridge, which has maximum depths near 730 m (2395 ft; 400 fathoms). The adjacent channels shallow to depths of 145–185 m (475–606 ft; 80–100 fathoms) to limit the influence of the Arctic Ocean to those waters found above those depths. Sea surface temperatures reach a maximum in August with a high of 8.9 °C (48 °F) in the Labrador Sea. Temperatures above 4.5 °C (40.1 °F) are found along the Greenland coast to 73°N and in the "North Water" off the entrance to Lancaster Sound. In the channels, temperatures are near 0 °C (32 °F) and warm to 1.7 °C (35 °F) as the waters approach Baffin Bay. The Arctic influence of the Canadian Current along the east coast of Baffin Island is reflected in the depressed temperatures, 0.3–3.3 °C (32.5–37.9 °F). Summer salinities are as low as 25 parts per 1000 in the islands near the ice edge. Along the coasts they are near 30 parts per 1000 in the Labrador Sea.

Tides in the area are of Atlantic origin and course through Davis Strait. Only in the extreme north are Arctic tides in evidence. Time of high waters at the head of the Gulf of Boothia occurs 12 h after that in the southern limits of the area. Tidal ranges are near 2–3 m (6.5–9.8 ft) over most of the region but go to 4 m (13.1 ft) on the southern side of Davis Strait and are over 12 m (39.3 ft) at the head of Frobisher Bay. Tidal currents are extremely variable throughout the region, being light in open waters and as high as 7 kn (12.9 km/h; 8 mi/h) in areas of large tidal ranges.

The general surface circulation, like the temperature and salinities, reflects the influences of the Arctic and Atlantic Oceans. The West Greenland Current flows northward from Kap Farvel at speeds of 0.5 kn (0.9 km/h; 0.5 mi/h) in Davis Strait. This current carries around the head of Baffin Bay where it is joined by waters flowing southward through the islands to form the Canadian Current, which flows along the east coast of Baffin Island. After crossing the Davis Strait Ridge, the Canadian Current is joined by the west-flowing branch of the West Greenland Current to form the Labrador Current. In the channels between the islands, currents are about 0.5kn (0.9 km/h; 0.5 mi/h), while the Canadian Current has velocities of about 0.2 kn (0.37 km/h; 0.2 mi/h). In the Gulf of Boothia, the waters flow southward and exit through Fury and Hecla Strait into Foxe Basin. Because eastern Arctic America extends through more than 20 degrees of latitude, there are extensive climatic differences. In general, the Canadian portion is colder than the Greenland coast in winter, with many storms in the southern sector. During the summer season there is a greater uniformity of weather over the whole area. The cold sea has a strong influence on coastal weather, creating much fog and cloud and keeping temperatures below 15 °C (49.2 °F).

3.4.2 Bathymetry

The bathymetry of the northeast American Arctic consists of the shallow areas to the north, in which are located the islands of the Canadian Archipelago with deep channels, and the deep Baffin Bay basin (2300 m; (7545 ft; 1257 fathoms) which is separated from the Atlantic by the Davis Strait Ridge. At the entrance to the area from the Atlantic, the bottom rises gradually from the depths of the Labrador Basin (4,000 m; 13,123ft; 2,200 fathoms) to the Davis Strait Ridge where depths are less than 800 m (2624 ft; 440 fathoms). This ridge separates the Labrador Basin from the Baffin Bay Basin where depths exceed 1825 m (5987 ft; 1000 fathoms). On the Canadian side of the Labrador Sea, the 200 m (656 ft) line is very narrow and is cut by the deep fiords of Frobisher Bay and Cumberland Sound. The Greenland shelf is well defined and narrow near Kap Farvel, and broadens northward to form Fylla, Little Hellefiske and Great Hellefiske Banks. North of Disko Island the continental shelf narrows and all but disappears near Cape York. Along the Baffin Island shore the shelf is narrow. The channels between the Arctic islands are deep where they join the major basins in the area. However, in every case they have shallow ridges which limit the depth and hence the types of waters that may be drawn through them. The narrow channel that separates Canada from northwest Greenland consists of two basins, Hall and Kane, joined by Robeson and Kennedy Channels and Smith Sound. In the north, at the narrowest par the two shores, are separated by only 11 nm (20 km; 12.4 mi) of water. The shallowest part is in the north part of Kane Basin where the sill depth is 90–185 m (295–606 ft; 50–100 fathoms). The floor slopes away in both directions to 365–450 m (1197–1476 ft; 200–250 fathoms) in the Lincoln Sea, and to about 650 m (2132 ft; 355 fathoms) at the south end of Smith Sound. In the central part of Jones Sound, depths of 730 m (2395 ft; 400 fathoms) have been reported. Lancaster Sound has a wide flat floor that drops eastwards to 825 m (2706 ft; 450 fathoms) where it enters Baffin Bay. As in Jones Sound, submerged valleys enter the main trough from the side fiords and channels. Lancaster Sound reaches a minimum depth of 90 m (295 ft; 50 fathoms) near Cornwallis Island in Barrow Strait. Prince Regent Inlet (Gulf of Boothia) and Admiralty Inlet have similar bathymetric features. Both are extremely shallow at the southern end. Depths increase northwards in what are considered submerged glacial troughs, to greater than 825 m (2706 ft; 450 fathoms) in the Gulf of Boothia. Sills with depths of 330 m (1082 ft; 180 fathoms) separate them from Lancaster Sound.

3.4.3 General Circulation

At Kap Farvel the waters of the West Greenland Current follow the Greenland coast to the north. Over the continental shelf the waters are colder and fresher than those just beyond the shelf edge. On approaching Davis Strait, the warm waters are deflected westwards toward Baffin Island. The freezing waters are depleted through mixing in their northward flow so that the extension of the West Greenland Current over the Davis Strait Ridge is warm. This warmth is reflected in weather conditions and ice coverage as described in Chaps. 3 and 5. By the time the north-flowing waters have reached the vicinity of Cape York, a substantial portion of the heat content has been dissipated. However, they are still warmer and more saline than the waters flowing southward from the Arctic Ocean, which are below freezing even in mid-summer. The joining of the waters of the extension of the West Greenland Current, with those flowing through the Arctic islands, forms the Canadian Current, which flows southward along the Baffin Island coast. This current is of frigid character, carrying with it sea ice and icebergs. In the area south of Davis Strait, it is joined by the warm waters of the West Greenland Current to form the Labrador Current. The waters in the three sounds bordering the northwest corner of Baffin Bay have a strong set towards the Bay, with their freight of cold low-salinity water. At greater depths there is a draw-in of the warmer high-salinity water from Baffin Bay to replace the water consumed in the mixing process with the waters from the Arctic Ocean. This effect is most noticeable in Lancaster Sound. In the northern part of Baffin Bay and Smith Sound there is an area described as “North Water.” In this region summer sea-surface temperatures have been reported as consistently warmer than in surrounding areas. In winter, various sources attribute it to be ice-free during all or most of the year.

Several reasons for the existence of the "North Water" have been advanced. The most recent one considered is that it is due to the nature of the circulation, which keeps it ice-free and hence allows the surface waters to be warmed through insulation.

3.4.4 Thermal Structure (Summer)

In general, below 35 m (114 ft; 20 fathoms) temperatures are uniformly cold with gradual warming noticeable in the extreme north (15, 16) due to warmer Atlantic waters in the Arctic Ocean entering the deep channel, and in the south (2) due to warmer waters of the adjacent Atlantic. In winter, temperatures are near −1.5 °C (29.3 °F) throughout the area north of Davis Strait, and between 2 and 4 °C (35.6–39.2 °F) at the southern boundary, except near the coast where temperatures are below 0 °C (32 °F). The influence of the East Greenland Current as it rounds Kap Farvel (1) is shown by the cooler waters in the upper 50 m (164 ft; 27 fathoms). Above 35 m (114 ft; 20 fathoms) strong negative gradients occur, and these are coupled with strong salinity gradients. In one instance (12) a strong positive gradient at the surface suggests cooling due to ice movement. Depressed sound channels are present throughout the summer but are weak due to the small contrast in temperatures in the water column.

3.5 Northeast Atlantic Waters

3.5.1 Bathymetry

The Norwegian and Greenland Seas are bounded on the west by the Greenland continental shelf, and on the east by the Norwegian, Barents and Svalbard continental shelves. The depth of the edges of the shelves varies greatly. A large area of the shelves is deeper than 185 m (606 ft; 100 fathoms) and frequently extends to 400 m (1312 ft; 220 fathoms). Seaward of the shelves are the continental slopes, with gradients of 1:15 to 1:40 leading down to the abyssal plains. The Norwegian shelf is 50–140 nm (90–260 km; 55.9–161 mi) wide in its southern and central parts but narrows to 10 nm (19 km; 11.8 mi) off the Lofoten Islands. It is more irregular than normal continental shelves and is crossed by a number of trough–like gullies; these are associated with the major fiord systems of Norway. The Svalbard shelf is narrow and is only 11–40 nm (20–75 km; 12.4–46.6 mi) wide. It is traversed by deep canyons originating in the fiords of Vestspitsbergen such as Kongsfjord and Isfjord. The Svalbard shelf stretches to the banks south of Bjornoya, and from there to the banks off northern Norway is the shelf area of the western Barents Sea. This shelf area is deeper and less irregular than the shelves off Norway and Svalbard. The Greenland continental shelf north of the Denmark Strait is broad (60–170 nm; 110–315 km; 68.3–195 mi) but narrows to about 15 nm (28 km; 17.3 mi) in the extreme north. It is widest at the Belgica Bank. Typically, the bottom is irregular and rough, but local smooth areas are common and there are a number of shoal areas less than 100 m (328 ft; 55 fathoms) in depth. Troughs are present in the shelf and are both parallel and normal to the shoreline. South from the Denmark Strait the Greenland Shelf narrows from 170 nm (315 km; 195 mi) in width to 25 nm (45 km; 27.9 mi) at Kap Farvel. It is crossed by prolongations of the fiords of east Greenland. Around Iceland the continental shelf is up to 80 nm (150 km; 93.2 mi) wide in the west and north and 50 nm (90 km; 55.9 mi) in the east, but off the south coast it narrows to 10 nm (19 km; 11.8 mi). The dominant feature of the central parts of the Norwegian and Greenland Seas is the mid-oceanic mountain range forming the Iceland-Jan Mayen and Mohn Ridges. This is the continuation of the Mid-Atlantic Ridge, which runs the entire length of the North and South Atlantic Oceans, and which rises above sea level in Iceland. The Mid-Atlantic Ridge is characterized in its most elevated region by a central rift zone and by the fact that earthquake epicentres tend to lie on or near to its axis. These features are continued in the Iceland and Jan Mayen Ridges. A narrow, deep-rift valley, 2–3 nm (4–5 km; 2.4–3.1 mi) wide at a depth of 3300 m (10,826 ft; 1800 fathoms), extends along their axial lines and has well-developed structural benches on its walls. It ends at 78° 30′ N where it meets the Vestspitsbergen block. The width of the rift mountains and flanks on each side of this rift is about 35 nm (65 km; 40.3 mi) and the crests lie at depths less than 2200 m (7217 ft; 1200 fathoms), but the mountains to the east of the rift are not easily discernible in the Greenland Sea and may have been buried. On both flanks of the ridges are many seamounts. The mountains of the ridge systems rise above the sea as the island of Jan Mayen. A massive, almost unbroken ridge (the South Jan Mayen Ridge) runs south from the island with least depths between 800 and 1100 m (2624–3608 ft; 440–600 fathoms).

The Greenland Sea Basin was, formerly considered to be separated from the North Polar Basin by the Nansen Sill, but recent Russian expeditions have shown that a continuous sill does not exist and that is cut through the middle at about 1°E longitude by the deep Lena Trough, the minimum depth of which is 3100–3500 m (10,170–11,482 ft; 1700–1900 fathoms). The mid-oceanic ridge turns in a north westerly direction from a region to the west of Prins Karls Forland (Vestspitsbergen) and runs in what is known as the Spitsbergen Fracture Zone to the northeastern tip of Greenland. This results in a bottom topography, with a complicated ridge and trench structure. To the east of this zone is the gently undulating Yermak Plateau, which extends for 130 nm (240 km; 149 mi) from the northwestern corner of Vestspitsbergen with its crest at depths shallower than 900 m (2952 ft; 490 fathoms). The basin of the Norwegian Sea is separated from the abyssal plains of the northernmost Atlantic, with depths down to 3000 m (9842 ft; 1640 fathoms) by the Scotland-Greenland Ridge. The eastern part, between Scotland and the Faroes, is mainly less than 600 m (1968 ft; 330 fathoms) in depth, but between Faroe Bank and the Faroe Islands is a narrow channel through the ridge with a sill depth of about 800 m (2624 ft; 440 fathoms) and a least width of about 13 nm (24 km; 14.9 mi). The Faroes-Iceland Ridge lies at a depth of 400–500 m (1312–1640 ft; 220–275 fathoms) with the central part shallower than 400 m (1312 ft; 220 fathoms) and having several peaks. The deeper parts of the Norwegian Sea can be divided into three regions. The Iceland basin lies between the Iceland-Jan Mayen and South Jan Mayen Ridges and has depths of 2000–2560 m (6561–8398 ft; 1100–1400 fathoms) in its deepest part. The Norwegian Abyssal Plain, to the east of the South Jan Mayen Ridge, is the most extensive of the three regions and has a considerable area deeper than 3500 m (11,482 ft; 1900 fathoms). To the east, it is bounded by the Norwegian Plateau, a flat-topped feature shallower than 2000 m (6561 ft; 1100 fathoms). An extension of this plateau to the northwest separates the Norwegian Abyssal Plain from the Lofoten Abyssal Plain. This third region is extensive, and it has a large area greater than 3300 m (10,826ft; 1800 fathoms) in depth. It is bounded on the north by the Mohn Ridge. Beyond the Mohn Ridge is the basin of the Greenland Sea. It consists in the south of the Greenland Abyssal Plain at a depth of 3660 m (12,007 ft; 2000 fathoms).

3.5.2 Water Masses and Currents

The water masses of the area together with their temperature and salinity characteristics are listed in Table 3.1. The distribution of these water masses, as far as the upper layers of the ocean are concerned, is shown in Fig. 3.4. The temperature distribution at 200 m (656 ft; 110 fathoms) depth, and the sea surface temperature isotherms at 5 °C (41 °F) intervals for one of the coldest months (February) and one of the warmest months (August). The trends of the 200 m (656 ft; 110 fathom) isotherm indicate the direction of the current speed; the steeper the gradient, the greater the speed. A comparison of the 200 m (656 ft; 110 fathom) isotherms with those for the surface in summer yields some indication of the amount of thermal stratification in the warm season, whilst comparison with those for winter shows the effect of the convective stirring of the water column brought about by winter cooling of the surface layer. The two dominant features are the warm currents which form the ends of the Gulf Stream system and the cold currents that derive from the North Polar Basin. Taking the warm currents first, the North Atlantic Drift divides at the Mid-Atlantic Ridge at about latitude 51°N. One arm moves northwards parallel to the ridge towards Iceland where it becomes the Irminger Current. This latter carries the Irminger Atlantic Water, and it bifurcates to the west of Iceland, one branch proceeding northwards and then turning east along the north coast of Iceland, and the other turning first west and then south to flow along the East Greenland Slope, eventually to round Kap Farvel and flow northwards along the West Greenland Slope in the Davis Strait.

A second arm of the North Atlantic Drift carries warm Northeast Atlantic Water eastwards towards the British Isles and passes through the Faroes-Shetlands Channel and to the west of the Faroe Islands into the Norwegian Sea where it is known as the Atlantic or Norwegian Current. This current moves northwards along the Norwegian coast and gradually becomes cooler and less saline as it does so. Off northwest Norway it divides into the North Cape Current, which flows eastwards into the Barents Sea, and the West Spitsbergen Current, which continues northwards past Vestspitsbergen to enter the North Polar Basin where the Atlantic water forms

Table 3.1 Temperature and salinity characteristics of the water masses of the European and American Arctic and Sub-Arctic Seas

Northeast Atlantic Water	9.5 °C; 35.35 parts/1000
Irminger Atlantic Water	4.6 °C; 34.95–35.10 parts/1000
Arctic Water	< 0 °C; < 34 parts/1000
Labrador Sea Water	3.4 °C potential temperature; 34.89 parts/1000
Northeast Atlantic Deep Water	3.0 °C potential temperature; 34.95 parts/1000
Northeast Atlantic Bottom Water	0.8–1.5 °C; 34.91 parts/1000
Norwegian Sea Deep Water	< 0 °C; 34.92 parts/1000
Arctic Intermediate Water	< 0–2 °C; 34.8–35.0 parts/1000
North Icelandic Winter Water	2–3 °C; 34.85–34.9 parts/1000
Skagerrak Water	3–16 °C (iaw season) < 34 parts/1000

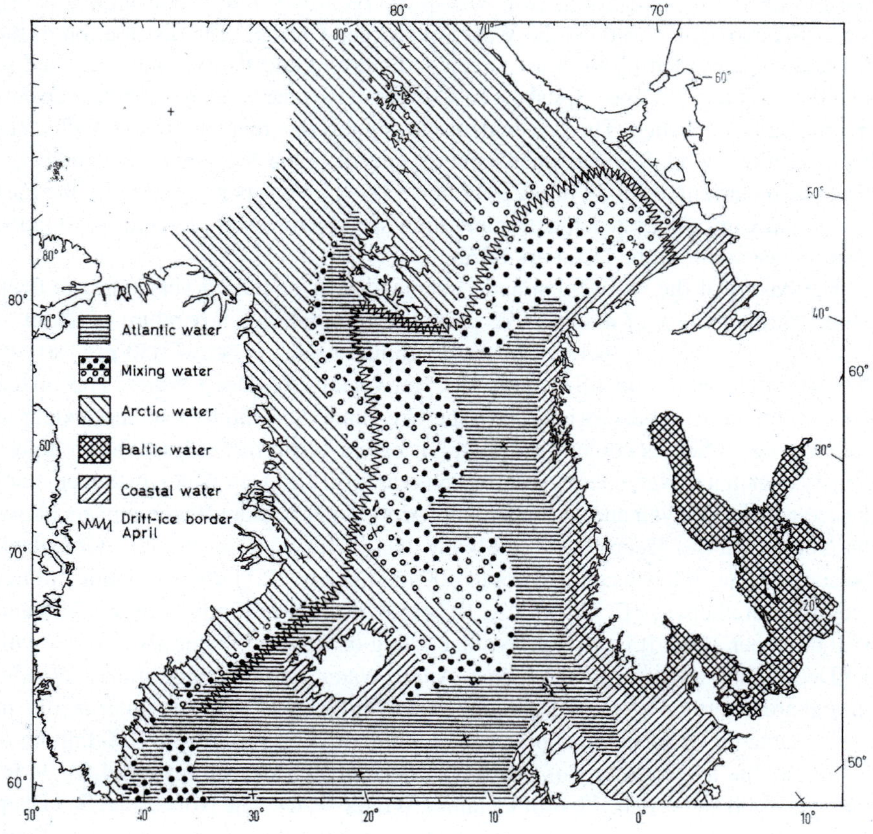

Fig. 3.4 Water masses (Norwegian and adjacent seas)

an intermediate warm layer. Part of the West Spitsbergen Current turns westwards off the northern part of Vestspitsbergen and flows towards the Greenland Shelf. It then turns south and proceeds below the East Greenland Current as a warmer, more saline, intermediate layer at 175–350 m (108–217 ft; 95–190 fathoms) depth. The main cold current of the region is the East Greenland Current. This carries ice and Arctic water, with subzero temperatures and low salinity, from the North Polar Basin along the entire length of the East Greenland Shelf to round Kap Farvel and enter Davis Strait. There are two branches of the East Greenland Current: the first, the Jan Mayen Current, flows eastward to the north of Jan Mayen in the region of the Mohn Ridge. The second, the East Icelandic Current, flows south eastwards past the northeast coast of Iceland to sometimes reach the north coast of the Faroe Islands and beyond. A smaller source of Arctic water and ice are the Bjornoya and East Spitsbergen Currents. These originate in the northeastern part of the Barents Sea and move south westwards over the Svalbard Shelf, the former to reach Bjornoya and the latter to round Sorkapp in Vestspitsbergen and flow northwards between the Vestspitsbergen coast and the West Spitsbergen Current. The distribution of the sea ice is determined by the East Greenland Current and its two branches, and to a lesser extent by the East Spitsbergen Current. The speeds of the various currents are not well established. There is evidence to suggest that there are frequent changes because of the wind. Some estimates put the speeds of the Atlantic Current and East Greenland Current at 12–24 nm (22–44 km; 13.6–27.3 mi) per day. Locally the East Greenland Current can reach extremely high speeds, for example, 3 kn (150 cm/s; 59 in/s) just south of the Denmark Strait.

The basins of the Norwegian and Greenland Seas contain a very uniform deep water with a salinity of about 34.92 parts per 1000 and a temperature of about −1 °C (30.2 °F). Seventy percent of the combined basins is below 550 m (1804 ft; 300 fathoms) depth and filled with this Norwegian Sea Deep Water. The mixed water in the upper layers, primarily in the Greenland Sea and to a smaller extent in the Norwegian Sea, is cooled in winter, but before it can freeze it reaches a higher density than that of water below it and so sinks to form the Deep Water. In the region of the Iceland-Jan Mayen and Mohn Ridges, Arctic Intermediate Water appears above the Norwegian Sea Deep Water and below the Arctic Water of the East Greenland Current system. It has its core at about 400 m (1312 ft; 220 fathoms) depth and has a temperature between 0 °C and 2 °C (32–35.6 °F) and a salinity between 34.8 and 35.0 parts per 1000. It is formed by the cooling of Atlantic Water and mixing with the Deep Water and, to a lesser degree, Arctic Water. Further south in the Icelandic coastal area, vertical mixing of Atlantic Water and Arctic Water in winter results in a homogeneous water in the uppermost 175–350 m (574–1148 ft; 95–190 fathoms). This water has a temperature of 2–3 °C (35.6–37.4 °F) and a salinity of 34.85–34.90 parts per 1000 and is called North Icelandic Winter Water. The Norwegian Sea Deep Water escapes into the North Atlantic Basin through the channel between Faroe Bank and the Faroe Islands. It flows at the bottom of this channel at a speed in excess of 2 kn (100 cm/s; 137 in/s) and as it does so the Northeast Atlantic Water, which lies above it, is entrained into the flow. The resultant mixing produces the Northeast Atlantic Deep Water. This water mass has minor constituents because, at times, the Arctic

Intermediate Water flows through the same channel and the Norwegian Sea Deep Water, Arctic Intermediate Water and North Icelandic Winter Water overflow the Faroes-Iceland Ridge, particularly near Iceland, proceed down its southern flanks, entraining overlying Northeast Atlantic Water as they do so, and eventually join the outflow from the Faroe Bank Channel as it flows westwards at the foot of the ridge.

The Northeast Atlantic Deep Water eventually turns south when it meets the Reykjanes Ridge; at about latitude 63°N it breaks through this ridge and flows northwards along its western flanks to fill most of the Irminger Sea at a depth of 1750–2500 m (5741–8202 ft; 950–1360 fathoms). Above it, in the Irminger Sea, is Labrador Sea Water with its core at 500–1275 m (1640–4183 ft; 275–700 fathoms). This water is formed as the result of the vertical mixing from the surface to a 1500 m (4921 ft; 820 fathoms) depth of low-salinity water in the Labrador Sea in winter. Below the Northeast Atlantic Deep Water in the Irminger Sea is the Northwest Atlantic Bottom Water, which originates with the overflow of water from the Norwegian Sea across the Iceland-Greenland Ridge, in the region of the narrow deep channel in the Denmark Strait. The overflowing water is at times Norwegian Sea Deep Water and at times Arctic Intermediate Water. In both cases the overflow proceeds at high speed down the East Greenland Continental Slope and entrains, first, overlying Irminger Atlantic Water and, later, Labrador Sea Water and Northeast Atlantic Deep Water, to produce a water mass of high density which fills the bottom part of the basins of the Irminger and Labrador Seas. Thus, in addition to there being a counter-clockwise horizontal circulation in the upper part of the water column, with the Northeast Atlantic Water entering our area and the East Greenland Current leaving it, there is also a circulation in the vertical plane with the inflow of the Northeast Atlantic Water being compensated by deeper outflows, over the Scotland-Greenland Ridge, of Norwegian Sea Deep Water and Arctic Intermediate Water.

3.6 Eurasian Coastal Waters

3.6.1 Bathymetry

The Barents Sea occupies 1,400,000 km^2 (8,699,919 mi^2) on the continental shelf of Eurasia. It has free contact with the Norwegian Sea on the west and with the Arctic Ocean on the north, and it is deeper than the other peripheral seas, much of it being greater than 185 m (606 ft; 100 fathoms) deep. The bottom is more like a continental borderland than a shelf, as it has both exceptionally shallow and deep areas scattered through it. The flat shelf areas are east and southeast of Svalbard and in the southeastern part of the sea. An east–west ridge at a depth of 185 m (606 ft; 100 fathoms) connects the shore areas around Z. Frantsa Iosifa with Svalbard, and a north–south ridge at 300 m (984 ft; 165 fathoms) separates the western Bjornoya basin with depths of over 350 m (1148 ft; 190 fathoms) from the eastern basin. The eastern depression extends southwest between Z. Frantsa Iosifa and Novaya

Zemlya with general depths over 275 m (902 ft; 150 fathoms) and occasional depths over 350 m (1148 ft; 190 fathoms). In the extreme north there is a third depression between Svalbard and Z. Frantsa Iosifa. The White Sea is a large arm of the Barents Sea that projects more than 270 nm (500 km; 310 mi) into the European mainland. The approach from the Barents Sea is appropriately called Voronka (Funnel) and is about 60 nm (110 km; 68.3 mi) wide. The entrance narrows to the southwest until between the Kol'skiy P-ov. and the mainland it becomes a strait, called Gorlo (the Throat), 27 nm (50 km; 31 mi) wide, where ice obstructs navigation even when the western, inland and wider part of the White Sea is ice-free. The White Sea is mostly less than 100 m (328 ft; 55 fathoms) deep.

The Kara Sea is exceptionally shallow in the east but deep for its position on the continental shelf in the west. Off the east coast of Novaya Zemlya there is a 43 nm (80 km; 49.7 mi) wide basin that in some places is 550 m (1804 ft; 300 fathoms) deep. A ridge in the north of the Kara Sea separates the basin from the Arctic Ocean. Between Yamal P-ov and Novaya Zemlya there is another trough of deeper water. There are two more basins in the north of the sea, which have maximum depths of about 550 m (1804 ft; 300 fathoms) and are separated by a ridge. O. Uyedineniya, Vize and Ushakova form the highest elevations of this ridge. In the southeast of the Kara Sea depths average only 50 m (164 ft; 27 fathoms) 50–135 nm (93–250 km; 57.7–155 mi) from the shore. The water adjacent to the Ob and the Yenisey rivers is also exceptionally shallow.

3.6.2 Currents

The surface characteristics of both climate and sea ice in the Barents Sea result primarily from an influx of warm water from the Norwegian Sea. This warm current flows north along the coast of Norway; a southern branch enters the Barents Sea along the north coast of Norway and Kol'skiy P-ov. as the North Cape Current, the northern branch flows north of Bjornoya and then turns northwest, passing along the south and west coasts of Svalbard as the West Spitsbergen Current. Off Varanger Halvoya, the North Cape Current flows at 8 nm (15 km; 9.3 mi) a day but the velocity decreases to the east. The current splits at Varanger Fjord, one-part flows in a belt 40–50 nm (75–90 km; 46–55 mi) wide to the entrance of the White Sea, the other curves north eastwards across the Barents Sea and passes north of Novaya Zemlya into the Kara Sea. This branch is weak in the north and south of its course, but between 36°E and 44°E it runs at approximately 18 nm (33 km; 20.5 mi) a day. The major inflow of freezing water into the Barents Sea is between Novaya Zemlya and Z. Frantsa Iosifa. This current also branches into two parts: one-part flows southwest of the archipelago, and the other west, as the Bear Island Current. In the southeast the general movement of water is towards the Kara Sea except for the Litke Current, which has the reverse direction. It moves west through the northern half of Proliv Karskiye Vorota, and then northwest along Novaya Zemlya, joining the general northerly movement there. In the White Sea is a weak outward current in the spring and summer, and an equally weak anti-clockwise eddy within the basin. The only important current in the Kara Sea forms a closed anti-clockwise circulation in the west. The gyre begins in the east with Ob and Yenisey waters which broaden as they leave the estuaries. One branch flow into Novaya Zemlya where it turns southwest to Proliv Karshiye Vorota. Within the main circulation are two small weak anti-clockwise eddies. Water also enters the Kara Sea around the north of Novaya Zemlya from the Barents Sea and eventually mixes with the Ob-Yenisey waters. Tides in the Kara Sea are semi-diurnal and weak. They come from the Barents Sea along eastern Novaya Zemlya and from the Arctic Ocean along western Severnaya and progress southwest. The average amplitude is 0.5–1 m (1.6–3.2 ft). Winds commonly increase the tidal range by 1 m (3.2 ft) or more.

3.7 Eastern Siberian Coastal Waters

3.7.1 Bathymetry

Both the Laptev and the East Siberian Seas are shallow basins with gentle shores. The edge of the continental shelf is up to 430 nm (800 km; 497 mi) offshore. Only in the northwest Laptev Sea, off Severnaya Zemlya, are depths greater than 90 m (295 ft; 50 fathoms). The western sector of the East Siberian Sea south of the Ostrova Novosibirskiye and east to the Kolyma River, is exceptionally shallow with many shoals. Between the Indigirka and the Kolyma Rivers, and almost continuous shorebank, defined by the 5.5 m (18 ft; 3 fathom) curve, extends about 24 nm (44 km; 27.3 mi) out from the shore. From the Kolyma east to Mys Shmidta, the coastal water is deeper. There are only a few islands, and these, with the exception of Ostrova Medvezh'i, are close to the shore. The sea deepens slowly to the northeast; maximum depths are 45–55 m (147–180 ft; 25–30 fathoms).

3.7.2 Currents

The general flow of water in both the East Siberian and Laptev Seas is counter-clockwise. There is a weak, easterly coastal current which is modified by water from the large rivers which forces it offshore in a northeasterly direction at 1 kn (2 km/h; 1.24 mi/h); anti-clockwise eddies develop when it is caught in coastal indentations. The major current entering the Laptev Sea comes through Proliv Vil'kitskogo between M. Chelyuskin and Severnaya Zemlya. It is joined by a cold current, flowing south eastward along Severnaya Zernlya, at 0.2 kn (0.4 km/h; 0.1 mi/h) and the combined waters move along the Taymyr coast into the shallow part of the Laptev Sea. At the Lena Delta, the current splits. One part, flowing along the west side of the Ostrova Novosibirskiye at 0.5–1 kn (1–2 km/h; 0.6–1.24 mi/h), sets to the north of the archipelago and joins the main Arctic drift. The other part flows through Proliv Dmitriya Lapteva and other straits into the Eastern Siberian Sea. The waters that pass through the straits separating Ostrova Novosibirskiye and the mainland spread out on reaching the East Siberian Sea. The main branch near the coast flows at 0.3–1 kn (0.5–2 km/h; 0.3–1.2 mi/h). A branch of this current is believed to pass north and west of Ostrov Vrangelya. In summer, a current flow out through the Bering Strait and flows northwest to the middle of Proliv Longa; its direction may be reversed in the winter. North of the coastal currents in both the Laptev and the East Siberian Seas, the water flows in large anti-clockwise eddies. Father north still is a west-northwest current, which runs northwest at the Ostrova De-Longa and passes north of the Ostrova Novosibirskiye into the Laptev Sea. It continues northwest across the northern margin of the sea and flows north of Severnaya Zemlya. Tidal progression is southwards from the Arctic Ocean in both seas. Tides are semidiurnal and their range is 30 cm (11.8 in), although it may be raised to 3 to 3.5 m (9.8–11.4 ft) with an onshore wind. Tidal currents flow from 0.5 to 0.8 kn (1–1.5 km/h; 0.6–0.9 mi/h) in the Laptev Sea but are weaker in the East Siberian Sea.

3.8 Central Polar Basin

3.8.1 Bathymetry

The Arctic Ocean is a true ocean. Beneath the deepest parts, the crust of the earth is about 10 km thick, a figure that is fairly typical of the other oceans. The Arctic Ocean also shares with other oceans the characteristic that relief occurs at two predominant levels: the continental shelves with depths of a few hundred metres, and the deep basins with depths of several thousand metres. A sharp break in slope between the continental shelf and the continental slope marks the boundary between the shelves and basins and defines the edge of the Central Polar Basin. In most oceans, the shelf break is at about 185 m (606ft; 100 fathoms) depth, but in the Arctic, it is deeper in some localities (365–475 m; 1197–1558 ft; 200–260 fathoms). First recognition that the Arctic Ocean was not a shallow sea but contained a true deep basin came from the soundings of the Norwegian North Polar Expedition during the drift of the Fram (1893–1896). The North Pole has been reached twice by ice-breakers and Arctic exploration has been conducted with aircraft, drifting ice stations, and nuclear submarines. Oceanographic research proceeded slowly after the successful voyage of the Fram, and it has only been in the last two decades that renewed efforts by the USA and Russia have resulted in our present level of understanding. Early bathymetric maps of the oceans usually showed smooth and unrealistic contours based on wire soundings. With the introduction of the continuous echo-sounder, it became clear that the contours often cut across regions with quite different bottom characteristics. Recently there has been a trend to analyse the ocean floor in terms of physiographic provinces which cover areas of similar topographic textures and which contrast with surrounding areas. All the various provinces found in other oceans, except deep-sea trenches, are known in the Arctic Ocean. The first-order features are the continental margins and the ocean basins. The continental margin is made up of the continental shelf, continental slope, continental rise, and marginal plateaux. The ocean basin consists of ridges, rises, and abyssal plains. A map of the major physiographic provinces of the Arctic Ocean is shown in Fig. 3.5.

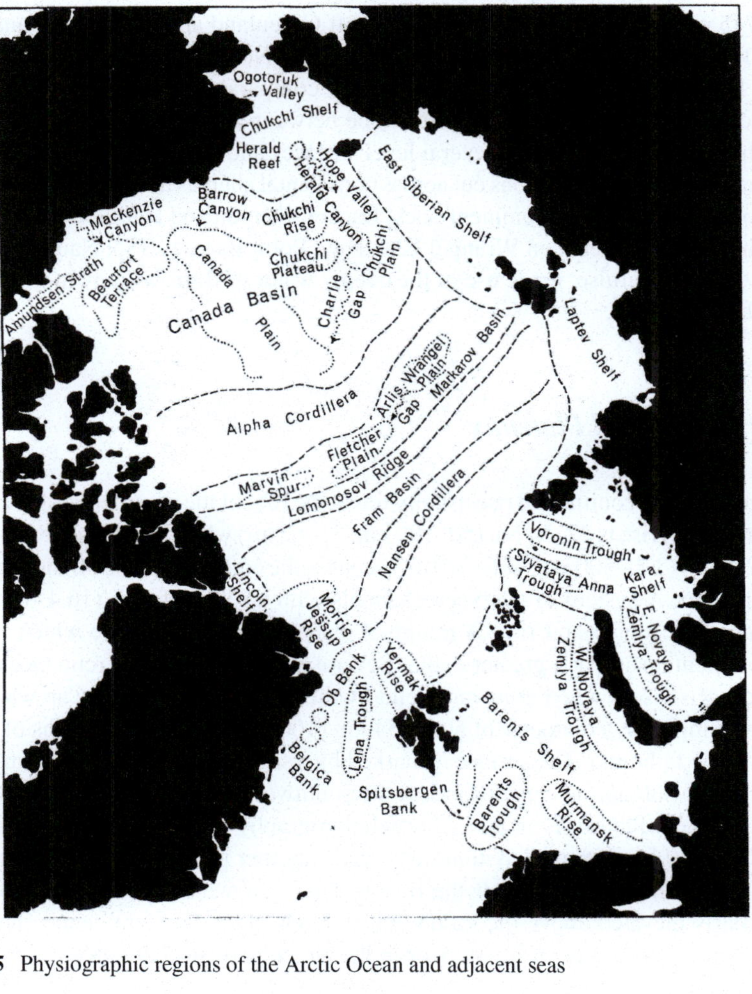

Fig. 3.5 Physiographic regions of the Arctic Ocean and adjacent seas

3.9 The Continental Margin

3.9.1 Shelves, Slopes and Canyons

The continental margin on the Eurasian side contains some of the broadest continental shelves in the world. The East Siberian Shelf is 270–485 nm (500–900 km; 310–559 mi) wide and the Barents and Kara shelves are 380–50 nm (700–1200 km; 434–745 mi) wide. In contrast the shelves off Greenland and northwest Canada are only 16–38 nm (30–70 km; 18.6–43.4 mi) wide. North of Alaska the continental slope has a gradient (1:15 to 1:40) which is comparable to that in other parts of the world. In contrast, the continental slope between the New Siberian Islands and the Chukchi Sea is formed of several level surfaces and has been likened to a "giant staircase." Submarine canyons cut across continental shelves and slopes in the Arctic as in other oceans. Two prominent ones, Barrow Canyon and Harald Canyon, incise the Chukchi Shelf between Wrangel Island and Point Barrow. SSN Nautilus, during her 1958 Arctic cruise, made use of the deeper water of Barrow Canyon to cross the shelf below the ice.

3.9.2 Rises and Plateaux

A gently sloping continental rise usually lies at the foot of the continental slope. North of Alaska, this rise is 27–54 nm (50–100 km; 31–62 mi) wide. Much greater widths of 160–270 nm (300–500 km; 186–310 mi) occur in the continental rise off the Canadian Archipelago. A system of deep-sea channels with a relief of 5 m (16.4 ft) crosses the Canadian continental rise. A marginal plateau is a level feature which borders the continental shelf at a greater depth. Several are known in the Arctic Ocean. The Chukchi Rise is crowned at its outer end by the Chukchi Plateau or Cap which has a flat summit with a diameter of about 54 nm (100 km; 62.1 mi) at depths of 300 m (984 ft; 165 fathoms). The surface is marked by a small-scale relief of 5-30 m (16.4–98.4ft). Two submarine canyons indent the southwest side of the plateau. Southeast of the Chukchi Plateau is an area of rough topography which has been described as a "continental borderland." Within this area is another plateau, the Northwind Cap. Other marginal plateaux include the Beaufort Terrace, which on its outermost edge, is similarly elevated above the saddle which connects to the continental shelf, and the Morris Jesup and the Yermak rises in the Greenland-Svalbard area.

3.10 The Ocean Basin

3.10.1 Lomonosov Ridge

The Central Polar Basin is crossed by three submarine mountain ranges. The ridges and rises are parallel to one another and span the basin from the Eurasian to the Canadian side. A bathymetric profile (Fig. 3.5) based on SSN Nautilus soundings, shows two of these ranges, the Lomonosov Ridge and the Alpha Cordillera. The Lomonosov Ridge stretching 970 nm (1800 km; 1118 mi) between the New Siberian Islands and the Greenland-Ellesmere Shelf, was discovered by Russian scientists in 1948. It is a single continuous feature, 54–108 nm (100–200 km) in width. Available echograms show a steep-sided ridge with a smooth profile. Minimum depth reported is 950 m (3116 ft; 520 fathoms). Saddles along the crest have depths of 1500–1600 m (4921–5249 ft; 820–875 fathoms). A flat surface near the crest of the ridge has been noted on two different crossings. An offshoot of the Lomonosov Ridge is known as the Marvin Spur.

3.10.2 Other Ridges

The Alpha Cordillera is about the same length as the Lomonosov Ridge, but it is much broader, ranging from 135 to 430 nm (250–800 km; 155–497 ft) in width. The crest of the Cordillera is 1500–1975 m (4921–6479 ft; 820–1080 fathoms) deep. Topography is much rougher than that of the Lomonosov Ridge. The magnetic fields over these two features also differ. The field over the Alpha Cordillera is rough, with many anomalies exceeding 1000 gammas, while there is little disturbance over the Lomonosov Ridge. Neither of these ranges is seismically active. The Nansen Cordillera is an extension of the Mid- Atlantic Oceanic Ridge into the Arctic Ocean. Where the topography has been sampled, it is rough, as it is on the Mid-Atlantic Ridge. The most distinctive characteristic of the Nansen Cordillera is the narrow earthquake belt along its crest (Fig. 3.6). The belt of earthquake epicentres crosses Iceland and then changes direction abruptly north of the island, where a large east–west fracture zone intercepts the mid-oceanic ridge near Jan Mayen Island. Between northeastern Greenland and northern Siberia, the earthquake belt is narrow and straight for a distance over 1080 nm (2000 km; 1242 mi). Within Siberia, the earthquake zone spreads out and disappears. In the Atlantic a similar earthquake belt coincides with a central rift valley at the crest of the mid-oceanic ridge.

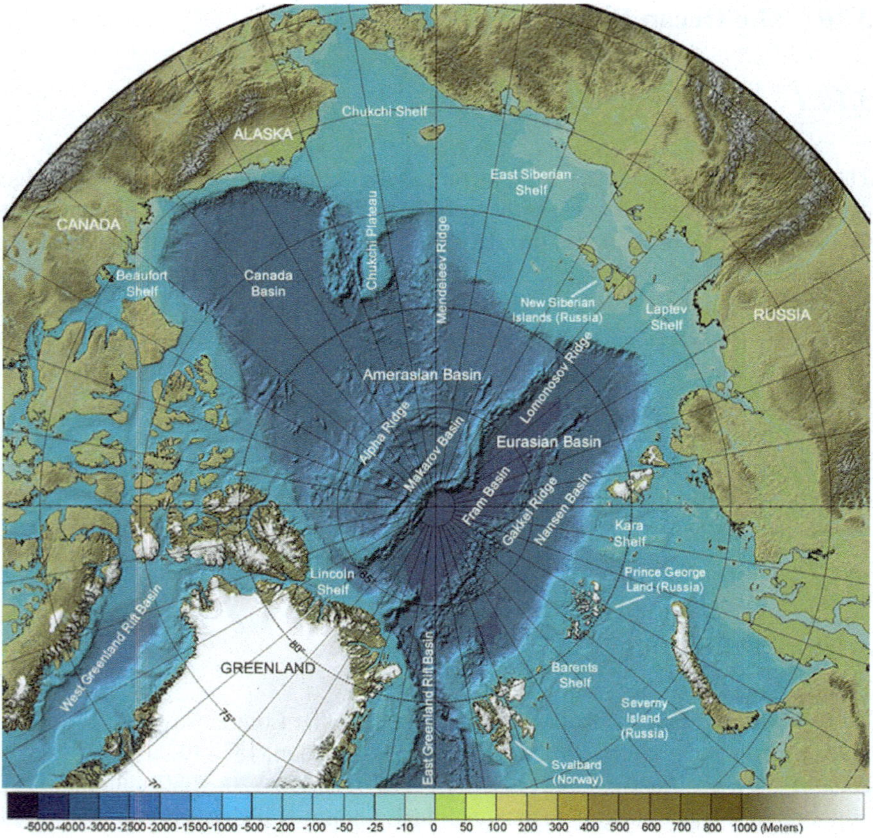

Fig. 3.6 Topographic profile across the Arctic Ocean

3.10.3 Abyssal Plains

These are the ultimate repositories for sediments, which have been transported across the continental shelves and down the continental slopes via the submarine canyons. In the deep basins the sediments collect to form the most extensive level surfaces on the globe, with gradients of 1:1000 or less. Four of the abyssal plains in the Arctic Ocean are arranged in step-like pairs. Each pair is connected by an abyssal gap through which sediments are transported from the upper to the lower plain. The Charlie Gap connects the Canada and Chukchi Abyssal Plains. The complete route of sediment flow is from Herald Canyon to the Chukchi Abyssal Plain and then through the Charlie Gap to the Canada Abyssal Plain; Wrangel and Fletcher Abyssal Plains are connected through the Arlis Gap. Abyssal gaps are commonly named after the ship of discovery. In this case, the discovering "ships" were drifting ice stations, Charlie and Arlis, respectively. Seismic reflection profiles show that a prominent sub-bottom basement ridge exists in the vicinity of the Arlis Gap. Sediments move from the Siberian Shelf to the Wrangel Abyssal Plain and then through the Arlis Gap to the Fletcher Abyssal Plain. A system of inter plain channels funnels the flow across the plain and into the gap. The right bank of these channels is higher than the left bank, due to the influence of the earth's rotation.

The most extensive plain in the Arctic Ocean is the Canada Abyssal Plain which covers an area of 254,000 km^2 (157,828 mi^2). It is remarkably flat, with depths ranging from 3750 m (12,303 ft; 2050 fathoms) in the north to 3850 m (12,631 ft; 2100 fathoms) in the south. On its northern and western edges, the scarps of the Alpha Cordillera and Chukchi Rise bond it. The eastern and southern boundaries grade smoothly into the continental slope. The Pole Abyssal Plain is deeper than the four plains previously mentioned. In the neighbourhood of the North Pole, it is flat and smooth with a depth of 4085 m (13,402 ft; 2230 fathoms). Away from the Pole the depth of the plain increases to 4575 m (15,009 ft; 2500 fathoms).

3.10.4 Bottom Sediments

Light-brown foraminiferal oozes are intermixed with sands and gravels on elevated topography such as ridges and rises. This is known as, a glacial marine sediment. The ooze represents normal pelagic sedimentation. The sands and gravels are ice-rafted material, which has been conducted from shore on ice floes and ice islands to be dumped when the ice melted or broke up. Rocks are often observed strewn over the bottom in deep-sea photographs. The unsorted ice-rafted rocks range in size. A cobble weighing over 7 kg (15.4 lbs) has been dredged. In bottom photographs, rocks with dimensions of about one meter have been seen. These rocks show faceting and striation, which are typical of glacial deposits. On the ridge and rise areas, sedimentation is extremely slow, only a few millimetres accumulating in 1000 years. Thus, many thousands of years are required to bury an ice-rafted rock. Signs of bottom life are rare in these regions. Sediments on the Canada Abyssal Plain are greyer in colour and lack the ice- rafted debris found on the elevated areas. The sediments of the abyssal plains also contain the pelagic and glacial components, but these have been inundated and diluted by turbidity-current deposits. Turbidity currents originate on continental slopes when over steepened sediments slump and flow down submarine canyons a, a slurry of mud and water. They flow and spread out on reaching the abyssal plain, depositing the mud over the level surface. The coarsest material settles first so that the beds are usually graded. No rocks are found in photographs taken on the Canada Abyssal Plain. Signs of animal life, such as intricate patterns of tracks and burrows, are abundant. Seismic reflection profiles give a thickness of about one kilometre (0.62 mi) for the unconsolidated sediments near the northern edge of the Canada Abyssal Plain. The thickness must be greater near the centre of the plain. The same technique has shown at least a 3.5 km (2.1 mi) thickness of horizontally stratified sediments underlying the Wrangel Abyssal Plain. The layer of unconsolidated sediment is much thinner on the ridges and rises. Measurements on the Alpha Cordillera give thickness between 300 and 500 m (984–1640 ft).

3.11 Water Masses

3.11.1 Four Water Masses Recognised in the Central Polar Basin

The Arctic Surface Water lies between the surface and a depth of 200 m (656 ft; 110 fathoms). Salinity may be as low as 30 parts per 1000 in this layer at the surface, but it increases rapidly below 55 m (180 ft; 30 fathoms). Temperature is close to the freezing point. On the Alaskan-Canadian side of the ocean, there is a small temperature maximum (−0.7 °C; 30.7 °F) at about 70 m (229 ft; 38 fathoms), which is known as the Pacific Water. The temperature maximum decreases with distance from the Chukchi Shelf. The Atlantic Water is marked by positive temperatures between uniform salinities of 34.9 to 35.0 parts per 1000 are attained. The depth of the maximum temperature in the Atlantic Layer increases with distance from its source northwest of Svalbard. It is initially at 300 m (984 ft; 165 fathoms) but increases to 500 m (1640 ft; 275 fathoms) north of Alaska. The temperature maximum decreases with distance from Svalbard from an initial value of 3–0.5 °C (37.4–32.9 °F) north of Alaska. The Arctic Deep Water lies between the Atlantic Water and bottom. Temperatures are below 0 °C (32 °F) and salinities increase very slowly with depth, from 34.93 to 34.99 parts per 1000. Below 1375–1500 m (4511–4921 ft; 750–820 fathoms), the temperatures are 0.5 °C (32.9 °F) warmer on the Canadian side than on the Eurasian side. These deep waters are formed during cold winters in the Norwegian Sea. They first enter the basin on the Eurasian side of the Lomonosov Ridge. Some of the water flows over the saddles in the Lomonosov Ridge to form the warmer bottom water of the Canadian side.

3.11.2 Tides

Tides in the Arctic Ocean are small. For example, the mean spring tide range at Point Barrow, Alaska, is only 15 cm (5.9 in). The semi-diurnal tide is derived entirely from the Atlantic. It enters the Arctic Ocean between Svalbard and Greenland, travelling across the Arctic Ocean in about 12 h as a progressive wave. On the Siberian shelves the tidal currents usually rotate clockwise. Storm surges may exceed the height of the tides and can cause damage to low-lying Arctic coasts.

3.11.3 Underwater Sound

The propagation of underwater sound in the Arctic Ocean differs in several ways from that in non-polar oceans. In the Atlantic and Pacific Oceans, the SOFAR channel lies at depths of 1000–1375 m (3280–4511 ft; 550–750 fathoms), but in the Arctic it is at the surface. Low-frequency sound is propagated to great distances in the Arctic SOFAR channel. Sound rays are alternately refracted upward in the water and reflected downward from the base of the ice. At great ranges signals consist of low frequencies, 8–100 Hz. Roughness on the lower surface of the ice strongly attenuate the high frequencies but have a negligible effect on the low frequencies. Signals generated by small explosions have been recorded clearly at distances from 43 nm up to about 1620 nm (80–3000 km; 49–3452 mi). Beyond a range of 325 nm (600 km; 372 mi), waves above 100 Hz are very weak. The Arctic SOFAR signal is dispersed so that an impulsive signal increases in duration as the range is increased. At a range of 325 nm (600 km; 372 mi) the duration of a signal from a small explosion is about 5 s. At shorter ranges, and over smooth bottoms such as abyssal plains, bottom-reflected arrivals can be of importance. They are late signals and increase still further the duration of the signal. In the shallow water of the shelves, propagation characteristics depend strongly on bottom parameters. In general, long-range transmission is much more strongly attenuated along shallow-water paths than it is along deep-water paths. Dispersion is even more pronounced in shallow-water transmission. A shallow sound-scattering layer has been observed beneath the central Arctic Ocean during the summer months. This scattering layer appears on echograms made at a drifting station with a 12 Hz sounder. Information on the sound velocity structure of the ocean can be obtained from measurements of temperature, salinity and depth. For very precise measurements of these factors at specific depths, Nansen bottles and reversing thermometers are employed. Information thus obtained can be converted into sound velocity by various methods.

3.12 Arctic Ocean Fronts

3.12.1 General

Ocean fronts are boundaries separating one water mass from another like weather fronts in the atmosphere. Unlike atmospheric fronts however, ocean fronts do not move great distances, but rather remain in any given area. Because of this, "regional" characteristic ocean fronts can be classified by the region where they are found and by the characteristic water masses they separate (see Fig. 3.7). Ocean fronts themselves are usually associated with cold or warm currents and can be influenced by subsurface topographic features such as submarine ridges. In any case, they are boundaries between water masses of different temperatures, salinities or both.

Fig. 3.7 General position of ocean fronts

3.12.2 Iceland-Faeroe Front

The Iceland-Faeroe Front forms the boundary between the warm, saline Atlantic water and cold, fresh Arctic water. The front is located near the extension of the submarine ridge that runs between Scotland and Greenland with depths over the ridge shoaling from 360 to 550 m (1181–1804 ft). The ridge forms a wall separating the deep waters of the Atlantic Ocean and the Norwegian Sea. It blocks the flow of most of the homogenous Norwegian Sea Deep Water into the North Atlantic, thereby affecting the formation and dynamics of the front. Although the surface reflects a complex mixing pattern, sea-surface temperature gradients across the front are present during all seasons. Frontal intensity increases and shows more definition with increasing depth. At depths of 180–275 m (590–902 ft), the horizontal temperature gradient reaches its maximum of −8 °C (17.6 °F) in the winter and −12 °C (10.4 °F) in the summer. Below 275 m (902 ft), frontal features weaken and become less defined due to the influence of the homogeneous Norwegian Sea Deep Water. Although the position of the front is determined by the Iceland-Faeroe Ridge brief period meandering within a 45 nm (83.3 km; 51.7 mi) envelope is frequent.

3.12.3 Greenland-Norwegian Sea Front

The Greenland-Norwegian Sea Front is sometimes referred to as the North Polar Front and is associated with the Norwegian Current. The Norwegian Current is an extension of the North Atlantic Current. The current flows northward along the eastern Norwegian Sea at speeds averaging 0.5kn (0.9 km/h; 0.5 mi/h) and brings warm, saline Atlantic water with it. Like all ocean fronts the Norwegian Sea Front, is the boundary between two water masses of different temperature-salinity combinations. Atlantic water with temperatures of 6–7 °C (42.8–44.6 °F) and salinities slightly greater than 35 parts per thousand are found east of the front. The colder, less saline waters of the Norwegian-Greenland Seas are found to the west of the front. The Norwegian Sea Front is present year-round and shows significant short-term variability (2–3 days). This variability takes the form of cyclonic and anti-cyclonic meanders and eddies of about 16–32 nm (29.6 km; 18.4 mi–59.2 km; 36.8 mi). The surface expression of the front is difficult to locate due to wind mixing of the near-surface waters and the extensive cloud cover limiting the usefulness of satellite imagery. Because the maximum horizontal temperature gradient is found at a depth of 200–300 m (656–984 ft), SXBT traces can be extremely useful in determining position.

3.12.4 Bear Island Front

The Bear Island Front is located midway between Spitsbergen and the Norwegian coast and results from the interaction of the Bear Island Current and the eastern branch of the Norwegian Current. The Bear Island Current carries cold, low-salinity Arctic water down into the northeastern Norwegian Sea. The Norwegian Current, also known as the North Cape Current in the region, carries modified Atlantic water around the North Cape into the Barents Sea. The temperature of the modified Atlantic water ranges from 1 to 7 °C at a salinity of about 35 parts per thousand. The Bear Island Front lies in shallow water in the vicinity of the Bear Island shelf break near the 90 m (295 ft) curve. The front is intricately linked to the bathymetry and shows some light meandering of about 30 nm (55.5 km; 34.5 mi) from year to year. This phenomenon may be due to periods of northeasterly or south westerly winds that favour the transport of Arctic or Atlantic waters respectively, or tidal effects. Typically, temperatures across the front range from 4–5 °C (39.2–41 °F) and salinity changes about one part per thousand. The associated sound speed change is like

21 m/s (68.8 ft/s). Also, one may expect a shoaling off the bottom to 185 m in the vicinity of the front.

3.12.5 West Spitsbergen Front

The West Spitsbergen and Greenland Sea Fronts form the northern and western boundaries of the large cyclonic gyre. The West Spitsbergen Front forms the boundary between the modified Atlantic water found adjacent to the southwest coast of Spitsbergen and the colder, less saline water found in the interior of the Greenland- Norwegian Sea. The temperature gradient associated with the front is evident at the surface where the temperature changes from an average 7–1 °C (44.6–33.8 °F). The warmer temperature is associated with the modified Atlantic water, which has salinity greater than the 34.88 parts per thousand found in the Greenland-Norwegian Seas.

3.12.6 East Greenland Front

The East Greenland Front is the result of the East Greenland Current, which is the western extension of the Spitsbergen Current. The front separates the cold, less saline waters adjacent to the east coast of Greenland from the warmer, saline water of the Greenland Sea. The water off the coast of Greenland is characterized by seasonal temperatures ranging from −1 to 0 °C (30.2–32 °F) and salinities from 30 to 34 parts per thousand owing to ice melt. In the summer, strong vertical salinity gradients may be expected to depths of 15–23 m (49.2–75.4 ft). The intermediate and deep waters are made up of Atlantic Intermediate Water found to depths of 730 m (2395 ft), temperatures of 0 °C (32 °F), and salinities of 34.88–35 parts per thousand. Below 730 m (2395 ft), Norwegian and Greenland Sea deep water is found with temperatures less than 0 °C (32 °F) and salinities between 34.87 and 34.95 parts per thousand.

3.12.7 Kolbeinsey Front

In the region between Iceland, Jan Mayen, and Greenland known as the Iceland Sea, colder, lower-salinity Polar water exists. The Kolbeinsey Front is the boundary between the Iceland Sea Water and warmer, saline water adjacent to the west and northern coast of Iceland. The surface temperature of the Atlantic water ranges from 2 °C (35.6 °F) in the winter to 5 °C (41 °F) in the summer. Even in the winter, surface ducts and sound channels should be expected in the Atlantic Water along the Iceland shelf. North of the front in the Iceland Sea Water, either very weak surface ducting or half-channel conditions prevail.

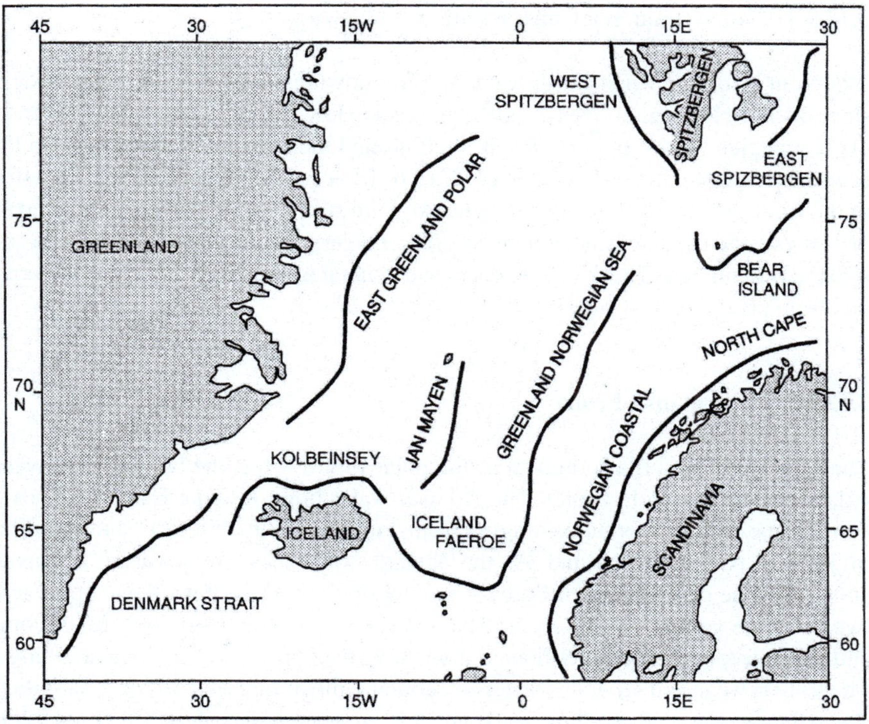

Fig. 3.8 Fronts in the Norwegian Sea

3.12.8 Denmark Strait Front

The Denmark Strait Front is the boundary between cold Polar water carried southward by the East Greenland Current and warm Atlantic Water carried north into the Irminger Sea along the west coast of Greenland. The Front follows the ice edge and continental shelf break fairly closely but is known to vary from 30 to 60 nm (55.5–111 km; 69–34.5 mi) (see Fig. 3.8).

3.12.9 Jan Mayen Front

The Jan Mayen Front is located south of the Jan Mayen Island and forms the boundary between two Arctic intermediate waters of slightly different temperatures and salinities. To the west of the Front, the water temperature is less than 0 °C (32 °F) and the salinity is less than 34.9 parts per thousand, and to the east of the front temperatures are greater than 2 °C (35.6 °F) and salinity is greater than 34.9 parts per thousand.

3.12.9.1 Norwegian Sea Coast Front

This front is formed when North Atlantic water flowing northward along the western slope of the Norwegian Trench contacts colder, low-salinity waters flowing from the Norwegian Fiords and the Danish shelf around the Skagerrak. A decrease in the sea-surface temperature of about −13 °C (8.6 °F) occurs along an eastward crossing of the front. Although the front characteristics are strongest along the southern coast of Norway, the front is detectable on satellite imagery all the way to North Cape and into the Barents Sea. The front meanders and generates numerous eddies on the scale of 54 nm (100 km; 62.1 mi).

3.12.9.2 Murmansk Front

The Murmansk Front, also known as the polar Front, forms the boundary between Atlantic water and Polar water. The Atlantic water flows into the Barents Sea over North Cape with temperatures ranging from 4 to 12 °C (39.2–53.6 °F) and salinities of 34.8–35 parts per thousand. As the Atlantic water moves eastward it begins to cool. Over the central/eastern Barents Sea, the flow turns northward and splits into two branches south of Franz Josef. The main branch curves west and further cools and sinks deeper. The smaller branch flows over the north coast of Novaya Zemlya. The front is weak and separates waters of similar temperatures but different salinities. The Atlantic water has a salinity of about 34.8 parts per thousand in this region while the Arctic surface water has a salinity of less than 34 parts per thousand. Intense, vertical circulation of water has been observed in the frontal zone. This mixing provides a good supply of nutrients to the surface waters, resulting in increased biological activity in the vicinity of the front.

3.12.9.3 Novaya Zemlya and Kara Sea Fronts

The extension of the North Cape current that flows over the north coast of Novaya Zemlya carries highly modified Atlantic water with temperatures around freezing and salinities of 34–34.8 parts per thousand into the Kara Sea. The Atlantic water then flows southwest along the coast of Novaya Zemlya and into the southern Kara Sea. Coastal freshwater input from the Obskaya and Yenisey Rivers amounts to about 940 km^3 (584 mi^3) annually. This water is warm and has salinities of from 2 to 20 parts per thousand. The general flow for this water is northerly into the central Kara Sea. In the southern Kara Sea, the Novaya Zemlya Front is the boundary between the Atlantic water and the river runoff. A diffuse front, the Kara Sea Front, exists over the central Kara Sea and forms the boundary between the Arctic and river runoff waters. Both the Novaya Zemlya and Kara Sea Fronts are seasonal features that begin to develop soon after the ice melts in the late spring. They reach their peaks in the summer and persist through the fall.

3.12.9.4 Laptev Sea Front

A weak front similar to the Kara Sea Front exists in the Laptev Sea. Fresh water input into the southern Laptev Sea from the Lena and Olenek Rivers amounts to 554 km^3 (344 mi^3) per year. This amount is augmented by the inputs of the Khgatanga and Anabar Rivers into the western Laptev (no data available on their output). The river runoff water is warm and has salinities that range from two parts per thousand coastally to 28 parts per thousand near the central Laptev Sea. The Laptev Sea Front forms a diffuse boundary between the southern fresh water and the northern Arctic water and is more or less a continuation of the Kara Sea Front. Like the Kara Sea Front, the Laptev Sea Front is seasonally diffuse and therefore not tactically significant.

3.12.9.5 East Siberian Sea Front

There are two general types of water masses found in the East Siberian Sea: Arctic and Siberian Coastal waters. The Arctic water is the same basic water mass found in the northern Laptev and Kara Seas and has a temperature near freezing and salinities of from 28 to 33 parts per thousand. The Siberian Coastal water begins to develop after the spring melt and is composed of input from the Indigirka and Kalyma Rivers. This water is fresher than the Arctic water, with salinities less than 20 parts per thousand. The East Siberian Sea Front separates the two water masses and extends eastward from the New Siberian Islands to Ostrov Vranglya (Wrangel Island). The front continues east of Wrangel Island into the western Chukchi Sea where it serves as a boundary between the Siberian Coastal water and the Bering Sea water flowing into the Chukchi from the Bering Strait. The East Siberian Sea Front is a weak seasonal front, sharing the same basic characteristics, which are found in both the Laptev and Kara Sea Fronts.

3.12.9.6 Alaskan Coastal and Beaufort Sea Fronts

The Alaskan Front is the boundary between central Chukchi and Alaskan coastal waters flowing north along the eastern side of the Bering Strait and coastally in the eastern Chukchi Sea to Point Barrow. Alaskan coastal water originates with freshwater input from Katzebue Sound. The central Chukchi Sea water is composed of Bering Sea and Arctic waters. The central Chukchi water is colder and more saline than the Alaskan Coastal water. The Alaskan Coastal Front extends from the Kotzebue Sound coastally to Point Barrow and is a weak front (even compared to the East Siberian Sea Front). East of Pont Barrow, the Colville and Mackenzie Rivers provide freshwater input into the coastal water of the Beaufort Sea. The Beaufort Sea Front forms the boundary between the coastal water and the Arctic water of the Central Beaufort Sea. The Beaufort Sea Front is a continuation of the Alaskan Coastal Front and is a weak front.

Chapter 4
Preparing the Vessel for Arctic Operations

4.1 Introduction

Weather conditions in the North Atlantic, particularly during winter, can be more severe than those experienced above the Arctic Circle during the short navigation season. Indeed, the Arctic summer can be quite pleasant, with warm sunshine, temperatures well above freezing, dear skies, little wind, and, because of the dampening effect of ice, calm seas. There is, in fact, a threat of sunburn and a definite need for sunglasses. Climatic and other environmental conditions affecting ships and their equipment during Arctic operations includes low surface air temperatures at certain seasons; sudden changes in air temperatures; high winds; low seawater injection temperatures; low humidity; ice conditions ranging from slush and brash to solid pack; snow, sleet and freezing rain; fog and overcast, occurring at the ice/water interface; heavy seas with attendant spray in areas clear of pack-ice; the possibility of heavy and rapid ice accretion, with consequent loss of stability; abnormal magnetic conditions and low directivity of magnetic compasses; and the possibility of gyro compass errors.

4.2 Topside Preparations

4.2.1 Prevention of Slippery Decks

Ensure that deck tread ladders, and deck non-skid areas meet safety standards (renew if necessary). Non-skid areas can be expanded to enhance traction. Thin ice can be removed most effectively from decks and other flat surfaces by the use of dry chemicals such as sodium chloride (rock salt), calcium chloride and urea. These materials are simply spread over the frozen surfaces in a thin layer as required. Rock

A. Olsen, *Ship Operations in Extreme Low Temperature Environments*, Springer Series on Naval Architecture, Marine Engineering, Shipbuilding and Shipping 19,
https://doi.org/10.1007/978-3-031-52513-1_4

salt is the most economical material and is effective above −9 °C (15.8 °F). Calcium chloride gives off heat when mixed with water, so it acts faster than rock salt. A mixture of one part calcium chloride and three parts rock salt will be effective to temperatures of −18 °C (−0.4 °F). Urea (granular or pellets) is effective for melting thin layers of ice above −9 °C (15.8 °F) and for preventing ice accumulation in freezing rain. Sand can be used alone or in combination with any de-icing chemicals to improve traction on ice-covered decks. Develop procedures for entry, egress, and safety of topside personnel:

- Topside personnel must use two-man buddy system.
- Topside personnel will need to be tethered with tended safety lines during heavy weather.
- Temporary lifelines and guide-ropes can be run along flight-decks attached to tie downs (during non-flying hours) to permit watch personnel crossing during helicopter operations.

Lubricate all topside fittings with appropriate cold-weather greases. When selecting covers for equipment, the most important characteristics to look for are strength, durability and water resistance. Non-porous, fire-retardant covers are recommended for (at a minimum):

- Ship's boats (complete boat must be covered.)
- Davit winches.
- Capstan/windlass and associated controls (because of exposure to severe weather forward, extra covers will be required).
- Unheated combat system equipment.
- Sound-powered phone boxes.
- All outside (exposed) command, control, and communications stations.

Precautions taken to protect hydrostatic release mechanisms on life rafts should include the fitting of polyethylene sleeves over the devices and sealing them. Develop and promulgate ice accretion removal procedures and instructions to avoid damage to equipment or undue hazard to personnel during removal operations. Obtain and install, when necessary, temporary shelters or windscreens for exposed personnel and topside watchkeepers. Rig, when necessary, additional life and safety lines for protection of personnel. Heavy-weather lifelines should be rigged well in advance to facilitate early identification and correction of deficiencies. Set up cargo lines on lifelines at UNREP stations to prevent line handlers from falling overboard. Ice removal equipment and de-icing materials should include those items listed in Tables 4.1 and 4.2. Fire main valves topside will have to be cracked sufficiently to prevent freezing. Run an old hose section over the side to prevent ice buildup. Securing the risers from below deck is effective, as is filling piping and stations with antifreeze.

Table 4.1 Ice removal equipment

Broomstick handle Hose (for steam deicing) Mallet (plastic/wooden) Portable hair dryer Portable heat gun	Sharp sand Shovel Special ice footwear Steam lance

Table 4.2 Deicing materials

Calcium chloride Denatured ethyl alcohol Ethylene glycol Isopropyl alcohol Rock salt	Safety equipment for handling calcium chloride: • Rubber gloves • Rubber boots • Safety goggles

4.2.2 Topside Damage Control Equipment

Fire hoses and nozzles will perform satisfactorily at freezing temperatures and below, provided water is kept flowing and a good pressure is maintained. If the pressure is reduced or the hose is secured, the nozzle and plug may become frozen. If a long lead of hose is to be secured at temperatures below −12 °C (10.4 °F), stop the flow only for the time necessary to disconnect each length of hose. After securing, hoses and nozzles should be taken below decks and completely dried prior to returning to topside stowage. Duplex proportioners should be drained after use, dismantled, dried, oiled, and the chamber change-over valves reassembled. Oxygen-breathing apparatus (OBA) with spare canisters should be returned to below-decks stowage as soon as no longer needed. Portable water pumps should be stowed below decks. External connections on the fire main will need to be isolated, drained and backfilled with ethylene glycol, or kept on a low trickle flow. Dead ends or low flow spots such as magazine or cargo storeroom sprinkles on the fire main should be watched closely to avoid freezing.

4.2.3 Ship's Boat Preparation

The use of boat engine heaters (engine block, oil system or cooling system) will ensure easier starting. Equipment from auto parts stores such as dipstick heaters are effective. Other means to keep boat engines warm include heat lamps, flood lights, drop lights and insulation blankets. Use antifreeze in engine cooling water system to prevent freezing. Freezing weather lubricants and oils should be used for engines and transmissions. Circulating water heater and heat strips around engine block and oil pan is effective. Ethylene glycol (60/40) in the bilges and saltwater pumps will prevent ice buildup. Install and utilize boat jump-start connections. A constant trickle charge is recommended. Obtain spare boat batteries. Lubricate and protect davits and winches. Boat falls can be prevented from bird caging by keeping them clean and running the winch machinery to heat lubricants prior to use. Fabricate boat covers for each boat. The cover must protect the entire boat down to the water-line and should be made out of heavy canvas, not herculite. Secure the cover with multiple lines so parts of the boat can be uncovered to conduct routine maintenance without removing the entire cover. Ensure that procedures are developed to drain salt-water cooling systems and top off fuelling systems daily. Ensure that adequate repair parts are available. During very freezing conditions, provisions must be made for periodically starting and warming boat engines if no use, is expected to be made of them. Time varies from once daily to once per watch, depending upon the weather. Lubricate entire boat throttle cable to prevent freezing and binding. Not only may boats become damaged by ice, but they can also be cut off from their parent ship by poor visibility or by drifting ice floes brought down by a change of wind or tide. Thus, all boats should be equipped with emergency rations and survival kits including first-aid supplies, sleeping bags, firearms, and a suitable selection of hand-held pyrotechnics. In addition, all boats ought to be radio-equipped. Boats' crews should always have their cold-weather clothing with them while away from their ship. For larger boats, an inflatable life-raft ought to be carried, and all boats should be fitted with radar reflectors. Additional suggestions include:

- Hoisting slings may need reinforcement for rough-weather handling of boats.
- Wooden boats should be copper-sheathed along the water-line, especially forward.

Foam flotation material (e.g., Styrofoam), which must remain impervious to both water and fuel, should be applied to boats to provide an additional measure of buoyance in the event they are holed by ice. Boats in excess 9 m (29.5 ft) should have some form of watertight subdivision for the same reason. In Arctic service, outboard motors are dependable and efficient. While damaged propellers can be easily repaired or replaced, consideration should also be given to the fitting of some form of propeller guard. Fuel tanks should be kept topped up to minimise condensation. Fill drinking water containers in boats to only 75% capacity to avoid bursting the containers. Provide anti-exposure suits to small boat crews for floatation and the best protection from freezing weather.

4.3 Mooring Lines and Anchor Gear Preparations

Mooring lines should be kept dry and stowed under cover when not in use. Manilla lines can freeze and dry rot when exposed to the cold for prolonged periods. Polypropylene and polyethylene lines absorb little water but are stiff and brittle in the cold. Nylon and Dracon lines, particularly braided type, absorb more water but remain easy to work with. Particular attention should be paid to the small sizes of wire rope where the component wires are of small diameter, since frost and ice can cause them to break. When proceeding in waters calling for constant use of soundings, steps should be taken to ensure that anchors and cables are ice-free and ready for use at short notice.

4.4 Engineering Preparation

4.4.1 Window Heaters

All heaters must be checked, and their operability ensured, before leaving port. Overheating may cause lamination to separate on windows, so ensure that temperature controlling mechanism is functioning. Determine means to clear ice from area in front of bridge windows. Using the window wash system, with a water/antifreeze (50/50) fluid mix, is an effective de-icing method. Ensure wash system is operated after addition of water/anti-freeze solution to purge fresh water from system piping.

4.4.2 Turbine and Ventilation System Intakes

Provisions must be made to ensure that intakes are cleared of snow and ice. Accumulation can be controlled by the use of low pressure (LP) air to blow snow from demister pads. Ensure that turbine de-icing systems and inlet heaters are operable and effective.

4.4.3 Sea-Chest Inlet and Outlet Blockage

Use steam blow-out connections or LP air to de-ice engineering intakes. Monitor freezing of overboard discharges, such as CHT, and develop an ice removal plan for these systems.

4.4.4 Main Engineering Plant Spaces

Check all air reducers by operating them at their proper pressure and ensure that they can be adjusted so that a drop in air pressure in freezing weather can be overcome. Add sufficient antifreeze to diesel freshwater system to preclude freezing. Obtain extra batteries for all requirements. Battery locker temperature should be maintained above 15 °C (59 °F) or store batteries elsewhere. Monitor water/steam usage during ice removal operations. Inefficient ice removal may lead to excessive water usage.

Install insulation behind and above main switchboards where condensation may form. Ensure that a full allowance of damage control and repair material such as shores, plates, clamps, wedges and plugs are on board. Monitor temperature of idle machinery. Prepare evaporators for cold-weather operations. Prior to sailing, hydro the freshwater side of the evaporators. Locate and fix all leaks. Use new gaskets when reassembling the units. Check air ejector nozzles for proper operation. The best way to maintain capacity is to start with a "tight plant." Extremely high vacuums occur during cold-weather operations. Throttling the main condenser overboard valve to control condensate depression does not work well. Use of the masker belts causes air to be entrapped, and air may bind the main condenser. Venting of the condenser regularly has proven effective in maintaining vacuum and efficient engine operations.

4.4.5 Interior Space Heating and Ventilation

Ventilation heater controls should be adjusted to maintain space temperatures of 19–20 °C (66.2–68 °F) in lieu of 21–27 °C (69.8–80.6 °F). This will minimise the drying effect in living compartments, since the relative humidity of the compartment air will be greater at the lower space temperature differential for personnel going to and from topside or exposed locations. Machinery space ventilation fans should be operated to give a slight positive pressure within the area. This will avoid creating drafts through the ship proper and will conserve the heat in the interior of the ship. All living space ventilation supply fans should be operated on low speed to reduce the amount of outside air taken into the ship. To avoid freezing of ventilation heaters, it is essential that condensate lines are kept open and that traps are in proper operating condition. To ensure that blowers are operated at low speed, that thermostats are not tampered with, and that thermostatic traps or pre-heaters are operative, it is suggested that a heating patrol be established as part of each watch. This procedure is particularly recommended for large ships engaged in low-temperature operations. A roving watch, checking berthing spaces, fan rooms and heating boundaries, has proven effective.

4.5 Towing Considerations

4.5.1 Prolonged Stay

Ships operating in Arctic waters should be equipped and ready to tow or be taken in tow at short notice. Towing at prolonged stay can be difficult because ice, if there is any present, can get between the ships involved. Some ice-breakers are equipped with a notched stern, suitably padded, in which the stem of the ship to be towed can be secured. Ships with high freeboards, however, are unsuited for this method.

4.5.2 *Short Stay*

Towing at short stay can be undertaken though the ship towed should not use its engines because of the risk of overrunning and striking the towing vessel.

4.6 Ice Preparations/Considerations

4.6.1 *Removal Techniques*

(1) Topside icing can result in a dangerous loss of stability, reduction in reserve buoyance, and critical impairment of a ship's ability to withstand damage. Other adverse effects are the increased load on decks, masts, and other structures, and the danger that men will slip on icy decks or be struck by falling ice from aloft.

(2) There are various methods by which ice can be removed:

   (a) Manual. Ice is broken away and chipped off using mallets, clubs and scrapers. Caution must be exercised to avoid damage to metal surfaces, electric cables (including degaussing cables) and equipment. Ships should lay in a stock of wooden mallets, shovels, wire brooms and scrapers suitable for the removal of ice and snow. Table 4.1 provides a list of ice removal equipment.

   (b) Steam jet. A steam lance can be used to undercut layer ice as, for example, in freeing an anchor which has become frozen in its hawse pipe. It can also be used for spot heating.

   (c) Anti-icing coatings. These coatings can be used to protect comparatively small areas or items. Anti-icing compounds retard the formation of ice and contribute to the ease of its removal.

   (d) Heated salt water under pressure. This is sea water heated to at least 50°C, or higher if possible. It is used in high- pressure steams to slot, undercut, and break up large accumulations of ice. No attempt is made to melt ice directly, but rather to break it up or weaken it to facilitate removal. When using water to clear ice, scuppers and overboard drains must be clear.

   (e) De-icing. Use of Ethylene Glycol, Methanol or other de-icing chemicals can be used to remove ice build-up on exterior surfaces.

4.6.2 *Superstructure Icing Considerations*

When severe topside icing occurs and ice continues to accumulate, despite all attempts to remove it, it may become necessary to alter course, heave to, or make for an area of more moderate weather. Icing can occur when heading into fierce winds and heavy seas and in conditions of low air and seawater temperatures. Smaller ships having low free boards and reserve stability are particularly vulnerable. The rate of ice accretion from ocean spray is related to surface wind speed roughness of the sea, air temperature (below 1.67°C (35°F), and duration of exposure. Figure 4.1, the Overland Monograms, offer the most current prediction method for icing conditions. The smaller the vessel, the more serious the problem of ship icing. Any prolonged exposure to gale-force winds and below-freezing air temperatures could result in the vessel capsizing. Fishing trawlers approximately 52 m (170 ft) long have capsized and sank as a result of rapid ice accumulations on the hull and superstructure; freezing spray accumulating at a rate of two tons an hour over a period of 24 h has been reported. Meteorological observations taken in the vicinity of Iceland indicate that storms of gale force accompanied by freezing air temperatures and lasting as long as three days may occur as often as three times a year. Sparseness of synoptic meteorological information in this area makes forecasting of such prolonged periods of gale-force winds and below-freezing temperatures quite difficult. The use of aircraft engines can be effective in removing snow from around closely parked aircraft. Rock salt should not be used because of its corrosive effect on metals.

4.7 Cargo Handling Considerations

Cargo should be stowed in such a manner that it cannot shift as a result of repeated impacts between the ship and the ice. It should be stowed far away from the sides of the ship thus allowing easy access to the side plating in case of damage and to permit passage of water to the bilges. Ships operating in ice, should be so loaded that they will be trimmed by the stern. If in ballast, consideration should be given to flooding to immerse the rudder and propellers to minimise the risk of damage to them by ice. Helicopters should be fitted with hoisting gear and quick-release hooks. There may be occasions when ice floes block beach unloading sites, or ice-pack prevents a ship getting close enough to employ boats and landing-craft effectively.

4.8 Equipment Safety General

Safe handling of equipment cannot be over-emphasised, particularly in a freezing weather environment. The dangers associated with operating much equipment during routine evolutions in a moderate climate are sometimes formidable. In a cold, harsh climate there is no such thing as a routine evolution. Evolutions typically take 2–3 times the normal period to perform. Every operation requires effective planning and extra attention to safety for successful completion. This section addresses some particular safety concerns when operating equipment in freezing weather. The purpose is to encourage a thoughtful, safety-conscious approach to hazards involved.

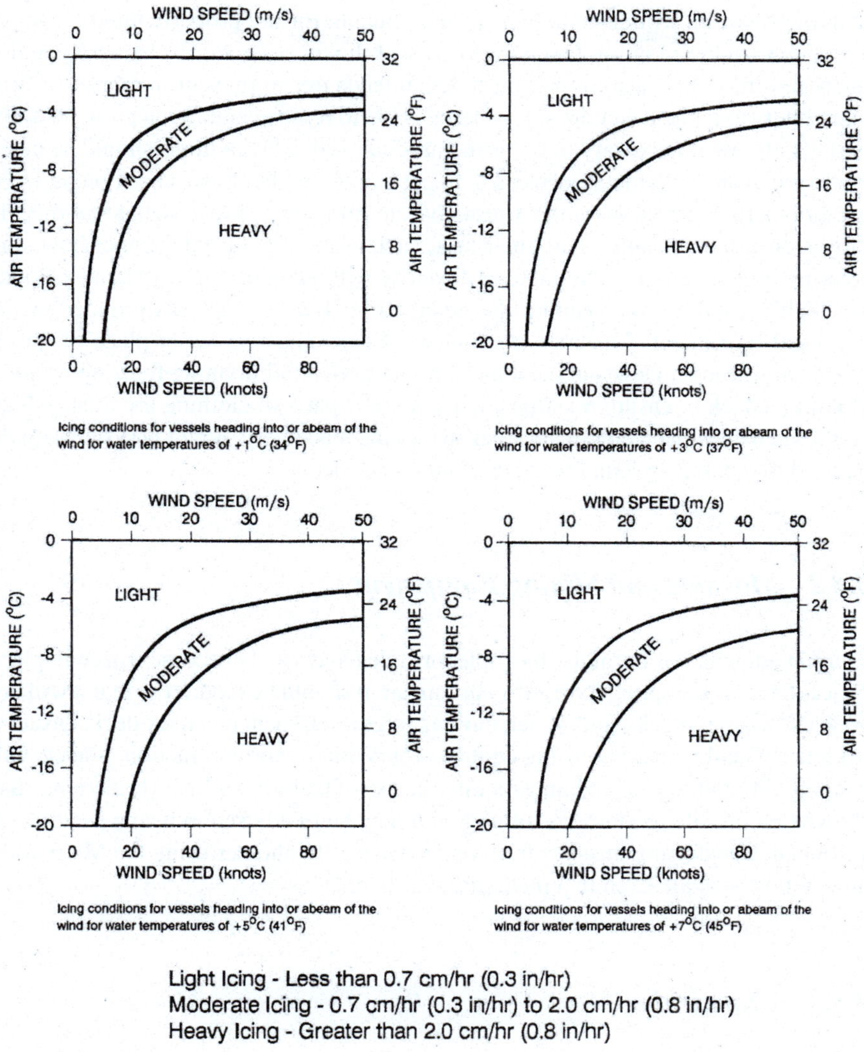

Fig. 4.1 Nomograms (Overland et al. 1986)

4.8.1 Equipment Handling with Weather Clothing

Equipment handling may be difficult while wearing freezing weather clothing because manoeuvrability and dexterity will be reduced as the thickness of clothing increases. Mittens and gloves make it more difficult to grasp some objects, such as spanners and screwdrivers. It will be impossible, for example to fine-tune electronic equipment with small potentiometer knobs while wearing large gloves. The following safety precautions should be observed while wearing freezing weather clothing. Be extremely careful while handling heavy equipment. Ensure properly fitted gloves of suitable warmth are worn. Leather gloves with liners are good for handling equipment and lines. It is also important to keep hands dry to prevent numbness. Carry extra liners to replace wet ones. It is permissible to remove hand protection for short periods of time if necessary to perform particular tasks. Precautions should be taken not to touch any cold metal surfaces or to get hands wet. Put gloves back on between evolutions to maintain warmth. Alternatives to removing gloves include modifying gloves or mittens to suit a particular task, such as cutting a trigger finger hole, and wearing light glove liners to improve dexterity without exposing bare hands. Hoods are highly effective for keeping the head warm, but they can also hinder vision and head movement. Consider wearing wool caps and scarves in place of hoods when unobstructed view and movement are needed. After donning freezing weather clothing, check to ensure no straps, flaps or other parts of clothing are loose which could catch on appendages or get stuck in moving machinery. A buddy system should be used to conduct freezing weather clothing checks.

4.8.2 Movable and Sliding Equipment

Equipment, which is normally movable, or slides may easily become frozen in place by ice. Before attempting to free the equipment, it should be secured so that it will not go adrift when it breaks free. When moving large equipment on an icy deck, ensure it is securely held to prevent losing control of it. Without normal traction, objects will be more difficult to stop and may require the use of restraining lines. Experience has shown that missile rearming is difficult and may be ill advised in heavy weather. In particular, transferring missiles from one magazine to another using the Mk 6 dolly may not work at all, even in a moderate sea state.

4.8.3 Chemicals

Certain chemicals will be used onboard ships operating in freezing weather conditions that are not used under normal conditions. These chemicals will be used for such purposes as:

- Preventing ice build-up or melting ice which has accumulated on decks and equipment.
- Removing or preventing fog or frost from forming on windows and optical equipment.
- Preventing freezing of cooling water systems.
- Starting cold engines.

Safety precautions, should be observed during the use of these chemicals. Because calcium chloride is so widely used to melt ice on decks, everyone aboard ship will come into contact with it at some time. For this reason, and because its potential hazards are not known, the precautions for handling this chemical are presented in more detail. Calcium chloride is an effective de-icer at temperatures below −17 °C (1.4 °F) because it gives off heat when mixed with water. A concentrated solution may get hot enough to boil. This same property may cause irritation or burns to the eyes, mucous membranes (nose and mouth) and skin. In case of contact with the eyes, promptly flush with plenty of water for at least 15 min and call medical personnel. For skin contact, flush with water and wash thoroughly. Remove and wash contaminated clothing. Because regular and anhydrous flake calcium chloride can absorb moisture from the surroundings, it can dry out leather and damage fabrics. Calcium chloride should always be stored in a cool dry place. Protective clothing which should be worn by personnel handling bulk calcium chloride includes:

- Rubber boots or well-oiled leather shoes.
- Rubber or latex canvas gloves with gauntlets.
- Rubberised raincoat, goggles.
- Dust respirator or face mask.

4.9 Cold Weather Preparations to Be Performed Well in Advance of Arctic and Polar Operations

Check all heating and cooling systems (HVAC systems). Check coolers for leaks. Weather leakage can occur because of contracting of the tubes in the cooler tube sheet/plate due to colder salt-water inlet temperatures. Identify unheated or inadequately heated spaces. Air, cooling-water, salt-water, and other lines running through spaces should be insulated for protection. Heaters should be installed. Portable items subject to freezing should be moved to heated spaces. Other ad hoc measures to improve temperature control include fans and other ventilation techniques to draw warm air from other areas. Additional heating will be needed to maintain crew comfort. Check areas where condensation and frost may form on interior metal surfaces adjacent to exterior bulkheads.

(1) Hatches, doors and lagging buttons are likely to be most frequently affected. If interior ambient temperature is cool, heavy frost will form. If interior spaces warm up, condensation will drip onto deck and equipment.
(2) More insulation may be needed on exterior bulkheads, overheads, and especially on doors and hatches.
(3) Check all watertight doors' gaskets and knife/sealing edges.

Where it is determined that spaces require portable heaters, check wiring set-ups. Ensure that vital electronic equipment (navigation gear, etc.) is not wired to the same circuit breaker as a high-amperage portable heater.

4.10 Exposed Piping and Scuppers

Where overflow pipes from freshwater gravity tanks should be led inside the ship to a warmed scupper and not be allowed to drain onto an open deck. Deck-edge waterways should be kept clear of obstructions so that a clear passage will exist when snow is being washed down. Fill tanks to a level that precludes routine overflow on decks. Lagging by itself will not keep liquid in a pipe from freezing if the ambient temperature falls below the freezing point. Liquid must be warmed and kept in continuous motion; failing this, heating of all the piping lying outside the insulated structure will be necessary where piping systems cannot be internally isolated. Where possible isolate and drain external liquid piping.

4.11 Stowage Space Requirements

Additional stowage space will be required to accommodate Arctic equipment and material. Space will also be required for the stowage of anti-ice compounds and liquids. Crew members will need extra stowage space for their gear. Space should also be set aside for Arctic clothing in living spaces or in compartments immediately adjacent thereto. Similarly, suitable facilities should be provided for the drying of heavy Arctic clothing. Allot additional stowage space for food storage since a 10 percent increase in food consumption may be experienced and underway resupply may be limited.

4.12 Fresh Water Supply

Consideration should be given to freshwater tanks that are not fitted with heating or insulation systems as fresh water will freeze at nominal polar sea water temperatures.

4.13 Saltwater Supply

For fire mains potentially exposed to below freezing temperatures, leak-offs should be fitted at the ends of the fire main to assure a continuous circulation of water. These leak- offs must be of sufficient diameter, be well lagged, and discharge directly overboard. Fire main risers can have water secured at the main deck, or for flight deck supplies use antifreeze in the riser itself. Salt-water hydrants fitted in exposed positions should be isolated inside the main structure, and arranged so that the exposed length of the hydrant supply pipe can be drained and left empty in freezing weather. Careful attention must be paid to the lagging of salt-water supply pipes which, although not directly exposed to weather, can be subject to low temperatures, e.g., those fitted in the hangar of aircraft-carriers near lift openings. Careful attention must also be paid to spraying arrangements in ready-use magazines, to similar compartments in exposed positions, and to magazine-flooding lockers if these are located on weather decks.

4.14 General Instructions for Machinery

Machinery space temperatures should be maintained at or above 5 °C (41 °F). In systems containing a mixture of glycerine and water, a check on the homogeneity of the mixture should be made by taking samples from both the top and the bottom of the mixing tank and testing with a glycosometer. A homogeneous 50/50 mixture will remain completely fluid down to −27 °C (−16.6 °F). A 60/40 mixture will remain fluid down to −40 °C (−40 °F). When engines which have saltwater cooling systems are not required for immediate use, the circulating water system should be kept empty or heated. Engines with a freshwater cooling system should have the system charged with an approved antifreeze solution. The raw water side of the heat exchanger system should be drained, or circulation maintained. Glycerine and water should not be used as an antifreeze solution. Because of the fine clearances in the water pumps of both types of cooling systems, drops of water remaining after draining may cause seizure of the pump, so pumps should be warmed before turning the engines.

4.15 Exposed Equipment

Particular attention should be given to equipment on deck, to guard against congealing of lubricant or formation of ice. Covers should be used, where possible. Pressure gauges in exposed positions should be disconnected when non-essential. When removing ice from any mechanism by use of steam jet, care should be taken to avoid ingress of steam which will condense and freeze on interior working parts. Gas engine (consideration extended to CODOG propulsion systems). Anti-icing equipment should be on before starting the engine even when ambient temperature remains over 0°. As the air gets accelerated, it becomes colder because of the venturi effect. This is where an icing risk exists on the leading edge of the blades.

4.16 Miscellaneous Systems

4.16.1 Fire Hoses

Hoses should be drained, dried, and stowed below deck.

4.16.2 Fire Extinguishers

Foam-type and soda-acid extinguishers should be stowed in a warm place. Arrangements should be made to ensure that the supply from the fire main for continuous foam is warm and that foam compound is stowed in a position protected from the cold.

4.16.3 Arrestor Gear and Safety Barrier System

Heating should be provided to air- reducing valves fitted in the arrestor gear and barrier system to prevent freezing.

4.16.4 Hydraulically Operated Mechanisms

These should be charged with a 50/50 mixture of glycerine and water, or a national approved liquid as directed and be assessed periodically.

4.16.5 Motorboats

To prevent damage to the engines and to ensure that they can be started when required, the following precautions should be taken:

(1) An external source of heat should be provided to maintain the engines and pipe systems at a reasonable temperature. It may also be necessary, under conditions of extreme cold, to warm the lubricating oil and diesel fuel tanks.
(2) When engines are not in use, or at "short notice," the seawater circulating system should be kept empty. Engines with freshwater cooling systems should have them charged with an approved anti-freeze solution. To facilitate draining the system, drain cocks should be fitted where necessary to drain water from any pockets in the system. On account of the fine clearances in the water pump, small drops of water remaining after draining may be sufficient to cause seizure of the pump if the temperature has not been above −2.2 °C (28 °F). To avoid damage to the pump, specific measures should be taken to warm the pump before turning the engine.
(3) Full use should be made of devices fitted to assist cold starting of diesel engines, such as heater plugs, ether, external battery chargers etc.
(4) Outboard motor precautions: Boats and the engines stored on open deck should be covered when not in use. Engines should be started daily and flushed with fresh water after each use. When launching care should be taken not to immediately release boat's bowline as sudden stopping of engines that are started before entering the water can occur. Spare engines should be carried and kept in warm storage.

4.17 Insulation

Ensure that insulation is intact at the shell and bulkheads behind and above switchboards if such areas suffer from condensation. Check that all cold-weather piping and equipment such as fire, sanitary and freshwater systems, soil pipes and freshwater tanks are properly insulated where damage to equipment or discomfort to personnel may result from dripping condensate. All ventilation supply and exhaust ducting and heaters, including flanges and joints, must be insulated to prevent sweating and heat loss.

4.18 Gas Cylinders

Pressure in compressed gas cylinders subjected to cold temperatures will drop considerably and may be insufficient for practical use. Therefore, bottled gases should be stowed inside if safety regulations permit.

4.19 Batteries

Cold temperatures drastically reduce the output of all types of batteries (both dry cells and storage batteries). For example, at −18 °C (0.4 °F), the ampere-hour capacity of a typical dry cell battery is reduced to about 25 percent of the 20 °C (68 °F) rated capacity. At this temperature, capacities of lead-acid and nickel–cadmium storage batteries are down to about 35 percent and 50 percent, respectively. Li-ion can operate to a temperature of −40 °C (−40 °F) but only at a reduced discharge rate; charging at this temperature is out of the question. The lowest temperature for reliable cranking is about −18 °C (0.4 °F). The output of all batteries reaches zero at about −34 °C to −40 °C (−29.2 °F to −40 °F). The rate at which storage batteries can accept a recharge is also reduced in cold temperatures. To obtain a good recharge in a reasonable amount of time, the temperature of the battery should be about 15 °C (59 °F) or higher. The sulphuric acid electrolyte in a discharged acid battery can freeze at −15 °C. If the battery is fully charged, the electrolyte freezing point is depressed to −60 °C (−76 °F) or below. Freezing may damage the plates, crack the battery case, and split the cover-to-case seal or the terminal-to-cover seal, thus leading to electrolyte spillage. Freezing of electrolyte may also form crystals which can pierce separators, eventually leading to internal short-circuits and premature failure of the battery. The potassium hydroxide electrolyte in a nickel–cadmium battery does not vary significantly with the state of charge and the freezing point is constant at −60 °C (−76 °F). Freezing of this type of battery is not a problem. Storage batteries should be kept fully charged and stowed in a heated space or equipped with heaters. Flashlight and other dry cell batteries should be kept warm when not in use.

4.19.1 Battery Maintenance and Safe Handling

- *Daily*. See that all parts of the ventilation system in battery rooms and battery lockers are in proper condition. Clean battery hydrometers.
- *Weekly*. Observe the height of electrolyte in cells and measure readings and record cell specific gravity and temperature readings for all batteries (when batteries are in warm rooms). Fill water batteries (only with distilled water) if the height of electrolyte is at the low mark or will drop below the low mark before the next weekly inspection. Check the charging rate of engine battery charging generators and voltage at which batteries are being floated.
- *Monthly*. Clean batteries and grease the terminals with petrolatum, as necessary. Examine battery connections and correct any faulty condition such as breaks, frayed insulation or grounds. Inspect for broken or cracked battery cases or jars. Give all batteries an equalising charge, except those charged from their own generator or being floated. Take a complete set of voltage, temperature and specific gravity readings on all batteries, which have been given an equalising charge.

- *Quarterly*. Give all batteries which are charged from their own generators or are being floated an equalizing charge and take a complete set of voltage, temperature and specific gravity readings.
- *Semi-annually*. Give each battery a test discharge at a 5-to-10-h rate or as specified on the battery nameplate. A test discharge is the most reliable means of determining storage battery conditions, but functional testing may be done in lieu of a test discharge if authorised by the ship's maintenance requirement cards.

Functional testing of maritime-type portable storage batteries for various shipboard applications varies with usage, size of battery and load. Battery test schedules typically means engine starting batteries should be capable of starting an engine at least once a week. Portable lantern batteries (using storage lead-acid batteries) should be capable of providing sufficient light for a period of one minute without dimming and should be assessed at least once a week. Gyrocompass batteries should be functionally assessed for a 20-min period on battery power alone, once a month. Telephone batteries should be functionally assessed during a peak load period for 4 h on battery power alone, once a month. In the event of battery failure, give the battery an equalising charge, then retest. If the retest fails, replace the failed battery. Batteries subjected to the cold should be heated, if possible, and must be kept fully charged to prevent freezing. If using heaters or a warm air supply to heat boat engines, heat the batteries by the same method. Battery chargers should be provided for all boat batteries, if possible. The battery charger should be placed in the same temperature environment as the battery being charged since they have a temperature compensation feature for charging voltage. Battery chargers located in heated spaces, used to charge batteries by long leads outside the spaces, should have their charging voltage output adjusted for the exterior temperature. Any acid batteries, which are required to operate consistently in temperatures below 44 °C (111.2 °F) may have the average specific gravity raised to 1.280 (between limits of 1.270 to 1.285) by adding diluted acid.

4.19.2 Batteries and Electrolytes

Due to the fact that the performance of batteries is severely degraded at cold temperatures, ship's personnel who use battery-powered equipment are more likely to become involved with the handling of batteries during a freezing weather cruise than under normal conditions. This handling may include recharging batteries, warming up batteries, which have become too cold or replacing batteries, which are no longer serviceable. In recognition of the potential serious safety hazards involved in handling batteries and electrolytes, this section describes some of the safety precautions which should be enforced at all times. Emphasis is given to considerations, which are especially important under freezing weather operating conditions.

4.19.3 Rechargeable Batteries

The most common storage battery used onboard ship is the lead-acid type, which contains an electrolyte solution of sulfuric acid and distilled water. Sulfuric acid is dangerous to personnel and highly corrosive to equipment or materials. Another type of storage battery, which will be encountered aboard ship is the alkaline type of battery containing nickel–cadmium and an electrolyte solution of potassium hydroxide. This electrolyte is also dangerous to personnel and is corrosive to many materials including aluminium and glass. Modern Lithium and other chemical composition batteries also present challenges to operations in cold regions. Operators should review technical publications and guidance when using or charging these devices in cold temperatures. There are severe limitations for charging many modern battery types in cold temperatures.

4.19.4 Hydrogen Hazards

Sparks and flames. Both hydrogen and oxygen gas are given off from a storage battery, especially during recharging. Because hydrogen mixed with oxygen or air is very explosive, sparks, smoking or flames of any kind must never be allowed in the vicinity of any rechargeable battery. *Battery compartment ventilation.* Extra care is necessary after opening a battery compartment, which has been sealed. Because hydrogen is colourless, odourless and tasteless, such compartments should always be thoroughly ventilated before they are entered. To avoid the possibility of an explosion, no light switches should be turned on and no electrical connections should be made or broken in the compartment until it has been well ventilated. When preparing to recharge a battery located in a battery compartment, verify that the ventilating system is operating properly before starting the charge. Stop the charge if the ventilation is interrupted. *Charging rate.* Charge a battery only at the rates given on its nameplate. Reduce the charging rate if the battery electrolyte begins to evolve bubbles of hydrogen and oxygen or if the battery temperature reaches 35 °C (95 °F). Stop charging if the battery temperature approaches the upper limit of 52 °C (125 °F). *Sparks.* To prevent dangerous sparks, ensure that no current is flowing into or out of the battery before disconnecting or connecting battery terminals. When batteries are used with one terminal grounded, the grounded terminal should be disconnected first when removing the battery and connected last when replacing the battery. Verify that all terminal connections are tight to preclude sparks due to lose connections. Use only tools with insulated handles to prevent short-circuiting the battery terminals.

4.19.4.1 Handling Electrolytes

Protective equipment. Personnel engaged in handling any electrolyte (or cleaning up a spill) should wear rubber aprons, rubber boots, rubber gloves and goggles or a full-face shield so that the electrolyte cannot come in contact with clothing or skin. *Treatment of electrolyte burns.* Should acid or electrolyte come in contact with the skin, immediately wash the affected area freely with a large quantity of freshwater for about 15 min. Should acid or electrolyte come in contact with the eyes, flush with freshwater for a minimum of 15 min ensuring that both upper and lower lids are pulled sufficiently to allow water to flush under them. In either case, the sickbay must be notified of the accident as soon as possible and must be requested to come to the location of the casualty. If the medical officer cannot be notified, wash or flush with water for 15 min before transporting the casualty to the sickbay. In an extreme emergency, where freshwater is not available, seawater may be used, but only as a last resort. Clothing that may have been splattered with electrolyte should be promptly removed. Skin areas touched by electrolyte beneath contaminated clothing, should be promptly treated by flushing with water as described above. *Replacing electrolyte.* Nothing but distilled water should be added to a battery except when it is necessary to replace spilled electrolyte. When replacing spilled electrolyte, use only premixed electrolyte, sulfuric acid of specific gravity greater than 1.350 should not be added to a battery.

4.19.4.2 Concentrated Sulfuric Acid

Only premixed electrolyte is to be used or stored onboard the ship. The use and storage of concentrated acid for the purpose of preparing electrolyte or for adjustment of specific gravity should be specifically authorised. Two of the most important precautions to be followed in handling concentrated sulfuric acid, especially in freezing weather, are:

(1) In making electrolyte, always pour acid slowly into water and never water into acid. The addition of even a small quantity of water to a container of strong sulfuric acid may cause an explosion due to the sudden evolution of heat.
(2) Containers of concentrated sulfuric acid must be stored in a place where freezing cannot occur. Under certain conditions, concentrated acid can freeze at temperatures as high as 4 C. Freezing can cause the container to break with consequent grave danger of serious acid burns to personnel.

4.19.4.3 Prevention of Battery Damage

Precautions to prevent battery damage also serve to protect the personnel handling the batteries. The following are precautions which should be followed to protect batteries from damage caused by freezing weather and related conditions:

(1) *Discharged batteries*. The sulfuric acid electrolyte in a discharged battery can freeze at -15°C (5°F) whereas, in a fully charged battery, it will remain liquid down to −51 °C (−59.8 °F) or below. Freezing may damage the battery internally and/or crack the case causing a hazardous spillage of electrolyte. All batteries exposed to freezing weather conditions should be kept fully charged.
(2) *Adding water.* When distilled water is added to a battery to replace that lost during normal operation, the water should be added just before the battery is placed on charge. The water remains on top of the electrolyte until mixed with it by charging. In freezing weather, the unmixed water may freeze, causing the battery case to crack and leak electrolyte.
(3) *Battery heaters*. Equipment which may be installed to warm batteries, which are exposed to freezing weather should be controlled to keep the battery compartment temperature below 35 °C (95 °F) and the battery below 52 °C (126 °F) or damage to the battery may occur.

4.19.5 Other Battery Hazards

4.19.5.1 Gasoline Fumes

Batteries used for engine starting are frequently located near the engine itself. Care should be taken to avoid sparks when removing or replacing batteries located in compartments, which may contain gasoline fumes.

4.19.5.2 Salt-Water

Care should be taken not to allow salt-water to splash or leak onto an acid battery because salt-water entering a battery cell will produce chlorine gas which is extremely toxic.

4.19.5.3 Dry Batteries

The following safety precautions should be observed for dry-cell type batteries:

(1) *Hydrogen hazards*. Never continue to use a multi-cell dry battery after its closed-circuit voltage has dropped below a value equal to 0.9 V per cell. Discharging a battery beyond this point will force current through some cells, which may be completely discharged. This will result in the generation of hydrogen and oxygen gases caused by the electrolysis of electrolyte. When this happens, there is danger of a hydrogen explosion and injury to personnel and damage to equipment.
(2) *Shock hazards*. Some ships have high voltage dry batteries. These are capable of imparting a profoundly serious, if not fatal, shock to anyone coming in

contact with their terminals. When disconnecting them, the current flow should be stopped before disconnecting the plug. It is possible for sufficient gaseous hydrogen to accumulate in this battery to produce a serious explosion if ignited (by the spark caused by pulling the plug while the current is flowing).

(3) *Short-circuits.* In order to prevent a short circuit, wire leads should be kept insulated when the battery is not connected to apparatus. Short circuits may result in sufficient heat to cause a fire. In addition, a discharge caused by a short circuit causes the cells to burst, spilling corrosive electrolyte, which can damage equipment and cause injury.

(4) *Mercury cell batteries.* These batteries may explode if improperly used. Never discharge a mercury cell battery after the battery fails to operate the equipment or the voltage falls below 0.9 V per cell. Do not leave the battery switch on when the equipment is not in use or after the battery fails to operate the equipment. Never impose a dead short-circuit on a mercury cell or allow it to become overheated. A temperature of about 204 °C (399 °F) will cause such a cell to explode. Discard exhausted mercury cell batteries as soon as possible. Dead single and multi-cell batteries with steel jackets should have holes punched in the jackets before being discarded to release any gas which might have formed. Follow proper disposal procedures.

4.20 Aircraft Ground Support Equipment

On board ships operational aircraft ground support equipment in freezing weather is particularly important. Failure of vehicles, which move aircraft equipment, supplies or ammunition could interrupt the ship's efficient operation. Normal greasing, filter cleaning, electronics and hydraulic checks are extremely important. Just as essential are the use of fuel, oil and lubricants designed to allow the vehicles to operate effectively and efficiently in freezing weather. The use of ether starting systems already installed on the vehicles or manual applications of ether to the air intake regions will provide the necessary freezing weather starting assistance. Ether canisters should be stocked in sufficient quantities to ensure quick starts throughout the deployment. Excessive use of ether damages the internal seals and gaskets of the engine. An engine may need to be overhauled in warmer temperatures after being started with ether. Air tanks on all aircraft ground support equipment vehicles should be drained daily. The air drain should be left open to allow all condensate to drain and to prevent freezing the airlines. Use of air dryers and their maintenance will be covered in vehicle maintenance manuals. Freezing weather affects pneumatic equipment as much as diesel or gas equipment or vehicles, so preventive maintenance and daily checks for moisture in the system needs to be accomplished. Pneumatic systems should be fitted with dryer units which use methanol. Freezing weather can cause severe stress on metal. Inspections of the aircraft ground support equipment stress points, the forklift carriage and points where hydraulic cylinders attach to the frame need to be completed to ensure no corrosion or stress fractures are evident prior to the deployment.

4.21 Underway Replenishment

In certain Arctic waters, underway replenishment is precluded due to concerns for environmental impact. Where operations are allowed careful preparations must be made.

4.21.1 Training

Before a freezing weather cruise, personnel should receive training on special aspects of freezing weather underway replenishment (UNREP) operations, including:

(1) Freezing weather hazards of hypothermia and frostbite.
(2) How to function in special freezing weather clothing.
(3) Proper relieving procedures.
(4) General freezing weather UNREP operations.

4.21.2 Preparation for UNREP

As a preventive measure, covers for deck equipment should be provided, here possible, to reduce the penetration of water and subsequent formation of ice. Because of freezing weather effects on UNREP equipment, more time and effort must be devoted to getting the equipment ready to begin UNREP operations:

(1) Ice and snow must be removed from covered and uncovered equipment, including decks and hatches.
(2) De-icing chemicals will be needed to remove ice from equipment and decks.
(3) Equipment must be started and operated at low speed to check for proper operation and to allow warm-up prior to full speed operation.
(4) The non-skid surface should be extended to cover as much deck area as possible.
(5) The technical manuals for all UNREP equipment exposed to the weather should be reviewed to determine what freezing weather lubricants and hydraulic fluids will be needed.
(6) Some specific preparations for UNREP include:

- Grease wire ropes and sheaves.
- Put freezing weather hydraulic fluid in winches and freezing weather lubricating oil in the winch gear boxes.

- Drain the water and purge the moist air from compressed air systems to prevent freezing of on-deck pneumatic equipment. Check air dryers to ensure that the dew point temperature of the compressed air is maintained within the proper range.
- Make room below decks to store all small equipment and fibre ropes used during UNREP operations. Break equipment out when needed and replace it below when the operation is completed.
- Procure winch covers to protect UNREP equipment from spray and to facilitate de-icing without damage to equipment by ice removal tools. These covers should be part of the ship's "Cold Weather Kit."
- Check the gaskets in refuelling probes and bellmouth to preclude leaks during refuelling operations. Stock extras because gaskets may become brittle and crack in freezing weather.
- Stock a large supply of chemlites because they will lose intensity after an hour or two in freezing weather, and because of the reduced number of daylight hours.
- Fibreglass reinforced plastic traction mats may be obtained which give effective traction even when partially iced over.
- Stock alkaline flashlight batteries rather than the ordinary cells. Alkaline are degraded less by cold temperatures.

(7) Conduct an electrical check to ensure traffic lighting and other necessary control measures are operational.
(8) Provide for portable, topside heaters or enclosures for personnel involved in extended topside evolutions. Hangar space heaters have proven effective.
(9) Anticipate that routine evolutions could take twice as long in freezing weather.
(10) Periodically inspect berthing spaces for warmth and ventilation. Ensure that adequate bedding materials are available.
(11) Provide for additional heaters for particularly exposed areas, such as the Signal Bridge and Pilot House.
(12) Hot beverages and soup should be available on the mess deck 24 h a day and distributed to look-outs, UNREP and flight-deck crews.
(13) Stripping of waxed decks and the addition of non-skid strips is recommended.
(14) Remove ice manually from the king-post and sliding pad eye prior to use. Self-de-icing by means of system operation damages the sliding pad eye and should be avoided.

4.22 Electronics and Characteristics of Electronic Emissions

The following guidance provides an outline of some of the requirements for electronic equipment applicable to most types of ships:

- All exposed shipboard equipment should be capable of operating at temperatures down to −50 °C (−58 °F).
- All exposed electronic equipment should be capable of withstanding wind velocities of 100kn (at 185.2 km/h; 115 mi/h) any temperature down to −50 °C (−58 °F).
- All mechanically operated equipment must be lubricated with low-temperature lubricant and adequate space heating fitted in spaces where this is feasible, e.g., inside radar pedestals.
- Electric heaters should be installed in inside covers of electronic equipment fitted on weather decks.
- Insulators should be lengthened approximately 10% or the next higher voltage rating over the normal standard as extra protection against shorting caused by ice formation.
- Antennas should be designed to withstand an ice load, up to 5 cm (1.9 in) thick around the component parts of each antenna. The diameter of wire rope antennas should not be less than 1 cm (0.39 in). All whip antennas should be capable of withstanding wind velocities of 100 kn 185.2 km/h; 115 mi/h) when coated with ice 5 cm (1.9 in) thick.
- Cables and cords should be handled with care to prevent fracturing. At extremely low temperatures, cables should be unreeled in a warm space before use.

4.22.1 Exposed Electronic Equipment

With the foregoing requirements in mind, all equipment should be appropriately winterised and pre-heaters used wherever possible. The practice of lengthy warm-up periods under no-load or light-load conditions is also recommended. Equipment not in regular use needs to be heated periodically to drive out moisture. Waveguides (and other non-pressurised lines where condensation can accumulate) should be fitted with drain cocks and drained at regular intervals. Planned maintenance schedules for ships should include checks of all drain cocks. Waveguide dryers should be used as directed. Equipment used in exposed positions, then returned to warm spaces for stowage, must be thoroughly dried to remove all moisture. After drying, the equipment should be allowed to reach the ambient temperature of the space prior to stowage. Ice formation can be a hazard on all types of external antennas, causing breakage or changes in antenna characteristics and reduction of operating range. Ice, which has formed should be knocked off with a thin pole. Ice and snow should also be removed from strain insulators to prevent excessive transmission losses. Wire antennas should be secured loosely, so any slight whip action can crack ice build-up.

4.22.2 Freezing Weather Preparations

Antennas. Radio equipment requires preparation prior to deployment for cold-weather operations. Of particular note is the problem of antenna icing. Antennas should be coated with a silicone/oil-based substance with sufficient on-board stock available for additional coating, if required. Antennas suffer sea-spray icing in the northern latitudes. The thicker the ice on the antenna, the greater the loss imposed on the signal. Factors such as air temperature, salinity of the water, structural shapes, and wind velocity play key roles in the antenna icing process and should be taken into consideration when operating in the area. Using proper preventive-maintenance methods can aid in reducing the amount of ice adhesion that occurs on an antenna. Wiping the antenna down, ensuring that it is free of dirt, and checking the smoothness of the antenna's surface, are procedures, which can be used by radio personnel to prevent icing. Additionally, all radio equipment (transceivers/receivers) should be brought up to technical manual standards prior to deployment It is imperative that all radio equipment be checked, and the necessary preventive maintenance be performed on all communication systems. Test antenna drive motors. Apply cold-weather grease, after purging old grease, about one week before getting underway. Make sure grounding straps were properly insulated. Obtain spare antennas and insulators to replace those which may be damaged from heavy weather/ice. A complete inspection and Planned Maintenance Schedule (PMS) of antennas/topside equipment is strongly advised prior to Arctic operations to minimise need for personnel to work aloft.

(1) Check rotating antenna heater circuits and cooling systems.
(2) Verify weather-tightness of exposed antenna components.
(3) Check operating mechanism on all communications antenna safety switches.
(4) Check operating mechanisms on all antennas with lowering systems installed.
(5) Insulate anemometer as much as possible. Hand-held anemometers should be available.
(6) Inspect wire antennas (HF broadband fans) and slacken if required.

The satellite communications (SATCOM) antenna pedestal may freeze, preventing the antenna from tracking in azimuth. The following actions may be helpful:

(1) Place temporary heat strip around pedestal
(2) Place fibreglass insulation and herculite cover around pedestal.
(3) Place desiccant bags inside pedestal to absorb moisture.
(4) Rotate directors/illuminators at low speed every eight hours. During extreme icing conditions rotate antennas in azimuth and elevation every 30 min to prevent the pedestal from freezing.

Fill antenna cooling systems with an anti- freeze solution. Fire control radar antenna scanners (the stinger on the director) may ice over on the inner rotating axis, holding the scanner in the conical scan pattern and preventing the spiralling motion in automatic-track mode.

(1) Keep a coating of antifreeze on the areas between rotating and non-rotating areas.
(2) Train the director aft with elevation depressed during freezing conditions.
(3) Install a lightweight, non-reflective cover on the antenna when Arctic operations are planned.
(4) Rotate directors/illuminators at their slowest speeds every eight hours.

Operate variable-spaced antennas at low speed for a brief period before increasing rotation speed. Do not change the direction of rotation or accelerate/decelerate rapidly, until gear trains and drive motors are thoroughly warmed up. Installation of electric heaters in antenna pedestals is recommended. Replace oil in the pedestals of rotating radar antennas with oil that will not congeal at low temperatures. Consideration should be given to the installation of oil heaters. Connect the radar antenna pedestal heater to a circuit that will remain energised when the radar set itself is disconnected. Drain wave guides and other non-pressurized lines fitted with drain cocks at regular intervals to eliminate accumulated condensation. Use wave guide dryers as directed.

4.22.3 Slow-Moving Mechanical Parts

Slow-moving mechanical parts, such as shafts and bearing surfaces, will operate more satisfactorily if the surfaces are buffed and polished. Such surfaces should not be lubricated unless this is essential, and, even then, it should be done sparingly.

4.22.4 Covered Cables

Rubber-covered, flexible cable becomes stiff at temperatures lower than −6 °C (21.2 °F); the insulation becomes brittle and can crack and shatter rather than bend. Synthetic materials (e.g., polyethylene) are replacing rubber for insulating purposes because of much improved cold-weather characteristics.

4.22.5 Electric Installations

Electric installations should be made with particular care to minimise the effect of cold, moisture, high humidity and stresses due to icing, as follows:

- Increase wire size of radio antennas to compensate for high winds and ice loading of long horizontal spans. Receiving antennas using small-size wire are vulnerable to damage by ice loading.

- Install double antenna suspension insulators for transmitting antennas to increase leakage path and to prevent arc-over.
- Increase the size of mounting brackets and foundations of equipment in ships which may become involved in ice-breaking operations.

4.22.6 Communications Equipment

Communications equipment for high-latitude operations should include a transmitter with a high-frequency operating range of at least 26 MHz, and another, which will operate on frequencies as low as 175–195 kHz. Both of these transmitters must have substantial power capabilities. All transmitters and receivers should be calibrated on the several frequencies expected to be used. To capitalise on the advantages of low-frequency operation and very low-frequency reception during blackout periods, antenna systems additional to those already installed may be required.

4.22.7 Telephones and Microphones

When using telephones or microphones in exposed locations, care should be taken to prevent the speaker's breath from freezing the microphone. In general, the use of portable microphones supplied with a transmitter station, which is in a protected space is preferable to those located in the open. Cover sound-powered telephone boxes.

4.23 Care of Electrical Equipment in Low Temperatures

In general, special procedures will be required to ensure satisfactory operation of electronic equipment at temperatures lower than −2 °C (28.4 °F). Dry-cell batteries and electrolytic condensers will not work at low temperatures but will recover and resume normal operation when warmed. Maintenance will be required to prevent corrosion from condensation caused by temperature changes. Components may require frequency readjustment of critical circuits because of changes in electrical characteristics. When temperatures fall below 2 °C (35.6 °F), heat should be applied to electronic equipment if it is to be ready for immediate use. Batteries and electrolytic condensers, in particular, must be warm if they are to operate efficiently. Continuous operation in the "standby" or "filament" position is good practice as it keeps equipment at the correct operating temperature before plate voltage is applied, and it eliminates condensation inside the equipment.

4.24 Underwater Electronic Equipment

Echo-ranging equipment with retracting transducers requires an "ice-gate" to prevent entry of ice into the transducer sea chest. Radar waveguide and ECM antenna masts on submarines should be filled with pressurized nitrogen and the heaters should be continuously energised.

4.25 Portable Power Sources (For Electronic Equipment)

A small aerosol-type can of ether, should be included to facilitate cold-weather starting of portable internal-combustion power-generated equipment.

Warning

- **Extreme caution should be exercised to limit the quantity of ether injected in both gas-engine carburettors and diesel engine air intakes.**
- **To avoid the possibility of frostbite, do not allow ether to come in contact with exposed body areas.**

If equipment is to be shut down or left turned off for extended periods, the lubrication oil should be drained while the engine is hot, then stored inside the ship or personnel shelter along with the starting batteries, if so equipped. Often it is more advantageous to leave engine-starting batteries inside and extended cables outside to the generator.

4.26 Maintenance (Personnel Precautions)

No maintenance aloft should be attempted unless an emergency condition exists, and then only after rigging an appropriate canvas windscreen protection cover against wind-chill and installing portable heaters inside to allow maintenance personnel an opportunity to remove gloves, masks and attach safety harness, prior to attempting any electronic-type repairs. (Solder will not blow at −20 °C (−4 °F) using conventional irons.) Gloves should be donned prior to leaving the shelter, to keep hands from freezing to ladder rungs or handrails. Safety belts/harnesses must be worn when working aloft. Positive footing is never available aloft in northern waters and becomes more pronounced when personnel are required to wear cumbersome clothing and footwear. The installation of communication whip antennas in the vicinity of smokestacks or boiler room uptakes usually prevents the formation of ice. However, periodic cleaning is necessary to remove soot accumulations and prevent degradation of antenna performance. The cleaning aspect does not present a problem when the whips are of the retractable type and access to the top of the stack is gained through an internal trunk-and-scuttle arrangement. Breathing masks and safety belts/harnesses must be worn at all times to prevent smoke and stack gas inhalation and to conform to "safe working practice aloft" criteria.

4.27 Communication Characteristics

Radio communications beyond "line-of-sight" limits in the Arctic pose certain problems that occur only rarely at lower latitudes. These problems are caused by ionospheric disturbances which affect the behaviour of radio waves in the low-frequency (LF), medium frequency (MF), high frequency (HF), and extremely high frequency (VHF) bands. The standard shipboard transmitter required for high-latitude operations usually covers a frequency range down to 175–195 kHz, depending on the model. At the lower end of the range, the radiated power obtained may be small, being something of the order of 0.6–1.5% of the rated output power. This is caused by the inability to load the antenna properly because of its short effective length. Radiation is further decreased by the necessarily short vertical section of the antenna, upon which radiation depends. Optimum radiation with current shipboard antennas is usually realized at around 500 kHz, and effective radiation decreases as the operating frequency is decreased. It has been found, too, that in operating the low-frequency transmitter at frequencies in its lower range at substantial power (one or two kW), using a standard shipboard antenna, severe arc-overs and large losses in antenna trunks will be experienced due to high standing-wave ratios, with high voltage occurring at the insulators caused by inadequate effective lengths. For these important lower frequencies to be used during periods of ionospheric disturbances, arrangements may have to be made to install temporary transmitting antennas of the greatest length and height practicable. The possibility of paralleling existing antennas in series might be considered, and emergency use of balloon suspension should not be overlooked.

4.28 Arctic Anomalies

4.28.1 Auroral Zones and RF Interference

Zones of extreme RF interference and "electronic storms" become increasingly prevalent and severe in higher latitudes. Such "auroral zones" are believed to be due to charged particles ejected from the sun and deflected by the earth's magnetic field. It has been noted that the frequency and severity of electrical storms vary directly with the number of sunspots observed on the sun. Communications in high latitudes are affected both by electronic storms and ionospheric disturbances. In HF/MF bands, propagation becomes erratic, the highest usable frequency decreases and the lowest usable frequency increases. Transmitters and receivers will fade. At times, communications black out completely and may remain out from a few minutes to several days. On the other hand, communications on VHF/UHF circuits may extend beyond the line of sight. Radar may similarly be rendered useless for brief periods, or its range extended by ducting. A communication coordinating circuit should be included in all communications plans. The coordination circuit should be covered using narrow-band and wide-band secure voice equipment.

4.28.2 Antenna Icing

Radio equipment requires preparation prior to deployment for cold-weather operations. Of particular note is the problem of antenna icing. Antennas should be coated with a silicone/oil-based substance with sufficient on-board stock available for additional coating if required. Antennas suffer sea- spray icing in the northern latitudes. The thicker the ice on the antenna, the greater the loss imposed on the signal. Factors such as air temperature, salinity of the water, structural shapes, and wind velocity play key roles in the antenna icing process and should be taken into consideration when operating in the area. Using proper preventive-maintenance methods can aid in reducing the amount of ice adhesion that occurs on an antenna. Wiping the antenna down, ensuring that it is free of dirt, and checking the smoothness of the antenna's surface, are procedures which can be used by radio personnel to prevent icing. Additionally, all radio equipment (transceivers/receivers) should be brought up to technical manual standards prior to deployment. It is imperative that all radio equipment be checked, and the necessary preventive maintenance be performed on all communication systems.

4.28.3 UHF Communications

Effective UHF ranges are significantly reduced. During heavy snow, absorption and fading are common, further reducing UHF ranges. With unrestricted visibility conditions, maximum UHF ranges are 15–16 nm (28–30 km; 17.3–18.6 mi); when visibility is reduced to 1–2 nm (2–4 km; 1.2–2.4 mi), UHF ranges are reduced to 10–12 nm (19–22 km (11.8–13.6 mi). Few incidents of apparent UHF black-out have been noted. However, on some occasions UHF communications have been good with a ship 10 nm (19 km; 11.8 mi) away, while having no communications with a closer ship.

4.28.4 Satellite Communications

A major problem is the potential for the satellite receiver antenna to freeze. Communications dependent upon satellites in geostationary, equatorial orbits will be degraded above 70°N because the satellite is near or below the horizon.

4.28.5 Static

Continuous high-level static is rarely experienced at Arctic latitudes, but sporadic noise is common. Irregularly occurring steady rushes of increasing noise frequently signify auroral disturbances on the frequency employed. LF is less affected by this type of atmospheric noise than HF. Flakes or pellets of highly charged snow are occasionally experienced in the north during periods of high winds. Charged particles of snow driven against metal vehicles, masts, and antennas, discharge with a high-pitched static sound that can be heard on all received frequencies. This form of noise is more severe on aircraft radios than on ground or vehicle stations.

4.28.6 Ground Conductivity

Ground conductivity in northern areas is poor. At HF and below, effective ground-wave ranges are reduced over ice, permafrost or snow-covered terrain. Moreover, the efficiency of an antenna, especially when propagating in the sky-wave mode, is lessened when installed over this kind of ground. Great care must be taken in sitting to secure the best available ground, and considerable effort may be needed to build an artificial ground system.

4.28.7 Communications Plans

Because of the foregoing, the following communication factors will influence the communications plan:

(1) Extended communication ranges, which must be achieved are normally accompanied by a reduction in communication capacity. Normal scales of communication equipment cannot usually be maintained under these circumstances.

  (a) Depending on the nature of terrain and other factors, ground rebroadcast stations for voice operation may have to be deployed at very moderate distances. Airborne rebroadcast may be required for greater ranges.

Finally, Carrier-Wave (CW) Operation may be necessary for point-to-point communications at still greater ranges or during disturbed conditions.

(b) Range or terrain may dictate the use of sky-wave transmission. Although usually dependable, it is not entirely so due to the unpredictability of atmospheric conditions.

(c) Where a higher degree of reliability is required at sky-wave transmission ranges, normal HF links must be guarded, by parallel LF circuits. Where this is necessary, two stations must be established at each circuit terminal. Extensive antenna systems may be required for continuous day/night operation.

(2) Because of the vagaries of sky-wave transmission, much greater attention must be paid to protecting information than is usually the case. When voice codes are inadequate, ciphers must be employed, and these are likely to be of the manual type. Manual enciphering is time-consuming and handling times are in proportion to the length and number of messages being transmitted.

(3) In northern areas, the siting of operational headquarters to gain the extended communication ranges is critical. Detailed ground reconnaissance and testing will be necessary to find adequate transmission sites. In the absence of suitable retransmission facilities, a major headquarters or base should not be deployed to a remote locality until a satisfactory site has been determined.

Chapter 5
Operating in Arctic Conditions

5.1 Introduction

Given the propensity for freezing conditions, it should come as no surprise to the seasoned mariner that waterways and harbours are often frozen over. This means ice-breaking vessels are needed to clear a pathway through the ice to permit safe and unimpeded passage. Broadly speaking there are three categories of ice-breakers:

(1) Harbour.
(2) Non-Polar; and
(3) Polar.

The harbour ice-breaker category typically covers small ice-breaking vessels, generally not much larger than tugs, engaged in keeping rivers and small harbours clear of thin ice. The second category embraces large ice-breakers employed in keeping areas of sea clear of ice, freeing ships trapped in areas such as the Baltic, the Gulf of St. Lawrence and Greenlandic waters. The third category covers ice-breakers capable of operating in very heavy and hard Arctic ice. In addition, ships of this last type usually support oceanographic, hydrographic and other related scientific activities in the Polar regions. The guidance, which is provided in this book, applies in the main, to the Polar category of icebreakers. Should an ice-breaker become stuck in the ice, fitted heeling and trimming tanks may be used to create motion. Ice-breakers lacking a heeling system, and fitted with a centre line crane, can rock the ship by swinging a heavy weight from side to side. Draught should be sufficient to accommodate large propellers whose tips, when in the upper position, will be at least 2 m (6.5 ft) below the surface of the water. There are five factors which govern the ability of a Polar ice-breaker to do the job she was designed to do; and moreover, to do it effectively. These are:

(1) The power, displacement, and strength of the ship.
(2) The nature and extent of the ice being attacked.

A. Olsen, *Ship Operations in Extreme Low Temperature Environments*, Springer Series on Naval Architecture, Marine Engineering, Shipbuilding and Shipping 19,
https://doi.org/10.1007/978-3-031-52513-1_5

Fig. 5.1 Nuclear ice-breaker *Yamal* arriving in Murmansk

(3) The shape of the hull, particularly the bow form.
(4) The capability of rapid heeling and trimming; and
(5) The skill, experience, and intelligence of the ship handler (Fig. 5.1).

5.1.1 Ships

As a general rule, a full-powered ice-strengthened ship should be able to make relatively timely progress through 6/10 winter ice. A careful assessment should be made of every aspect of wind and current in relation to ice drift when a ship is moving alone through waters where pack ice extends over large areas. Every effort should be made to avoid being caught between an extensive area of pack-ice and the shore, or between the pack and a danger such as a shoaling area, when the wind is blowing strongly onshore.

5.1.2 Air-Cushioned Vehicles

Air-cushion vehicles have an ability to operate over Arctic terrain and possess certain advantages. These advantages include:

- In case of engine failure, air-cushion vehicles are safer than light aircraft or helicopters.
- The reservoir of air holding the craft up leaks away comparatively slowly enabling the vehicle to settle on the surface.

- Air-cushion vehicles can float on water; and
- Emergency stops can be made on snow from speeds of 26 kn (48 km/h; 29.9 mi/h) without suffering damage.

Air-cushion vehicles are capable of traversing a wide variety of terrain at speeds up to 60 kn (111 km/h; 69 mi/h) or more. In addition to travelling over water, land and ice, air-cushion vehicles can negotiate muskeg, soft snow and mixtures of thin ice and open water with equal facility. By their nature they are best suited to lands of gentle relief and flat non-ridged ice. Present designs can pass over irregularities (rocks, ridges of ice, etc.) up to 120 cm (3.9 ft) high. Operation over water at temperatures below the freezing point can lead to spray freezing onto the craft and loading it down with ice.

5.1.3 Diving

5.1.3.1 Diving Conditions

Arctic waters are cold, the visibility excellent, and the bottom flat. Shifting ice, driven by wind and current, is the greatest danger to divers. Most diving work is conducted in shallow water, clearing beach obstructions and surveying approaches to landing sites. Divers must be warned and taught about the specific conditions when diving under the ice or in very freezing water.

5.1.3.2 Equipment

A dry suit with compressed-air breathing apparatus, both of which are easily transported and self-sufficient, are suited to diving requirements in the Arctic. An air compressor is necessary for charging bottles and an inflatable rubber boat or raft, equipped with a small outboard motor, is handy and safe for transporting divers and their equipment.

5.1.3.3 Explosives

Explosives used in demolition of ice and rocks must have extremely high shattering properties. Ice is extremely hard to dispose of due to its density, and large charges are often necessary. A 60/40 ratio of Forcite and Nitrone has proved to be effective.

5.1.3.4 Ship Repair Activities

Underwater cutting and welding equipment should be carried to facilitate fixing ship's plating, rivets, propellers, and rudders.

5.1.3.5 Safety

Rules regarding the conduct of diving should be more restrictive in the Arctic. Given the long distances to medical attention regarding decompression sickness, i.e., a hyperbaric chamber, it is recommended that divers in Arctic waters are especially attentive to the risk management of the dive.

5.2 Arctic Navigation, Navigational Aids, and Pilotage

With multiple global navigation satellite systems being available, navigation in the Arctic is achievable by all properly equipped platforms.

5.3 Charts and Sailing Directions

Requirements for charts and related publications can be met from the following sources:

(1) Catalogue of Maps, Charts and Related Products, published by the Defence Mapping Agency, Washington, DC, 20315-0010.

(2) Catalogue of Nautical Charts and Publications, published by the Canadian Hydrographic Service, are available from the Hydrographic Chart Distribution Office, Department of Fisheries and Oceans, 1675 Russell Road, PO Box 8080, Ottawa, Ontario Canada K1G 3H6.
Telephone (613) 998-4931
Fax (613) 998-1217.

(3) Catalogue of Admiralty Charts and Other Hydrographic Publications, published by the UK Hydrographic Office, Admiralty Way, Taunton, Somerset, TA1 2DN. Email:
customerservice@ukho.gov.uk
Customer Service +44 (0)1823 723366.
American and British authorities also publish classified catalogues describing selected classified charts and publications embracing Arctic areas.

5.4 General Reference Books

A number of Arctic navigational reference books are available. The Arctic navigator must procure all necessary references prior to sailing for northern waters.

5.5 Chart Projections

The Mercator projection satisfies the navigator's needs to the 70th parallel of north latitude. In latitudes higher than 70°N, however, the usefulness of the Mercator projection decreases rapidly, primarily because the value of the rhumb line becomes progressively less, and because there is an increasing rate of change in chart scales. Chart producers opt for a more appropriate projection for Arctic use. The four most commonly used are:

(1) Transverse Mercator projection, particularly when considering areas extending North-South.
(2) Modified lambert conformal projection.
(3) Polar stereographic projection; and
(4) Polyconic projection.

5.6 Polar Grid

Because of increasing convergence of the meridians near the pole, the true directions of an oblique course line will vary depending upon the length of the course line and its proximity to the pole. A polar grid provides the navigator with the changing true direction. The polar grid is described in the HO Publication No.9, Chapter XXV-Polar Navigation, and shown in Fig. 5.2.

5.7 Magnetic Compasses

The directive force of the magnetic compass is derived from the horizontal component of the earth's magnetic field. Although the total intensity of the earth's field remains fairly constant in all latitudes, the horizontal component decreases as the magnetic poles are approached, until the directive force becomes so weak that the compass is insensitive and unresponsive. Conversely, the vertical component increases and may give rise to large heeling errors. Although the horizontal component of the earth's fields and hence the induced magnetism in horizontal soft iron, decreases as the magnetic poles are approached, the field of sub-permanent magnetism of the ship's structure retains its absolute value and therefore becomes much more important in causing deviation. Small uncompensated deviations due to sub-permanent

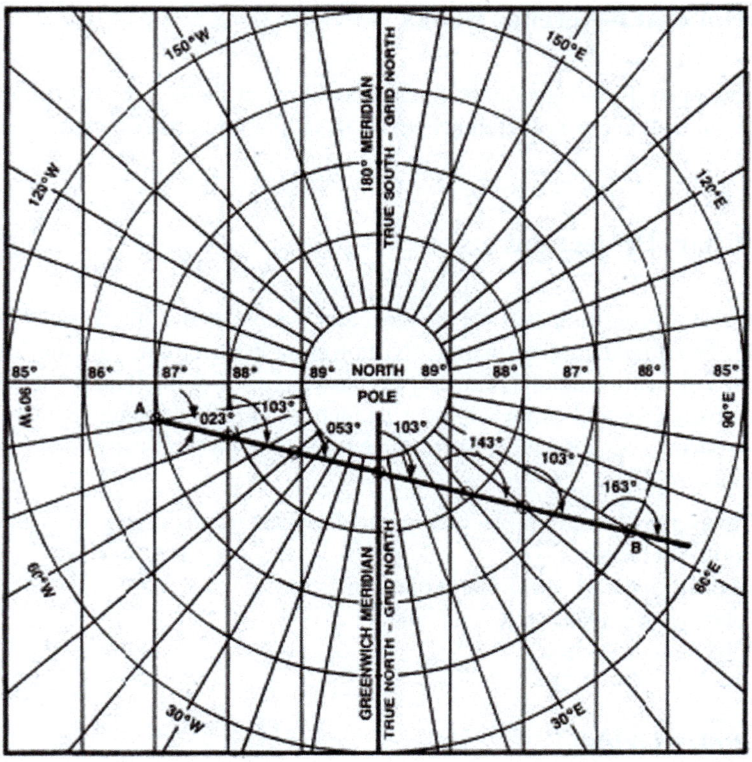

Fig. 5.2 Polar grid navigation

magnetism may attain exceptionally large values in high latitudes. Furthermore, magnetic disturbances or magnetic storms cause fluctuations in magnetic variations. For example, a severe magnetic disturbance can shift the effective position of the magnetic pole as much as 150 km (93.2 mi) or more. This shift has significant effects near the magnetic pole, becoming less of a problem with distance from the magnetic pole. To obtain the best performance from the magnetic compass in Arctic waters, a ship should be swung, and the compasses adjusted in high latitudes, preferably before entering the pack-ice. If the Flinders bar has not been permanently set at the magnetic equator, it must now be adjusted to the position indicated by computation, and the horizontal and heeling magnets carefully placed to produce minimum deviation. The US Defence Mapping Agency publication NVPUB226, "Handbook of Magnetic Compass Adjustment" is a useful reference; AVPUB9V1, "American Practical Navigator", also applies. Even if this recommended procedure is followed, changes in magnetic latitude may cause large deviations to reappear.

Likewise, the magnetic variations will change rapidly with locality and may undergo large diurnal changes, particularly if auroral activity is present, so that the navigator must undertake frequent azimuth determinations. If large compass errors are found, and if it is uncertain whether these are due to variation or deviation, swinging the ship again to see whether the error persists on all headings will establish the cause. Provided that precautions have been taken (e.g., burning the binnacle light continuously and keeping the binnacle itself covered by canvas when not in use), the liquid in modern magnetic compasses is capable of withstanding Arctic temperatures. The flux-gate compass has proved to be quite sensitive and has given fairly accurate and reliable results. The Admiralty gyro-magnetic compass with a pivoted card has proved to be serviceable for navigation up to 300 km (186 mi) from the north magnetic pole and has the additional advantage of being available as a simple magnetic compass should power fail. Both wet (floating) compasses and fluxgate (digital) compasses are subject to error due to local magnetic anomalies of geologic origin. These are reported in "Notices to Mariners." Any suspicious compass behaviour should be carefully documented and reported to the appropriate defence or national charting agency.

5.8 Gyro-compasses

When operating north of 70°N, particular attention should be paid to the operation of gyro-compasses and associated equipment. Manuals on military gyro-compasses usually specify 85°N as the highest latitude at which a gyro-compass will function reliably as a direction-indicating device. Between the latitudes of 70°N and 85°N, gyro errors may be considerable. Therefore, when weather conditions permit, frequent azimuths of heavenly bodies are recommended. For certain gyrocompasses, nomograms are available in operating manuals, and certain gyrocompasses are capable of being fitted with high-latitude correctors which improve performance above 70°N.

5.9 Azimuths

Every opportunity should be taken to determine the errors of compasses, particularly by azimuths of the sun. An azimuth attachment for a telescopic alidade is recommended; it may be of value in obtaining accurate azimuths for determining gyro error when the sun is not brilliant enough to obtain an azimuth by way of an azimuth circle. The present azimuth tables for high latitudes can be used only during a certain portion of the day, but azimuths for use at any time can be computed.

5.10 Celestial Compasses

Where magnetic and gyro-compasses cannot be relied upon to provide trustworthy directional references, a navigator should consider using:

(1) The Sun compass. This indicates direction by means of a shadow cast by a shadow pin. This instrument is, of course, of use only when the sun is visible, and the observer knows his position.
(2) The celestial (Astro) compass. This is similar in principle to the sun compass but may be used for any celestial body; and/or
(3) The sky (or twilight) compass. This compass indicates direction by means of the polarizing effect of the earth's atmosphere on sunlight. Its usefulness arises principally from the fact that twilight periods in high latitudes are of several hours duration and during this time no celestial body is visible unless the moon or a bright planet is above the horizon.

5.11 Dead Reckoning

When steaming in poorly charted waters, a ship can run aground or be exposed to other unexpected hazards. Because of inadequate tidal information, as well as other shortcomings, the most accurate estimated position may not result in an exact position. But careful reckoning, in accordance with the suggestions outlined in the following paragraphs, may be of help to the navigator. Inaccurate charts, together with limited fixing marks, may compel the navigator to resort to "relative navigation," i.e., fixing relative to selected charted landmarks or other objects. Repeated alterations of course to avoid ice make it difficult to plot a ship's track on a chart, especially a small-scale one. Plotting the mean course and distance made good once or twice each watch is recommended. Plotting the positions of large icebergs, especially if they are known to be aground, can sometimes be most useful as an aid to navigation when identifiable landmarks are not available. This is a good example of "relative navigation." Bottom soundings should be relied upon to assist in maintaining a dead-reckoning position or for fixing the ship.

5.12 Astronomical Observations

For a great part of the navigation season, cloud or fog obscures the sun in the Arctic, while continuous daylight prevents stellar observations. Ice horizons and abnormal mirage may also complicate the task of obtaining a precise altitude of the sun during periods of good visibility. Standard refraction tables are not accurate in high latitudes. HO Publication 229 does not allow for solution of the celestial problem for observed altitude of less than 5, a common condition in the early and late summer periods. The use of a bubble sextant, bubble attachment to the standard sextant, or artificial horizons can be helpful. The following chapters of HO Publication No.9 American Practical Navigator, are recommended reading during "in harbour" preparations:

- Chapter XVI Sextant altitude corrections.
- Chapter XX Sight reduction.

- Chapter XXV Polar navigation.
- Chapter XXIX Navigation errors.

5.13 Sunrise, Sunset, and Twilight Phenomena

Tabulated local mean times of Sun/Moonrise, Sun/Moonset, Nautical Twilight and Civil Twilight listings go no higher than 72°N in nautical Almanacs. In latitudes higher than 72°N, the graph in the Air Almanac should be consulted.

5.14 Abnormal Refraction

Generally speaking, abnormal refraction at sea is caused by an inversion of temperature in a layer of air. The variations in density, thus produced, cause light rays to be bent in excess of normal conditions. The most favourable conditions for excess refraction, when the more fantastic forms of mirage and distortion take place, occur when a layer of warm air is in contact with cooler water. The air next to the surface of the sea is cooled, and consequently the upper layers are warmer than the lower, so instead of the usual decrease, there is an increase of temperature with height. Most refraction phenomena are formed at the boundary between this layer of cold, dense air at the surface of the sea, and the less dense warm air above. This condition is identical with that which is responsible for the formation of most sea fog, and the presence of fog is therefore an indication that excessive refraction can be expected. Similar inversions may be caused by the presence of frigid air over warm water. A marked difference between air and sea temperature is thus a guide to the presence of excessive refraction. Although abnormal refraction is not restricted to particular geographical areas, certain regions are so situated with respect to general meteorological conditions as to be more favourable than others for the occurrence of abnormal refraction phenomena. In this respect the Arctic coasts are ideal because of the marked difference between sea and air temperatures. In the Arctic, excessive visibility, or some form of mirage, is often manifest when comparatively warm light winds blow over cold ice surfaces, or when cold winds blow over open water. A temperature milder over open water than over a nearby ice-clad shore also leads to refraction phenomena.

One form of abnormal refraction is "looming," which is the apparent raising of an object above the horizon. It is quite common at sea, especially in high and middle latitudes, and results in the appearance of distant objects which, in many instances, may actually be below the normal horizon at the time of observation. There are two types of looming. On the one hand an object (island, iceberg, ship) is increased in elevation though not in size; on the other hand, an object is enlarged and brought much nearer to the observer. Superior mirage is another form of abnormal refraction and is the apparent reflection from a mirror-like atmospheric condition where a pronounced temperature inversion exists about a metre above the surface. This inversion introduces an abnormal change in density resulting in extraordinary refraction. Its most frequent appearance is that of an inverted image above the object, but under suitable conditions a second image is seen erect, close above the inverted one. Sometimes the object is not observed directly and the inverted image, or the upper erect image of an object below the horizon, may be seen. As with looming, the condition needed for superior mirage is a warm layer of air existing over the sea at a suitable height, i.e., an inversion of temperature. The only difference between this and the condition necessary for looming is that for superior mirage there must be a more sudden change from cooler to warmer air at a certain height. At sea, ships and icebergs are the mirage subjects usually sighted. Ocean fog is also associated with mirage, since the temperature and humidity variations which favour condensation of moisture as fog in the air, are factors in causing mirage. Mirage is not visible, of course, in dense fog, but mock fog, or the typical refraction band, is often seen under such conditions and may lead to the erroneous report of true fog.

5.15 Echosounder

In Arctic waters the echosounder is primarily a warning device. In poorly charted waters it is one of the navigator's most valuable aids and should be manned and operated continuously in dangerous waters. A ship's echosounder will not always give a reading when ice is under the ship, or when water beneath the ship is disturbed by propeller swirl when the engines are put astern, or by turbulence caused by ice floes being shoved around. A ship having to proceed in uncharted coastal waters may minimise the risk of grounding by sending a boat away when ice conditions permit, equipped with a portable echosounder, to scout ahead. If continually operating in uncharted waters, consider embarking a forward-looking echosounder.

5.16 Electronic Aids to Navigation

There are several electronic aids to navigation that are useful in one or more areas of the Arctic. Detailed information on the availability of different aids to navigation systems is available in the HO Publication 117. GPS is the best and most accurate electronic aid for Arctic navigation. GPS accuracy is independent of latitude. The read out is in latitude/longitude with no corrections required. Inertial navigational systems may accumulate a small error over a period of time and require recalibration using an external source. These systems are not as accurate as GPS; however, they are passive and cannot be affected by enemy action. A combination of external navigational positioning systems (GPS) and an internal INS is recommended for Arctic operations.

5.16.1 *Radar as a Fixing Aid*

Radar-range navigation in polar waters has some limitations in accurately and reliably fixing the ship's position. Compared to other regions of the world, charts from polar regions are substandard making fixes obtained from the radar ranges inaccurate. Also, it is often difficult to distinguish between points of land and ice on the radar. Suitably spaced fixes using radar ranges will reveal variable gyro-compass errors and the influence of unknown tidal streams. However, the precision of radar range fixing depends upon the correct selection of radar conspicuous points, and on their correct interpretation, from an accurately calibrated radar plan position indicator (PPI). Plot radar ranges to as many points of land as possible. Continue to plot and study the ranges to determine which of the charted points of land plot accurately for navigational purposes if ranges fail to intersect with a fix. The technique of radar ranging for fixing in the Arctic has the following advantages:

(1) It is available under all conditions of visibility and using all types of land targets, i.e., shorelines that have both low-lying and steep-to features.
(2) Gyro errors are usually variable whereas a PPI index error for a given range scale can be ascertained and removed.
(3) If, as is often (but not necessarily) the case, the centre of a "cocked hat" is taken as a ship's position, the small, neat triangle produced by a radar range error (assuming there is one) gives a better fix than would be the case with an enormous "cocked hat" resulting from an unknown gyro error.
(4) At night, the bearing method of fixing is impossible; few shore objects are visible and lights for navigation purposes in Arctic waters are non-existent.
(5) Visual fixing in conditions of fog and snow is impossible.
(6) By fixing at frequent intervals, a practical ship's track can be derived, and adjusted, to avoid known shoals. The radar ranging technique is a rapid and simple method of ascertaining a ship's position and movement over the ground.

Radar should be calibrated whenever an opportunity occurs if it is to provide the navigator with accurate ranges. Range errors are not obvious at the lesser ranges but become increasingly evident at ranges over 20,000 m (12.4 mi).

5.17 Radar and Sonar for Ice Detection

5.17.1 *Radar*

It is important to keep radars operating at peak efficiency. As sea states increase, so does the minimum size of icebergs that can be detected. In rough seas, icebergs as tall as 15 m (49.2 ft) cannot always be detected in the sea return. Only in smooth seas can radar be relied upon to pick up growlers. Meteorological conditions in some areas affect radar propagation in a manner which, under certain conditions of fog and rain, may reduce or obscure returns from ice. Targets may become lost due to the ducting effect of the beam caused by a decrease of moisture content which is often accompanied by a temperature inversion. Such ducting happens occasionally, but seldom to the extent that a target is completely lost. Another form of ducting has the opposite effect, i.e., bending the radar beam so that it follows the curvature of the earth, permitting detections at very great ranges. The blending of sea returns and returns from growlers pose a severe problem in ice detection, but one which can sometimes be overcome by an alert operator. In moderate seas, growlers alternately appear and disappear from the PPI but in the same position at each sweep of the antenna. Sea returns, however, will fail to appear in the same relative position. Therefore, the relative trails function should be used here to be able to better ascertain if an echo is a growler. Large iceberg returns can be distinguished from adjacent pack-ice returns at ranges of 3300 m (10,826 ft) or more, but can be obscured by returns from pack-ice at lesser ranges. What are actually shadows cast by large icebergs can easily be mistaken for leads or open water. Anti-jamming controls are of some value in differentiating between pack-ice and large icebergs at reduced ranges but should not be relied upon. Floes up to 11,000 m (36,089 ft) from a ship are well patterned on PPIs, therefore radar can be of considerable assistance in showing up leads in the ice.

5.17.2 Sonar

Experience in the Arctic indicates that icebergs can be detected more reliably by passive sonar than by active sonar. Icebergs produce a loud noise similar to high-speed propellers in a ship, caused by the release of air bubbles under pressure. Echo ranging has proved to be less dependable, often failing to indicate the presence of icebergs at ranges where they could present a hazard.

5.18 Ship Handling in Ice and Ice Seamanship

The first principle of ice seamanship is to maintain freedom of manoeuvre in the presence of ice. In general, navigators should accept one but not willingly two of the following conditions:

- Ice.
- Fog/darkness/low visibility due to precipitation; and/or
- Fierce winds.

Three thoughts should be uppermost in the mind of the Master whose ship is working in ice:

- Keep the ship moving-even slowly.
- Try to work with, not against, the ice; and
- Be patient; circumnavigating ice can often be the quicker and safer course of action.

5.19 Non-ice Strengthened Ships

Ice is an obstacle to any ship, and it is dangerous to those ships which, by their construction, were never really intended for ice navigation. Nevertheless, it is possible for non-ice strengthened ships to navigate through regions of open pack-ice. The long hours of summer daylight combined with a vigilant use of radar and look-outs in high latitudes facilitate such operations.

5.20 Indications of Ice

Iceblink, which is the reflection of ice on low cloud, is an indicator much used by experienced navigators. It appears as a diffuse whitish glare above an accumulation of distant ice and is especially noticeable when observed on the horizon. Isolated fragments of floating ice often presage the approach of larger quantities of ice or warn of the presence of icebergs nearby. In late spring, and during the Arctic summer, there is frequently a thick bank of fog over the edge of the pack-ice. In fog, patches of whiteness can indicate the presence of ice at short range. Absence of sea or swell, especially in a fresh breeze, can be a reliable sign of ice to windward. A drop in temperature of the surface of the sea, or a drop in the air temperature, can indicate that a ship has entered waters where ice is likely to be encountered.

5.21 Signs of Open Water

Dark patches on low clouds, sometimes almost black in comparison with the general overcast, can indicate open water, and is known as "water sky." In fog, dark areas discerned through the murk can give an indication of open water. A dark band of cloud at high altitude can indicate the presence of a lead giving access to open water further away. A surging action in the pack-ice can indicate the presence of open water nearby. If, when approaching ice, there is darkness on the horizon beyond a light sky, this is good evidence of open water or land lying beyond the ice, in some cases 40 nm (75 km; 46.6 mi) or more beyond the visible horizon. If dark streaks are observed in the sky, the presence of leads is indicated. If there are no dark streaks, a ship should steer for the place where the iceblink is dullest. Iceblink is increased after a fresh fall of snow, since its reflection in the sky will be whiter from this snow than from the ice. In a cloudless sky and calm weather, abnormal refraction may raise the horizon enabling an observer to see ice at greater distances than normally would be possible. The image of ice or areas of open water, or a mixture of the two, may be seen as an erect or inverted image, or both images may be seen at once, one above the other. In this latter case the erect image is the higher of the two. Refraction also causes bergy bits to appear like icebergs. Where there is open water, a dark-blue colour will be seen towards which the ship should steer.

5.22 Icebergs

On dark clear nights, icebergs may be sighted at a distance of 1.8–3.6 km (1.13–2.2 mi), appearing either as white or black objects. The use of searchlights is paramount when navigating Arctic waters at night. In such conditions bergy bits or growlers constitute a greater hazard to ships. Such pieces of ice may also be difficult to distinguish in daylight, especially in a rough sea. A clouded sky at night, through which the moon appears and disappears, makes ice detection difficult. Heavy passing clouds may dim or completely obscure an object sighted ahead. Fleecy cumulus and cumulonimbus clouds often give the appearance of blink from bergs. Radar can usually detect large icebergs in ample time to avoid collision, but small bergs and growlers, capable of causing damage even to ice-strengthened ships, may remain undetected under quite moderate conditions of wind and sea. As the state of a sea increases, so does the minimum size of the iceberg that can be detected. The use of the "relative" trail function on the radar is recommended. Air and sea temperatures are not a reliable warning of the presence of icebergs. The presence of stationary icebergs may give some indication of the depth of water in their vicinity. For example:

(1) An accumulation of stationary icebergs often marks an isolated shoal.
(2) A line of grounded icebergs extending seaward from the shoreline, or between an island and the mainland, may indicate the presence of a submerged ridge or shoal.
(3) A shoreline fringed by glaciers or studded with icebergs inshore but free of ice to seaward, would indicate that the shoreline falls off steeply into deep water.
(4) A bay in which icebergs are found must have a deep channel leading into it. The sides of a channel which are bordered with icebergs, but with its centre clear, may be considered safe. Open water will usually be found during the summer months along a coast where offshore winds prevail.

5.23 Entering the Ice

Before committing a ship to the ice-pack, a complete reconnaissance of the area should be undertaken, using every means available-radar, look-outs, helicopter observations, notices to mariners, satellite-charts, and ice reports from long-range fixed-wing aircraft. The point of entry should be carefully selected, and the ice entered at right angles, after which the ship should make for the loosest areas in the pack. Preferably make the entry upwind, remembering that the windward edge of an ice-field will be more compact than its leeward edge. The violent motion of ice from the action of the waves will also be damped out on its leeward side. Avoid a lee shore on which ice is usually dense and hummocky. Favor a windward shore where an open channel may be found. Avoid manoeuvring so close to points of land that a combination of wind, ice, and unknown currents could force the ship aground. If the ice is drifting rapidly, wait (preferably in open water) for a change in direction of the ice movement, considering the times of ebb and flood tide; ice tends to compact on the flood and to loosen up on the ebb. An ice edge is usually not straight but often having tongues projecting between bights. Select a suitable bight and enter there where the surging action will be least. Enter at slow speed to reduce the initial impact on the stem. Once the bow is in the ice, cutting and pushing it aside, power should be increased to avoid losing headway. Give all icebergs and other forms of glacial ice in the pack a wide berth. If a collision with a floe cannot be avoided by a non-ice-enforced vessel, it is best to take the blow on the stem while going astern at full speed. Navigation in pack-ice after dark, or in fog, or when the ice is under pressure, should not be attempted. Propellers and rudders are the most vulnerable part of a ship, and conning officers must see to it that whenever heavy ice approaches the stern, action is taken to slow the shaft. If the bow of a ship rebounds off a floe, the stern may be swung into heavy ice with a risk of damage to rudder and propellers. When entering ice, it is recommended only to use 80% of the engine power available, so if the vessel gets stuck, it has the possibility to use additional power to move out of the ice, avoiding a potentially dangerous situation.

5.24 Speed of Ships Working in Ice

Speed through ice must be a matter of judgment and will depend, among other things, upon the amount of open water, the hardness of the ice, and the strength of the ship. Coasting into ice, with engines stopped, will result in loss of steerage-way. Ships must, at all times, be prepared to go "Full Astern," and keep the rudder amidships during this manoeuvre. In less than 6/10 ice, the speed of a ship passing through the ice, without ice-breaker escort, should depend on the distribution of leads and pools of water. Seven to nine-tenth ice should be negotiated throughout at slow speed, so that collision with floes will not damage the hull. When moving through very loose pack at night, or in poor visibility, continue with caution and at slow speed. In such circumstances, searchlights, preferably mounted in the eyes of the ship, can be of considerable help. Ice projectors with a powerful Halogen-light have proven an effective tool to locate ice. The white light emitted from a halogen source, is reflected much better than light from traditional light bulbs. An ice projector should be installed on each side of the vessel as well as amidships. Furthermore, it is recommended that the projectors have a function which allows the light beam to be focused, in order to concentrate the light to make observation of ice easier from further distance.

5.25 Working Through Ice

The extent of the ice should be studied from aloft, preferably from a helicopter. In this way distant leads and open water, invisible from the bridge, may be revealed. The character of the ice ahead, when viewed from aloft, may sometimes be assessed by comparing it with the ice through which the ship has just passed. Pressure ridges and ice which is greenish-blue in colour (polar ice) should be avoided. Sometimes pools of melt-water form on top of the ice. From the air, and even from a distance at sea level, these pools can resemble open pack. Upon closer examination it will be found that the ice is continuous under the pools and may even be un-navigable. Aerial ice observers should be thoroughly trained and familiar with the problems confronting the ships to which they are passing valuable information. To provide continuity, the same ice observers ought to be used, and such observers must have the confidence of the captains, who may have to rely heavily upon their reports. A short burst of full speed ahead, with the helm over, may be of assistance in speeding up a turn to avoid a floe. Propellers and rudder may be afforded some protection by trimming by the stern. Ships should go astern in ice with extreme care, always with their rudders amidships, and while keeping a sharp outlook for ice under the quarter. One system for working astern in ice is to:

(1) Allow propellers to wash the ice astern for a few minutes before going astern.
(2) Go full astern until just before contact with the ice debris, then stop and allow the momentum to carry the ship into the churned-up ice.
(3) When all ice has surfaced, give a short burst of ahead power and stop.
(4) Repeat this process until sufficient manoeuvring room has been produced.

Another effective system for working a twin-screw ship astern, when surrounded by heavy brash, is to:

(1) Go astern on one engine while going ahead on the other, setting up a current under the stern which opens up an area of clear water to one side depending upon the engine direction.
(2) Go astern on both engines. This should move the stern into the unsecured area until the ice eventually brings the ship to a halt.
(3) When stopped, repeat backing on one engine and going ahead on the other, but in the combination opposite to that used previously, until an unsecured area of water appears on the other side.

(4) Go full astern until just before contact with the ice debris, then stop and allow the momentum to carry the ship into the churned-up ice.

Cracks may form in ice-fields along a line of pressure perpendicular to the movement of the ice. Such cracks are sometimes ridged. At the least change in wind or current, these heavy masses may come together crushing and grinding anything caught between them. Fine weather in the pack often portends lower temperatures, close pack and little open water, whereas damp and misty weather signifies the presence of some open water and better conditions for manoeuvring. The presence of swell indicates loose pack is near at hand and open water is not extremely far off. Any offshore wind usually creates a channel between the coast and the pack. If this is exploited by a ship, she should be alert for an onshore wind driving the pack onto the coast if the latter is steep-to. In such a case, shelter should be sought in a bay or behind an island. Failing any such refuge, the only alternative would be to go out and meet the ice hoping to work through it to open water. Icebergs move at a different rate than the sea ice and in strong currents may even travel upwind. In these conditions, open water will exist to leeward, and piled-up pressure ice to windward of icebergs. In a strong wind the pack may overtake the icebergs, resulting in a heaping up of the pack to windward, while a lane of open water opens to leeward of the icebergs. This creates the illusion that the icebergs are traveling in a direction opposite to the pack. The movement of an iceberg through wind-compacted ice creates a lead which may remain open for a time. In traversing pack, advantage might be taken of such leads. If a ship enters a narrow strait or bay into which the prevailing winds are known to blow, she should be alert to the possibility of ice being driven in by a sudden change in weather and trapping her there. A ship should exercise caution when to windward of a prominent headland because a sudden increase in the wind may drive the pack down upon the vessel which, if set toward a lee shore, may become beset and subject to pressure. An exception to this occurs in the western Canadian Arctic and along the north Alaskan coast where ships successfully navigate close inshore, depending upon their draught, thus avoiding the heavy pack lying too seaward. When moving through ice in dense fog relying solely on radar for navigation, the navigator must be aware of giant floes with a smooth surface being presented as open water. The navigator may instinctively move towards this only to find the floe blocking the way. The use of bow thrusters in order to turn the vessel in narrow spaces between hummocks is not recommended due to the risk of small pieces of ice to enter the thruster tunnel and damage the blades. Recirculating cooling water to the main engines from the sea well is recommended, so the cooling system of the engine does not freeze up, and results in an overheated engine and subsequent shutdown.

Fig. 5.3 *USCGC Healy* breaking ice around the Russian flagged tanker *MT Renda*

5.26 Convoying in Ice

An ice convoy consists of one or more ships, some of which may be strengthened for ice navigation, accompanied by one or more ice-breakers. It is essential that such a convoy, while in ice, be under the direction of the master or commanding officer of the lead ice-breaker. Should the senior officer of a naval force happen to be embarked in a ship without ice-breaking capabilities, he must delegate tactical control to the senior ice-breaker master or commanding officer (Fig. 5.3).

5.27 Types of Convoys

There are two types of ice convoy:

(1) Simple convoy: one ice-breaker escorting a group of ships; or
(2) Composite convoy: two or more ice-breakers escorting several ships.

For a simple convoy, the master of the icebreaker will decide upon the number of ships they can manage. Their decision will depend not only on the number, but on the power and type of ship requiring escort and the ice conditions expected en route. If the ships are reinforced for ice navigation, and have sufficiently powerful engines, one ice-breaker can usually escort four of them through 7/10 to 9/10 ice. In 5/10 or 6/10 ice or less, the number of ships can be increased. If there is close pack (more than 9/10) only one or two ships can be managed. The first factor to be considered must be the power of the ships requiring escort. The weakest, as a rule, should be stationed immediately astern of the ice-breaker to avoid ice obstacles and to move in a comparatively clear channel. The most powerful ships with wide beams should be interspersed in the convoy so that less powerful ships can proceed in their wake. Consideration should also be given to whether a ship is loaded or in ballast. Finally, it is essential that one of the most powerful ships in the convoy be last in line. A composite convoy consists of two or three simple convoys. The number of ships allocated to each ice-breaker, and their place in the column, is determined in the same way as for a simple convoy. The difficulty of controlling from a position ahead is a drawback in this type of convoy, which frequently extends over a distance of 2 nm (3.7 km; 2.2 mi) or more. The customary procedure is for the most powerful ice-breaker to lead the convoy, breaking a channel in the ice without stopping to break out other ships. Following the leader, at a distance decided upon by the master, come two or three ships, the weakest and longest in the convoy. The second ice-breaker proceeds astern of the first group followed by two or three ships, etc. The assignment of the second ice-breaker is to break out the ships ahead so the leader will not have to return, thus delaying the whole party. The second ice-breaker, on receiving a "stuck" signal from any of the preceding ships, increases speed, leaves the column and breaks out that ship. When the latter is freed and moving, the ice-breaker resumes previous position in the column. Similar action is taken by the second ice-breaker upon receiving a signal from one of the ships astern, provided there are no more ice-breakers available.

5.28 Distance Between Ships

Before entering the ice, the masters of all ships should set the agreed distance between their ships and the ice breakers, and between other ships. It is unwise to have the convoy strung out in too long a line. At the same time, the distance between ships should be great enough for way to be checked and collision averted if a "stop" signal is originated by any one of the ice-breakers. At ice convoy speed, way in merchant ships of average tonnage can be checked in ice-free waters by going astern over a distance of three to three and a half ship lengths, provided an order for full speed astern is given. This distance should, therefore, be the minimum between ships when navigating in less than 7/10 ice. Depending upon ice conditions, exceptionally large ships (i.e., those displacing 100,000 tons or more) must keep farther apart. Until more experience is gained in convoying ships of this size through ice, at least 1 km (0.6 mi) is recommended as a minimum. A channel made by an ice-breaker will eventually fill with broken ice. The speed at which the channel closes will depend on the amount of ice pressure encountered. This will have a factor in determining the distance between the ships. The difficulties caused by this to a ship in a narrow channel increase when the distance between ships is increased, and even powerful ships may find their speed reduced. This makes it all the more important for ships

Fig. 5.4 Icebreakers *Ymer* and *Atle* changing convoys in the Bay of Botnia

to maintain the minimum prescribed distance apart. Signals from the leading ice-breaker must be obeyed promptly and correctly and all ships must be alert for any difficulties and delays caused by ice (Fig. 5.4).

5.29 Course and Speed of Convoy

The longest safe route in open water will be quicker than a more direct one in ice, and the selected track should pass through areas of thin ice or open water, regardless of the length of the voyage. Course changes should be gradual since most cases of ships getting stuck occur when sharp turns are made by the much more manoeuvrable ice-breaker. Speed over the ground through the pack usually varies between 4 kn and 7 kn (7.4 km/h; 4.6 mi/h–12.9 km/h; 8 mi/h). The higher speed is desirable due to better manoeuvrability of large ships, but ice conditions must be the governing factor. In a convoy composed of ships reinforced for ice navigation, a speed over the ground of 6–7 kn (11.11 km/h; 6.9 mi/h–12.9 km/h; 8 mi/h) can be maintained if the route lies through 5/10 open pack, and if the ships following the icebreaker will not meet with heavy ice. If a single-screw ship must suddenly go full astern without warning while passing through an ice-covered channel, the stern will kick to port and the bow to starboard. This could damage the propeller, rudder and starboard side of the ship. To avoid collision with a ship ahead, it is preferable to ram the ice on one side of the channel, bow foremost, rather than risk damage to the rudder and propeller by going astern in heavy ice. When navigating in close pack (7/10 to 9/10), speed over the ground should not exceed 5 kn (9.26 km/h; 5.7 mi/h). In such ice a convoy will be moving in a channel which will not remain navigable for exceptionally long after the passage of the ice-breaker. Therefore, the distance between ships must be reduced to enable them to move in as clear a channel as possible. Higher speeds not only increase the danger of hitting the ice, but also the possibility of colliding during unscheduled stops of the icebreaker or other ships of the convoy.

5.30 Conduct of a Convoy Through Ice

When following an ice-breaker, a convoy should keep in line. By looking for independent channels, ships break up formation and may become stuck. Since headway through heavy floes and ice-fields is more difficult than through "normal" pack, an ice-breaker increases speed and, by striking the ice, crushes or breaks it. Ships astern must maintain correct intervals and endeavour to enter the channel thus made before it closes. If an ice-breaker should encounter an obstacle where a glancing blow is struck by her stem, she will be thrown sideways. Ships following behind may be too unwieldy, or be unable to react quickly enough, and may suffer damage. This is particularly applicable to single screw ships. This sudden change of direction should be expected when moving through ice of varying structures and strength. In such circumstances an icebreaker should not make too rapid a return to her original course. In summer there are many signs indicating the state of the ice. Careful observations should be to determine whether the ice has been softened by the sun or if it still retains its winter hardness. Greenish or greenish-blue ice is the hardest to break and such ice should be outflanked. This type of ice is sometimes covered with pools of clear melt-water formed during the thaw of snow on the surface of the ice. If sections of dirty-looking ice are encountered in areas of light-coloured ice, the former should provide the easier route, since the darker ice absorbs more heat from the sun and melts sooner. The most navigable type of ice is brash, even though it may be devoid of leads. Although this ice usually closes as a result of tide and wind, it consists of separate cakes and does not present a serious obstacle. When the pressure is great, however, even though an ice-breaker can get through, a ship astern may be hindered as the channel behind the ice-breaker closes immediately. In brash, even under pressure, ships are in less danger than if they were being pressed by larger and heavier forms of ice. When hummocked ice is met, an ice-breaker should first attempt to outflank it. The outward characteristics of hummocky ice may indicate to what extent it will be navigable. If the hummocks consist of loose blocks not fused together into one solid piece, they may easily be overcome, but if they are composed of larger masses of ice many feet thick, they will be impassable even to an icebreaker. Usually an ice-breaker backs up one to three ship lengths, then goes full ahead until the stem attacks the ice. If ice is strong and extends over a great area, this process must be repeated, and progress will be slow. It may be necessary to make either a simple channel, equal to the beam of the ice-breaker, or a double or triple

Fig. 5.5 Russian icebreaker *Yamal* at close quarters to *50 Let Pobedy*

one, depending on the strength and character of the ice and on the size of the ships waiting to get through. While navigating in heavy ice, ships should be so loaded and trimmed that only the waterline plating will be in contact with ice. In the after part, the propeller is exposed to danger. It is often assumed that blades are damaged only when a ship is going astern. Blades can be damaged or lost while going ahead as well. Sometimes large blocks of ice pass under the ship's hull and turn on edge. Such ice is dangerous and can damage propellers (Fig. 5.5).

5.31 Signaling Between Ice-Breakers and Ships

Suitable signals which have been adopted for use are given in Chap. 13 of the revised (1969) International Code of Signals and may be used between icebreakers and ships navigating in their vicinity or under their escort. The signal "K" (-.-) by sound or light may be used by an ice-breaker to remind ships of their obligation to listen continuously on their radios. The use of the special signals from the International Code of Signals (1969) does not relieve any ship from complying with the International Regulations for Preventing Collisions at Sea (COLREGS). Whistle signals are limited in their value. Experience has shown them to be of questionable value in a convoy of several ships because of the time lag and the danger of misinterpretation. Whenever possible, voice radio should be used and, when ships are close aboard, loud hailers can be used very effectively.

5.32 Stopped Ice-Breaker (Red Warning Lights and Sound Signals)

Canadian ice-breakers escorting ships in ice make use of two special rotating red warning lights to indicate that ice-breakers are fouled or jammed in ice. These two red lights are disposed in a vertical line, one over the other about 2 m (6.5 ft) apart and they are visible all around the horizon at a distance of 2 nm (3.7 km; 2.2 mi). They rotate in a manner similar to the lights on air beacons and are unmistakable as warning lights. The rotating red lights operate in conjunction with a motor-driven siren, facing astern, audible up to a distance of about 5 nm (9 km; 5.5 mi) depending upon atmospheric conditions.

5.33 Breaking Out Ships

When a ship is beset, awaiting ice-breaker assistance to get moving again, it should keep propeller(s) turning slowly to keep the ice away. If engines are stopped, ice will move in around the stern making it dangerous to start turning the propeller(s). Ice-breakers usually prefer to drop back stern first from ahead until abeam of the stuck ship. They then move ahead, instructing the latter to follow. The open water left by the ice-breaker should be immediately occupied by the beset ship. Most ice-breakers dislike coming up from astern of a beset ship for fear of shoving ice under the latter's stern and jamming rudder and propeller(s). Ice-breakers should be careful not to push heavy floes against the sides of thin-skinned ships, which could rupture their plating. There are many ways in which an ice-breaker can free a beset ship and no attempt is made to cover them all here. Ice conditions have a bearing on the tactics to be used. Ice-breaker masters will always communicate their plan to the ship to be helped.

5.34 Replenishment in Ice

Replenishment in ice can be accomplished only when both ships are stopped and lying as close aboard each other as practicable. Prevailing ice conditions should be carefully studied to ensure that replenishment can be completed safely. Pack that is drifting into a lee shore must obviously be avoided. When approaching another ship in the pack there is a danger that the pressure generated by the approaching ship will force intervening ice blocks through the plating of one or both ships or will damage rudders and propellers of the ship approached. A bow-to-bow approach is safest for berthing alongside another ship. If an ice-breaker is available, it should proceed carefully through the ice ahead of the ship making the approach.

5.35 Anchoring

Anchoring in the presence of ice is risky. The minimum amount of cable should be paid out and the capstan must be available for immediate use in the event heavy pack-ice approaches the anchorage. When anchoring in rotten ice or in shallow water, a ship should first attempt to penetrate the ice as far as necessary to avoid any swell. If the water is deep and ice is present, anchoring should be avoided. In such circumstances it is preferable to heave to keep power available and manoeuvre as necessary to avoid floes which may approach and threaten the ship. Stopping in the ice instead of anchoring can be a better solution. This is especially useful in the bottom of frozen fjords where the ice does not move. The navigator should stop with the wind astern so that the wind pressure, which can be considerable, will only push the ship forward into unbroken ice, negating the possibility of the ship drifting astern in the just broken lane of ice. Caution must be taken that the ice system in which the ship is navigating is not drifting towards an obstacle.

5.36 Arctic Aviation

Cold-weather flying operations call for a high degree of professional competence among aircrew plus careful inspections of aircraft, more faithful adherence to prescribed procedures, and more frequent maintenance than is necessary for comparable operations in more temperate climates. Experience has also shown that successful air operations in Arctic conditions require more preparation than air operations in other areas. While routine tasks take longer because of difficulties posed by low temperatures, experience shows that aircraft and equipment can be maintained and serviced when exposed to very cold ground temperatures. Required and recommended technical procedures from the aircraft's AOI, TTP, SMM, National or Local Orders such as operating in winter or Arctic conditions must be adhered to.

5.36.1 Air Navigation

Air navigation in high latitudes require careful preparation for the following reasons:

(1) Electronic aids to navigation will be limited, as will meteorological information.
(2) Limitations may exist with some aircraft navigational equipment at high latitudes.
(3) Mapping is still inadequate in the more remote areas.

5.37 Servicing Aircraft in Cold Temperatures

Both wind and temperature affect the overall efficiency of technicians working in freezing weather. Except for minor tasks, it is worthwhile erecting a shelter and using a ground heater. Temporary shelters of tarpaulins can be put up over a work area to reduce windchill even when no heat is available.

5.37.1 Refuelling

When refuelling at low temperatures, care should be taken because objects can become charged with static electricity more readily than at normal temperatures. Explosive mixtures for JP4 exist down to a temperature of −23 °C (−9.4 °F), and for Avgas down to a temperature of −43 °C (−45.4 °F). All activities that could cause a buildup of static electricity, e.g., the sweeping of frost and snow from an aircraft, must be followed by complete dissipation of the static charge thus accumulated before fuelling is attempted. Refuelling should be conducted as soon as possible after shut-down to prevent condensation. Spilled fuel on the skin can result in quick freezing and severe frostbite of the affected area.

5.37.2 Propellers

Be aware of the implications that frigid conditions can have on propeller seals. Follow the guidance within the aircraft AOI.

5.37.3 Airframes

Slushy snow entering parts of an aircraft, where it can freeze overnight or during flight, can cause frozen brakes, undercarriage frozen “up” or “down,” flaps frozen “up,” or plugged vent lines. After take-off from slush, the landing gear and flaps should be operated several times to prevent freezing in the “up” position. Do not apply parking brakes while brakes are still warm.

5.38 Flight-Deck Operations

5.38.1 Flight-Deck Covering

A coated canvas-type cover can be used to keep the flight-deck clear of ice and snow. The trade-offs involved, speed in readying flight-deck versus the amount of ice/snow, must be considered. Ice-removal teams should be created to clear the flight-deck and tie-downs. The typical ice-removal sequence is as follows:

- Clean ice/snow off the flight-deck, all areas (use brooms to remove snow to prevent flight - deck non-skid damage).
- After snow removal, break up ice with steam.
- Remove water and loosen ice with high-pressure pressure air.
- Apply de-icer to pad eyes and any remaining frozen areas.

Lubricate safety nets and hangar fittings with appropriate cold-weather greases. Rotate flight-deck personnel every 15–20 min to minimise exposure and fatigue. If possible, two complete flight-deck crews should be trained. Additional auxiliary power cables may be necessary for helicopter starts. Additional heaters should be obtained for aircraft hangars. Prior to aircraft operations, flight-deck fire mains and other firefighting gear must be checked for normal operations. Fire extinguishers may need to be kept at a temperature above freezing for reliable operation. Flight-deck non-skid should follow current requirements. Aircrews should ensure that sufficient quantities of waterless cleaner, oil, greases, and other aircraft fluids suitable to freezing weather are available. Modify spotting arrangement and flight deck usage. Allow more space between aircraft to facilitate safe parking and work.

5.38.2 Flight Operations

Unfavourable weather restricts flying in any area. In the Arctic it is especially necessary to be familiar with conditions affecting flight operations, for storms can occur there with little warning. Icing conditions always exist, and as long as de-icing equipment remains only partially effective, icing can be overcome only by avoiding those altitudes at which it is most likely to occur. Fogs and low cloud are also natural products of polar areas in summer. At times dry surface snow raised by high winds tends to obscure the ground and the horizon, reducing visibility and obliterating all references to surface and sky ("white-out"). Any of these conditions makes flying dangerous, but a thorough knowledge of the hazards involved coupled with observance of proper precautions will help to reduce the risk. In the matter of maintenance, close diligence is necessary because of the strain placed on aircraft by low temperatures. Low temperatures increase the difficulty of performing routine tasks and also increase the time required to complete them. The time involved depends on the protection afforded personnel and the kind of facilities provided. Aircraft on-deck time prior to engine start must be minimised. Flight deck crews on all diversion decks must be ready. Boat crews should be suitably prepared and standing by during flight operations to enable immediate response as required. If possible, helicopters should operate in pairs so that one is immediately available to rescue downed personnel in the event of a forced water landing. All participating ship air controllers must keep close contact, by both voice and radar, with helicopters. Anticipate rapid loss of visibility and subsequent recovery of the aircraft.

5.38.3 Pre-flight

The Aircraft Operating Instructions (AOI) for any particular aircraft type provide the best source of information for cold-weather operation and must always be consulted.

5.38.4 Rotary Wing Aircraft Operations

Cold-weather operation of helicopters should be in accordance with the applicable aircraft AOIs for each individual aircraft (Fig. 5.6).

5.38.5 White-Outs

A “white-out” condition is usually indicated by a complete overcast and an indistinct or non-existent horizon. A complete lack of depth perception may be experienced, resulting in an extremely hazardous flight condition. Complete overcasts, hazy horizon, loss of shadow, and reduction in depth perception are conditions that normally exist. Helicopters should not be flown unless visual contact with the surface is assured unless flown infrared.

5.38.6 Landings in White-Out and/or Low-Visibility Procedure

If a pilot must land in white-out conditions, they should use all information available such as the nature of the terrain, wind conditions and, information from ground parties. Landings should be made in accordance with the published aircraft SOPs for snow operations.

Fig. 5.6 Canadian Coastguard helicopter onboard *USCGC Healy*

5.39 Meteorological Organisations

Various government organisations provide readily accessible climatic, environmental and meteorological data sets and historic atlases. Table 5.1 provides a list of these organisations along with their contact information.

Table 5.1 Meteorological organisations and contact details

Country	Address	Email	Web address
All Russian Research Institute of Hydrometeorological Information – World Data Centre	RIHMI-WDC 6, Korolev St. Obninsk Kaluga Reg., Russian Federation	wdcb@meteo.ru	www.meteo.ru/
Arctic and Antarctic Research Institute	38 Bering str. St. Petersburg Russian Federation 199397	aaricoop@aari.nw.ru	www.aari.nw.ru/default_en.asp
Argentine Navy Meteorological Service (SMARA) of the Naval Hydrographic Service	Av. Montes de Ocal 2124 1217 Buenos Aires Argentina	cnhsmara@rina.ara.mil.ar	
Australian Bureau of Meteorology	GPO Box 1289 Melbourne VIC 3001 Australia	webclim@bom.gov.au	www.bom.gov.au/
British Antarctic Survey	High Cross Madingley Road Cambridge CB3 0ET United Kingdom	basweb@bas.ac.uk	www.antarctica.ac.uk/
Canadian Ice Service	373 Sussex Drive Block E, Third Floor Ottawa, Ontario K1A 0H3 Canada	cis-scg.client@ec.gc.ca	www.ice-glaces.ec.gc.ca.
Climatic Research Unit University of East Anglia	School of Environmental Sciences University of East Anglia Norwich NR4 7TJ United Kingdom	cru@uea.ac.uk	www.cru.uea.ac.uk/
Cold Regions Research and Engineering Laboratory	72 Lyme Road Hanover New Hampshire 03755-1290 USA	contactcrrel@crrel.usace.army.mil	www.crrel.usace.army.mil/
Danish Meteorological Institute	DMI Lyngbyvej 100 2100 København Ø Denmark	vmarked@dmi.dk	www.dmi.dk/dmi/index/

(continued)

Table 5.1 (continued)

Country	Address	Email	Web address
Environment Canada	Meteorological Service of Canada Atmospheric and Climate Science Directorate (ACSD) 4905 Dufferin Street Toronto Ontario M3H 5T4 Canada	firstname. lastname@ec.gc.ca	www.climate.weatherof fice.ec.gc.ca/climat eData/canada.e.html
European Climate Support Network	KNMI PO Box 201 3730 AE De Bilt The Netherlands		www.eumetnet.eu/ ECSN_home.htm
Icelandic Meteorological Office	Bustadavegur 9 150 Reykjavík Iceland	climate@vedur.is	www.vedur.is/english/
International Oceanographic Data and Information Exchange (IODE)	IOC Project Office for IODE Wandelaarkaai 7 B-8400 Oostende Belgium		www.iode.org/
International Polarview Organisation		info@polarview.org	polarview.org/project/ index.htm
National Snow and Ice Data Center University of Colorado	449 UCB University of Colorado Boulder CO 80309-0449 USA	nsidc@nsidc.org	nsidc.org/data/ewg/ index.html
Network of European Meteorological Services	Refer to European Climate Support Network		www.eumetnet.eu.org/
Norwegian Meteorological Institute (DNMI)	PO Box 43 Blindern 0313 Oslo Norway	met.inst@met.no	www.met.no/english/ index.html
Norwegian Polar Institute (Norsk Polarinstitutt, Polarmiljøsenteret)	Norwegian Polar Institute Polar Environmental Centre N-9296 Tromsø Norway	postmottak@npolar. no	www.npiweb.npolar.no/

(continued)

Table 5.1 (continued)

Country	Address	Email	Web address
Russian Federal Service for Hydrometeorology and Environment Monitoring	Novovagankovsky Per., 12 Moscow 123995 Russian Federation		www.mecom.ru/
Swedish Meteorological and Hydrological Institute	Folkborgsvägen 1 SE-601 76 Norrköping Sweden	kundtjanst@smhi.se	www.smhi.se/en/index.htm
United States Government Department of Commerce National Oceanic and Atmospheric Administration National Climatic Data Center	National Climatic Data Center Federal Building 151 Patton Avenue Asheville NC 28801-5001 USA	ncdc.info@noaa.gov	www.ncdc.noaa.gov/oa/ncdc.html
United Status Government National Oceanic and Atmospheric Administration Cooperative Institute for Research in Environmental Sciences NOAA-CIRES - Climate Diagnostic Center Physical Sciences Division	325 Broadway R/PSD1 Boulder CO 80305-3328 OR CIRES/Climate Diagnostics Center University of Colorado 216 UCB Boulder, CO 80309-0216 USA	cdc.webmaster@noaa.gov	www.cdc.noaa.gov/
United States Government United States Navy, the National Oceanic and Atmospheric Administration (NOAA) United States Coast Guard, National Ice Center	National Ice Center Federal Building #4 4251 Suitland Road Washington DC 20395 USA	liaison@natice.noaa.gov	www.natice.noaa.gov/

Part II
Arctic Vessel Requirements

Chapter 6
Arctic Vessel Hull Structure Materials, Welding and Coatings

6.1 Introduction

In the first part of the book, we examined the general environment and climatic conditions that are often experienced in the Arctic and polar latitudes. This is important for mariners to know as it enables them to prepare for living and working in very inhospitable conditions. In Part 4 of this book, we will look at the health, safety and welfare of crew members onboard vessels operating in the Arctic region. Before that, however, we will examine the Class Rules which dictate the design and construction of vessels intended to operate in the Arctic. Needless to say, the challenging conditions associated with Arctic waters requires unique vessel designs as well as specially developed hull structure materials, welding techniques and coatings which can withstand below freezing temperatures and impact shocks. The materials used for exposed hull structures and deck machinery must be selected by taking into consideration their intended functions in freezing weather. As such, all hull and structural materials for ice class vessels must be selected in accordance with the Class specific rules governing the use and integrity of materials. For context, in this chapter, we will consider the Class Rules set by the American Bureau of Shipping (ABS) in the ABS Marine Vessel Rules (ABS/MVR). For vessels not classed by ABS, vessel designers, owners and operators are strongly advised to consult with their own Class Rules for the vessel in question. For vessels subject to the requirements set by Class for vessels operating in low temperature regions, including the Baltic Sea, or other recognised ice class Rules as may be applicable, the guidance on material requirements provided in this chapter are expected to be applied. For vessels subject to the requirements discussed in Chap. 15 *Baltic Ice Classes*, the materials selected for the construction of shipboard machinery and systems should be in accordance with the requirements set out below in the section *Materials and fabrication.* When the provisions of *Materials and fabrication* are applied, the definition of temperature provided in *Materials and fabrication* becomes applicable. For non-ice class vessels, the hull and structural materials should be selected in accordance with the requirements of section

A. Olsen, *Ship Operations in Extreme Low Temperature Environments*, Springer Series on Naval Architecture, Marine Engineering, Shipbuilding and Shipping 19,
https://doi.org/10.1007/978-3-031-52513-1_6

Materials and fabrication. When section *Materials and fabrication* is applied, the definition of temperature provided in section *Materials and fabrication* applies.

The guidance in this chapter provides additional information to the reader and, in some cases, recommendations on how to comply with the Class Rules. Material selection for vessels with ice class requirements should be in accordance with the provisions discussed in Chap.7, *Arctic vessel hull construction and equipment,* where applicable. If no ice class is specified or the ice class does not have material requirements, the guidance provided in the section on *Materials and fabrication* may be referred to instead. In any case, it is recommended that Grade A steel (or equivalent) should not be used in temperature conditions below −15 °C (5 °F). Successful application of Grade A and B steels for ships of Baltic ice classes is based on the correct selection of the combinations of structural thicknesses, steel grades and the appropriate operation of the vessel in question. For these cases there has been several reported damage incidents to such ships during winter navigation in the basin of the Baltic Sea. However, for operation in the Arctic region, under lower temperatures, there have been reported occurrences of brittle cracks of vessels' structures when constructed from Grade A steel. Crack development may take place even under the short-term impact of low temperatures and is related to the stress in the structural element or to its vibration level (bridge wing, bulwark, etc.). Accordingly, materials with resistance to brittle fracture are recommended for construction of exposed structural members.

6.2 Materials and Fabrication

The guidance provided in this section is adapted from the Class Rules set by ABS in the MVR. For vessels classed by alternative classification societies, the reader is directed to refer to those Rules for specific guidance. To apply the ABS/MVR Rules, the following conditions must be complied with:

(1) The requirements of ABS/MVR apply to vessels of all welded construction.
(2) For vessels with riveted hull construction, the applicable sections in the 1969 edition of the ABS MVR Rules must be complied with; and
(3) The design temperatures for unrestricted service are assumed to be not less than −10 °C (14 °F) for air temperature and not less than 0 °C (32 °F) for sea water temperature.

6.2.1 Materials

Steel. The Class Rules are intended for vessels of welded construction using steels complying with the requirements of Chap. 1 of the *ABS Rules for Materials and Welding (Part 2)*. The use of steels other than those stipulated in Chap. 1 of the *ABS*

Rules for Materials and Welding (Part 2) and the vessels' corresponding scantlings, must be considered with especial attention by the vessel designer in conjunction with Class.

Aluminium alloys. The use of aluminium alloys in hull structures may be considered upon submission to Class of a specification of the proposed alloys and their proposed method of fabrication. The use of aluminium alloys for deckhouses, helicopter landing platforms, masts, ladders and hatch covers are acceptable upon approval in accordance with the requirements stipulated in Chap. 5 of the *ABS Rules for Materials and Welding (Part 2)*. The connections between steel and aluminium structures must be made by interfacing bimetallic inserts. Details of interfacing bimetallic inserts must be in accordance with the requirements set out in section 3-2-13/3 of the *ABS Rules for Building and Classing High Speed Craft*. Furthermore, said aluminium structures and components must be fully and efficiently precluded from having direct contact with steel by approved non-wicking and non-water absorbing insulating materials, the purpose of which is to avoid incidences of electrolytic corrosion.

Design consideration. Where scantlings are reduced in association with the use of higher-strength steel or where aluminium alloys are used, adequate buckling strength must be provided. Where it is intended to use material of cold flanging quality for important longitudinal strength members, this steel must be indicated on the vessel's design plans and approved by Class.

Guidance for repair. Where a special welding procedure is required for special steels used in the construction, including any low temperature steel and those materials not encompassed in Chap. 1 of the *ABS Rules for Materials and Welding (Part 2)*, an additional set of plans showing the following information for each steel must be kept onboard the vessel:

(1) Material specification.
(2) Welding procedure; and
(3) Location and extent of application.

These plans are in addition to those documents and plans normally maintained onboard the vessel and must show all material applications.

6.2.2 Selection of Material Grade

Steel materials for locations must not be of a lower grade than those required by Table 6.1 for the material class given in Tables 6.2, 6.3, or Table 6.4 depending on the length (L) of the vessel. This provision is not required for vessels with L under 61 m (200 ft). For vessels under 61 m (200 ft) in length, the material used must comply with the *ABS Rules for Materials and Welding (Part 2)*. ASTM A36 steel that is manufactured by a Class approved steel mill, and which is assessed and certified to the satisfaction of Class, may be used in lieu of Grade A for a thickness up to and including 12.5 mm (0.5 in) for plate and 19 mm (0.75 in) for sections. In addition, the dimensions for ASTM A36 steel must comply with the Class permissible variations in dimensions. The responsibility for meeting the tolerance standards rests with the steel manufacturer, who is required to maintain a procedure that is deemed acceptable to Class. Where any tolerance (including over thickness tolerance) to be used is more stringent than normal commercial tolerance, Class should be advised before the steel is presented for acceptance to assure that the thickness measuring procedure is appropriate and meets Class requirements. When tensile stresses through the thickness (Z direction) exceed approximately 50% of the minimum specified yield stress (as defined in the applicable Class Rules), consideration should be given to applying Z grade steel (refer to the *ABS Rules for Materials and Welding (Part 2)* for further guidance). Alternatives to applying Z grade may be proposed—provided it is demonstrated by ultrasonic testing before and after welding—that no through thickness tearing has occurred, and/or the welding preparation, weld size and bead sequence is such that damaging through thickness loads induced by weld shrinkage are avoided. It is worth noting that when fatigue loading is present, the effective strength of higher-strength steel in welded construction may not be greater than that of ordinary-strength steel. Therefore, precautions against corrosion fatigue may also be necessary.

Also, each plate is to be UT inspected in accordance with either ASTM A578 Level B or another equivalent recognised standard to evaluate the internal soundness.

Table 6.1 Material grades

Plate thickness t mm (in)	Material class		
	I	II	III
$t \leq 15$ ($t \leq 0.60$)	A[(2)], AH	A, AH	A, AH
$15 < t \leq 20(0.60 < t \leq 0.79)$	A, AH	A, AH	B, AH
$20 < t \leq 25(0.79 < t \leq 0.98)$	A, AH	B, AH	D, DH
$25 < t \leq 30(0.98 < t \leq 1.18)$	A, AH	D, DH	D[(1)], DH
$30 < t \leq 35(1.18 < t \leq 1.38)$	B, AH	D, DH	E, EH
$35 < t \leq 40(1.38 < t \leq 1.57)$	B, AH	D, DH	E, EH
$40 < t \leq 100(1.57 < t \leq 4.00)$	D, DH	E, EH	E, EH
$100 < t \leq 150(4.00 < t \leq 6.09)$[(3)]	E, EH	E, EH	E, EH

Notes:
[(1)]Grade D, of these plate thicknesses, is to be normalised.
[(2)]ASTM A36 steel otherwise manufactured by a Class approved steel mill, assessed and certified to the satisfaction of Class may be used in lieu of Grade A for a thickness up to and including 12.5 mm (0.5 in) for plate and up to and including 19 mm (0.75 in) for sections.
[(3)]Rolled plates over 100 mm (4 in) thick are to be specially considered. When Class grade and non-Class grade rolled plates of over 100 mm (4 in) thickness are used for vessel hull structural application, in addition to chemical analysis the following test data is to be obtained at one quarter and mid thickness locations:

- Tensile properties; and
- Impact properties in the longitudinal or transverse directions.

6.2.3 Dissimilar Materials

Problems have been experienced in the past where a vessel is constructed or outfitted in a warm location then moves to operate in low temperature environments. Typically, these problems are caused by stresses related to thermal expansions or contractions. If an assembly is constructed in a warm ambient temperature, then subsequently

Table 6.2 Material class or grade of structural members for vessels ≥ 90 m (295 ft) in length

Serial	Structural members	Within 0.4*L* amidships	Outside 0.4*L* amidships
		Material class(8) or grade	Material class or grade
A	**Secondary**		
A1	Longitudinal bulkhead strakes, other than those belonging to the Primary category	I	A(10)/AH
A2	Deck plating exposed to weather, other than that belonging to the primary or special category		
A3	Side plating(12)		
B	**Primary**		
B1	Bottom plating, including keel plate	II	A(10)/AH
B2	Strength deck plating, excluding that belonging to the special category(13)		
B3	Continuous longitudinal plating of strength members above strength deck, excluding hatch coamings(13,14)		
B4	Uppermost strake in longitudinal bulkhead		
B5	Vertical strake (hatch side girder) and uppermost sloped strake in top wing tank		
C	**Special**		
C1	Sheer strake at strength deck(1,9,13)	III	II (I outside *0.6L* amidships)
C2	Stringer plate in strength deck(1,9,13)		
C3	Deck strake at longitudinal bulkhead(2,9,13)		
C4	Strength deck plating at outboard corners of cargo hatch openings in container carriers and other ships with similar hatch opening configurations(3,13)		
C5	Strength deck plating at corners of cargo hatch openings in bulk carriers, ore carriers, combination carriers and other ships with similar hatch opening configurations(4,13)		
C6	Trunk deck and inner deck plating at corners of openings for liquid and gas domes in membrane type liquefied gas carriers(4,13)		
C7	Bilge strake(5,6,9)		
C8	Longitudinal hatch coamings of length greater than 0.15*L* including coaming top plate and flange(7)		
C9	End brackets and deck house transition of longitudinal cargo hatch coamings(7)		

(continued)

Table 6.2 (continued)

Serial	Structural members	Within 0.4L amidships	Outside 0.4L amidships
		Material class[8] or grade	Material class or grade
D	**Other categories**		
D1	Plating materials for stern frames supporting rudder and propeller boss, rudders, rudder horns, steering equipment[15], propeller nozzles, and shaft brackets	–	II[11]
D2	Strength members not referred to in A to C and D1[17]	A[10]/AH	A[10]/AH
D3	Equipment foundations, non-hull structure, appurtenant structures	(16)	(16)

Notes

[1]Not to be less than grade E/EH[9] within 0.4L amidships in ships with length exceeding 250 m (820 ft).

[2]Excluding deck plating in way of inner-skin bulkhead of double hull ships.

[3]Not to be less than class III within the length of the cargo region.

[4]Not to be less than class III within 0.6L amidships and class II within the remaining length of the cargo region.

[5]May be of class II in ships with a double bottom over the full breadth and with length less than 150 m (492 ft).

[6]Not to be less than grade D/DH within 0.4L amidships in ships with length exceeding 250 m (820 ft).

[7]Not to be less than grade D/DH.

[8]Special consideration will be given to vessels of restricted class.

[9]Single strake required to be class III or E/EH are to have breadths not less than 800 + 5Lmm (31.5 + 0.06L in), but need not exceed 1800 mm (71 in), unless limited by the geometry of the vessel's design.

[10]ASTM A36 steel otherwise manufactured by a Class approved steel mill, assessed and certified to the satisfaction of Class may be used in lieu of Grade A for a thickness up to and including 12.5 mm (0.5 in) for plates and up to and including 19 mm (0.75 in) for sections.

[11]For rudder and rudder body plates subjected to stress concentrations (e.g., in way of lower support or at upper part of spade rudders), class III is to be applied.

[12]Single side strakes for ships exceeding 150 m (492ft) without inner continuous longitudinal bulkheads between bottom and the single strength deck are not to be less than grade B/AH within cargo region in ships.

[13]Not to be less than grade B/AH for members contributing to the longitudinal strength within 0.4L amidships in ships with length exceeding 150 m (492 ft) and single strength deck.

[14]Not to be less than grade B/AH for inner deck plating and plating between the trunk deck and inner deck for members contributing to the longitudinal strength within 0.4L amidships in membrane type liquefied gas carriers and other similar ship types with a double deck arrangement above the strength deck and with length exceeding 150 m (492 ft).

[15]Steering equipment components other than rudders.

[16]ASTM A36 steel otherwise assessed and certified to the satisfaction of Class may be used in lieu of Grade A for a thickness up to and including 12.5 mm (0.5 in) for plates and up to and including 19 mm (0.75 in) for sections.

[17]Deck plating below the strength deck at corners of cargo hatch openings immediately forward and aft of the engine room and/or deck house in container carriers and other ships with similar hatch opening configurations, is not to be less than Class I.

brought into a cold ambient temperature, the various expansion/contraction rates of the materials can cause binding or excessive internal stress. An example of this is an external light fixture. If the bulb burns out and needs to be replaced, the cage and glass globe must be removed from the steel base. If the cage and globe were installed in a warm location it is possible that the threads are locked due to thermal contraction. It should therefore be assumed that the component was built and installed at the normal temperature for the shipyard (+20 °C (68 °F) or more is suggested). Maintenance should therefore be performed at temperatures similar to the Design Service Temperature (t_D). The guidance provided in this section is considered mandatory for systems and equipment which form part of the vessel's essential services (refer to Tables 6.1

Table 6.3 Material class or grade of structural members for vessels 61 m (200ft) $\leq L < 90$ m (295 ft) in length

Structural member	Material class(1)	
	Within 0.4L amidships	Outwith 0.4L amidships
Shell		
Bottom plating including keel plate	II	A(4)/AH
Bilge strake	II	A(4)/AH
Side plating	I	A(4)/AH
Sheer strake at strength deck(2)	II	A(4)/AH
Decks		
Strength deck plating(3)	II	A(4)/AH
Stringer plate in strength deck(2)	II	A(4)/AH
Strength deck plating within line of hatches and exposed to weather, in general	I	A(4)/AH
Strength deck strake on tankers at longitudinal bulkhead	II	A(4)/AH
Longitudinal bulkheads		
Lowest strake in single bottom vessels	I	A(4)/AH
Uppermost strake including that of the top wing tank	II	A(4)/AH
Other structures in general		
External continuous longitudinal members and bilge keels	II	A(4)/AH
Plating materials for stern frames supporting rudder and propeller boss, rudders, rudder horns, steering equipment(5), propeller nozzles, and shaft brackets	–	I
Strength members not referred to in above categories and above local structures	A(4)/AH	A(4)/AH

Notes
(1) Special consideration is typically given to vessels in restricted service.
(2) A radius gunwale plate may be considered to meet the requirements for both the stringer plate and the sheer strake, provided it extends suitable distances inboard and vertically. For formed material, refer to section 2-4-1/3.13 of the *ABS Rules for Materials and Welding (Part 2).*
(3) Plating at the corners of large hatch openings may be specially considered.
(4) ASTM A36 steel otherwise manufactured by a Class approved steel mill, assessed and certified to the satisfaction of Class may be used in lieu of Grade A for a thickness up to and including 12.5 mm (0.5 in) for plates and up to and including 19 mm (0.75 in) for sections.
(5) Steering equipment components other than rudders.

and 6.2). Alternatively, for systems and equipment not listed in Tables 6.1 and 6.2, the guidance provided in this section is optional, but recommended. The application of dissimilar materials should be avoided wherever possible. Where it is impractical or impossible to avoid the use of dissimilar materials, tolerances should be such that thermal expansions or contractions will not cause the components to bind at the t_D. This is applicable when components may need to be subject to maintenance and

Table 6.4 Application of material classes and grades (structures exposed at low temperatures)

Structural member category	Material class	
	Within 0.4L amidships	Outwith 0.4L amidships
Secondary	I	I
Deck plating exposed to weather, in general		
Side plating above BWL		
Transverse bulkheads above BWL(5)		
Cargo tank boundary plating exposed to cold cargo(6)		
Primary	II	I
Strength deck plating(1)		
Continuous longitudinal members above strength deck, excluding longitudinal hatch coamings		
Longitudinal bulkhead above BWL(5)		
Top wing tank bulkhead above BWL(5)		
Special	III	II
Sheer strake at strength deck(2)		
Sheer strake at strength deck(2)		
Stringer plate in strength deck(2)		
Deck strake at longitudinal bulkhead(3)		
Continuous longitudinal hatch coamings(4)		
Deck strake at longitudinal bulkhead(3)		

Notes

(1)Plating at corners of large hatch openings to be specially considered. Class III or Grade E/EH to be applied in positions where high local stresses may occur.

(2)Not to be less than Grade E/EH within 0.4L amidships in ships with length exceeding 250 m (820 ft).

(3)In ships with breadth exceeding 70 m (230 ft) at least three deck strakes to be Class III.

(4)Not to be less than Grade D/DH.

(5)Applicable to plating attached to hull envelope plating exposed to low air temperature. At least one strake is to be considered in the same way as exposed plating and the strake width is to be at least 600 mm (24 in).

(6) For cargo tank boundary plating exposed to cold cargo for ships other than liquefied gas carriers, refer to footnote (1).

consist of dissimilar materials. The control of dissimilar materials used in construction is the responsibility of the final assembly shipyard and should be covered by the shipyard's quality management system (QMS). Components or assemblies may be enrolled in a Class Type Approval programme with specific approval on material applicability ranges provided for low temperature service.

6.2.4 Material of Machinery

Machinery subject to Class requirements includes any machinery associated with the function or purpose of the vessel. Exposed machinery must be suitable for operation at the minimum anticipated temperature where the vessel owner/operator intends to operate the machinery in low temperatures. Moreover, safety critical systems are required to be suitable for operation at the minimum anticipated temperature unless otherwise stipulated by the machinery/equipment manufacturer and confirmed by Class. Any machinery component/equipment can be excluded from applying the requirements if the machinery component/equipment is not intended to be operated in low temperatures. In that case, a note will be entered in the vessel's official Record to indicate that the specific machinery component/equipment is not reviewed in accordance with the low temperature environment requirements. Table 6.1 lists the design temperature requirements for various systems, equipment and components.

6.2.5 Machinery, Structural Members, and Components Exposed to the Cold Temperatures

The material class and temperature of steel products used for exposed machinery foundations and load bearing components should be designed and manufactured to the standards in Table 6.1. Material grade selection should be selected in accordance with Class Rules.

6.2.6 Vessels Intended to Operate in Low Air Temperatures

For ships intended to operate in areas with low air temperatures [i.e., below −10 °C (14 °F)], the materials in exposed structures should be selected based on the t_D. Materials in the various strength members above the lowest ballast water line (BWL) exposed to air (including the structural members covered by *Note 5* of Table 6.4) and materials of cargo tank boundary plating for ships carrying cold cargoes other than liquified gas carriers[1] must not be of lower grades than those corresponding to Classes I, II and III, as given in Table 6.4, depending on the categories of structural members

[1] For ships other than liquefied gas carriers, intended to be loaded with liquid cargo having a temperature below -10 °C, e.g., loading from cold onshore storage tanks during winter conditions, the material grade of cargo tank boundary plating is defined in Table 6–5 based on the following:

- t_c design minimum cargo temperature in °C (°F).
- steel grade corresponding to Class I as given in Table 6–4.

The design minimum cargo temperature, t_c is to be specified in the loading manual.

(secondary, primary and special). For non-exposed structures (except as indicated in *Note 5* of Table 6.4) and structures below the lowest BWL, the guidance provided in the section *Selection of material grade* applies. The material grade requirements for hull members of each class depending on thickness and t_D are defined in Table 6.5. For $t_D < -55$ °C (−67 °F), materials are to be specially considered. Single strakes required to be of Class III or of Grade E/EH or FH are required to have breadths not less than $800 + 5L$ mm ($31.5 + 0.06L$ in), maximum 1800 mm (71 in). L is as defined as the distance in m (ft) measured on the waterline at the scantling draught from the fore side of the stem to the centreline of the rudder stock. To comply with Class Rules, L must not be less than 96% and need not be greater than 97% of the extreme length on the waterline at the scantling draught. The forward end of L is to coincide with the fore side of the stem on the waterline on which L is measured. In ships without rudder stock (e.g., ships fitted with azimuth thrusters), L is to be taken equal to 97% of the extreme length on the waterline at the scantling draught. In ships with an unusual stern and bow arrangement the length, L, is to be specially considered. Plating materials for stern frames, rudder horns, rudders and shaft brackets are not to be of lower grades than those corresponding to the material classes given in the section *Selection of material grade*.

6.2.7 Design Temperature

The t_D is to be taken as the lowest mean daily average air temperature in the region of operation; accordingly:

- Mean: statistical mean over observation period (at least 20 years).
- Average: average for one day and night; and
- Lowest: lowest during year.

Table 6.5 Material grade requirements for Classes I, II and III at low temperatures Class I

Thickness, in mm (in)	−11 to −15 °C (12 to 5 °F)	−16 to −25 °C (4 to −13 °F)	−26 to −35 °C (14 to −31 °F)	−36 to −45 °C (−32 to −49 °F)	−46 to −55 °C (−50 to −68 °F)
t ≤ 10 (t ≤ 0.39)	A, AH	A, AH	B, AH	D, DH	D, DH
10 < t ≤ 15 (0.39 < t ≤ 0.60)	A, AH	B, AH	D, DH	D, DH	D, DH
15 < t ≤ 20 (0.60 < t ≤ 0.79)	A, AH	B, AH	D, DH	D, DH	E, EH
20 < t ≤ 25 (0.79 < t ≤ 0.98)	B, AH	D, DH	D, DH	D, DH	E, EH
25 < t ≤ 30 (0.98 < t ≤ 1.18)	B, AH	D, DH	D, DH	E, EH	E, EH
30 < t ≤ 35 (1.18 < t ≤ 1.38)	D, DH	D, DH	D, DH	E, EH	E, EH
35 < t ≤ 45 (1.38 < t ≤ 1.80)	D, DH	D, DH	E, EH	E, EH	–, FH
45 < t ≤ 50 (1.80 < t ≤ 1.97)	D, DH	E, EH	E, EH	–, FH	–, FH
t ≤ 10 (t ≤ 0.39)	A, AH	B, AH	D, DH	D, DH	E, EH
10 < t ≤ 20 (0.39 < t ≤ 0.79)	B, AH	D, DH	D, DH	E, EH	E, EH
20 < t ≤ 30 (0.79 < t ≤ 1.18)	D, DH	D, DH	E, EH	E, EH	–, FH
30 < t ≤ 40 (1.18 < t ≤ 1.57)	D, DH	E, EH	E, EH	–, FH	–, FH
40 < t ≤ 45 (1.57 < t ≤ 1.80)	E, EH	E, EH	–, FH	–, FH	–, –
45 < t ≤ 50 (1.80 < t ≤ 1.97)	E, EH	E, EH	–, FH	–, FH	–, –

For seasonally restricted service the lowest value within the period of operation applies. For the purpose of issuing a Polar Ship Certificate in accordance with the IMO Polar Code, the t_D must be no more than 13 °C (23.6 °F) higher than the Polar Service Temperature (PST) of the ship. In the polar regions, the statistical mean over observation period is to be determined for a period of at least 10 years. Materials in the various strength members above the lowest BWL exposed to air (including the structural members covered by *Note 5* of Table 6.1) and materials of cargo tank boundary plating for which footnote (1) is applicable are not to be of lower grades than those corresponding to Classes I, II and III, as given in Table 6.1, depending on the categories of structural members (secondary, primary and special). For non-exposed structures (except as indicated in *Note 5* of Table 6.4) and structures below the lowest BWL, the section *Selection of material grade* applies. The material grade requirements for hull members of each class depending on thickness and t_D are defined in Table 6.5. For $t_D < -55$ °C (−67 °F), materials are to be specially considered. Single strakes required to be of Class III or of Grade E/EH or FH are to have breadths not less than $800 + 5L$ mm ($31.5 + 0.06L$ in), maximum 1800 mm (71 in).

6.2.8 Deck Machinery, Piping, Valves and Fittings

Deck machinery materials are required to comply with the material specifications set by Class and of any national or international material standard. The proposed materials must be approved in connection with the design. All such materials are to be certified by the material manufacturer(s) and must be traceable to the manufacturer's certificates. Materials for piping, valves and fittings for minimum anticipated temperatures lower than −18 °C (0 °F) should be selected in accordance with the vessel's Class rules relating to materials and welding standards. For contextual purposes, section 2-A1-1 of the *ABS Rules for Materials and Welding (Part 2)* contains a *List of Destructive and Non-destructive Tests required for Materials*, and *Responsibility for verifying*. This list (provided below as Table 6.6) indicates which tests require a Class approved surveyor witness, manufacturer provided data without Class approved surveyor witness, and those tests for which data must be provided by the manufacturer and subject to auditing by a Class approved surveyor.

6.2.9 Cranes, Lifting Appliances, Vehicle Ramps, and Boat Davits

Cranes and lifting appliances. In respect of vessels operating in low temperature environments, specific certification relating to lifting appliances is optional. Section 6.2–6.1/7.21 of the *ABS Guide for Certification of Lifting Appliances (Lifting Appliance Guide)* defines the "Design Service Temperature" as "the minimum anticipated temperature at which the crane will operate as specified by the owner, crane manufacturer or builder." For vessels seeking lifting appliance certification, the minimum anticipated temperature as defined in this chapter should be applied. Having established the minimum anticipated temperature for each component, the following may then be applied:

- The material class and minimum anticipated temperature of steel products used for crane and foundation/pedestal are in accordance with Chapter 2, section 3 of the *ABS Lifting Appliance Guide.*
- Crane structural components are to be evaluated in accordance with the requirements of section 2-3 of the *ABS Lifting Appliance Guide.*

Table 6.6 List of destructive and non-destructive tests required for materials and responsibility for verifying

Key to test and test data	
Witnessed tests. The designation (W) indicates that the surveyor is to witness the testing unless the plant and product is approved under the appropriate Class Quality Assurance programme	
Manufacturer's data. The designation (M) indicates that test data is to be provided by the manufacturer without verification by a surveyor of the procedures used or the results obtained	
Other tests. The designation (A) indicates those tests for which test data is to be provided by the supplier and audited by the surveyor to verify that the procedures used, and random tests witnessed comply with the Class Rule requirements	
General • Through thickness properties (W)	**Steel plates for intermediate and lower-temperature service** • McQuaid-Ehn (M) • Chemical composition (M) • Tensile properties (W)
Ordinary-strength hull structural steel • Ladle analysis (M) • Product analysis (M) • McQuaid - Ehn (M) • Tension test (W) • Charpy V-notch impact test (W)	**Seamless forged-steel drums** • Tension tests (W)
Higher-strength hull structural steel • Ladle analysis (M) • Tension test (W) • Charpy V-notch impact test (W) • Product analysis (M) • McQuaid - Ehn (M)	**Seamless-steel pressure vessels** • Tension test (W) • Flattening test (W) • Hydrostatic test (W) • Thickness test (W)
Tension test (W) • Magnetic particle inspection (A) • Dye Penetrant inspection (A) • Ultrasonic inspection (A)	**Boiler and superheater tubes** • Chemical composition (M) • Product analysis (M) • Tensile properties (W) • Flattening test (W) • Reverse flattening Test (W) • Flange test (W) • Flaring test (W) • Crush test (W) • Hardness test (W) • Hydrostatic test (W) • Non-destructive electric test (NDET) (A) • Thickness test (A)
Hull steel castings • Materials for low temperature service • Charpy V-notch impact test (W) • Drop-weight test (NDTT) (W)	**Boiler rivet and staybolt steel and rivets** • Tensile properties (W) • Bending properties (bars) (W) • Bending properties (rivets) (W) • Flattening test (W)

(continued)

Table 6.6 (continued)

Hull steel forgings • Ladle analysis (M) • Tension test (W) • Brinell hardness test (BHN) (W)	**Steel machinery forgings** • Chemical composition (M) • Tensile properties (W) • Surface inspection of tailshaft forgings (W) • Ultrasonic examination of tail shaft forgings (A) • Hardness test (W)
Ordinary and higher strength steels with enhanced corrosion resistance properties for cargo oil tanks • Ladle analysis (M) • Tension test (W) • Charpy V-notch impact test (W) • Product analysis (M) • McQuaid-Ehn (M) • Corrosion tests (A)	
• **Extra high strength quenched and tempered steel** • Ladle analysis (M) • Tension test (W) • Charpy V-notch Impact Test (W) • Product analysis (M)	**Steel Castings for Machinery, Boilers, and Pressure Vessels** • Chemical composition (M) • Tensile properties (W) • Magnetic particle or dye penetrant inspection (W)
Anchors • Proof test (W) • Product test (W)	**Ductile (nodular) iron castings** • Tension tests (W) • Chemical composition (M)
Anchor chain • Ladle analysis (M) • Tension test (W) • Bend test (W) • Charpy V-notch impact test (W) • Breaking test (W) • Proof test (W) • Magnetic particle inspection (A) • Brinell hardness test (W)	**Gray-iron castings** • Tension test (W)
Unstudded short-link chain • Ladle analysis (M) • Tension test (W) • Bend test (W) • Breaking test (W) • Proof test (W)	**Steel piping** • McQuaid-Ehn (M) • Chemical composition (M) • Product analysis (M) • Tension tests (W) • Bend test (W) • Flattening test (W) • Hydrostatic test (W) • Thickness test (A)

(continued)

Table 6.6 (continued)

General requirements for all grades of steel plates for machinery, boilers, and pressure vessels • Ladle analysis (M) • Product analysis (M) • Test specimens (W) • Tensile properties (W)	**Piping, valves and fittings for low temperature service [below -18°c (0°f)]** • McQuaid-Ehn (M) • Chemical composition (M) • Mechanical test (M) [(W) for Piping] • Impact properties (M) [(W) for Piping]
Steel plates for intermediate temperature service • Chemical composition (M) • Tensile properties (W)	**Valves on vessels intended to carry liquefied gases in bulk for low temperature service [at or below -55°c (-67°f)]** • McQuaid-Ehn (M) • Chemical Composition (M) • Mechanical Test (W) • Impact Properties (W)
Steel plates for intermediate and higher-temperature service • Chemical composition (M) • Tensile properties (W)	**Valves on vessels intended to carry liquefied gases in bulk for low temperature service [above -55°c (-67°f)]** • McQuaid-Ehn (M) • Chemical composition (M) • Mechanical test (M) • Impact properties (M)
Bronze castings • Chemical composition (M) • Tensile properties (W) • Dye penetrant inspection (W)	**Austenitic stainless steel propeller castings** • Dye penetrant inspection (W) • Chemical composition (M) • Tensile properties (W)
Seamless copper piping • Chemical composition (M) • Tension test (W) • Expansion test (W) • Flattening test (W) • Hydrostatic test (W) (M) • Thickness test (A)	**Seamless red-brass piping** • Chemical composition (M) • Expansion test (W) • Flattening test (W) • Mercurous nitrate test (M) • Bend test (W) • Hydrostatic test (W) (M) • Thickness test (A)

(continued)

Table 6.6 (continued)

Seamless copper tube • Chemical composition (M) • Tension test (W) • Expansion test (W) • Flattening test (W) • Hydrostatic test (W) (M) • Thickness test (A)	**Monel pipe and tube** • Chemical composition (M) • Tension test (W) • Flattening test (W) • Flare test (W) • Flange test (W) • Hydrostatic test (W) (M) • Non-destructive electric test (NDET) (A) • Dimensions (A)
Condenser and heat exchanger tube • Chemical composition (M) • Tension test (W) • Expansion test (W) • Flattening test (W) • Non-destructive electric test (NDET) (A) • Hydrostatic test (W) (M) • Dimensions (A)	**Copper-nickel tube and pipe** • Chemical composition (m) • Tension test (W) • Expansion test (W) • Flattening test (W) • Non-destructive electric test (NDET) (A) • Radiographic examination (A) • Hydrostatic test (W) (M) • Dimensions (A)

- The machinery components are to be assessed as per section 2-6 of the *ABS Lifting Appliance Guide*. Therefore, it would have to be demonstrated by way of testing or analysis that such components will operate satisfactorily at the minimum anticipated temperature. In this regard, the testing requirements of section 2-3 of the *ABS Lifting Appliance Guide* may also be applied to such components.

Vehicle ramps. Vehicle ramps and boat davits should be designed, constructed and maintained in accordance with the provisions outlined above as far as is practicable. Where it is neither desired nor necessary to design, construct and maintain vehicle ramps and boat davits in accordance with the provisions outlined above, the following provisions may be applied instead:

- For the large structural components of vehicle ramps exposed to the weather, materials should be selected in accordance with the requirements in the section *Ships intended to operate in low air temperatures* for material class II. T_{DIT} may be used instead of t_D if the structures are contained within an enclosed space.
- The material used in the construction of machinery components for vehicle ramps, such as locks, guides, or rollers, should be in accordance with the section *Machinery, structural members, and components exposed to the cold temperatures.*

Boat davits. Boat davit materials should be selected based on section 2-3 of the *ABS Lifting Appliance Guide*. The temperature to use for material selection is to be the Minimum Anticipated Temperature as defined in this chapter. Davits are normally not certified in accordance with the *ABS Lifting Appliance Guide.*

6.2.9.1 Material of Exposed Outfitting

Steel products material class used for deck machinery should be selected from Table 6.1. Material selection must be in accordance with guidance contained in the section *Ships exposed to low air temperatures,* or specific Class conditions where required.

6.2.9.2 Insulated Members

The design service temperature for insulated members will be considered by Class upon submission of substantiating data for review by Class Engineering.

6.2.9.3 Criteria for Other Steels

Steels with specified minimum yield strength below 410 N/mm^2, (*42* kgf/mm^2, *60* ksi). Where steels other than those listed in 2-1-2/15.9 Table 5 or 2-1-3/7.3 Table 5 of the *ABS Rules for Materials and Welding (Part 2)* are intended, their specifications are to be submitted for approval. These steels are to comply with the following impact test requirements:

Yield strength			CVN (longitudinal)		
N/mm^2	kgf/mm^2	ksi	J	kgf-m	ft-lbf
235–305	24–31	34–44	27	2.8	20
315–400	32–41	45.5–58	34	3.5	25

At the following temperatures:

Class I: design service temperature
Class II: 10 °C (18°F) below design service temperature
Class III: 20 °C (36°F) below design service temperature.

Steels with specified minimum yield strength in the range of 410–690 N/mm^2 (*42–70* kgf/mm^2, *60–100* ksi). Where steels having this yield strength range are intended, their specifications are to be submitted for approval. These steels are to comply with the impact test requirements of 34 J (3.5 kgf-m, 25 ft-1bf) at the following temperatures:

T_D	Test temperature
0 °C (32 °F)	−30 °C (−22 °F)
−10 °C (14 °F)	−40 °C (−40 °F)
−20 °C (−4 °F)	−40 °C (−40 °F)
−30 °C (−22 °F)	−50 °C (−58 °F)
−40 °C (−40 °F)	−60 °C (−76 °F)

6.2.9.4 Alternative Requirements

As an alternative to these requirements, higher strength steels may comply with the following. For transverse specimens, $^2/_3$ of energy values shown in Chap. 17, in the section *Criteria for other steels.* For longitudinal specimens, lateral expansion is not to be less than 0.5 mm (0.02 in). For transverse specimens, lateral expansion is not to be less than 0.38 mm (0.015 in). Nil-ductility temperature (NDT), as determined by drop weight tests, is to be 5 °C (9 °F) below the temperature specified in Chap. 7, in the section *Criteria for other steels.*

Additional requirements for plates over 100 mm (4 in) thickness. In the application of the guidance provided in this book, rolled steel plates of thickness over 100 mm (4 in) for primary structural members are considered as thick plates and are subject to additional requirements as set out in 2-1-1/16 of the *ABS Rules for Materials and Welding (Part 2).* This requirement aims to avoid possible brittle failure in cold environments.

Cast irons. Ordinary cast iron or grey cast iron must not be used. Nodular (ductile) iron may be considered acceptable where specifically approved for use by Class.

Cast steels. Cast steel components which are not intended to be welded in construction or fabrication, should comply with the following impact test requirements:

Yield strength			CVN (longitudinal)		
N/mm^2	kgf/mm^2	ksi	J	kgf-m	ft-lbf
235–305	24–31	34–44	20	2.0	15
305–410	31–42	44–59	24	2.4	18
410–690	42–70	59–100	27	2.8	20
> 690	70	100	27	2.8	20

At the following temperatures:

Class I: t_D
Class II and Class III: 10 °C (18 °F) below t_D.

For cast steels with yield strengths exceeding 690 N/mm^2 (70 kgf/mm^2, 100 ksi), the test temperature for all classes should be a minimum of 10 °C (18 °F) below t_D. The impact test requirement is subject to agreement between the designer/manufacturer and Class.

Steel forgings. Forged steel components which are not intended to be welded in construction or fabrication, are to comply with the following impact test requirements:

Yield strength			CVN (longitudinal)		
N/mm^2	kgf/mm^2	ksi	J	kgf-m	ft-lbf
235–305	24–31	34–44	27	2.8	20
305–410	31–42	44–59	34	3.5	25
> 410	42	59	42	4.3	31

At the following temperatures:
Class I: t_D.
Class II and Class III: 10 °C (18 °F) below t_D.

For steel forgings with yield strengths exceeding 690 N/mm^2 (70 kgf/mm^2, 100 ksi), the test temperature for all classes should be a minimum of 10 °C (18 °F) below t_D. The impact test requirement is subject to agreement of the designer/manufacturer and Class.

6.3 Weld Metal

Class grade hull steels. When the ABS ordinary and higher strength hull steels of 2-1-2/15.9 Table 5 or 2–1-3/7.3 Table 5 of the *ABS Rules for Materials and Welding (Part 2)* are applied in accordance with the vessel's ice class requirements or section *Ships exposed to low air temperatures*, approved filler metals appropriate to the grades as shown in section 2-A3-1 of the *ABS Rules for Materials and Welding (Part 2)* may be used. For vessels not classed by ABS, refer to the appropriate specific Class Rules.

Criteria for other steels. For the welding of hull steels other than the Class approved grades listed in Section 2-1-2 or 2-1-3 of the *ABS Rules for Materials and Welding (Part 2)*, the weld metal should exhibit a Charpy V-Notch toughness value at least equivalent to the transverse base metal requirements ($^2/_3$ of longitudinal base metal requirements).

6.3.1 Repair Welding in Low Temperatures

Some welding machines can be damaged by operating in low air temperatures. The operating manual of the welding machine must be referenced before outside operation. Post weld NDT is more important in low air temperature welding due to the rapid cooling, but the low temperature may cause difficulties with the NDT. Most liquids and/or gels will tend to freeze on contact with the cold steel, the NDT equipment and any consumables used for the procedure should be suitable for use in the low air temperature at the time of examination. As with any outdoor welding, shielding from the effects of wind are important and essential where shielding gas is used in the welding process. The procedure used for welding in low temperatures should be agreed in advance with the attending Class approved surveyor prior to the work taking place. The procedure is to specify the minimum ambient temperature, as well as pre and post weld heating temperatures. Welding materials with an ambient temperature below −10 °C (14 °F) are not recommended and should be avoided where possible. Where welding in low temperatures cannot be avoided, the work area should be protected from low ambient temperatures and wind as far as is possible by a heated temporary shelter. The weld base materials are best pre-heated in accordance with the applicable procedure. Post weld heating and insulation are to be applied unless otherwise specified in the applicable procedure. It should be noted that a vessel's hull is a large heat sink and heating of the local area to be welded will be required on a continuous basis during the repair. Subsequently, the weld area should be subject to NDT in accordance with the Class approved procedures for conducting non-destructive inspections once the weld area has cooled to the ambient temperature.

6.4 Coatings

Coatings in low temperatures applied to those areas of the hull subject to contact with ice are expected to be durable and resistant to peeling, abrasion or other degradation. Coating product information and testing results must be submitted to Class for assessment (Table 6.7).

With reference to the coatings that may be applied, the purpose of this section is to provide ship owners and operators with general information to assist in the selection of coatings for those vessels intending to operate in low temperature environments. Whenever possible, ship owners and operators are encouraged to obtain assistance from coating manufacturers or suppliers in the selection of the most appropriate coating(s). The physical properties of coatings applied to vessels intending to operate in low temperature environments varies and is dependent on the intended area of operation and the specific location on the vessel where the coating is to be applied. For vessels intended to operate in low temperature environments, Class will usually take note of the following coating types:

Table 6.7 Material class of machinery structural members/components

Machinery members	Material class	Temperature
Exposed load bearing structural and machinery components directly exposed to the weather	II	MAT[(1)]
Load bearing structural and machinery components attached to and within 600 mm from the other parts which are directly exposed to the weather	II	MAT[(1)] + 10 °C
Load bearing structural and machinery components attached to but at over 600 mm from the other parts which are directly exposed to the weather in enclosed spaces		
Unheated space	II	DIT[(2)]
Heated space	I	0 °C

Notes
[(1)]Minimum anticipated temperature.
[(2)]Design internal temperature.

- External hull coatings.
- Ice release coatings.
- Maintenance or repair coatings; and
- Interior coatings.

External hull coatings. External hull coatings on ships operating in ice face different conditions than hull coatings on vessels operating in warmer, ice-free waters. Therefore, the selection of hull coatings on vessels operating in ice requires a uniquely different approach. The main purpose of the coatings applied to ship hulls is (1) to protect against corrosion whilst the secondary purpose is (2) to provide and maintain a smooth hull which provides as low friction as possible when the ship is sailing. For ships operating in warmer waters, it is a known phenomenon that their hulls are prone to bio fouling. Therefore, several types of anti-fouling coatings have been developed which are applied on top of the anti-corrosive coating underneath (i.e., a "dual coat system"). For ships operating in ice, however, the fouling problem is minor. Additionally, most designated anti-fouling coatings are quickly destroyed and torn from the anti-corrosion coating underneath when operating in ice conditions. Most of the commonly used types of anti-corrosion coatings are also easily damaged by the forces of ice hitting against or scrubbing along the hull's surface. The industry answer to this problem is the application of a "single coat system" which has the benefit of the two main functions combined, i.e., (1) corrosion protection and (2) providing a smooth surface without surface irregularities. In today's market, a limited number of specific ice coatings are available of which the majority are based on epoxy resins, most of which are glass flake reinforced. In general, epoxy-based coatings perform relatively well, but over time they tend to become brittle which may end up in cracking, chipping and finally disbonding from the steel. In general, epoxies tend not to last as long as coatings based on polyester or vinyl ester resins.

Hull coatings can be applied at a wide range of film thicknesses. For ice hull coatings in general, the thinner the coating film, the more often it must be repaired or replaced. Typically, a dry film thickness (DFT) of 500 microns (half a millimetre) or less will not provide long term protection. Particularly in the ice belt areas DFT's up to 1000 microns may be considered, although coating system manufacturers should be consulted for information on the specific system under consideration. It is important to apply coating systems strictly in accordance with the recommendations of the manufacturer. The most important parameters to consider during the pre-treatment of the steel are roughness and cleanliness. Roughness, which is provided by proper grit blasting and the use of the correct type of grit, provides the required 'anchorage.' Cleanliness helps to prevent poor attachment to slippery oil/grease or foreign particles underneath, as well as to prevent soluble salts to eliminate osmotic blistering. Additional coating selection considerations include environmental restrictions on coatings. This also includes the environmental character of the coating. In case the coatings are to be applied in areas known to be prone to cavitation damages, an additional test on this phenomenon may be considered. Finally, for the selection of hull coatings for vessels operating in ice, the following Test Standards are routinely suggested (Table 6.8):

Table 6.8 Suggested coating test standards

Property	Test standard
Abrasion resistance	ASTM D 4060 taber abrasion
Impact resistance	ISO 6272-93
Hardness	ISO 2815
Scratch resistance	ISO 1518
Adhesion	ISO 4624
Friction	ASTM D 4518-91
Anti-corrosion properties	As advised by NORSOK M-501
Environmental Character	A High Solids Content (preferably Solvent Free or Ultra High Solids), as a minimum to follow the VOC Regulations applicable at location of application, and not containing biocides, tin or copper compositions is to be considered
Resistance to Cavitation damage	ASTM G 32

*Election to be in harmony with the expected operation sequence and ice conditions such as type of ice and content of highly abrasive volcanic lava, gravel or sand as existing in Antarctic regions

6.4.1 Ice Release Coatings

The accretion of ice or snow onboard vessels can have a significant adverse effect on a vessel's operation, specifically the potential impacts of the added weight on the vessel's stability and machinery systems. Various spray-on fluids to prevent ice build-up are readily available on the market. Prior to use, it is advised to precheck the chemical compatibility with the coating underneath. To lessen or mitigate the accumulation of ice and snow, the application of ice release coatings above the abrasion resistant hull coating and all other externally exposed surfaces (including decks, deck houses and superstructure ends and sides, guardrails, bulwarks and deck machinery) may be considered. Because of their inherent slickness, they should **not** be used on walkways. These ice release coatings, which are sometimes referred to as icephobic coatings, create a slippery surface. The hydrophobic (not capable of uniting with or absorbing water) surface causes the water to roll off the surface before it is able to solidify and form solid ice. The development of these ice release or icephobic coatings within the coating industry is ongoing and is currently being explored by some coating companies as a niche product. Acceptable testing standards are under development (as of 2023).

6.4.2 Maintenance or Repair Coatings

It should be realised that prior to operating in Arctic conditions, in most cases, the vessel will have been built under cover and under air-conditioned conditions with favourable temperatures and humidity which enable the coating(s) to cure properly. It should be realised that most coatings used in the shipbuilding industry, the majority being so called epoxies, cure at temperatures as low as 5 °C (41 °F). In the event maintenance or repair coatings need to be applied at lower temperatures, specially formulated coatings are needed with volatile ingredients that can function and evaporate at such lower temperatures. Therefore, so called 'Winter Type' or 'Low Temperature Type' (often designated as "LT Type") coating versions are readily available. It is important to know that even some of these LT coatings must be heated prior to application. Under freezing temperatures, it may be difficult eliminate all ice crystals on the steel prior to the application of the coating(s). It is strongly advised to follow the recommendations provided by the coating manufacturer by referring to the coating data sheet. Coating manufacturers should confirm the compatibility of the coating(s) in case a winter type or LT type coating is to be applied over a normal type coating. Finally, it should be kept in mind that maintenance coatings are best stored cool and dry but above freezing temperatures. For vessels operating in the Arctic, it may be wise to keep the sea stock at a minimum and store the maintenance coatings ashore wherever possible.

6.4.3 Interior Coatings

With respect to interior coatings, the 2014 adopted *IMO Code on Noise Levels* (Res. MSC.337(91)) should be kept in mind. Various coating manufacturers already deliver spray-on coatings specially developed to reduce noise and/or vibration levels inside ship accommodations. Simultaneously, other coatings have been developed with an especial focus on warmth insulation. Some coatings also prevent condensation, which on its own helps to prevent corrosion. The use of insulating coatings may at least minimise the thickness or even eliminate the use of insulation materials. Satisfactory results have been realised in eliminating corrosion under insulation by applying such coatings while it may pay back itself quickly when used on deck in case of heated cargoes being transported in Arctic conditions. When selecting such insulating or noise cancelling coatings, the coating should be checked to determine its compliance with the *IMO Fire Test Procedures* (FTP) *Code*.

6.4.4 Coating Application and Maintenance

The ability of the crew or service technicians to perform repairs or routine maintenance onboard vessels operating in low temperatures is extremely limited. This is especially true for those coatings which are directly exposed to the weather or are on a plate or structural member that forms a boundary with the weather. As a result, the application and inspection of coatings during the construction and shipyard periods takes on an even greater importance when compared to those vessels that operate in more temperate zones. It is recommended that vessel owners and operators consider employing coating consultants/inspectors to confirm that all coatings have been properly applied.

Note: The static friction measures the coefficient of friction from a stopped position to some relative speed between the ice and the steel. The breakaway friction accounts for specimen to ice contact, under load for a specific length of time. The kinetic friction coefficient is the value obtained during continuous movement between ice and test specimen.

Chapter 7
Arctic Vessel Hull Construction and Equipment

7.1 Introduction

In the previous chapter we examined some of the main Class Rules relating to the classification and use of steel, welding and coatings. In this chapter, we will turn our attention to the methods of construction and the equipment used for vessels intended for extreme low temperature service. For vessels intended to navigate in ice waters in Arctic regions, the applicable requirements for hull strengthening are provided later in this book. If the intended routes through which the vessel is expected to operate are ice free, then ice strengthening classification is not required. The materials used in the hull structural members must be in accordance with the requirements discussed previously in Chap. 6.

7.2 Water, Fuel and Lube Oil Tanks

7.2.1 Freshwater Tanks

Freshwater tanks with a space between the tank boundary and the side shell may be susceptible to the contents freezing. Consideration should be given to providing the tanks with turbulence-inducing systems, for example, bubble systems or heating coils. Freshwater tanks with boundaries including the side shell or the deck exposed to the weather must be provided with appropriate arrangements to prevent freezing (refer to Chap. 8 for further details). Freshwater tanks arranged such that the tank boundaries do not include the side shell, or the deck exposed to the weather and are located in a heated area such as the machinery space usually do not require heating arrangements to be provided.

A. Olsen, *Ship Operations in Extreme Low Temperature Environments*, Springer Series on Naval Architecture, Marine Engineering, Shipbuilding and Shipping 19,
https://doi.org/10.1007/978-3-031-52513-1_7

7.2.2 Fuel Oil Tanks

The location of the fuel oil tanks will determine the heating requirements. For tanks located below the waterline, the seawater temperature can be assumed to be – 3 °C (27°F). For tanks located above the waterline in contact with the shell, the t_D is to be assumed. Fuel oil tanks must be provided with heating arrangements in all circumstances where the fuel will cool to below its pour point. Fuel oil tanks for residual fuels should be maintained to at least 10 °C (18°F) above the fuel's pour point. Tank heating calculations must be provided to demonstrate sufficient heat transfer capacity for the t_D.

7.2.3 Ballast Water Tanks

For ballast water tanks located below the waterline, the seawater temperature can be assumed to be – 3 °C (27°F). For ballast water tanks located above the waterline in contact with the shell, the t_D is to be assumed. The required heating calculations for determining heat transfer should assume the seawater will be maintained at a temperature of 2 °C (36°F). Ballast water tanks on vessels with t_D equal to or above – 30 °C (– 22°F) but lower than – 10 °C (14°F) and arranged such that the top of the tank is located above the lightest operating draught of the vessel must be provided with arrangements to prevent the freezing of the ballast water. Acceptable arrangements include turbulence and convection-inducing systems, for example, bubble systems or heating coils. Class may consider other arrangements where there is a justifiable reason for having alternative arrangements. Vessels with t_D less than – 30 °C (– 22°F) should be provided with some means of heating. Tank heating calculations must be provided to show sufficient heat transfer capacity for the t_D. For ice class vessels, the capacity and arrangement of the water ballast tanks are to be sufficient to immerse the top of the propeller edges at least to the maximum thickness of level ice in the anticipated area of navigation. Further guidance may be sought from IMO Resolution MEPC.163(56) *Guidelines for Ballast Water Exchange in the Antarctic Treaty Area.*

7.3 Forward Areas, Navigational Bridge and Vessel Superstructure

7.3.1 Vessel Bow

The bow and forward area of the vessel must remain accessible for personnel in all weather conditions. Forecastles are recommended so as to deflect waves and water spray away from the deck area immediately aft of the bow. If a forecastle is not provided, the shell plating should be flared so to deflect waves and spray away from the main deck area immediately aft of the bow. The bow area is to be designed to protect anchoring and mooring equipment and operating personnel by some means of sheltered or enclosed spaces. A sheltered area is considered a structure enclosed on three sides with the side facing aft open to the weather.

7.3.2 Forecastle

Forecastles are recommended for vessels less than 50,000 DWT as well as for all bulk carriers, ore carriers and combination carriers in accordance with IACS UR S28, *Requirements for the Fitting of a Forecastle for Bulk Carriers, Ore Carriers and Combination Carriers.*

7.3.3 Navigation Bridge and Bridge Wings

Contemporary design practice is to enclose the bridge wings of vessels operating in low temperatures. If a designer elects not to enclose the bridge wings, all equipment (for example, the binnacles, propeller speed repeaters, etc.) are to remain fully functional. Heating for personnel working on the bridge wing is typically required. Access to the navigation bridge windows may be provided from the navigation bridge. The navigation bridge windows may be defrosted by the use of hot air. If all non-protected windows are considered blocked with snow and ice, the vessel's arrangements must still meet SOLAS Chap. 5, Regulation 22. All doors on emergency escape routes, and the navigation bridge doors, must be kept free of ice at all times. This can be achieved the by installation of heat tracing equipment or some similar method. The location of heated deckhouses is to be determined by the vessel designer or owner. Here, the intent is to provide personnel working on deck a place to shelter and/or warm up so as to be able to continue to perform their duties. The bridge wings should be enclosed or designed otherwise to protect navigational equipment and operating personnel and to permit operating personnel observation of the ice/hull contact (refer to Chap. 5) in accordance with the requirements of SOLAS, Chap. 5, regulation 22. Furthermore, external access to the navigation bridge windows must be provided to facilitate their cleaning and maintenance. Further information regarding the installation of catwalks and other navigational bridge structures to aid cleaning and maintenance are provided in *Human Factors in Ship Design and Marine Operations* (Olsen and Karkori: 2023). At least 50% of the total number of bridge windows must be heated and provided with window wipers and cleaning systems for de-icing purposes. The heated windows should be located in such a position so to maximise visibility assuming all non-heated windows are unusable. The window wiper blade material must be suitable for operation at the minimum anticipated temperature. Access to the accommodation spaces and work areas are to be provided with some means, such as two doors, which minimises heat loss from the heated space. Windows and side scuttles should be so designed as to also minimise heat loss. An effective example is the installation of double or triple glass panes. All doors and the adjacent deck areas on emergency escape routes and navigation bridge doors and the adjacent deck areas must be designed and arranged such that they can be maintained free of snow and ice so to be readily functional in accordance with the vessel's winterisation plan. Further guidance on escape routes is provided in Chap. 9. When personnel are required to perform functions such as a lookout when underway, or security at the gangway when in port, a heated deckhouse should be provided for sheltering against the elements.

7.3.4 Exterior Stairs

These requirements apply only to inclined stairs. They are not intended to apply to vertical ladders providing access to foam monitors, deck equipment, foremasts, etc. Many smaller vessels or vessels with restricted deck space require the use of stairs with higher angles than 35°. Exterior stairs that are adequately protected from the effects of spray or freezing rain as applicable for the area of operation may be considered at angles over 35°. All exterior stairs must be installed at a lower angle of about 35°. Stair step and landing material is to be technically consistent with the solution selected to maintain stairways free of ice and snow. Where no heat is applied to the stair step, a grating type of material is to be used. In all cases the walking surface of the stairs must be of high traction material.

7.3.5 Operating Platforms for Deck Equipment

Platform material is to be selected based on the intention to keep the platforms free of ice and snow. Grating will allow most snow to fall through and makes ice removal by manual impact means easier. However, grating may not be appropriate for heat tracing application. Solid top platforms will accumulate more snow and ice. Additionally, a rigid solid platform surface may be more difficult for the manual removal of ice but may offer significant improvement when heat tracing is used. Platforms for deck equipment material are to be technically consistent with the solution selected to maintain the platforms free of ice and snow. Where no heat is applied to the platform, a grating type of material is to be used. In all cases the walking surfaces must be of high traction material.

7.3.6 *Railings*

It is preferable to heat railings internally. When this is not possible, heating may be accomplished by installation of exterior heat tracing. For vessels with a **CCO** notation, or equivalent, there are currently no additional requirements (check with the vessel specific Class rules to confirm this is applicable). For **CCO-POLAR** notations, or equivalent, internal heating of railings exposed to low temperatures should be provided where practicable on all escape routes (refer to Chap. 5 for further discussion on emergency escape routes).

7.3.7 *Towing Fittings for Ice Class Vessels*

All ice-class vessels are required to have arrangements for being towed. Substantial fittings such as bollards should be fully incorporated into the vessel's hull structure and should be sized for the high dynamic loads that can be experienced when operating in ice. The towing fittings must be suitably marked with the maximum allowable load, the manufacturer, and any restrictions associated with their operation. Reference may be made to MSC Circ.1175 "*Guidance on shipboard towing and mooring equipment*". In some maritime areas (for example, the Baltic Sea which is subject to Finnish and Swedish Maritime Administration requirements), vessels may be provided with arrangements to secure the anchors onboard the towed vessel. This may be accomplished by swinging the anchor aft to prevent contact with the towing vessel, and particularly so for ice breakers with a notch stern. This may be accomplished with shackles and cables. In either case, emergency towing arrangements must be provided together with some form of suitable arrangements or equipment to provide de-icing capability.

7.3.7.1 Towing Notch

The design and characteristics of the towing notch and towing devices installed to ice breakers and ice class vessels are typically selected on the basis of the anticipated operation of the vessel when navigating in ice. The towing notch is a design feature that has been incorporated into the design of many Russian, Finnish and Swedish icebreakers but has not—to date—been included on any North American vessel. Icebreaking cargo vessels may also be equipped with a towing notch. The towing notch is installed on all Russian ice class vessels with the **ULA** notation, or equivalent. The towing notch arrangement enables a towed vessel to be drawn into the stern of the towing vessel and together the vessels work in tandem during ice breaking. This method has been used extensively in the Baltic and Russian Arctic and its effectiveness is well documented. Because of the possibility of damage to the towing notch itself, it is common to construct the notch as an appendage to the main hull

Fig. 7.1 Towing notch on *MV Polaris*

structure. In some operations, it is necessary for the icebreaker to press its bow against the towing notch of the cargo vessel. Such a method of joint operation of the icebreaker and icebreaking cargo vessel has proved to be successful in Russian practice and may be considered especially for large ice class cargo vessels as towing from the bow can prove problematic. To reduce the possibility of damage at the towing notch, including the bow of the towed vessel, various methods of fendering have been employed inside the towing notch itself. Both rubber and wood have been used successfully, whereas bumper wheels have been found to wear quickly. Due to the wide variety of hull shapes that are likely to be towed, the notch should be designed to accept both deep and shallow V shapes (Fig. 7.1).

7.3.7.2 Stern Towing Notch for Ice Breakers

When provided, towing notches may be constructed as appendages to the main hull structure. Fendering should be used to reduce the possibility of causing damage to the vessel's stern at the towing notch and should include provisions for protecting the bow of the vessel under tow. The notch must be designed to accept various bow shapes.

7.3.8 Cargo Handling

Refer to Chap. 8 for guidance relating to the requirements for hatch covers, ramp doors and side doors, as well as the requirements for deck cargo securing equipment.

7.3.9 *Deck House Insulation*

Additional insulation will typically be required in the deck house superstructure to afford some level of protection and comfort against the elements.

7.3.10 *Access to Machinery on Deck*

Access to the deck for personnel (for example, doors, hatches, etc.) and to other equipment located on the main deck must remain accessible at all times. Access by the operating personnel must not be obstructed by deck plate level ice accumulation or otherwise endangered by overhead ice accumulation.

7.4 Stability

Ice accumulation is most probable in the forward sections of the vessel; however, smaller vessels can experience substantial icing thicknesses across all areas. This can present a significant safety hazard, as the increase in topweight (aggravated by changes in trim) reduces the stability of the vessel. Stability information onboard should contain sufficient data on the vessel's stability taking account of potential icing to allow masters to recognise such risks as they arise and to respond appropriately to address these situations. The *IMO Code on Intact Stability* requires an allowance for ice accretion for certain vessel types. Other Flag State Administrations (and Class) may have additional requirements for addressing ice accretion. One effective solution is the use of ice release coatings which can be applied to exposed vessel structures to minimise ice accretion and thereby reduce the effects of ice accretion on vessel stability. Whether or not the vessel is equipped with additional ice and snow removal facilities, as a minimum, the vessel must be provided with equipment for de-icing. Any freeing ports in accordance with the *International Load Line Convention 1966* must also be equipped with some means to remain operational irrespective of the risk of icing. Furthermore, weather decks are to be designed to avoid the accumulation of stagnant water such as in wells or in the form of small pools of water between structural stiffening. Should icing conditions develop, personnel should take all steps to reduce and remove ice. This may require a continuous procedure until icing conditions subside. If ice build-up cannot be controlled or removed, the vessel must be moved to a sheltered area whenever possible. Figure 7.2 is a grouping of icing nomographs developed by the United States National Oceanic and Atmospheric Administration (NOAA) from actual icing reports from fishing, US Coast Guard and towing vessels operating in Alaskan waters. These reports were based on icing events that lasted anywhere from 1 to 26 h but averaged 3 to 6 h. The NOAA National

Weather Service Environmental Modelling Centre provides an online forecast for ice accretion at http://polar.ncep.noaa.gov/marine.meteorology/vessel.icing/

Figure 7.3 provides guidance for ice accretion versus wind velocity for air temperatures ranging from – 34 °C (– 30°F) to – 7 °C (20°F). Icing may only occur when there is a source of water for wetting the deck, superstructure and other exposed parts of a vessel. Some vessel factors to consider are vessel's speed, heading (with respect to wind, waves and swell), vessel length, freeboard height, handling and the cold soaking aspect (cold soaking occurs when a vessel has been in cold conditions for two to three weeks and the hull structure remains cold for a period of time after the vessel operates in warmer temperatures). For the same environmental conditions there will be more sea spray reaching the vessel deck, superstructure, etc. when the vessel is traveling faster, into the wind and waves, and for smaller vessels and ships with less freeboard. Water waves result from the wind blowing over a vast enough stretch of the ocean. Some waves in the oceans can travel thousands of miles before reaching land. The threshold significant wave height, $h_{1/3}$, and associated

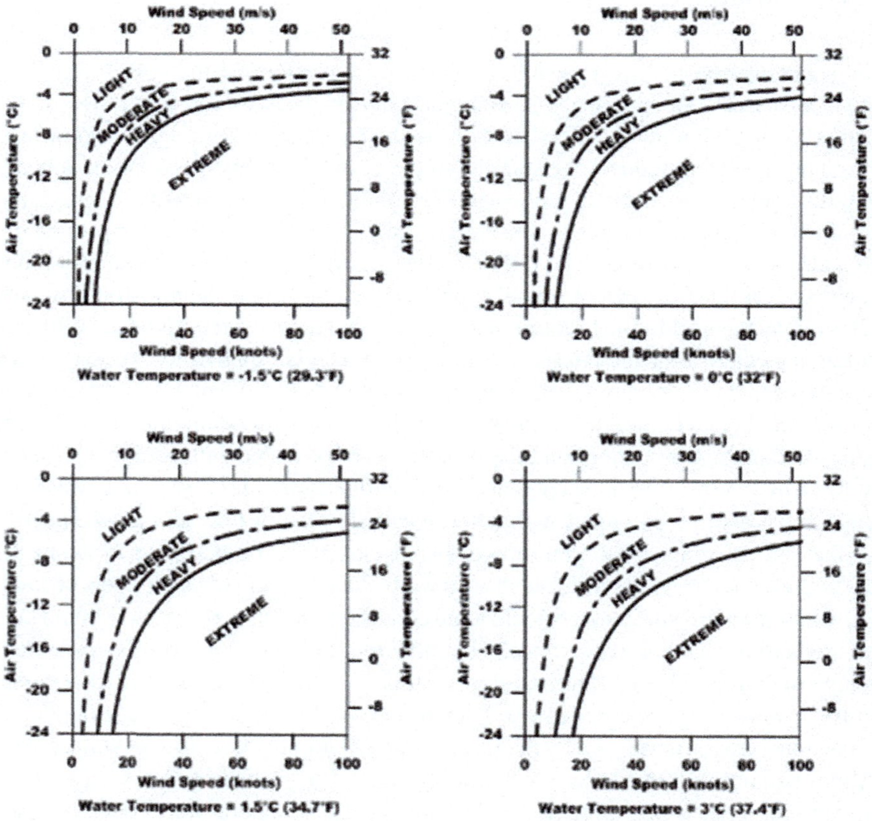

Fig. 7.2 Icing conditions for vessels into or abeam of the wind. *Source* Guest P et al. (2005) Vessel Icing, Mariners Weather Log, Volume 49, No. 3

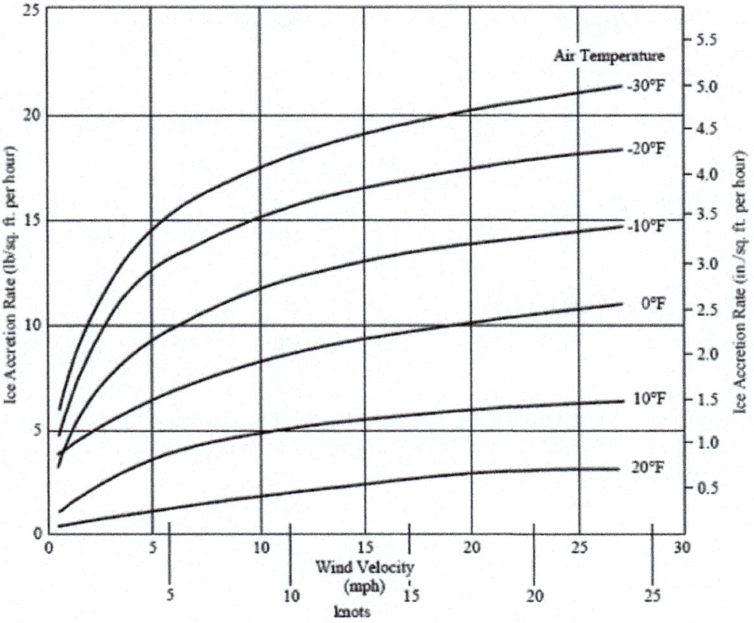

Fig. 7.3 Ice accretion versus wind velocity for six air temperatures. *Source* U.S. Navy Cold Weather Handbook for Surface Ships (May 1988) Accreting Surface: Flat Panel; Water Spray Temperature: 41–48°F

wind speed, for a 200 km (124 mi) fetch (the distance over water which the wind blows to generate water waves) at which enough sea spray reaches the decks and superstructures to cause severe icing, assuming frigid air and water temperatures are also present, are listed in Table 7.1.

Icing class	None	Light	Moderate	Heavy	Extreme
Icing rates					
(cm/hour)	0	< 0.7	0.7–2.0	2.0–4.0	> 4.0
(inches/hour)		< 0.3	0.3–0.8	0.8–1.6	> 1.6

7.5 Anchor Chain

The anchor chain is to be delivered to the vessel manufactured with impact resistant properties based on the design service temperature in compliance with the applicable requirements discussed previously in Chap. 6.

Table 7.1 Threshold wind speeds for icing to occur on various length ships

Parameter						
Vessel length						
Metres	15	30	50	75	100	150
Feet	49	98	164	246	328	492
Significant wave height—$h_{1/3}$						
Metres	0.6	1.2	2.0	3.0	4.0	6.0
Feet	2.0	3.9	6.6	9.8	13.1	19.7
Wind speed at 200 km (108 nm) *fetch*						
Metres/second	5.0	7.4	9.8	12.5	15.0	20.0
Knots	9.7	14.4	19.0	24.3	29.3	38.9

Source Guest P et al. (2005) Vessel Icing, Mariners Weather Log, Volume 49, No. 3, Table 1. Based on Overland, J.E., (1990), Prediction of Vessel Icing for Near-Freezing Sea Temperatures, Weather and Climate, pp. 5, 62–77

Chapter 8
Arctic Vessel Systems and Machinery

8.1 Introduction

In the previous two Chaps. (6 and 7) we examined the main Class Rules associated with the structural design and construction of vessel hulls operating in Arctic and extreme low temperature environments. In this chapter, we will instead consider guidance relating to the management and safe use of Arctic vessel systems and machinery when operating in extreme low temperature environments.

8.2 Systems and Machinery Care

8.2.1 Anti-icing and De-icing

Icing can be a major issue for vessels that operate in low temperature but open water conditions. These conditions are common around northern Norway, off the east coast of Canada, and on several other shipping routes and fishing grounds. In ice-covered waters, icing is less of a problem, as wave-induced spray is the main source of thick ice accumulations. Ice build-up is most probable in the forward parts of a vessel, however smaller vessels can experience significant icing thicknesses across all areas. Spray, fog, freezing rain and snow can all cause sufficient icing to render equipment inoperable or extremely difficult to operate. Subsequently, measures should be taken to manage icing issues. Icing of safety-related equipment exposed working areas and access routes can also lead to safety hazards, while icing of deck machinery, valve manifolds and other systems can interfere with cargo handling operations, leading to delays and economic loss. Depending on the operations of the vessel, de-icing or more permanent heat tracing equipment should be considered. Vessels intended to operate for extended periods of time in Arctic climates should be designed with

A. Olsen, *Ship Operations in Extreme Low Temperature Environments*, Springer Series on Naval Architecture, Marine Engineering, Shipbuilding and Shipping 19,
https://doi.org/10.1007/978-3-031-52513-1_8

permanently installed equipment. Those vessels which normally operate in open oceans and only occasionally experience extreme conditions should be provided with de-icing equipment. Smaller vessels have been equipped with pneumatically operated rubber pillows to break ice away from superstructure fronts and masts, reducing the risk of loss of stability due to ice build-up. However, for larger vessels, portable de-icing equipment is normally preferred. Care is to be taken to provide that any de-icing equipment used is suitable for the location, particularly in hazardous locations. It is imperative to account for the required heating load early in the design cycle. Determining the level of protection required to maintain an area ice free requires an understanding of the weather conditions the vessel is expected to encounter such as temperature, wind velocity and humidity. Equipment vendors can provide design manuals and engineering advice for the design of de-icing and heat tracing equipment. Unfortunately, it is not practical to design a heat tracing system that is suitable for all weather conditions as the power requirements would be excessive. Likewise, the system should not be designed solely on averages as it will likely be underpowered. Accordingly, a decision must be made by the designer about what percentage of snowfall/icing hours the system can keep heated surfaces clear.

8.2.1.1 Design Approaches

Design approaches for de-icing and heat tracing equipment include steam, thermal oil, electric resistance heating or hot air. Combinations of these approaches are installed for some applications. Often the desire is to employ anti-icing systems designed for the extreme condition but, this is often impractical. An optimal design would be one that can function as an anti-icing system most of the time. A risk assessment may be used to establish the time intervals acceptable for the specific vessel. This will enable the crew to focus on other duties most of the time rather than ice removal. In severe conditions operations may be suspended and crew resources can be applied to manual ice removal and the anti-icing system used as an aid to de-ice. A time-based definition for de-icing has been considered by several regulatory bodies in the past but has proven an ineffective measure. It is therefore suggested that de-icing systems and procedures be evaluated based on the level of risk involved with the system being de-iced.

8.2.1.2 Steam Systems

Equipment for de-icing includes steam generators, steam hosing, hot water and even saltwater spray. Steam tracing may be considered if steam will be used onboard for cargo heating as in an oil carrier. Steam is extremely effective but requires training to be used safely. As with all fluid lines, de-icing systems should be kept heated or drained when not in use, particularly for the more remote branches. Availability of personnel experienced in operation of steam boilers may be another factor in the decision to install a steam system. Pumps used for de-icing systems should be provided with redundant arrangements. Any equipment used for de-icing must be suitable for operation onboard, particularly when operating in hazardous areas on oil carriers, chemical carriers and liquefied gas carriers.

8.2.1.3 Thermal Oil

Thermal oil is a substitute for steam systems as the oil will not freeze in low temperatures if suitable oil is chosen. However, the thermal capacity of oil is less than water, so a greater volume of oil is required to be pumped. Protection from release of oil into the environment must be considered with thermal oil systems.

8.2.1.4 Electric Resistance Heating

Electric resistance heating can be used for several applications such as exposed piping, tank heating, stairways, deck access and handrails. It can be used in cabin reheat units and areas where personnel regularly perform vessel operations. Portable units can be provided for localized areas where maintenance is performed or to meet the needs of personnel. Handrail heat tracing system design philosophy considers de-icing of those handrails in areas subject to emergency evacuation or high foot traffic to avoid wasting heating resources. There are standards available such as IEEE 515.1, "*IEEE Standard for the Testing, Design, Installation, and Maintenance of Electrical Resistance Heat Tracing for Commercial Applications*" and ASHRAE's handbook—HVAC Applications, Chapter 45 "*Snow Melting*". ASHRAE 926-RP "*Development of Snow Melting Load Design Algorithms and Data for Locations Around the World*" is another resource. For applications in which the standards are unable to provide information, computer modelling and analysis may be performed.

8.2.1.5 Hot Air

Portable hot air generators can also be used for de-icing purposes. These units, if used, can be electric, or can operate on diesel fuel oil with a self-contained fuel system. These systems have also been effectively used for heating enclosures or shelters when exterior work is required.

8.2.1.6 Manual Removal

Even where specialised de-icing equipment is carried onboard, manual means to remove ice and snow build-up should also be considered. This includes the use of mallets, shovels, axes, etc., and in some cases snow blowers and other mechanical equipment. Care must be taken in the operation of all these so that onboard fittings, valves and electrical components are not damaged. Storage of all de-icing equipment should be such that items can always be accessed, and that if located within exterior compartments, their access is itself protected from freezing.

8.2.2 Ice Prevention and Limitation

There are also many methods that can be used to prevent icing or to limit its effect. Some common anti-icing methods that have been used successfully include protective locations/covers with or without heating; bow form; electric heat trace wiring; heating coils; steam generators; ice resistant coatings; self-draining piping; and the circulation of media (e.g., hydraulic oil). Arrangements should be made to prevent the freezing of all exterior moving equipment. This includes installing heat tracing on bridge and critical windows, davits, deck equipment, lifeboats and other personnel safety and rescue equipment, hatch and door seals, etc. Tracing systems may include electric wired tape, hot water, steam and heated glycol, although water-based systems must be kept in operation to avoid freezing themselves. Alternatively, protective covers can be a cost-effective option to protect fire hydrants, mooring equipment and other operational components exposed on open decks. Although the cover itself does not prevent ice accumulation onboard, it may provide local protection around fittings, valves and controls where ice removal may be more difficult or may risk damage to the equipment. Special consideration may be warranted for exterior electronics equipment. Communication transmitters and receivers may require anti-icing features to provide continual functionality, although whip type antennas can usually be de-iced with the strike of a wooden mallet or else shaken to remove ice build-up. Other communications antennas with horizontal surfaces or dish shaped configurations may require built-in heat elements. Exposed rotating radar scanners normally require no specific measures, even at extremely cold temperatures, due to internal heating elements. However, the smaller enclosed type arrays can become encrusted with ice and can be difficult to de-ice due to their inherent fragility of construction. Consultation with the manufacturer is strongly recommended.

8.2.3 Heat Tracing

Heat tracing installations must comply with the requirements discussed in the section *Fire safety system requirements*. However, requirements related to heat tracing of various systems referenced elsewhere in this chapter (and throughout this book in general) are optional but are to be in accordance with the vessel's winterisation plan. The vessel's superstructure surfaces and systems are expected to be heat traced or de-iced in accordance with the vessel's winterisation plan, considering all open deck areas; gangways; stairways; superstructure areas subject to icing; and railings. A test programme for de-icing and heat tracing is usually required to be submitted to Class for engineering review and subsequent use by the Class approved surveyor.

The capability of the de-icing and heat tracing systems to function at the minimum anticipated temperature is to be demonstrated by onboard testing or calculations. Calculations should be in accordance with—as a minimum—recognised standards or the manufacturer's code of practice. These must account for anticipated snowfall in operation; ambient temperatures (applied as a minimum anticipated temperature); wind velocity in the vessel's area of operation; and water source. The factors, for which the heating power capacity is selected, should be clearly identified.

8.3 Hydraulic Systems

Hydraulic oil sumps are required with heaters or other suitable means for heating purposes, where necessary and should be capable of circulating to avoid cold spots. The use of steam heating coils is not permitted. Where used, the hydraulic oil must be suitable for the minimum anticipated temperature. If the power unit and hydraulic lines are in a heated space, a heater is not considered necessary. The effect of thermal expansion and the contraction of fixed lines must be considered and incorporated into the system design.

8.3.1 Overboard Discharges and Drainage

The use of grey type cast iron material for piping, valves and fittings is prohibited. Moreover, drains must be protected from freezing over. Direct steam or compressed air connections to the sea chests should be provided for de-icing. For vessels with a t_D below – 30 °C (– 22 °F), steam or heat tracing should be provided on all overboard discharges to prevent freezing. Deck drains should be adequately protected from freezing over or else provided with some means to enable blockages caused by ice and snow to be quickly cleared.

8.3.2 Lubricating Oil Systems

Lubricating oil must be maintained at the proper minimum temperature in accordance with the manufacturer's recommendations to allow for safe equipment and machinery start-up. The use of steam heating coils within lubricating oil tanks is prohibited. The lubricating oil for the main propulsion and auxiliaries, particularly for idle equipment, should be maintained at a minimum temperature to permit equipment start-up when demanded. Current practice is to operate a standby pump along with the lubricating oil purifier, which is fitted with a heater to keep the oil at the proper temperature. Steam coils in lube oil sump tanks are difficult to inspect regularly and have been known to leak, rendering the equipment inoperable. Synthetic lubricants, which have the correct viscosity at low temperatures (i.e., – 45 °C (– 50 °F)) without the use of heaters, may be considered by Class on a case-by-case basis as an alternative or supplement to heating coils.

8.4 Prime Movers

The vessel's propulsion plant must be capable of continuous operation at low power outputs and/or low vessel speeds. If icebreaking vessels are likely to operate at or near their continuous ice breaking limit, their operation will be adversely affected by ridges, rafted ice and pressured ice conditions, meaning they may have to back and ram. The propulsion system should therefore be capable of rapid and repeated variation of power and direction; and should be able to develop a significant percentage of ahead thrust when moving astern. Ice class vessels following icebreakers should also be capable of rapid and repeated variation of power and direction to avoid colliding with the icebreaker or other vessels when operating in convoy. Manoeuvring mode is some fraction of the full ahead speed of the vessel. Ice-strengthened and ice breaking vessels have been designed with all the commonly fitted propulsion plants, including direct drive and geared diesel, steam and gas turbines, and electric transmission versions. Russia operates a nuclear ice breaker fleet, but nuclear power is considered highly unlikely for a commercial application in the near future and therefore is not specifically discussed within this book. The overriding consideration for the propulsion plant for any ice-capable vessel is that its power and thrust are matched to the ice conditions in which the vessel is intended to operate. In some areas, currently including the Baltic and the Russian Northern Sea route, minimum powering requirements are imposed on vessels as a condition of ice class. In particular cases, model tests may be used as an alternative to demonstrate adequate capability to an Administration. For non-ice class vessels, this book assumes that the propulsion plant will be operated continuously with few changes in speed and direction when operating in extreme low temperature environments.

8.4.1 Prime Mover Operating Characteristics for Ice Class Vessels

Much ice breaking takes place at extremely low speeds, and so it is necessary to consider bollard as well as free running conditions in selecting propeller properties during design. Bollard and low-speed thrust can be augmented by fitting nozzles, which can also provide protection to the propeller blades. However, nozzles can be blocked or clogged by ice, in which case, rapid reversal of thrust is needed to clear the blockage. Vessels following icebreakers will also have to reduce their speed during ice breaking operations. Low power outputs for propulsion plants are often in the range of

15 to 50% of maximum continuous rating (MCR). During ice breaking, power output would be expected to be in the lower part of this range. There are various operational issues associated with low load operation to consider such as: incomplete combustion of fuel oil and subsequent effect on turbochargers and exhaust piping system; continuous operation of auxiliary blowers for prolonged periods of time; effects on internal components of the prime mover; prime mover emissions; and functioning of auxiliary systems. Icebreakers, ice breaking support vessels, and higher ice class vessels as well as other smaller vessels are most likely to see frequent propeller-ice interaction. This effect places a range of demands on the propulsion system in addition to ensuring adequate component strength. Propellers may be slowed or jammed by ice blocks caught against the hull which is mitigated by increased propeller blade thickness and propulsion shaft diameter. The high torque available from certain electric transmission systems can be very advantageous for fixed-pitch (FP) systems. For controllable-pitch (CP) systems, rapid pitch changes may be required.

8.4.2 Combustion Air Systems for All Vessels

The combustion air system is to be based on the minimum anticipated temperature of the outside air. Some form of means should be provided to pre-heat the combustion air to ensure the proper functioning of the main propulsion, auxiliary and emergency generator internal combustion engines in accordance with the engine manufacturer's recommendations. For **CCO-POLAR** notations and equivalent, the combustion air is to be led directly from the exterior of the vessel to the engines (where having a rated power of 100 kW (135 HP)) and over by way of ducting or other suitable means such to prevent the combustion air from being drawn directly from the machinery space. The combustion air intakes must be located on both sides of the vessel and arranged in such a manner so to prevent the recirculation of engine exhaust gases into the combustion air intakes; ice accumulation; and blockage of the duct. The installation of a heat tracing system at the combustion air intakes is acceptable provided calculations confirming the adequacy of the heat tracing system are submitted and approved by Class.

8.4.3 Combustion Air for Internal Combustion Engines

Vessels operating in low temperature environments are at risk that the engine may not operate as the low temperature of the combustion air-fuel mixture may fail to enable auto ignition when compressed in the engine cylinders. Engine cylinder pressure limits may be exceeded when using low temperature air because of the cold air's higher specific density. Protection can be achieved by installing a waste gate (blow-off valve) for either scavenge air or exhaust gas. This may lead to operating restrictions for the engine. Specific features are incorporated within the combustion air system to preheat this frigid air. Some vessel designs are equipped with electric/steam heating to preheat the combustion air. Recent designs now use the diesel jacket water waste heat. Another strategy is to have an emergency air intake directly from the machinery space. It is not practical to have all combustion air taken from the machinery space because the space's temperature will become too low, affecting equipment function and personnel. For typical diesel engines operating at reduced loads (e.g., low cylinder compression temperature), it is likely that preheated combustion air will be required. Two-stage charge air cooling/heating systems have been employed to raise the combustion air temperature for acceptable engine operation. For engine loads above 30% MCR, jacket water can be used as the heating medium. For engine loads below 30% MCR, external preheat sources may have to be installed. Typically, at engine loads near 50% MCR, charge air temperatures will rise above 0 °C (32 °F) and the charge air cooler can be employed. The local operational conditions and outside temperatures should also be considered for the valve settings. Additionally, a relief valve may be installed, designed to activate (blow off) at a specific gauge pressure (typically around 2 bar). Some engine manufacturers allow the charge air blow off to be re-circulated back into air inlet ducting.

8.4.4 Combustion Air for Other Prime Movers

Combustion air for other prime movers, such as steam plants and gas turbines, should be provided in accordance with the manufacturer's recommendations. The details and arrangements will usually be subject to special consideration by Class. The manufacturers of steam plants and gas turbines should be consulted regarding any operational restrictions imposed by low temperature conditions and for recommendations related to system arrangements and features to mitigate these restrictions.

8.4.5 Turbochargers for All Vessels

Turbochargers and the associated combustion air system for internal combustion engines are to be designed to obtain surge-free operation throughout the range of ambient air temperatures in which the vessel is expected to operate. The turbocharger's pressure-volume curve with reference to the diesel engine's operating characteristics must be available. If surging is expected, special blow-off air intake systems may have to be engineered to obtain surge-free operation of the turbochargers over the entire range of operations for low design ambient temperatures. A blow-off valve installed in the charge air manifold of the engine can be used to blow off excess air to prevent turbocharger surging.

8.4.6 *Emergency Generator Starting*

In low temperatures the emergency generator is essential for survival. A self-contained emergency source of electrical power is to be provided so that in the event of the failure of the main source of electrical power, the emergency source of power will become available to supply power to services that are essential for safety in an emergency. The two sources of energy for starting the generator are to be selected based on minimising the single failure mode attributed to cold temperatures. Further provisions apply to passenger vessels. A self-contained emergency source of electrical power includes the prime mover and its starting equipment, generator, fuel tank, emergency switchboard, associated transforming equipment, if any, transitional source of emergency power, if applicable, and emergency lighting switchboard and associated transformers, if applicable. Attention is directed to the requirements of governmental authority of the country, whose Flag the vessel flies, for emergency services and accumulator batteries required in several types of vessels.

8.5 Propulsion and Manoeuvering Machinery

The guidance provided in the section *Strengthening for navigation in ice* applies a progressive strength approach to the propeller and propulsion shaft system. The philosophy of this approach is for the propeller blade to be the weakest link and the tail shaft the strongest. The relative strength of the components is diagrammatically shown as follows:

Propeller blade	Weakest
Propeller blade palm bolts	
Propeller hub	↕
Propeller hub flange and flange bolts	
Tail shaft	Strongest

Propeller damage to ice class vessels is common and can result from both encounters with heavy ice and/or from operator error. The progressive strength approach is intended to lessen the frequency of catastrophic failures of tail shafts and propeller hubs from occurring. The transmission system that delivers the power from the prime movers to the propulsors must also be able to cope with the demands of ice operation, which can include high dynamic loads due to ice impact and ice milling, vibration, and frequent reversals or rapid changes in rotational speed. These considerations are most important for smaller, higher powered and higher ice class vessels, but require some attention in any ice class vessel. Supporting structure for thrusters should be designed to withstand ice impact in addition to other operating loads imposed on the structure. When thrusters are used for dynamic positioning, either azimuthing thrusters or variable pitch thrusters may be installed. The maximum power and maximum thrust may be developed at zero speed for station keeping purposes. For conventional hull type vessels, dynamic positioning systems are commonly direct drive. Typical drillships power systems are diesel-electric with available power capabilities usually exceeding the power requirements for transit conditions.

8.5.1 Propulsion Shafting Bearing Lubrication

Oil-lubricated bearings. Vessels with mineral-based oil-lubricated stern tube bearings should be provided with a pollution free oil-seal gland.
Biodegradable oil-lubricated bearings. Vessels with stern tube bearings lubricated with biodegradable lubricant may be fitted with a pollution-free oil-seal gland. The requirements relating to *Lubricating oil systems* discussed earlier for providing lubricating oil heating arrangements are also applicable.
Water-lubricated bearings. Vessels may be fitted with water-lubricated bearings. The bearings should be suitable for continuous operation in low temperatures.
Shaft seal type. The final selection of shaft seal type, radial or axial, must consider the maximum diameter of the shaft itself and the speed of rotation.

Damage statistics indicate that problems with tail shaft seals have immobilised more vessels than any other single failure associated with vessel propulsion systems. Both radial lip and axial face type seals have been used onboard vessels transiting through ice-covered waters. Radial lip type seals have been the preferred installation to date. Although axial face seals can accommodate larger radial and axial movements of the shaft without losing seal efficiency, the enormous size of shafts found onboard some icebreaking vessels tends to make these more complicated and expensive compared to radial lip seals. Radial lip seals are, however, more sensitive to shaft speed, surface finish, shaft eccentricity, shaft vibrations and pressure differences across the seals themselves. The allowable rubbing speed limit is to prevent high lip surface temperatures that could result in seal failures and is commonly set to below 6 m/s, although slightly higher allowable speeds have been recently installed on some vessels. Oil-lubricated bearings with pollution free sealing arrangements are seals with more sealing surfaces, or operate at reduced oil pressures, or a combination of the two. Final selection of any lubricant used for stern tube bearings should always be verified with the bearing and sealing manufacturer, as the material characteristics of the contact surfaces is of paramount importance for correct operation.

Water-lubricated stern tube bearings have also been proven as a pollution mitigation strategy and have been successfully used onboard Canadian and Russian ice breaking vessels. The selection of the shaft sealing arrangement should also consider the possibility of damage and the repair or replacement of the unit itself. Consideration should be given to the installation of split type seals.

8.5.2 Fixed Pitch Propellers for Ice Class Vessels

Whilst solid propellers are permitted for use on vessels operating in ice waters, damage to ice class vessel propellers is common, therefore consideration should be given to the use of propellers with detachable blades for ease of repair. To reduce the probability of damage to the propeller blades, especially for CP propellers, it is advisable to operate with the top of the propeller edges below the maximum thickness of level ice.

8.5.3 Controllable-Pitch Propellers for Ice Class Vessels

The CP system should be designed to account for the numerous pitch reversals to be expected when operating in ice conditions. Calculations confirming the adequacy of the CP system to withstand the additional pitch changes or service experience from the manufacturer are to be submitted for information. For any hydraulic components exposed to low air temperatures, the requirements discussed earlier under *Hydraulic systems* are applicable. Table 8.1 lists some of the pros and cons for the consideration of installing a CP propeller versus a FP propeller.

8.6 Deck and Other Machinery

The deck and other machinery discussed in this section which are exposed to low temperatures or located in unheated spaces must be deemed suitable for operation at the minimum anticipated temperature. Note that piping and other machinery in cargo holds will be considered exposed to the t_D when hatch covers, ramp doors and side doors are opened to load and/or discharge cargo.

Table 8.1 Considerations for installation of CP versus FP propellers

Pros	Cons
Operating conditions vary widely, and maximum thrust is desired throughout these operating conditions	Higher initial cost for more complex equipment
Extensive low-speed manoeuvring is required for a diesel-powered vessel. Thrust can be varied continuously from ahead to astern, including zero thrust while operating in the minimum speed range of the diesel	When constant shaft rpm is required over a wide range of operating powers the propeller pitch can be adjusted. However, a propeller efficiency penalty is incurred when operating the propeller off the design point
Unidirectional rotation of a CP propeller subjects the blades to less ice damage because the leading edges of the blades are thicker and stronger than the trailing edges. Fouling of propeller blades by ice blocks is less likely with continuous propeller rotation	Greater equipment complexity (e.g., hydraulic system, controls systems, additional moving parts in propulsion shafting and hub) requires additional maintenance and increases frequency of failure
Improved manoeuvrability and a minimum vessel stopping distance. Variable thrust capability in either direction and more rapid response to thrust reversal commands improve vessel manoeuvrability and reduce the vessel's head reach	
Individual blades can be replaced when damaged	

Source Harrington, Roy L. Editor, (1992) Marine Engineering, Society of Naval Architects and Marine Engineers

8.6.1 Anchoring Arrangements for Ice Class Vessels

Ice breaking vessels. The locations of the anchors are to be considered with respect to the waterline and anticipated ice conditions. Consideration is to be given to the height of anticipated ice ridges and rubble; and whether close towing operations will be conducted.
Ice class vessels. The locations of the anchors are to be considered with respect to towing operations with the ice breaker.

Ice breaking vessels. While operating in ice-covered waters, the towing of vessels is frequent, either to supplement powering capability or when a vessel suffers propulsor damage. Ice breakers and other ice breaking support vessels are most likely to conduct towing operations, but all ice class vessels should be designed to be towed on either a longer line or 'in the notch,' depending on practices in the anticipated area of operation.

Ice class vessels. When ice class vessels are towed 'in the notch' of the ice breaker, the towed vessel's anchors must be lowered, and the anchors moved aft and secured to be clear of the ice breaker's stern. Other vessels have arrangements in which the anchors are in an enclosure over the hawse pipe.

8.6.2 Anchor Windlass for All Vessels

The anchor windlass design should include these additional considerations:

- It should be suitable for operation at the minimum anticipated temperature.
- Anchor releasing arrangements are to be provided in such a manner as to reduce the effects of icing.
- The Class requirements which apply to all electrical components of the windlass.
- The Class requirements which apply to all hydraulics associated with the windlass; and
- All windlass components subject to tensile stresses are to be constructed of materials subjected to impact testing.

The deck machinery for the anchoring systems of ice breakers and ice class vessels can be either windlasses or capstans. Windlasses are fitted with a horizontal shaft used for raising and lowering the anchor while capstans are fitted with a vertical shaft to perform the same function. Anchor windlasses exposed to the weather can be expected suffer from coating of ice from saltwater spray and rendered inoperable if heating or some other protective device or design feature is not provided. It is preferable to have machinery fitted under cover; hence capstans offer advantages inherent in their design. It is easier to have the machinery associated with the capstan located below deck and the wildcat or wildcat and barrel above deck. Both windlasses and capstans can be located entirely below decks, but the height required for the wildcat is chosen so that the cable is self-stowing in the chain locker, requiring that the wildcat itself be located on an open deck. Power units located within heated spaces do not require oil sump heating. If windlasses or other equipment are located on the exposed deck, the remote release of the cable and anchor from the wheelhouse should be considered, shortening the time that crew members must spend on the open deck. The effects of icing on anchor release must be addressed and mitigated. For example, can a build-up of ice in the hawse pipe affect the ability of the anchor to be released into the water? Can icing of any of the rotating components on the anchor windlass affect release? What design modifications or operational changes would be required to mitigate the effects of ice? One way of responding to these questions is to fit a closed-circuit TV (CCTV) camera for monitoring the functioning of anchor and cable, released length of cable and monitoring of sea conditions. However, the crew must still prepare the anchor for release. Sufficient supplies of suitable cold-weather clothing and, where possible, protected access and shelter stations should be provided. Steam-driven machinery is not recommended due to inherent problems associated with the piping systems; the need to provide for proper draining of all lines and the consequent freezing and cracking of pipes. Stressed parts of machinery should consider the t_D if exposed on open decks. It is always preferable to encase moving parts, such as reduction gears and brakes, to prevent icing or freezing spray build-up. Wildcats may be fitted with covers. In some cases, the rotating parts may be provided with heaters. The bearings installed in motors/hydraulic pumps should be protected with suitable packing to prevent water/moisture entrapment (Figs. 8.1 and 8.2).

8.6.3 Towing Winch

If a vessel is expected to undertake regular towing operations, a towing winch is usually required. The design, construction and testing of the towing winch should conform to an acceptable standard or code of practice. To be considered acceptable, the standard or code of practice is to specify the criteria for stresses, performance, and testing. Moreover, towing winches are to be of the electric or electro-hydraulic type, single drum, constant tensioning, and fitted with an automatic spooling gear. The winch must be controllable from the navigation bridge and an emergency override is to be positioned near to the winch itself. The design load of the towing winch is to be based on the available total thrust of the vessel. The sizing of towing winches should be based on the available thrust for the ice breaker concerned. In some cases, ice-strengthened vessels are capable of towing other vessels. Standard equipment today for towing in ice is the use of a traction winch. Depending on the ice breaker supporting the escorted vessel, the observed pull on the towing line will be some fraction of the ice breaker's available thrust. Historically, Russian ice breakers tended to have high brake traction settings in the magnitude of 80% of available thrust. However, most other ice breakers operate within the 20–50% range. To limit dynamic loads originating during close towing of a vessel in ice, the towing gear is to be provided with a damping device. On Russian ice breakers there is a hydraulic damper which serves this purpose. If a vessel is expected to undertake regular towing operations, it should be outfitted with a towing winch. Protection for the winch from the elements is recommended, including arrangements for the passage of the tow line. During towing, the winch, pre-set for a given tow pull, will reel in or pay out on the drum as the pull on the cable varies. This line action may occur quite suddenly if ice breaks, and the vessel's crew should be trained to respond to this specific hazard. The haul in rate for most ice breakers is in the order of 9m/min (29.5 ft/m), with reel out approaching double that of the haul in (Fig. 8.3).

8.6.4 Towing Lines

Each vessel is to be capable of passing an emergency towing line to the towing vessel by a line throwing appliance or similar device. A "slip" should be provided to allow the towing hawser to be disconnected rapidly from the vessel in the event of imminent danger. The towline should, as for all mooring arrangements, break before damage to the towing winch takes place, notwithstanding the constant tension features of the winch. Towlines in ice-covered waters are shorter than those often seen for open water tows. These shorter tows are conducted to prevent the channel formed by the ice breaker or ice-strengthened vessel from quickly closing and increasing the resistance of the towed vessel. Towlines fall between 200 and 500 m (656 and 1640 ft) in length, keeping the wire length to a minimum on the drum. This minimises cable wear resulting from the outer windings depressing on the inner windings during a towing operation. The number of windings on a drum should not exceed 4 or 5 for any vessel expected to conduct regular towing operations.

8.6.5 Towing Fittings for Ice Class Vessels

Any vessel that is intended to operate in ice-covered waters should have some arrangements for being towed. Whether towed in close tandem configuration or at a distance in a more separated mode, substantial fittings such as bollards should be well incorporated into the hull structure and should be sized for the high dynamic loads that can be experienced in ice. Towing fittings are to be marked with their Safe Working Load (SWL). IMO MSC/Circ.1175, *Guidance on Shipboard Towing and Mooring Equipment* provides additional information. Anchoring systems should be provided with an independent means of securing the anchor so that the anchor cable can be disconnected for use as an emergency towing bridle. When this is accomplished, the anchor chain is brought on deck and led out the forward chocks. The storage of the anchor has always been an issue with the tow hook-up. Previously, the anchors were lowered down onto the deck of the towing vessel. However, recent recommendations from Baltic administrations indicate that a vessel should have arrangements to secure the anchors onboard the towed vessels. For some vessels, it may become necessary to swing the anchors aft to prevent contact with the towing vessel for ice breakers with a notch stern. This may be accomplished with shackles and cables. Another method to secure a towed vessel is to simply secure the towing line onto the forward bollards. Most vessels are fitted with capstans or other systems such as warp ends for line hauling. However, if towing in ice is anticipated, it is advised to have a bow roller fitted to accept the towline. A bridle is preferred to a single line hauled onboard and it is recommended that this be included within the vessel's own equipment.

8.6.6 Cargo Handling for All Vessels

Cargo handling equipment that may be affected by low temperature environments include cranes, hatch covers, ramp and side doors, cargo pumps and deck cargo securing equipment.

8.6.6.1 Cranes

Cranes and any other lifting devices exposed to low temperatures are to be certified in accordance with the *ABS Guide for Certification of Lifting Appliances* (or vessel specific Class Rules) at the minimum anticipated temperature. In addition, the operator's cab is to be able to be heated to 20 °C (68 °F) at the minimum anticipated temperature. If the crane is hydraulically actuated, the hydraulic oil should be suitable for the design service temperature, including an electric heater within the oil sump. If the crane is not controlled from an operator cabin, it is recommended to have, as a minimum, nylon or other synthetic material covers for all controls to keep ice/snow off the controls.

8.6.6.2 Hatch Covers, Ramp Doors and Side Doors

A thin layer of ice on the seal in way of the hatch cover contact has been known to cause trouble during opening of the cover. The ice bonding the cover to the seal may be broken by mechanical shock such as mallets, or by thermal means such as heat tracing the seal, steam or hot water spray. Arrangements or equipment is to be provided to efficiently remove ice build-up around hatches and doors. If the hatches or doors are hydraulically operated, the requirements discussed under *Hydraulic systems* are applicable. Materials selected for hatch and door seals are to remain pliant at the minimum anticipated temperature. If covers are to be opened in low temperatures, a means is to be provided to de-ice the seals prior to opening.

8.6.6.3 Cargo Pumps

Cargo pumps and their auxiliary equipment (e.g., valves) exposed to the weather are to be suitable for operation at the minimum anticipated temperature. If the cargo pumps are hydraulically operated, the requirements discussed under *Hydraulic systems* are applicable. Electric motors driving cargo pumps must be provided with space heaters.

8.6.6.4 Deck Cargo Securing Equipment

Deck cargo securing equipment (e.g., container lashings) is to be suitable for use in the minimum anticipated temperature (Fig. 8.4).

8.7 Piping Systems

Piping of auxiliary machinery systems and equipment subjected to low temperatures, ice accumulation and/or ice ingestion in areas exposed to the weather or in unheated enclosed areas of the vessel are expected to meet the requirements discussed in this section. Piping systems are to be operable for the minimum anticipated temperature. All piping systems throughout the vessel should be designed to minimise exposure of crew to low temperature environmental hazards during normal operation and routine maintenance. If possible, valves and controls should be automated, and pipe tunnels should be considered if feasible.

8.7.1 General

8.7.1.1 Materials

Materials subjected to the minimum anticipated temperature are to be resistant to brittle fracture and are to meet the requirements discussed in Chap. 6.

Materials subjected to the t_D are to be resistant to brittle fracture because, on occasion, it may be necessary to use hammers or axes to clear ice accumulation, making fracture a possibility. Materials used in piping systems and equipment should be suitable for operations at the design service temperature, considering also specific circumstances such as the lay-up of the vessel for extended periods. Particular attention should be paid for those components required for the prevention of pollution or the safety of the vessel. Areas of concern are the areas external to the hull and the unheated areas inside the hull and above the waterline. In all cases, these materials should not be susceptible to brittle fracture. Refer to the Class Rules for the material requirements deemed acceptable by Class for piping components. For components manufactured from grey cast iron or other cast steels exhibiting brittle characteristics in low temperatures, refer to Chap. 6. Consideration may be provided for applications where a heating device is provided for the component. Piping accessibility can be improved by providing walkways, platforms, ladders, etc., keeping in mind there could be obstructions from ice/snow. To reduce/eliminate snow/ice, heating would need to be provided. On some vessel designs, piping is enclosed in a tunnel, so personnel are not exposed to the weather. Providing a tunnel may not be suitable for vessel designs where flammable fumes are present.

8.7.1.2 Flexible Hoses

Seals and hoses must remain flexible at the minimum anticipated temperature. Lubricants and working fluids must also be determined as suitable for the t_D.

8.7.1.3 Valves

Valves and closures in exposed primary and secondary essential service systems listed in Tables 8.2 and 8.3 should be selected and located or otherwise protected to avoid freezing of fluids on either side of the valve and to prevent the accumulation of snow and spray ice. Moving parts may require to be heated, continuously or prior to operation. Canvas covers with electric heating device may be considered. If a hazardous area is present, the covers must be of a certified safe type.

Table 8.2 Primary vessel essential services

Azimuth thrusters which are the sole means for propulsion/steering with lubricating oil pumps, cooling water pumps, etc. Control, monitoring and safety devices/systems of equipment for primary essential services. Electric generators and associated power sources supplying primary essential equipment Electrical equipment for electric propulsion plant with lubricating oil pumps and cooling water pumps Fire pumps and other fire extinguishing medium pumps Forced draft fans, feed water pumps, water circulating pumps, vacuum pumps and condensate pumps for steam plants on steam turbine ships, and for auxiliary boilers on vessels where steam is used for equipment supplying primary essential services Fuel gas supply pumps, low duty gas compressor and other boil-off gas treatment facilities supporting boil-off gas usage as fuel to main propulsion or electric power generation machinery	Hydraulic pumps supplying primary essential equipment Internal safety communication equipment Lighting system Navigation lights, aids and signals Oil burning installations for steam plants on steam turbine vessels and for auxiliary boilers where steam is used for equipment supplying primary essential services Pumps for controllable pitch propellers Scavenging air blower, fuel oil supply pumps, fuel valve cooling pumps, lubricating oil pumps and cooling water pumps for main and auxiliary engines, turbines and shafting necessary for propulsion Steering gears Ventilation necessary to maintain propulsion Viscosity control equipment for heavy fuel oil

Table 8.3 Secondary vessel essential services

Ambient temperature control equipment	Hydraulic pumps supplying secondary essential equipment
Bilge, ballast and heeling pumps	Lubrication oil transfer pumps and lubrication oil treatment equipment
Control, monitoring and safety devices/systems of equipment for secondary essential services.	Methods used to comply with the class rules for liquefied gas carriers
Control, monitoring and safety systems for cargo containment systems	Pre-heaters for heavy fuel oil
Electric generators and associated power sources supplying secondary essential equipment	Services considered necessary to maintain dangerous spaces in a safe condition (inert gas system of an oil carrier, ventilation for Ro-Ro cargo spaces, etc.)
Electrical equipment for watertight and fire-tight closing appliances	Starting air and control air compressors
Fire detection and alarm system	Ventilating fans for engine and boiler rooms
Fuel oil transfer pumps and fuel oil treatment equipment	Watertight Doors
	Windlass

Table 8.4 Required number of starts for propulsion engines

Engine type	Single propeller vessels		Multiple propeller vessels	
	One engine coupled to shaft directly or through reduction gear	Two or more engines coupled to shaft through clutch and reduction gear	One engine coupled to each shaft directly or through reduction gear	Two or more engines coupled to each shaft through clutch and reduction gear
Reversible	16	20	20	20
Non-reversible	6	8	8	8

Table 8.5 Minimum number of required EEBDs

A.	In machinery spaces for category A containing internal combustion machinery used for main propulsion[a]
	1 × EEBD in the engine control room, if located within the machinery space
	1 × EEBD in workshop areas. If there is, however, a direct access to an escape way from the workshop, an EEBD is not required; and
	1 × EEBD on each deck or platform level near the escape ladder constituting the second means of escape from the machinery space (the other means being an enclosed escape trunk or watertight door at the lower level of the space)
B.	In machinery spaces of category A, other than those containing internal combustion machinery used for main propulsion
	1 × EEBD should, as a minimum, be provided on each deck or platform level near the escape ladder constituting the second means of escape from the space (the other means being an enclosed escape trunk or watertight door at the lower level of the space)
C.	In other machinery spaces
	The number and location of EEBDs are to be determined by the Flag administration

Note [a]Alternatively, a different number of locations may be determined by the Flag Administration, taking into consideration the layout and dimensions or the normal manning of the space

8.7.1.4 Pipes

Components, such as valves, valve control units, manifolds, vents, fittings, sounding pipes, and other piping components exposed to the minimum anticipated temperature should be protected from ice accumulation or be provided with a means to provide for continued functionality such as positioning in sheltered locations, installing trace heating or heated covers. Horizontal runs of piping may be unavoidable on certain vessel types, such as a tanker. Effective drainage of these pipes can be accomplished by trimming the vessel aft or installing drain lines in the piping.

8.7.1.5 Tank Vents

The blockage of vent pipes by ice accumulation at the deck or by the freezing of plugs inside the pipe from the vapours rising from the tank contents or ingress of air into the tank can result in safety hazards, for example, due to over-pressurisation. It is common to fit, or provide for, heat tracing. If the there is a heated liquid in the tank being vented, anti-icing equipment may not necessarily be installed provided supporting calculations are provided.

8.7.1.6 Pipe Drainage

All piping within areas that can be expected to freeze should be designed to prevent freezing or should be able to be drained of contained fluids. This applies to exposed pipes, valves, pumps and fittings, including those found in areas onboard with no heating available such as voids, cofferdams, etc. Additional heat tracing and insulation should be installed on all safety-related piping exposed to extreme cold temperatures unless other means to provide for operability have been introduced. Drain cocks (valves) should be provided as these are quick-acting devices. Drain plugs may be provided; but drainage of the piping will take much longer because the operator will have to remove each one with a tool.

8.7.2 Ship Piping Systems and Tanks

8.7.2.1 Ballast Water Tanks

Cargo vessels operating in cold regions are likely to make one leg of any voyage in ballast. Ice class vessels are likely to have significant ballast capacity to reduce the range of waterlines at which ice must be broken. Local regulations regarding the discharge of ballast water are to be considered. Ballast water systems, therefore, need both design and operational consideration. Ballast water inside or hopper tanks above the waterline may freeze, starting at the top of the tank and at the side walls.

In contemporary designs, it is advisable to minimise the amount of ballast carried high in the vessel, especially in stand-alone tanks. Even where the tank itself does not freeze completely, valve and suction line freezing can occur. In extremely cold regions, thick ice formation or complete blockage within air and vent pipes has been noted. The extent of freezing will depend on the temperatures encountered and on the duration of the voyage, and on the salinity of the ballast water. Fresh or brackish water will freeze more easily, and so higher salinity sea water should be used where voyage routing and local or regional environmental regulations allow. It is highly unlikely that any sizeable tank will freeze solid, as the ice itself acts as an insulating layer, reducing the rate of heat transfer. However, ice represents a weight that may not be dischargeable when the vessel is loading, reducing deadweight capacity. If ice chunks fall from the tank sides after the discharge of the liquid ballast, they may damage coatings or components. Ballast tanks should not be pressed full in any conditions when freezing is possible, as expansion during the freezing process can damage structure and pipes. Ballast tank freezing can be prevented or minimised by adding heat. This will be most practical for oil tankers, where cargo heating systems can be adapted to be operated on both loaded and ballast voyages by designing the system to be able to heat either the cargo tanks or the ballast tanks depending on the voyage. In some arrangements, the heating coils in the cargo tanks operate during the ballast voyage and the radiation from these coils may be sufficient to warm the ballast water. These arrangements may be considered, provided calculations indicating sufficient heat transfer at the design service temperature are submitted for review. Circulating the water within the tanks is only effective over the short term, and on longer voyages can aggravate problems by increasing heat transfer from ballast water to air.

Ballast water tank freeze prevention requirements are discussed in Chap. 7 however the piping requirements under the section *Freshwater systems* must also be applied. Convection-inducing systems are to extend the entire length of the tank. *Ballast water tanks for t_D below* – 30 °C (– 22 °F). For t_D below – 30 °C (– 22 °F), each ballast water tank is to be fitted with steam heating coils.

8.7.3 Fuel Oil Systems for Prime Movers

8.7.3.1 Fuel Oil System

Fuel oil heating. Fuel oil heating systems should be designed to provide uninterrupted service of fuel oil to the prime movers.

The fuel oil heating system must be designed for uninterrupted service so that heavy and intermediate fuels are always available. Steam systems are standard equipment on most oceangoing vessels. However, problems associated with freezing of the steam system have occurred. Additional problems with condensate return and failure of heaters exposed to frigid air can be eliminated with the newer thermal fluid heating systems. It may also be necessary to add heating coils to distillate fuel tanks to prevent clouding or waxing of these fuels at low temperatures. These systems would normally be secured and only put into use in extreme conditions.

8.7.3.2 Fuel Oil Filling Stations

Fuel oil filling stations are to be in an area sheltered from the weather for access and operation. For fuel oil filling pipes located on the deck, alternative arrangements to protect the filling pipes may be provided such as having a removable or hinged cover installed.

8.7.4 Cooling Systems

8.7.4.1 Seawater Systems

> Seawater systems draw their supply from the sea around the vessel, and therefore must be designed to reduce the risk of the inlets becoming blocked by ice or ice forming within them.

The most cost-effective solution will depend on the type of vessel and the nature of the service. As an alternative, box coolers have been installed in recent years. In these arrangements, cooling water is forced through a U-tube-bundle, which is placed in a sea-chest having inlet- and outlet grids. Cooling is achieved by natural circulation of the cooling water in the sea-chest or by circulation because of the vessel speed. However, ice may still accumulate during periods alongside in frigid conditions, and so it is still advisable to provide heating/water recirculation to deal with possible freeze-up and associated problems during start-up. Box coolers may also become subject to damage during heavy ice conditions for vessels with "standard" hull configurations. This damage could occur during either transit or when in near-stationary dynamic positioning mode. Shallow water depths on the order of 10 m (32.8 ft) in the Arctic Ocean are common in many areas. Sea chests may be subject to mud intrusion for vessels operating in shallow waters and therefore should be provided with a means to adequately clean the sea chest of the mud.

8.7.4.2 Sea Box/Bay Arrangement

Sea inlets should be located to prevent ingestion of ice. Locations offering the best protection depend solely on the hull form and the size of the vessel. For example, larger tankers or bulk carriers of conventional ice breaking form will see little ice at the aft end below the turn of bilge. If sea suctions are in this area of the hull, ice ingestion during voyages is unlikely to be a severe problem. Smaller vessels with limited draft may locate the sea inlets closer to the vessel's centreline, thereby offering better protection to ice floes projected downwards along the length of the hull. However, since ice may still accumulate during periods alongside in frigid conditions, it is advisable to provide additional protection by way of heating/recirculation to avoid problems during equipment start-up. The sea boxes should be configured to run the maximum vertical extent possible. A sea box designed with its associated tank top well above the load waterline offers a region for any ice floes that may pass through the sea grid to float clear of the sea suctions. The suctions in turn should be located as low as possible to decrease the likelihood of ice being ingested. To improve the protection of any sea suction inlet, the use of weirs or baffle plates have been used successfully in the past. A weir type sea box will reduce the possibility of suction pipe clogging. As the suction is isolated from the sea inlet grill by way of a solid vertical plate or a perforated baffle, ice that may enter the sea box will float to the top of the sea box and in turn be prevented from being dragged down to the suction inlet. Refer to Fig 8.5 through 8.6 for example configurations. Perforations for a baffle plate are recommended to be nominally 20 mm (0.7 in) diameter to prevent large ice particles from being ingested yet still provide sufficient cooling capacity. There should be the capability of manually clearing the systems of all ice blockages. This can be accomplished by designing efficient access points at strainers and designing the sea box for access above the load waterline.

8.7.4.3 Piping and Valve Arrangements

Sea bays should be supplied from at least two (independent) sea suctions. For sufficient cooling water to enter the system, the area of the inlets themselves between the sea box and the sea bay should be a multiple of the total suction sectional areas of all seawater pumps. The selection of the multiple can be based on the ice class of the vessel. As a minimum, four times is required, however, where vessels are **PC1** through **PC5**, six times is required. As noted previously, the location of sea suctions is critical for adequate cooling water to the vessel's systems. Ice blockage can result in overheating of the vessel systems and has caused shutdowns in the past. Sea bays can be protected by extra capacity strainer boxes. These sea bays should have additional isolation valves installed to allow access to the strainer boxes for removing debris without the need of shutting down the vessel's systems.

Fig 8.1 Enclosed anchor windlass (Polarstern)

Fig 8.2 Open deck anchor windlass (MV Algarve)

Fig 8.3 Enclosed towing winch

Fig 8.4 Container securing, MV Pollux, Torshavn

8.7.4.4 Cooling and Heating

Ice-classed vessels operating in extremely cold temperatures must consider the effects that the temperature itself may have on the vessel. Although the design of the sea bays,

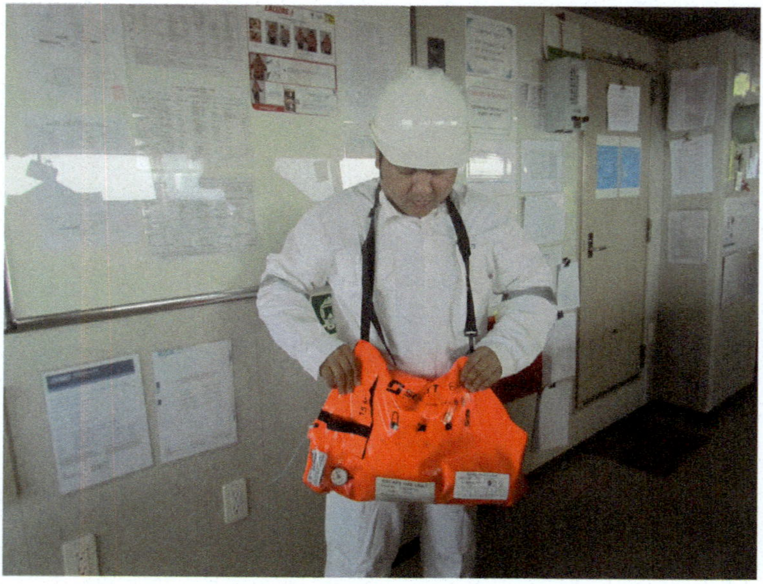

Fig 8.5 Third officer demonstrating the use of EEBD

strainers and inlets may reduce the effects of ice ingestion or prevent some icing, these will not eliminate the possibility. There are proven practices and functional designs for heating and cooling arrangements to reduce this potential inherent for vessels operating in cold regions. Most high ice class vessels have an emergency supply of seawater separate from the standard engine room sea bay. Ballast tanks or an entirely separate emergency sea bay have been used in the past. Many vessels have a recirculated seawater cooling system back to the sea bay. This limits the amount of cold seawater entering the vessel system and aids in reducing icing of the sea bay itself. Recirculating cooling water through ballast tanks provides an additional level of reliability. If sea boxes become blocked, these tanks may be used for cooling purposes. De-icing systems within the sea boxes and at cooling system suctions are another feature which may be incorporated to eliminate ice accumulation and blockages. Additionally, these sea boxes and bays may be provided with steam de-icing systems. The steam system requires isolating valves, strainers, and pressure gauges sized for adequate steam capacity and supply characteristics. Hot water has also been used effectively in the past. An alternative to steam is installation of a thermal oil heating coil inside the sea chest and bay provided (1) the heating coil is of extra heavy pipe; (2) the heating coil is all welded joints; (3) the heating coil is arranged to absorb the expansion and shrinkage by heat variation; (4) an isolation valve is fitted at the sea chest/bay inlet and outlet inside the engine room in an accessible location. Most of these designs are taken from ice breaker design practice, as these vessels face the most severe challenges due to small size, high power and aggressive operation.

8.7.5 Seawater Piping for All Vessels

Standard piping, valves and fittings may be used in the sea water system. However, the use of cast iron material is not advised for any vessel operating in heavy ice conditions. Previous experience has shown that heavy vibrations during propeller/ice interactions may result in system failures because of fracture of this material. Overboard discharges should be kept to a minimum to reduce the possibility of ice forming in the vessel's valves, and wherever possible, water drains should not be discharged above the load waterline. In areas of extreme temperatures, steam or heat tracing is advised on all overboard discharges. Alternative means to de-ice sea chests and sea bays such as through electric heating will usually be considered by Class.

8.7.6 Engine Cooling

Main and auxiliary engine cooling lines, where exposed to freezing temperatures, are to be suitably insulated or heated. Alternatively, antifreeze may be used in the cooling water.

Main and auxiliary engine cooling is the most frequent cause of operational problems for vessels operating in cold regions and ice-covered waters. This results from the inlet blockage problems discussed above but may be caused by other effects such as the freezing of cooling lines serving harbour and emergency generators. Insulating, heating, and/or adding antifreeze to any lines exposed to freezing temperatures may be required for any engine, and particularly for those using freshwater cooling systems.

8.7.6.1 Cooling System for Non-ice Class Vessels

The above requirements are applicable exclusively for ice classed vessels; accordingly, there are no requirements for non-ice classed vessels.

8.7.6.2 Freshwater Systems

> Refer to Chap. 7. In addition, freshwater pipe runs through void spaces must be insulated and heat traced. All water lines exposed to freezing temperatures are to be fitted with drain cocks at the lowest part of the piping arrangement. Furthermore, freshwater generation systems' inlet temperatures are to be designed and operated always in accordance with the manufacturer's requirements.

Freshwater tanks should be located away from the vessel's sides to prevent freezing of contents. As with all water-filled systems onboard, pipe runs within void spaces should be insulated and heat-traced to prevent freezing. Assume the void space temperature is the minimum anticipated temperature. The lowest part of the pumping arrangement and pipe runs are to be fitted with drain cocks. Freshwater generation systems may be sensitive to low sea water (inlet) temperature. The system designer must consult with the manufacturer to determine inlet temperature requirements and any other requirements necessary for satisfactory operation. Reverse osmosis systems are not efficient at low temperature and will require preheat arrangements. Other auxiliary equipment may need to be considered.

8.7.6.3 Other Piping Systems

Exhaust gas outlets and pressure/vacuum arrangements (e.g., for prime movers and/or their associated spaces) are to be protected from ice build-up that could interfere with effective operation.

8.7.7 Waste Storage and Disposal Systems

> Regulations in the Baltic, Arctic and Antarctic regions vary among the Maritime Administrations. Refer to the appendices for contact information. Waste storage systems (i.e., for grey and black water) including piping are to be protected from freezing by either placement into a heated compartment or insulation and heat tracing. The system is to be protected enough to prevent freezing at the minimum anticipated temperature.

Many voyages in ice-covered waters are prolonged by the need to maintain safe (low) operating speeds and by the long distances involved. For wastes normally stored for subsequent disposal ashore, this may impose additional space requirements. Few ports in polar regions have adequate waste disposal arrangement for shipboard wastes. These must be carried back to ports with better infrastructures. Many Arctic and Antarctic waters are also covered by restricted discharge requirements for waste streams such as oily, grey and black water, leading to additional requirements for onboard storage or to the installation of more capable treatment and disposal systems. Waste compactors and incinerators can be used to reduce storage volumes and may be essential for vessels with a high number of persons onboard, including both passenger and science vessels. For other waste treatment systems, care should be taken that these will function adequately at low temperatures where both chemical and biological reactions may slow down markedly. Consult with the designer/manufacturer of these systems.

8.7.8 Compressed Air Systems

8.7.8.1 General

The compressed air supplied to the control air system is to have a dew point 20 °C (36 °F) below the design service temperature. Extremely dry control air is required to avoid condensation forming in the control air piping and ship's service air piping from freezing. Noting that the low air temperature can result in a drop in the vessel's service air pressure, consideration is to be given to either insulating or protecting with heat tracing the service air system piping.

8.7.8.2 Starting Air System (Ice Class Vessels)

The increase in the required number of starts for propulsion engines addresses the propulsion reliability and faster reaction time required for ice class vessels following an ice breaker. Loss of starting air after an engine shutdown increases the risk of collision when navigating in escorted formation (Table 8.4).

8.8 Fire Safety Systems

8.8.1 Firefighting Systems

8.8.1.1 Fire Extinguishing Design and Location

Fire extinguishing systems must be designed or located so that they are not made inaccessible or inoperable by ice or snow accumulation or the minimum anticipated temperature. System configuration should consider the need for operators to be wearing bulky and cumbersome freezing weather gear. Alternative firefighting system designs suitable for use in a low temperature environment, [e.g., aerosol firefighting system for enclosed spaces (engine room, pump room, etc.)], will be considered upon submission to Class of substantiating data. The fire extinguishing system is to be designed such that all equipment, appliances, extinguishing agents and systems are always protected from freezing and remain operable at the minimum anticipated temperature; and suitable precautions are taken to prevent nozzles, piping and valves from freezing or becoming clogged due to internal or external ice build-up. Refer to Olsen: *Firefighting and fire safety systems on ships*, Routledge, 2023) for further information and guidance.

Many fixed and portable firefighting systems can suffer from performance degradation due to low temperatures or to ice and snow accumulation at locations ranging from sea water intakes to hydrants and hoses. It is extremely important that the selection and configuration of firefighting systems for vessels operating in low temperature environments consider potential scenarios under extreme operating conditions. Equipment, appliances, extinguishing agents and systems should always be protected from freezing. It is common to provide drain cocks or valves in the low points of the piping system. For water piping, if draining the system is not possible, options include charging the system with low temperature fluids (such as glycol) or maintaining continuous flow through the system. Maintaining operability of hydrants can be accomplished by locating them in protected areas to avoid accumulation of snow and ice. Alternatively, the hydrants can be provided with a heated cover or strip heater to keep them free of snow and ice. Some of the firefighter's outfits may be stored in the Fire Control Room.

8.8.1.2 Water or Foam Extinguishers

Portable and semi-portable water and foam extinguishers should be positioned in locations that are adequately protected from freezing temperatures, as far as is wholly practicable. Locations subject to freezing must be provided with extinguishers capable of operation at the minimum anticipated temperature.

8.8.1.3 Fire Pumps and Associated Equipment

Where fixed firefighting systems or alternative fire extinguishing systems are situated in a space separate from the main fire pumps and using their own independent sea suction, this sea suction is to be also capable of being cleared of accumulations of ice. Fire pumps, including emergency fire pumps, must be in heated compartments. The pumps and their auxiliaries in the compartment are to be adequately protected from freezing at the minimum anticipated temperature. Isolating valves must be in accessible locations. Isolating valves located in exposed locations should be protected from freezing spray. The fire main is to be arranged so that external sections can be isolated and adequate means of draining are provided. Hydrants are to be positioned or designed to remain operable at the minimum anticipated temperature. The risk and hazards associated with ice accumulation and freezing over must be accounted for. All hydrants are to be installed with a two-handed valve lever or hand wheels and provided with quick connects for hoses. Valves and hydrants exposed to minimum anticipated temperatures less than or equal to – 30 °C (– 22 °F) are not permitted to be of cast iron.

8.8.1.4 Protection Against Ice Build Up

Components of the firefighting system that may be exposed to icing that could interfere with the proper functioning of that component are to be adequately protected to remain operable at the minimum anticipated temperature. Portable firefighting equipment such as portable foam applicators are to be stored in a protected space and are to be readily available.

8.8.1.5 Firefighter's Outfits

Firefighter's outfits are to be located within accommodation areas and other spaces, as appropriate, suitably protected from low temperature conditions. The outfits are to be stored in positions as widely separated as practicable and readily accessible by the vessel's crew. In addition to the firefighter's outfits required in the main ship's spaces, for vessels operating in remote areas, one spare firefighter's outfit is to be provided. The spare outfit is to be stored in a location protected from low temperatures where access is provided to the open deck, such as a steering gear room.

8.8.1.6 Components of the Firefighter's Outfit

Firefighter's outfit. The firefighter's outfit is required to consist of a set of personal equipment and breathing apparatus. The personal equipment is to comprise of protective clothing manufactured from material that is rated to protect the skin from heat radiating from the fire and from burns and scalding by steam. The outer surface must be water-resistant. The wearer must don boots of rubber or some other electrically non-conducting material and a rigid helmet which provides effective protection against falling objects and head on impact. An electric safety lamp (hand lantern) of an approved type with a minimum burning period of three hours may be carried manually or attached the person. Onboard tankers, an electric safety lamp rated to be used in hazardous areas must be used as these are of an explosion-proof type. Firefighters may also carry an axe with a handle fitted with high-voltage insulation. *Breathing apparatus.* Breathing apparatus is to be of a self-contained compressed air-operated type. It is common for the capacity of the cylinders to contain a minimum of 1200 l (317 gal.), capable of functioning for at least 30 min. The compressed air breathing apparatus should be fitted with an audible alarm and a visual or other device which alerts the wearer before the volume of air in the cylinder has reduced to 200 l (53 gal.). Two spare charges are to be provided for each required breathing apparatus set. All air cylinders for breathing apparatus must be interchangeable. Vessels that are equipped with suitably located means for fully recharging the air cylinders free from contamination need carry only one spare charge for each required apparatus. *Lifeline.* For each breathing apparatus, a fireproof lifeline of at least 30 m (98.5 ft) in length is to be provided. The lifeline is to successfully pass an approval test by static load of 3.5 kN (360 kg f, 787 lb f) for 5 min without incurring failure. The lifeline is to be capable of attachment by means of a snap hook to the harness of the apparatus or to a separate belt to prevent the breathing apparatus becoming detached when the lifeline is operated. *Two-way portable radiotelephone.* A minimum of two two-way portable radiotelephone apparatus for each fire party for firefighter's communication should be carried onboard. Those two-way portable radiotelephone apparatus should be of a certified safe type suitable for use in zone 1 hazardous areas, as defined in *IEC Publication 60079*. The minimum requirements in respect to the apparatus group and temperature class are to be consistent with the most restrictive requirements for the hazardous area zone onboard which is accessible to the fire party.

8.8.1.7 Required Quantity of Firefighter's Outfits (All Vessels)

All vessels are required to carry a minimum of two complete firefighter's outfits. Additional sets of personal equipment and breathing apparatus may be required, having due regard to the size and type of the vessel. In addition to the requirements outlined above, onboard tankers, two additional firefighter's outfits must be provided (i.e., four in total).

8.8.1.8 Storage of Firefighter's Outfits

The firefighter's outfits or sets of personal equipment are to be kept ready for use in an easily accessible location that is permanently and clearly marked. Where more than one firefighter's outfit or more than one set of personal equipment is carried, these are to be stored in widely separated locations.

8.8.2 Emergency Escape Breathing Devices (EEBD)

All ships are mandated to carry at least two emergency escape breathing devices (EEBD) and one spare device within the vessel's accommodation spaces. On all vessels, within the machinery spaces, EEBDs are to be situated ready for use at easily visible locations, which can be reached quickly and easily at any time in the event of fire. The location of EEBDs is to consider the layout of the machinery space and the number of personnel ordinarily working in said spaces. (See the *Guidelines for the performance, location, use and care of emergency escape breathing devices*, MSC/Circ.849 and 1081). The number and locations of EEBDs must be clearly indicated in the vessel's Fire Control Plan. A summary of the MSC/Circ.1081 requirements are shown in Table 4.5. This applies to machinery spaces where crew are normally employed or may be present on a routine basis. The regulations differ for vessels with unmanned machinery spaces (UMS). Refer to the specific Class Rules where applicable (Table 8.5).

8.8.2.1 EEBD Specification

An EEBD is a supply-air or oxygen device only used for escape from a compartment that has a hazardous atmosphere and is to be of an approved type. EEBDs are not to be used for fighting fires, entering oxygen deficient voids or tanks, or worn by firefighters. In these events, a self-containing breathing apparatus, which is specifically suited for such applications, is to be used.

8.8.2.2 EEBD Particulars

The EEBD is required to have a minimum service duration of at least 10 min. The EEBD is to include a hood or full-face piece, as appropriate, to protect the eyes, nose and mouth during escape. Hoods and face pieces must be constructed of flame-resistant materials and include a clear window for viewing. Inactivated EEBD should be carried hands-free.

8.8.2.3 EEBD Storage

When stored, the EEBD is to be suitably protected from the environment.

8.8.2.4 EEBD Instructions and Markings

Brief instructions or diagrams clearly illustrating their use are to be clearly printed on the EEBD. The donning procedures are to be quick and easy to allow for situations where there is little time to seek safety from a hazardous atmosphere. Maintenance requirements, manufacturer's trademarks and serial number, shelf life with accompanying manufacture date and name of approving authority are to be printed on each EEBD. All EEBD training units are to be clearly marked.

8.9 Electrical Systems

Most aspects of electrical system design and operation can be identical to normal practice for ocean-going vessels. Some issues do, however, need to be considered. These are discussed below.

8.9.1 Main Source of Electrical Power

Where central electric power plants or shaft generators are used, speed fluctuations during ice interactions can create frequency instability, which can damage certain subsystems and components.

8.9.2 Emergency Source of Electrical Power Additional Systems, Equipment and Spaces

The electrical power available from the emergency source is to be sufficient to supply all those services that are essential for safety in an emergency, due regard being paid to such services as may have to be operated simultaneously. Where the sum of the loads on the emergency generator switchboard exceeds the power available, an analysis demonstrating that the power required to operate the services simultaneously is to be produced. The analysis is to be submitted for review in support of the sizing of the emergency generator. The emergency source of electrical power is to be capable, having regard to starting currents and the transitory nature of certain loads, of supplying simultaneously at least the services listed in Table 8.6 for the period specified.

The initial ambient temperature in the listed spaces below can be assumed to be 20 °C (68 °F) for calculation purposes. The time periods listed in Table 8.6 may need to be increased from 18 hours considering rescue times in the remote Arctic and Antarctic regions may be increased because of weather conditions, distances from nearby ports and proximity to other vessels operating in the area. Heat tracing of the piping systems for the various equipment and systems listed below are to be functional during the emergency period to permit continued system/equipment functionality, such as heat tracing of the steering gear equipment in a very cold steering gear compartment. In addition to those listed in Table 8.6, the following systems and equipment are also to be provided with an emergency source of power available for a period of 18 h:

- Battery room.
- Cargo control room.
- Centralised control station.
- Emergency generator room.
- Engine control room.
- Heated compartments for fire control station and firefighting equipment.
- Hospital.
- Machinery space workshop.
- Navigation bridge.
- Radio room; and
- Two shared areas (e.g., mess room and recreation room).

As an option, the time periods listed in Table 8.6 may be increased from 18 hours or greater for vessels anticipated to operate in remote areas. This increased time is optional and will be shown in the classification notation.

8.9.3 Navigation Lighting Systems

Certain lighting designs will fail frequently in low temperature environments. Lighting manufacturers should certify by way of testing or service history that the lights are suitable. It is recommended that navigation lights be of a greater intensity for improved visibility at an acceptable range. Dual filament lights would be acceptable for this purpose.

Table 8.6 Services to be powered by an emergency source and by a transitional source

Service	Gross tonnage	Duration (h)				
		500 GT or over		Under 500 GT		
	Electrical plant's capacity	–		75 kW or over		Under 75 kW
	Emergency power consumers	Emergency power	Transitional power	Emergency power	Transitional power	Emergency power
	At every muster and embarkation station for survival craft, their launching appliance, over the sides for launching	3	0.5	2	0.5	2
	At the area of water into which the survival craft is to be launched	3	0.5	2	0.5	2
Emergency lighting	In all service and accommodation alleyways, stairways, and exits, personnel lift cars, and personnel lift trunks	18	0.5	6	0.5	6
	In the machinery spaces and main generating stations including their control positions	18	0.5	6	0.5	6
	In all control stations, machinery control rooms, and at each main and emergency switchboard	18	0.5	6	0.5	6
	At all storage positions for firefighter's outfits	18	0.5	6	0.5	6
	At the steering gear	18	0.5	6	0.5	6
	At the emergency fire pump, the sprinkler pump, the emergency bilge pump, their starting positions	18	0.5	6	0.5	6
	In all cargo pump rooms	18	0.5	6	0.5	
	Floodlight and perimeter lights on helicopter landing decks	18	0.5	6	0.5	

(continued)

Table 8.6 (continued)

Service	Gross tonnage	Duration (h)				
		500 GT or over		Under 500 GT		
	Electrical plant's capacity	–		75 kW or over		Under 75 kW
	Emergency power consumers	Emergency power	Transitional power	Emergency power	Transitional power	Emergency power
Navigation lights	The navigation lights and other lights required by the International Regulations for Preventing Collisions at Sea	18	0.5	6	0.5	
Communication	Radio equipment as required by Chap 4 of SOLAS	18	0.5	6	0.5	
	All internal communication equipment as required in an emergency	18	0.5	6	0.5	
	Intermittent operation of daylight signalling lamp, ship's whistle, manually operated call points, all internal signals that are required in an emergency	18	0.5	6	0.5	
Navigational aids	Navigational equipment required by Regulation V/19 and V/20 of SOLAS	18(1)				
Alarm systems	Fire detection and alarm	18	0.5	6(4)	0.5	
	Gas detection and alarm system	18	0.5	6	0.5	

(continued)

Table 8.6 (continued)

Service	Gross tonnage	Duration (h)				
		500 GT or over		Under 500 GT		
	Electrical plant's capacity	–		75 kW or over		Under 75 kW
	Emergency power consumers	Emergency power	Transitional power	Emergency power	Transitional power	Emergency power
Fire pumps	Emergency fire pump; plus, one fire pump and fixed pressure water-spray system pump if dependent upon the emergency generator for its source of power	18				
Steering gear	Steering gears	$0.5^{(2,3)}$		$0.17^{(3)}$		
Navigation	Remote propulsion control and monitoring system for **ACC**, **ACCU** and **ABCU** notations	0.5		0.5		
Other emergency services	Free-full lifeboat secondary launching appliance, if not dependent on gravity, stored mechanical power or other manual means	0.5		0.5		
	Power-operated watertight doors	0.5		0.5		

Notes (1) Vessels less than 5000 GT may be waived if approved by the Administration. (2) Applicable for rudder stock diameter over 230 mm (9 in). (3) 10 min continuous operation on vessels of less than 10,000 GT. (4) where a fixed fire detection and fire alarm system is installed
The bold letters are indicative of the vessel's assigned ice class

8.9.4 *Electrical Equipment*

8.9.4.1 Rotating Machines

Motors and generators. Electric motors and generators in unheated spaces are expected to be fitted with some means for preventing moisture condensation in the machine when sitting idle; and, provided with lubricating oil suitable for the design internal temperature if the bearings require forced lubrication or pre-lubrication. *Motors.* Electric motors in exterior locations are to be fitted with some means to prevent moisture condensation in the machine when idle; and provided with lubricating oil or grease suitable for the minimum anticipated temperature. *Switchboards, motor controllers, etc.* Switchboards, power and lighting distribution boards, motor control centres and motor controllers and battery charging panels located in compartments that may be subject to low temperatures must be provided with space heaters. These components are to be suitable for continuous operation resulting from vibrations from ice impacts and vessel manoeuvring and must not trip inadvertently. The manufacturer's certification or specification sheet is required to be submitted to Class for review.

Vibration during ice breaking can cause circuit breakers and other components to trip.

8.9.4.2 Cables

Electric cables in exposed areas are to be suitable for operation at the design service temperature. Sheathing is to be provided to protect the cables from mechanical damage. Electric cables are to be in accordance with one of the recognised standards. Electric cables constructed of stranded copper conductors, thermoplastic, elastomeric or other insulation, moisture-resistant jackets, and, where applicable, armouring and outer-sheathing are to be in accordance with *IEC Publication 60092-350, 60092-352, 60092-353, 60092-354, 60092-360, 60092-370, 60092-376, IEEE Std-45* or other marine standards of an equivalent or higher safety level, acceptable to Class. Network cables are to comply with a recognised industry standard. Cables such as flexible cable, fibre-optic cable, etc., used for special purposes may be accepted provided they are manufactured and assessed in accordance with recognised standards accepted by Class. Cables on exposed decks are to be protected from the effects of mechanical ice removal.

8.9.4.3 Accumulator Batteries (Storage Space)

Conventional battery capacity is influenced strongly by temperature. Where batteries are installed, this should be considered, or battery banks should be provided with supplementary heating.

8.9.5 Electronic Equipment (Monitoring and Control)

Indicators and alarms for critical systems are to be duplicated as far as possible. The use of components that have demonstrated an elevated level of reliability operating in low temperatures should be used wherever possible. The intended operation of the vessel and the equipment must be considered. Vessels that are expected to enter a lay-up period in low temperatures are to have systems rated for survival at the minimum anticipated temperature while inactive. Systems that are required to operate in low temperatures are to be rated for operation at the minimum anticipated temperature. The systems to consider include—but are not necessarily limited to those systems used to meet the requirements for automatic sprinkler, fire detection and fire alarm systems and fixed fire detection and fire alarm systems; essential vessel services required to support the Class notation requirements; displays and monitors; computers and integrated circuits; systems and functions related to the human machine interface (HMI), including touch screen monitors; sensors; and cameras.

8.10 Heating, Ventilation and Air Conditioning

Standard heating, ventilation and air conditioning (HVAC) systems (coupled with typical levels of insulation) may not have the capacity to provide adequate comfort levels during operations in extreme frigid conditions. Furthermore, these standard systems may not take account of the temperature gradients that are likely to exist between the outer and inner areas of the accommodation block. An increase in the amount of approved insulation for the steel structure inside the accommodations area to support the efficiency of the HVAC system may be necessary. The associated increase in thickness and weight must be considered in the design. It is therefore not only important to calculate the overall heating demand accurately, but also that the layout of HVAC systems provides for a satisfactory distribution of heating. Another consideration is relative humidity. In the world's temperate zones, the formation of condensation on the interior surfaces of those boundaries between air-conditioned spaces and the weather may occur. Accordingly, vessels are insulated in such a manner as to mitigate the formation of condensation and are arranged to provide for the drainage of the condensation by leaving a channel to allow for the condensation to drain. For vessels operating in extreme low temperature environments moisture or humidity must be added to the make-up air so to increase the relative humidity to within a range of between 40% minimum and 70% maximum to provide for an acceptable comfort level and to prevent the build-up of static electricity. As indicated earlier in the section, an adequate amount of insulation must be provided to maintain an acceptable temperature and to reduce the energy demand required to support the heaters. In association with the increased thickness and density, the insulation is to be arranged to prevent the formation of condensation. Such an insulation arrangement may result in the need for additional space to be provided between the steel boundary and the interior lining. In general, the design of the HVAC systems for vessels intended to operate in extreme low temperatures requires a sound understanding of all the issues and therefore, must be designed accordingly. Heating can be a major issue for propulsion and auxiliary machinery spaces when combustion air is drawn into and from the compartment. At extremely low external temperatures, the machinery waste heat may not be adequate to provide acceptable temperatures. Additional heating will be required, both in operation and for periods alongside when machinery is shut down or idling. Two-speed fans offer flexibility within the system design. For extreme cold temperatures, consideration should be given to a cross-over between the supply air inlet and the exhaust vent ducting. Air ventilation ducting is to be insulated to prevent condensation, which leads to corrosion and mould growth. Closing apparatus for ventilation inlets and outlets must be able to be always operated, particularly for closing in the event of a fire.

The following requirements apply to the vessel's accommodation and other spaces that are normally manned or occasionally manned for certain vessel operations. The HVAC system(s) are to be designed for satisfactory distribution of heating at the minimum anticipated temperature. Supplementary electric heaters may be used in cabins and other manned spaces. High amperage permanently installed heaters must not be wired to the same circuit breakers as vital electronic equipment. Portable electric heaters are not permitted. The relative humidity of manned spaces should be maintained in a range of between 30 and 70%. Heating systems for accommodation spaces and those spaces listed above with the following exceptions are to have a rated capacity sufficient for maintaining the temperature to 20 °C (68 °F) at the minimum anticipated temperature. However, the machinery space workshop, emergency generator room, battery room and compartments for firefighting equipment are to be fitted with heating systems with a rated capacity sufficient to maintain the compartments to 10 °C (50 °F) at the minimum anticipated temperature. Recirculation of air in the accommodation spaces is to be in accordance with a recognised standard. The standard applied is to be provided for information. Air ventilation ducting is to be insulated with non-combustible insulation. The closing apparatus for all ventilation inlets and outlets are to be located to be protected from snow and ice accumulation that may interfere with effective operation of the closures and recirculation of exhaust gases. Alternatively, heat tracing may be provided in lieu of a protected location. All other spaces that are normally manned and/or have equipment installed within are to be provided with a heating system with a rated capacity sufficient to maintain 0 °C (32 °F) at the minimum anticipated temperature. Consideration must be given if it is unpractical to heat the space due to its arrangement or size.

8.11 Monitoring of Remote Propulsion Controls and Automation Systems

Standby propulsion and auxiliary machinery must be provided with either low temperature monitoring with an alarm installed on each machine, or low temperature monitoring with an alarm for the compartment. These alarms are to alert the operator in the Engine Control Room and/or Centralised Control Station when the temperature is too low for the machinery to start unaided.

Annex

For checking the results of calculated powering requirements, the table below presents input data for a number of sample vessels (Table 8.7).

Prop. No./type	1/CP	1/CP	1/CP	1/CP	1/CP	1/CP	1/CP	1/CP	1/FP
New ships kW (HP)	7840 (10514)	4941 (6626)	3478 (4664)	2253 (3021)	6799 (9118)	6406 (8591)	5343 (7165)	5017 (6728)	3872 (5192)
Existing ships kW (HP)	9192 (12327)	6614 (8870)			8466 (11353)	7645 (10252)	6614 (8870)	6614 (8870)	

Table 8.7 Parameters and calculated minimum engine power for sample vessels

Sample vessel number	1	2	3	4	5	6	7	8	9
Ice class	**I AA**	**I A**	**I B**	**I C**	**I AA**	**I AA**	**I A**	**I A**	**I B**
α, degrees	24	24	24	24	24	24	24	24	24
φ_1, degrees	90	90	90	90	90	90	90	90	90
φ2, degrees	30	30	30	30	30	30	30	30	30
L, m (ft)	150 (492)	150 (492)	150 (492)	150 (492)	150 (492)	150 (492)	150 (492)	150 (492)	150 (492)
B, m (ft)	25 (82)	25 (82)	25 (82)	25 (82)	25 (82)	25 (82)	25 (82)	25 (82)	25 (82)
T, m (ft)	9 (29.5)	9 (29.5)	9 (29.5)	9 (29.5)	9 (29.5)	9 (29.5)	9 (29.5)	9 (29.5)	9 (29.5)
L_{BOW}, m (ft)	45 (147.6)	45 (147.6)	45 (147.6)	45 (147.6)	45 (147.6)	45 (147.6)	45 (147.6)	45 (147.6)	45 (147.6)
L_{PAR}, m (ft)	70 (229.7)	70 (229.7)	70 (229.7)	70 (229.7)	70 (229.7)	70 (229.7)	70 (229.7)	70 (229.7)	70 (229.7)
A_{wf}, m^2 (ft^2)	500 (5382)	500 (5382)	500 (5382)	500 (5382)	500 (5382)	500 (5382)	500 (5382)	500 (5382)	500 (5382)
D_P, m (ft)	5 (16.4)	5 (16.4)	5 (16.4)	5 (16.4)	5 (16.4)	5 (16.4)	5 (16.4)	5 (16.4)	5 (16.4)

Note The bold letters are indicative of the vessel's assigned ice class

Chapter 9
Arctic Vessel Safety Systems

9.1 Introduction

In the previous chapter we discussed some of the primary safety systems associated with eliminating or reducing icing on critical equipment. This chapter discusses the requirements for various systems and equipment necessary for the protection and survival of the personnel onboard vessels operating in low temperatures. It should be noted that for vessels operating in low temperature environments, rescue and medical services may be significantly delayed due to the prevailing weather conditions or remoteness of the maritime area of operation. A list of Maritime Administrations with coasts on the Arctic and Antarctic Oceans and the Baltic Sea are provided at the end of the chapter.

9.2 Heating for Survival

Heating systems are considered essential for human safety and survival. Therefore, the emergency generator is to be sized to maintain crew survivability. The spaces listed in Chap. 8 are to be fitted with heating systems with a rated capacity sufficient for maintaining 10 °C (50 °F) at the design service temperature. An electric or steam pre-heater should be fitted in the air-handling unit for ventilation emergency use. Alternatively, individual space heaters for the emergency service spaces may be installed. Emergency generators should be sized for additional heating loads during an emergency. Special low temperature fuel may be required onboard for reliable start-up and operation of the emergency generator during an emergency. This is diesel oil that has been refined to remove "waxes" which precipitate out of the fuel resulting in fuel line clogging and subsequent engine malfunction. Emergency

A. Olsen, *Ship Operations in Extreme Low Temperature Environments*, Springer Series on Naval Architecture, Marine Engineering, Shipbuilding and Shipping 19,
https://doi.org/10.1007/978-3-031-52513-1_9

generator cooling should be preferably from an air-cooled radiator with sufficiently protected coolant for the design service temperature. Where batteries are used to provide power for emergency equipment, they should be suited and sized for low temperature operation.

9.3 Navigational Equipment in Ice-Covered Waters

Navigation in ice-covered waters is assisted by specialised equipment and services to support both strategic (route planning) and tactical (ice avoidance) decisions. All navigational equipment is to be capable of being operated in the environmental conditions at the design service temperature. All vessels should have sufficient communications capability to download ice charts and forecasts (where available) from local ice service providers. Prior to a voyage, arrangements should be made for this information to be available, as the most useful information is frequently provided on a fee-for-service basis. Systems are also becoming available to download and display higher-resolution imagery from satellites or aircraft-based radar systems. Typically, this information requires specialised equipment and training for the user before it can be interpreted effectively; but can be a cost-effective investment for vessels making regular polar voyages by permitting the personnel to navigate so as to avoid ice blockages, etc. Standard marine radar does not provide a good picture of ice conditions. As visibility is frequently limited by darkness, snow, or fog, other navigation aids should be considered. Cross-polarised radar systems can provide much better resolution of ice features, including bergy bits and other dangerous free-floating ice. These are now becoming available from specialised radar suppliers. Operations under escort or in a convoy can use standard navigation and communications equipment. Vessel masters should verify that procedures and protocols are understood prior to commencing such operations, and that other vessels are made aware of any vessel characteristics that may influence their conduct and safety.

9.3.1 Equipment

The following equipment is typically required to be installed and carried onboard: equipment capable of receiving high resolution ice and weather charts; radar systems capable of picking up ice targets; adequate communications and signalling equipment; high powered xenon arc searchlights remotely operated and positioned to provide 360° of lighting capability, if possible; sound reception system for navigation bridge with enclosed bridge wings for reception of exterior noises/signals; and navigation lights that are self-heating, or otherwise provided with some other adequate means to maintain ice-free lenses.

9.4 Life Saving Appliances and Survival Arrangements

This section discusses the requirements for various lifesaving appliances. They are based in part on the following International Maritime Organisation (IMO) documents:

- *IMO Guidelines for Ships Operating in Arctic Ice-Covered Waters.*
- *IMO Guidelines for Ships Operating in Polar Waters.*
- *IMO Life Saving Appliance Code (2003) (as amended);* and
- *IMO International Code for Ships Operating in Polar Waters (Polar Code),* MSC.385(94).

Flag State administrations and the administrations responsible for the coastal areas that the vessel is intended to operate in may have additional requirements to those listed. Adequate supplies of protective clothing and thermal insulating materials must be readily available to all vessels operating in low temperature environments. Moreover, lifesaving appliances are to be of a type that is rated to perform its functions at the lowest temperature of either:

- T_D; or
- A temperature of – 30 °C (– 22 °F).

Additional rations and drinking water must also be provided onboard all lifeboats and life rafts. Extra rations (viz. caloric intake) are required by personnel subjected to extremely cold temperatures. Drinking water containers must be suitable to allow for thermal expansion when frozen. Consideration should be given to storing rations and drinking water, in addition to those in the lifeboats and life rafts, inside the deckhouse/ accommodations to protect them from freezing. These additional rations and drinking water are to be stored in such a space or container so as to be convenient for placement into the survival craft should evacuation become necessary. Refer to Chap. 11 of the *IMO Guidelines for Ships Operating in Polar Waters.* It should be understood that most EPIRBs and Personal Floatation Device (PFD) lights are automatically activated by contact with water, which may not happen when vessels are evacuated in ice-covered waters. Crews need to be aware of this fact and are to be provided with equipment (for example, manually activated) that facilitates other ways of transmitting distress and location information. EPIRBs and PFD lights selected are to be rated for operation at the design service temperature. Note that most EPIRBs work only down to – 20 °C (–4 °F) according to manufacturers' instructions. Care must be taken to verify that the selected EPIRBs are suitable for the t_D. The manufacturer should be consulted for guidance.

9.4.1 Lifeboats

Ice build-up in way of all lifeboats, life rafts, cradles, davits and other launching gear should be regularly removed so that launching arrangements are not hindered. This may include ensuring a wooden mallet is available at each station or in the vicinity of the lifesaving appliances.

Partially or totally enclosed lifeboats are required for all vessels operating in low temperature environments, as they offer shelter to the occupants from the environment. Lifeboats are to meet the requirements of the *IMO Life Saving Appliance Code* and the Flag State Administration. The following additional requirements are mandatory to be complied with:

- The capacity of the lifeboats must be sized to accommodate a minimum of 125% of the crew size based on increased dimensions as crews are presumed to be dressed in bulky cold weather clothing. Lifeboat access and operations are to be based on these increased dimensions. Most lifeboats will not accommodate their SOLAS capacities when crews or passengers are dressed in bulky clothing.
- The lifeboat engine is to be suitable for cold starting and continuous operation at the design service temperature in lieu of the test temperature specified in the LSA Code. Engine block heaters may be considered for oil sumps. Cooling water, fuel and lubricating oils for engines are to be suitable for engine operation at the design service temperature.
- For ice class vessels, the engine power is to be sufficient to have ice-transit capability in thin ice with a low concentration.
- Lifeboat steering systems should be designed so that it does not directly transmit ice forces on the rudder to the coxswain's hands.
- For ice class vessels, the lifeboat's keel is to be adequately strengthened to withstand an impact with ice.
- Lifeboats are to be provided with radio equipment and batteries suitable for operation at the minimum anticipated temperature.
- Lifeboat releasing gear is to be shielded or protected from freezing for ready release or reattachment.
- Lifeboats are to be provided with heaters and heating fans to reduce humidity and prevent icing and blockage of entrance doors. For design service temperatures less than −30 °C (−22 °F), heater cables are to be installed for hatches and doors to prevent freezing. Lifeboat internal spaces are to be heated both when the lifeboats are onboard as well as when the boats are operating independently, the power for these heat sources need not be the same. Heating for lifeboats when deployed is to be capable of maintaining a temperature of at least 10 °C (50 °F) with no one onboard at the design service temperature and be capable of temperature adjustment. Air flow inside the lifeboat is to ventilate all areas including the bow area with fresh heated air.

- Lifeboat windows in way of the control station are to be provided with heating or other means to prevent icing and frost; and
- Lifeboats are to be equipped so to be capable of deterring native animal invasion (e.g., polar bears).

Air-cooled engines provide additional heating and can reduce problems associated with frozen valves, piping and water intakes. The lifeboat's propeller is susceptible to damage from ice, particularly when operating astern because of its relatively shallow draft. Lifeboat manufacturers typically provide an ice guard for the propeller. Lifeboat releasing gear may be provided with a heated cover. Alternatively, the hull structure can be designed in such a way that it protects the lifeboat from the accumulation of snow and ice. Further protection can be provided to the lifeboat by providing a cover over the lifeboat. To address the potential formation of condensation inside the lifeboat resulting from the exhalation of the lifeboat occupants, consideration should be given to installing supplementary ventilation or air circulation features and heaters within the lifeboat enclosure. The amount of condensation formed is related to the number of persons occupying the lifeboat, the physical exertion required to reach the lifeboat, and the outside air temperature. It is recommended that lifeboats on vessels operating in ice infested waters be of a type with enhanced manoeuvring and acceleration capability. Lifeboat engine distillate fuel should have a cloud point well below the design service temperature. Lifeboat engine lubricating oil should have the correct viscosity at the design service temperature without the use of a heater. For vessels trading between temperature extremes, it may be necessary to provide instructions to vessel personnel to replace lubricants and other fluids when operating in a different temperature. During lifeboat testing in ice an ice block struck the boats steering nozzle causing an uncontrolled spin of the steering wheel. This spin could cause serious injury to the coxswain. Most lifeboats are manufactured from a composite material that may change properties with changing temperatures. The hull of the lifeboat at any temperature between 10°C (50°F) and TDST, must be able to withstand impacts with large ice floes. Ice strength varies greatly depending on ice type and temperature, therefore the ice strength used in the hull calculations is to be justified based on operation season, location, and ice class. With respect to deterrence of native animal populations, this refers to polar bears, which are the greatest threat to personnel, particularly in the Arctic (Figs. 9.1 and 9.2).

9.4.2 *Life Rafts*

Life rafts are designed to be manned *after* deployment. Davit launched life rafts are manned *prior* to launching.

Fig. 9.1 Open top lifeboat (not ideal for Arctic conditions)

Life rafts are to meet the requirements of the *IMO Life Saving Appliances Code* and the Flag State administration. Inflation of inflatable life rafts is to be completed within a period of three (3) minutes at the lowest temperature of:

- Its design service temperature; or
- A temperature of – 30 °C (– 22 °F).

After inflation, the life raft is to maintain its form when loaded with its full complement of persons and equipment. A manual inflation pump is to be stored near the inflatable life rafts if not provided within the pack. The hydrostatic release mechanisms are to be protected from icing (Figs. 9.3 and 9.4).

9.5 Rescue Boat

A rescue boat, if provided, is to meet the requirements of the *IMO Life Saving Appliances Code* and the flag State administration. The following additional equipment is to be provided onboard the rescue boat:

- Radio equipment with batteries suitable for the minimum anticipated temperature is to be installed.
- A battery charger.
- An engine block heater; and
- Cooling fluid, fuel and lubricating oils for engines are to be suitable for engine operation at the design service temperature.

Fig. 9.2 Fully enclosed lifeboat (improved survivability in Arctic conditions)

9.6 Launching Stations and Arrangements

Lifeboats should be capable of being launched into the vessel's track because it is assumed there will be little room to transit between the vessel side and the ice cover. This is applicable when relatively thin, open, floating ice is present. The *IMO Life Saving Appliance Code (2003)* (as amended) requires free-fall craft be provided with a secondary means of launching. This secondary means may be by gravity or stored mechanical power whose function is not to be impaired by low temperature. Arrangements are to be provided to enable personnel to enter the lifeboat after it is lowered onto the ice. In some cases, the ice may be of sufficient thickness to allow the crew to descend upon the ice instead of launching lifeboats. Regarding the requirement to permit lowering life rafts onto the ice, it is better to do this with the life raft in an un-inflated condition to allow the raft to be dragged clear of the vessel. This may be accomplished through the use of a davit. An inflated life raft

Fig. 9.3 Life raft capsules (pre-deployment)

Fig. 9.4 Life raft (post-deployment)

can act as a tent on the ice surface, protecting survivors from the worst of the cold and wind. The requirements of this section are applicable for lifeboats, life rafts and rescue boats. For ice class vessels, launching appliances onboard need to be designed with the following special considerations. Free-fall lifeboats cannot be released onto ice-covered waters without risking the safety of occupants because of impact with the ice and/or surfacing under ice floes. Standard lifeboats have almost no ice-transit capability, and they should be launched into the vessel's track to maximise their ability to get clear of a sinking or burning vessel. Launching stations are to be located considering their suitability to facilitate abandoning the vessel during navigation in ice-covered waters. Arrangements are to be provided to permit lowering life rafts onto the ice rather than into the water. Life raft storage arrangements should be provided to permit this operation. For ice class and non-ice class vessels, the forward end of the lifeboat stations is to be shielded with a roof and wind walls to reduce ice and snow accretion on the lifeboat canopy, davits, sheaves and wires. The launching arrangements for boats and their retrieval are to be protected with a cover over the sheaves and wires. Grease or lubricant used for the sheaves is to be suitable and is not to exhibit degraded performance at the design service temperature (Figs. 9.5 and 9.6).

9.7 Ice Gangway, Personnel Basket and Escape Chutes for Ice Class Vessels

An ice gangway or personnel basket used in conjunction with a crane may be considered as an additional means of evacuation, subject to approval of the Flag State administration. Escape chutes, if provided, are to meet the requirements of the *IMO Life Saving Appliances Code* and the Flag State administration. The escape chute is to provide a means of landing safely on the ice pack. Equipment is to be provided for shelter of all personnel.

9.8 Immersion Suits and Life Jackets

The requirements in Chap. 11 of the *IMO Guidelines for Ships Operating in Arctic Ice-Covered Waters* or the *IMO Guidelines for Ships Operating in Polar Waters* with regard to immersion suits and life jackets and the Flag State administration are to be complied with. Adequate supplies of immersion suits and life jackets are to be provided on the vessel for all persons onboard at any time. Personal immersion suits may be stored in cabins to be used during lifeboat musters in order to avoid moisture accumulating in suits stored at lifeboat stations (Figs. 9.7 and 9.8).

Fig. 9.5 Lifeboat station

9.9 Alarms, Escape Routes, and Access Routes

The requirements in Chap. 4, Accommodation and Escape Measures of the *IMO Guidelines for Ships Operating in Arctic Ice-Covered Waters* or the *IMO Guidelines for Ships Operating in Polar Waters*, and the Flag State administration are to be complied with. In summary, all vessels are to have arrangements or procedures to maintain escape routes free of snow and ice. Public Announcement and General Alarms are to remain functional and audible considering snow and ice accumulation and operation at the minimum anticipated temperature. The foghorn is to remain

Fig. 9.6 Freefall lifeboat station

operational at the minimum anticipated temperature, and to be capable of being safely de-iced or designed to be anti-iced. In general, escape routes are to be clearly marked. External escape routes such as walkways are recommended to be kept free of ice and snow. This may be accomplished by heating the surface, protecting the route with an enclosure or a combination of the two. Escape routes through hazardous areas are to be well marked and lighted. The use of abrasive coatings on escape routes should be considered.

9.10 Lighting

The lighting and emergency lighting on open decks must be designed to prevent icing. Exterior lighting systems, including emergency lighting, are to be designed such that the decks are illuminated with consideration of fog or falling snow. Energy saving lighting sources such as fluorescent or light emitting diode (LED) typically do not generate as much waste heat as incandescent, therefore icing and snow accumulation is more likely. Some light sources such as fluorescent lights may not illuminate at lower temperatures.

Fig. 9.7 Immersion suit

9.11 Drills and Emergency Instructions

Emergency and evacuation procedures including instructions for at sea and ice-covered waters are to be developed with an appropriate emphasis to changes to standard procedures made necessary by operations in low temperature environments. Refer to the requirements of Chap. 13, *Operational Guidelines* of the *IMO Guidelines for Ships Operating in Polar Waters* (2009) (Resolution A.1024(26)). The procedures are to include the following topics:

Fig. 9.8 Life jacket

- For ice class vessels, the propeller is to be protected from contact with ice.
- Vessel evacuation.
- Operation of safety equipment and supporting vessel systems, such as cranes or lifting appliances during an emergency or vessel evacuation.
- Survival at sea.
- Survival on ice/ashore.
- Fire and damage control equipment and system; and
- Cross-training of crew members.

Crew members must be provided with proper onboard instructions and be regularly trained in the use of the equipment carried onboard for emergency situations and evacuations.

9.12 Provisions and Spares

Refer to the requirements for *Personal Survival Kits and Group Survival Kits* in Chap. 11 Life-Saving Appliances and Survival Arrangements and paragraph 13.4.5 of the *IMO Guidelines for Ships Operating in Polar Waters* (2009) (Resolution A.1024(26)). The Flag State Administration and the administrations responsible for the coastal areas that the vessel is intended to operate in may have additional requirements related to provisions and spares.

Various national Administrations have additional regulations for vessels operating in their territorial waters beyond those listed in this book. It is recommended that vessel designers, owners and operators contact these Administrations directly regarding these regulations. Table 9.1 provides a list of Administrations in areas subject to low temperatures.

Table 9.1 Coastal maritime administrations

Country	Address	Contact/telephone	Web address
ARGENTINA, REPUBLIC OF Prefectura Naval Argentina	Av. Eduardo Madero, 235 (C1106 CCA) Buenos Aires, Rep. of Argentina	Director Policia Seguridad de la Navegacion Jefe Departmento Reglamentacion de la Navegacion: (54-11) 4318-7475 Ext 2411 or (54-11) 4318-7467	http://www.prefecturanaval.gov.ar
AUSTRALIA, COMMONWEALTH OF Australian Maritime Safety Authority	GPO Box 2181 Canberra City ACT 2601 Commonwealth of Australia	Principal Adviser Technical: (61-2) 6279 5049 or (61-2) 6279 5966	http://www.amsa.gov.au/
CANADA Transport Canada Safety & Security	Tower "C", Place De Ville 11th Floor 330 Sparks Street Ottawa Ontario K1A 0N8 Canada	Director of Ships and Operations Standards: (1-613) 991-3131 or (1-613) 993-8196	http://www.tc.gc.ca/en/menu.htm
CHILE Direccion General del Territorio Maritimo y de Marina Mercante Armada de Chile	Avenida Errazuriz 537 Valparaiso Chile	Director General: (56-32) 2208102 or (56-32) 2208097 (fax)	http://www.directemar.cl/
DENMARK Danish Maritime Authority	Skibsregistret 38C, Vermundsgade PO Box 2605 DK-2100 Copenhagen O, Denmark	Chief Ship Surveyor (45-39) 17 44 00 or (45-39) 17 44 01 (fax)	http://www.dma.dk/
ESTONIA Estonian Maritime Administration	Valge 4 11413 Tallinn Estonia	Director General: (372) 6205 700 or (372) 6205 706 (fax)	http://www.vta.ee/atp/?lang=en
FINLAND Finnish Transport Agency Maritime Department	Finnish Transport Agency PO Box 33 FIN-00521 Helsinki Finland www.fma.fi	Technical Approval (358-20) 448 4249 or (358-20) 448 4336 (358-20) 63 7373 (Operator)	http://portal.fma.fi/sivu/www/fma_fi_en/services/winter_navigation

(continued)

Table 9.1 (continued)

Country	Address	Contact/telephone	Web address
GREENLAND Danish Maritime Authority	Skibsregistret 38C, Vermundsgade PO Box 2605 DK-2100 Copenhagen O, Denmark	Chief Ship Surveyor (45-39) 17 44 00 or (45-39) 17 44 01 (fax)	http://www.dma.dk/
ICELAND Icelandic Maritime Administration	Vesturvor 2 PO Box 120 202 Kopavogur Iceland	Head of Technical Regulations Section: (354) 560 0000 or (354) 560 0060 (fax)	http://www.sigling.is/Eng lish
LATVIA Ministry of Transport of Latvia, Maritime Administration	Trijadibas iela 5 Riga LV-1048 Latvia	Director: (371) 67062101 or (371) 67860082 (fax)	http://www.jurasadministra cija.lv/index.php?pid=211
LITHUANIA Ministry of Transport and Communications of the Republic of Lithuania	Gedimino Av.17 Vilnius 2679 Republic of Lithuania	Director: 370 46 469657	http://www.msa.lt/index. php/en/27717
NETHERLANDS Inspectorate Transport and Water Management	Netherlands Shipping Inspectorate Gebouw Prinsenpoort 's-Gravenweg 665 (PO Box 8634 3009 AP) Rotterdam The Netherlands	Director: (31-10) 266 8500 or (31-10) 202 3424	http://www.ivw.nl/en/water/ koopvaardij/
NEW ZEALAND Maritime New Zealand	Level 8 gen-i Tower 109 Featherston Street (PO Box 27006) Wellington New Zealand	General Manager, Maritime Operations: (64-4) 4730111 or (64-4) 4941263	http://www.maritimenz.gov t.nz/
NORWAY, KINGDOM OF Norwegian Maritime Directorate	PO Box 2222 N-5509 Haugesund Norway	Vessels and Seafarers 47 52 74 50 00 47 52 74 50 01	http://www.sjofartsdir.no/en/
POLAND Ministry of Transport and Maritime Economy	Department of Maritime and Inland Waters Administration U1. Chalubinskiego 4/6 00-928 Warsaw Poland	Ships inspection – Head Office (48-22) 621 1448 or (48-22) 621 9437	http://www.en.mi.gov.pl/2-482076faaa403.htm
RUSSIAN FEDERATION Ministry of Transport of the Russian Federation Russian Northern Sea Route Administration	Maritime Administration ul. Rozhdestvenka, 1/4 103759 Moscow Russian Federation	Head of Maritime Administration (7-495) 926 10 00 or (7-495) 978 91 70	Ministry of Transport http:// www.mintrans.ru/ Northern Sea Route http:// www.morflot.ru/
SWEDEN, KINGDOM OF Swedish Maritime Administration	Maritime Safety Inspectorate Slottsgatan 82 SE-601 78 Norrkoping Kingdom of Sweden	Director of Maritime Safety: (46-11) 19 10 00 or (46-11) 10 19 49	http://www.sjofartsverk et.se/

(continued)

Table 9.1 (continued)

Country	Address	Contact/telephone	Web address
UKRAINE Ministry of Transport of Ukraine	State Department of Maritime and Inland Water Transport 14, prospect Peremogy Kyiv 01135 Ukraine	Director General: (380 44) 461 5320 or (380 44) 461 5304 (Emergency Officer)	http://www.mintrans.gov.ua/# http://www.kmu.gov.ua E-mail: security@morrechflot.gol.ua
UNITED STATES OF AMERICA US Coast Guard	Commandant (G-MOC) 2100 Second Street, SW Washington DC 20593-0001 Federal Communications Commission Wireless Telecommunications Bureau 445 12th Street SW Washington DC 20554	Chief, Office of Vessel Activities (202) 372-1210 (202) 372-1918 (fax) Public Safety and Private Wireless Division (202) 418-2771 (202) 418-2643	http://www.uscg.mil

Chapter 10
Requirements for Specific Vessel Types

10.1 Introduction

Many vessel types have design and operational characteristics which require special consideration when intended, or called upon to, operate in Arctic or polar conditions. This chapter addresses some of those additional requirements for specific vessel types which are now operating or under consideration for operation in low temperature environment services, including liquid natural gas (LNG) carriers, bulk carriers, offshore support vessels (OSV), and oil tankers.

10.2 Vessels Intended to Carry Liquefied Gases in Bulk

LNG carriers, due to the nature of the cargos carried onboard, are constructed with containment systems specifically designed with cryogenic insulating characteristics. Many other systems and equipment are also designed with materials and equipment capable of operating at low temperatures. However, it is necessary that both external and internal low temperatures are taken into consideration. Safety systems such as venting arrangements need to be protected from ice and snow accumulation. These requirements are intended to apply to vessels with the classification ✠ **A1 LIQUEFIED GAS CARRIER** or ✠ **A1 LIQUEFIED NATURAL GAS CARRIER** or equivalent. These are additional requirements to those discussed throughout the other chapters of this book.

A. Olsen, *Ship Operations in Extreme Low Temperature Environments*, Springer Series on Naval Architecture, Marine Engineering, Shipbuilding and Shipping 19,
https://doi.org/10.1007/978-3-031-52513-1_10

10.2.1 Design Loads

The requirements of this section will only consider LNG carriers operating in waters without any significant ice cover. Loads resulting from interaction with ice and loads caused by navigation in ice are to be verified as part of the overall design development and considered when assigning a suitable Ice Class Notation. The following systems/ components are subject to the additional requirements in this section for operation at the t_D:

- Additional fire safety requirements, including water spray system.
- Ballast tank heating arrangements.
- Cargo piping diagram.
- Cargo pressure and temperature control systems.
- Cargo tank vent systems.
- Details and arrangements of re-liquefaction systems, if installed.
- Details of additional personnel protection arrangements.
- Details of the arrangements for the use of cargo as fuel and associated fuel gas piping diagrams and other arrangements for utilisation of boil-off gas (BOG).
- Electrical installations.
- Environmental control systems (e.g., nitrogen purging systems).
- Inert gas (IG) systems.
- Instrumentation for gauging, gas detection, and cargo handling controls.
- Inter barrier and insulation space venting systems; and
- Mechanical ventilation of the cargo area.

10.2.2 Material Selection

Materials entering the construction of LNG carriers and associated equipment and cargo containment systems must be in accordance with the requirements in Chap. 7 of this book. It may also be worth referring to the *ABS Rules for Materials and Welding (Part 2)* or alternative Class equivalent, where applicable.

10.2.3 Hull Construction and Equipment

Some concerns have been expressed about the robustness of different gas containment systems under the acceleration loads that may result from ice impact. There are no inherent reasons why any system capable of open water operation and meeting gas carrier damage tolerance standards cannot be used in vessels intended for navigation in ice. Ice breaking loads should be verified as part of the overall design development. Ice loads on the containment systems or any cover for the containment systems may be estimated by applying the allowance for ice accretion in accordance with the IMO

Code on Intact Stability or the requirements of the Flag State administration. *Ice loads.* Loads attributable to ice and snow accumulation on the containment system or any cover for the containment system are to be determined, considering both the static load and dynamic load from sliding or dropped ice. *Weather deck equipment.* Due to the nature of the cargo, many cargo-related systems and equipment are designed for operation at low temperatures. However, it is necessary to consider both external and internal low temperatures in the design and selection of materials. For cargo piping and piping supports installed on the main deck and exposed to the weather, including ventilation piping, the design loads are to include the expected amount of ice and snow accumulation. The supports for liquid and vapour cargo piping systems on deck are to be arranged so that free expansion and contraction of the pipes during cargo operations cannot be blocked by accumulated ice or snow. *Stability.* LNG carriers with cargo tanks or tanks covers protruding from deck and/or with many exposed piping systems and features in elevated locations are likely to accumulate substantial amounts of ice in the upper part of the vessel, adversely affecting stability. Intact and damage stability calculations are to consider the weight of ice accumulation in exposed areas by means of an allowance for ice accretion based on the specific configuration of decks, piping systems and features in the cargo area for the vessel.

Machinery and electrical equipment. Many LNG carriers utilise cargo boil-off as fuel for propulsion power. The normal rate of boil-off (NBOR) depends on the insulation of the cargo tanks and the external temperatures. For vessels capable of operating in the "gas only mode" and operating near the design service temperature, the expected boil-off volumes are to be determined to assure that the provisions for forced boil-off are adequate. Valves. Safety Relief Valves (SRV) and other valves exposed to the weather are to be arranged in such a way as to provide their local and remote activation under all circumstances. Means of ice removal and heat tracing of the valves are to be provided. Valve actuators with local position indicators are to be of the enclosed type. Valves for cryogenic service are more susceptible to packing gland leaks at low ambient temperatures. Data is to be submitted to confirm that these valves will be suitable for service at the minimum anticipated temperature. Cargo tank and inter-barrier space, venting and pressure regulating arrangements. All piping and pressure/vacuum preventing systems including valves are to be provided with a heat tracing system to always provide for their functionality. Operational procedures for local testing are to be provided to the operator. Heating media. The maximum temperature of heating media within the cargo area must take into consideration the temperature class associated with the auto ignition temperature of the cargoes being carried. The electrical equipment and instrumentation for heat tracing systems installed within the cargo area must be of an intrinsically safe type. Cargo control/ environmental systems. Equipment used for the environmental control of the cargo spaces, inter-barrier spaces and other spaces that require a controlled atmosphere, either by chemical composition, temperature or air quality, are to be designed to maintain a dew point below the lesser of the design service temperature or – 30 °C (–22 °F).

Mechanical ventilation. Spaces required to be ventilated before entering are to be equipped with suitable means of pre- heating the ventilation systems to provide for their proper functioning at the design service temperature. *Towing fittings.* The stowage compartment for emergency towing gear is to be provided with a means for de-icing and is to be always readily accessible. *Cargo manifolds.* The design of a dedicated watch station in way of the cargo manifolds should be considered for vessels that are expected to regularly load/discharge cargoes in extremely harsh environments. *Note:* emergency disconnection is performed by the terminals (without purging) to protect their terminal loading arms. The vessel provides the Emergency Shutdown System (ESDS). When an ESDS is activated, the manifold valves close (under all service conditions within 30 s), thus discontinuing loading or unloading operations. During the discharge/loading operation, some operators will pump seawater over the decks and manifold areas to act as a thermal barrier should there be an LNG leak. The cargo loading platform drip tray should be so designed that filling with seawater during the discharge/loading operation is not required. Seawater for the water curtain may be preheated to avoid excessive ice build-up at the side shell plating. Vent masts—It is extremely important that the vent mast is drained of any accumulation of water. The purpose of this is to ensure that the relief valves operate at their correct settings, which would otherwise be altered if any water were to accumulate in the vent mast and flow onto the valve assembly. Vapour heater—LNG carriers are fitted with vapour heaters for use during controlled venting of vapour. Controlled venting is rarely done. At low ambient temperatures, a vapour heater is mandatory, preventing the cold gas vapour to fall directly on deck. However, individual tank vents are not designed to be routed to the heater. In the instance of an individual tank relief valve lifting, uncontrolled venting occurs, and cold vapour will be released to the atmosphere.

10.2.4 Access to Deck Areas and Cargo Machinery

Electric motor room and cargo compressor room access. Access to the cargo machinery room must always be available. To facilitate this, heat tracing of the door coamings may be provided. *Cargo manifold and cargo vapour system.* If the cargo manifold and cargo vapour areas and associated valves are enclosed from the weather, the enclosures are to be arranged in such a way as to provide for sufficient ventilation of the space when the vessel is loading and unloading. When mechanical ventilation is provided, the ventilation capacity is to be at least 30 air changes per hour based on the total volume of the space. Additional gas detectors are to be provided. *Arrangement of vent mast.* The vent mast head is to be designed to prevent blockage by ice accumulation. Drip pans for the purpose of collecting moisture in the vent mast are to be provided. Drip pans with associated drainpipes inside the vent mast head and foot are to be provided with trace heating. *Vapour heater.* A steam-operated vapour heater for the purpose of preventing vapour descending to the deck is to be installed in the vapour line before the vent riser used for controlled venting. *Boil off gas piping.* Insulation of the boil-off gas piping is required in way of areas exposed or subject to the design service temperature.

10.2.5 Access to Machinery on Deck

Access to the deck for personnel (e.g., doors, hatches, etc.) and to other equipment located on the weather deck and the bow area are required to be always available. Access for the operating personnel is to be arranged in such a way that they are not obstructed by ice accumulation or endangered by overhead ice accumulation. Additional arrangements are to be made to prevent the crew from being hit by the release and sliding of accumulated ice and snow from inclined covers of the containment systems onto any walkways or passageways.

10.2.6 Monitoring Systems

Temperature monitoring systems. Additional temperature monitoring systems may be installed to monitor the efficiency of the heat tracing systems employed around the vessel. *Alarm systems.* Heat tracing and additional temperature monitoring systems must be arranged such that any failure will result in an alarm at the manned cargo control station and on the navigational bridge. *Instrumentation.* The following instrumentation is to be installed and fitted with heat tracing equipment as considered necessary by the vessel designer and/or Class:

- Level indicators for all cargo tanks.
- Overflow controls.
- Pressure gauges.
- Temperature-indicating devices; and
- Instrumentation for gas detection systems.

Other monitoring equipment. When remotely operated equipment, such as CCTV is utilised, any reduced movement of the CCTV enclosure must not impair its functionality to transmit images. Heated enclosures may be used to provide for their continued operation.

10.2.7 Fire and Safety Systems

Additional fire and safety systems are required to be installed onboard LNG carriers in accordance with the *International Gas Carrier Code* and are to be arranged and installed to provide for their continued operation at the minimum anticipated temperature. The water spray deluge system is to be equipped with heat tracing and sufficient drainage to prevent freezing. Alternative arrangements for water deluge systems and additives to the water may be accepted based on the approval of the Flag State administration. Dry chemical powder systems are to be arranged to prevent clogging of the nozzles and provide for the release of the medium at the minimum anticipated temperature.

10.3 Vessels Intended to Carry Ore or Bulk Cargoes/Cargo Ships

These requirements are intended to apply to vessels with the classification ✠ **A1 BULK CARRIER,** ✠ **A1 ORE CARRIER**, or ✠ **A1 OIL OR BULK/ORE (OBO) CARRIER** or equivalent. The requirements discussed in this section are also applicable to vessels engaged in the transportation of oil. These are additional requirements to those in the other sections/chapters of this book.

10.3.1 Material Selection

In frigid conditions, the chance of brittle cracking will be increased with frequent impact loading associated with cargo and by mechanical handling equipment. Accordingly, consideration may need to be given to using tougher materials in the cargo hold bottom and side plating, particularly if the vessel has the **GRAB** (or equivalent) notation. Materials entering the construction of Bulk Carriers are to be in accordance with the requirements in Chap. 2 and the *ABS Rules for Materials and Welding (Part 2)* or alternative Class equivalent.

10.3.2 Hull Construction and Equipment

Hatch cover sealing arrangements must be suitable for the minimum anticipated temperature. Arrangements are to be provided to prevent frozen water adhering to the seals. The selection of hatch sealing material should be specially considered considering their reduced ductility in extremely cold temperatures. Sealing problems of the main hatches may result in seawater entering the hold increasing risk of cargo contamination and release of products.

10.3.3 Machinery and Electrical Equipment

Ballast piping. If fitted, piping connecting the upper wing tanks and lower wing tanks is to be protected from freezing. If a steam heating coil is fitted on the surface of this connecting pipe, it is to be protected to avoid being damaged during cargo loading and unloading operations. If an air bubble system or heating coil system is installed and operating in the lower ballast tank to prevent freezing, the heated ballast water or air bubbles will be passing through the connecting pipe to the upper wing tank and no additional freezing protection will be required. *Ballast water tanks.* With the hatch covers open for cargo loading or discharge, the ballast tanks in the double bottom and lower wing tanks may be exposed to the design service temperature. Accordingly, the requirements in Chap. 4 are to be complied with.

10.3.4 Access to Deck Areas and Cargo Machinery

The requirements for cargo handling equipment are discussed in Chap. 8.

10.4 Offshore Support Vessels

These requirements are intended to apply to offshore support vessels (OSV) with the classification ✠ **A1 OFFSHORE SUPPORT VESSEL** or equivalent**.** Vessels with lengths more than the Class Rules relating to OSV are also generally held subject to these requirements. These are additional requirements to those in the other chapters of this book. OSV which operate in cold regions frequently double as ice breakers and ice management vessels and should be appropriately designed for these functions. This includes selection of ice class (strengthening of hull and machinery), and the selection of hull forms and appendages suitable for ice operations.

10.4.1 Material Selection

Materials entering the construction of support vessels are to be in accordance with the requirements in Chap. 2 and the *ABS Rules for Materials and Welding (Part 2).*

10.4.2 Hull Construction and Equipment

Tanks. Tanks are to be arranged or provided to accept waste fluids from the drill ship. *Stability.* About vessels' stability, it is important for masters to have a full understanding of the special challenges of freezing weather operations and of the operating limits of their vessels. If the hull form promotes ride-up onto larger ice features, stability when beached should be specially considered. Ice knives or skegs can be fitted both to reduce ride-up and to promote splitting of large floes. OSV can be particularly susceptible to reduced stability because of the effects of icing. These include both accumulation on equipment, making it difficult or impossible to operate, and impacts on heel, trim, and stability in general. Design features can be selected to reduce ice build-up, and de-icing equipment can be installed. Refer to Chap. 7. *Machinery and electrical equipment.* OSV typically conduct a wide range of operations that require the operation of deck machinery such as winches, cranes, etc. These, and their control stations, are to be protected from cold, ice and snow, as practicable. Drilling muds and other specialised bulk cargoes carried by OSV may be particularly susceptible to freezing and may flow less freely at low temperatures. This fact is to be considered at the design stage or in the selection of existing vessels for low temperature operations. Valve and control station locations are also to be considered. Where the transfer of fuel, drilling muds and other specialised bulk cargoes to drill ships at sea is anticipated, all equipment associated with the transfer such as, hoses, valves, fittings, etc. are to be suitable for operation at the minimum anticipated temperature.

10.5 Vessels Intended to Carry Oil in Bulk

Arctic maritime administrations and intergovernmental organisations recognise that pollution of polar waters can lead to severe consequences for the environment, aggravated by the limited availability of response infrastructure and the specific challenges of spill treatment in the presence of ice. Consequently, tankers are frequently subjected to additional design and operational requirements. Specific regulations and guidelines may include oil transfer requirements, restricted operations and routing, carrying dedicated ice navigators or observers and, in some circumstances, arranging for mandatory ice breaker support. Equipment and outfit issues particularly relevant to tankers include arrangements to prevent the freeze-up of vent lines, measures to protect valves and manifolds from icing, and special consideration of ballasting arrangements. Tankers intended for operations in ice may need additional ballast to immerse vulnerable elements such as propellers and rudders, which are near or above the waterline in some standard open water designs. These requirements are intended to apply to vessels with the classification ✠ **A1 OIL CARRIER**, ✠ **A1 FUEL OIL CARRIER**, or ✠ **A1 CHEMICAL CARRIER** and are in addition to those outlined elsewhere in this book.

10.5.1 Material Selection

Materials used in the construction of vessels intended to carry oil in bulk are to be in accordance with the requirements in Chap. 2 and the *ABS Rules for Materials and Welding (Part 2)* or alternative Class equivalent.

10.5.2 Hull Construction and Equipment

Vessels should be designed so that they are provided with double sides and a double bottom

For cargo piping and piping support details, including vent piping installed on the main deck and exposed to the weather, the design loads must include the expected amount of ice and snow accumulation. The supports for cargo and ballast piping systems on deck are to be arranged so that free expansion and contraction of the pipes during cargo operations cannot be blocked by accumulated ice or snow. Consideration is to be given to enclosing deck piping in a tunnel or similar enclosure.

10.5.3 Machinery and Electrical Equipment

Towing fittings. The stowage compartment for emergency towing gear is to be provided with a means for de-icing and is to be always readily accessible. For ice class tankers, emergency towing fittings at the forward end required as per SOLAS Chapter II-1, Regulation 3–4.

Pressure/vacuum valves

Acceptable means for de-icing pressure/vacuum valves are to be provided so that they always remain operable

Some vessel operators fill the pressure/vacuum valves with anti-freeze fluid. Other operators regularly de-ice pressure/vacuum valves using an on-deck steam supply. Some valve manufacturers permit a certain amount of icing on the valve before de-icing must be performed. The valve manufacturers should be consulted for instructions. *Liquid filled pressure/vacuum relief.* Where a liquid filled means of relieving pressure or vacuum is used, the liquid must be kept from freezing. Therefore, a liquid is to have a freezing point below the minimum anticipated temperature, or the unit is to be insulated and heat traced. If an anti-freezing liquid is used, the fluid level is to be adjusted accordingly with the density of the liquid. *Cargo tank purging and/or gas-freeing.* The cargo tank purging and/or gas-freeing system is to be designed to be operable at the design service temperature.

10.5.4 Inert Gas System

> The inert gas system deck seal is either filled with anti-freeze or heat traced to keep the seal operable

The inert gas system scrubber is to be in a compartment protected from the weather. The inert gas system deck seal is to be operable under temperature conditions specified in the vessel's winterisation plan. Where a liquid filled means of pressure relief is attached to the inert gas lines, measures are to be taken to from the inert gas lines, draining into the liquid filled means of pressure relief. This may be achieved by arrangements such as a trap and drain before the liquid filled pressure relief.

Piping

> Insulated piping should be properly installed and maintained to prevent moisture becoming entrapped between the insulation and the pipe and causing corrosion

All piping with contents with freezing points above the design service temperature is to be provided with a suitable means to prevent freezing. Acceptable arrangements include suitable insulation, heat tracing or continuous fluid circulation or a combination thereof.

10.5.5 Access to Deck Areas and Cargo Machinery

Cargo manifold. Protection from the weather or anti-icing equipment needs to be provided to permit hook-up, continuous operation or disconnection when scheduled. This may be accomplished by heated covers and/or hot water wash systems. An enclosure for the cargo manifold area is not required. If the cargo manifold area and associated valves are enclosed from the weather, the enclosures are to be arranged in such a way as to provide sufficient ventilation of the space when the vessel is loading and unloading. Additional gas detectors are to be provided. The design of a dedicated watch station in way of the cargo manifolds should be considered for tankers that are expected to regularly load/discharge cargoes in extremely harsh environments. At many offshore terminals, tankers may be forced off their moorings by ice movement and may need to shut down transfer operations rapidly to avoid spills. Quick disconnect and purging arrangements may need to be provided.

10.5.6 Monitoring Systems

Monitoring of heat tracing systems may be accomplished by installation of lamps which indicate the line current flow through detection of line current or voltage. *Temperature monitoring systems.* Additional temperature monitoring systems are to be installed to monitor the efficiency of the heat tracing systems. The monitoring system indicators are to be located at a manned control station. All systems provided with heat tracing are to be monitored. *Alarm systems.* Heat tracing and additional temperature monitoring systems are to be arranged in such a way that any failure will result in an alarm at the manned control station. *Instrumentation.* The following instrumentation is to be installed and fitted with heat tracing equipment as considered necessary:

- Level indicators for cargo tanks.
- Overflow control.
- Pressure gauges.
- Temperature indicating devices; and
- Instrumentation for gas detection systems.

10.5.7 Other Monitoring Equipment

When remotely operated equipment, such as closed caption television (CCTV), is utilised, any reduced movement of the CCTV enclosure is not to impair its functionality to transmit images. Heated enclosures are to be used to provide for their continued operation.

10.5.8 Fire and Safety Systems

If additional fire and safety systems are installed, they are to be suitable for continuous operation at the design service temperature. For example, if a water spray deluge system is installed, it is to be equipped with heat tracing and sufficient drainage to prevent freezing. If the water spray deluge system is in an enclosed space and the piping is normally dry until service is activated, heat tracing will not be necessary. Alternative arrangements for water deluge systems and additives to the water may be accepted based on the approval of the flag administration. Dry chemical powder systems are to be arranged to prevent clogging of the nozzles and release the medium at the design service temperature.

10.6 Ice Class Draught Marking

Subject to the provisions discussed in Chap. 15, Baltic ice classes, section *Maximum and minimum draught fore and aft*, the vessel's sides are to be provided with a warning triangle and with a draught mark at the maximum permissible ice class draft amidships (refer to Fig. 10.1). The purpose of the warning triangle is to provide information on the draught limitation of the vessel when it is sailing in ice for masters of icebreakers and for inspection personnel in ports.

The upper edge of the warning triangle is to be located vertically above the "ICE" mark, 1000 mm (39.3 in) higher than the Summer Load Line in fresh water but in no case higher than the deck line. The sides of the triangle are to be 300 mm (11.8 in) in length. The ice class draught mark is to be located 540 mm (21.2 in) abaft the centre

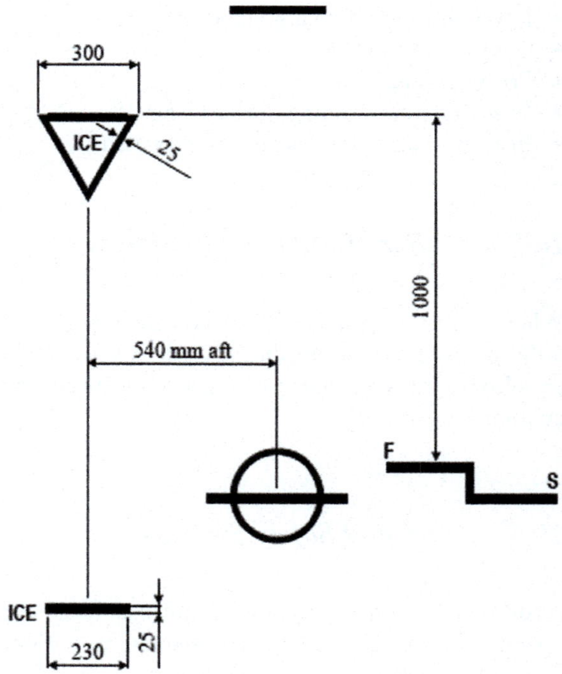

Fig. 10.1 Ice class draught marking

of the load line ring or 540 mm (21.2 in) abaft the vertical line of the timber load line mark, if applicable. The marks and figures are to be cut out of 5–8 mm (0.1–0.3 in) plate and then welded to the vessel's side. The marks and figures are to be painted in a red or yellow reflecting colour in order to make the marks and figures plainly visible even in ice conditions. The dimensions of all figures are to be the same as those used in the load line mark.

Part III
Polar Class Notations

Chapter 11
Structural Requirements for Polar Class Vessels

11.1 Introduction

In Part of this book, we examined the general Class requirements pertaining to vessels intended to operate in extreme low temperature environments (i.e., in Arctic or polar latitudes). In this Part 3, we will discuss the general requirements for ship designers and vessel owners/operators seeking optional ice strengthening class notations. This Chapter 11 provides important guidance on the structural requirements for vessels seeking (or are in the process of being assigned) a Polar Class (**PC**) notation. Chapter 12 concerns the machinery requirements for **PC** vessels based on based on IACS UR I1, I2, and I3. Chapter 13 sets out the requirements for enhanced **PC** notations, and Chap. 14 provides the requirements for vessels intending to navigate in first-year ice. In the final chapter of this Part 3, Chap. 15 examines the Class requirements associated with the Baltic Ice Class notation available to vessels intending to navigate the Baltic Sea during the winter season as provisioned in the *1985 Joint Finnish Swedish Ice Class Rules.* The guidance in these chapters is applicable to vessels of any length and are in addition to the guidance contained in Part 2 of this book, as appropriate. Vessels intended for navigation in the Canadian Arctic are required to comply with the requirements of the *Canadian Arctic Shipping Pollution Prevention Regulations.* Classification societies can issue an Arctic Pollution Prevention Certificate when duly authorised by the Canadian Flag State administration. It is the responsibility of the owner to determine which ice class is most suitable for the intended service. For guidance only, this chapter draws on the ABS rules for classing vessels operating in polar regions, however the IACS *Unified Requirements for Polar Class Ships* are incorporated into Sect. 11.1, which replaces the previous ABS general ice classes: **A5** through **A1**. Where the IACS requirements do not have a requirement comparable to the ABS requirement, the ABS requirement has been retained in Chap. 13 for vessels requiring the optional Enhanced notation. With these changes, the ice strengthening requirements for general ice classes **A0**, **B0**, **C0**, and

A. Olsen, *Ship Operations in Extreme Low Temperature Environments*, Springer Series on Naval Architecture, Marine Engineering, Shipbuilding and Shipping 19,
https://doi.org/10.1007/978-3-031-52513-1_11

D0 and for non-self-propelled vessels have been incorporated into Chap. 12. Furthermore, in addition to the guidance offered herein, ship designers and vessel owners/operators are strongly advised to acquaint themselves with the IMO statutory instruments having requirements specific to cold regions operations. The ice classes are as follows in Table 11.1.

The requirements for **PC** notations apply to those vessels constructed of steel and intended for independent navigation in ice-infested polar waters. Vessels that comply with the requirements of Chaps. 12 and 13 may be considered for a **PC** notation as listed in Table 11.1. The requirements of these chapters are in addition to the open water requirements set by Class. If the hull and machinery are constructed such to comply with the requirements of different polar classes, then both the hull and machinery will be assigned the lower of these classes in the Certificate of Classification. Compliance of the hull or machinery with the requirements of a higher polar class is also to be indicated in the Certificate of Classification.

Table 11.1 Ice class notations

Polar class	Polar class	Polar class, enhanced	First-year ice class	Baltic class
PC1	ICE BREAKER, PC1	PC1, ENHANCED		
PC2	ICE BREAKER, PC2	PC2, ENHANCED		
PC3	ICE BREAKER, PC3	PC3, ENHANCED		
PC4	ICE BREAKER, PC4	PC4, ENHANCED		
PC5	ICE BREAKER, PC5	PC5, ENHANCED		
PC6	ICE BREAKER, PC6	PC6, ENHANCED		I AA (Minimum engine output power XX kW/hp)
PC7	ICE BREAKER, PC7	PC7, ENHANCED	A0	I A (Minimum engine output power XX kW/hp)
			B0	I B (Minimum engine output power XX kW/hp)
			C0	I C (Minimum engine output power XX kW/hp)
			D0	
			E0	

Notes (1) The shaded ice classes are eligible for **ICE BREAKER** class notation, (2) This table shows the approximate correspondence between different ABS ice class notations. It is not to be interpreted to imply direct equivalencies between ice classes

Table 11.2 Polar class descriptions

Polar class	Ice description (based on WMO sea ice nomenclature)
PC1	Year-round operation in all polar waters
PC2	Year-round operation in moderate multi-year ice conditions
PC3	Year-round operation in second-year ice which may include multi-year ice inclusions
PC4	Year-round operation in thick first-year ice which may include old ice[1] inclusions
PC5	Year-round operation in medium first-year ice which may include old ice[1] inclusions
PC6	Summer/autumn operation in medium first-year ice which may include old ice[1] inclusions
PC7	Summer/autumn operation in thin first-year ice which may include old ice[1] inclusions

Note (1) "Old ice" means second year ice or multi-year ice

Vessels requiring ice breaker assistance are expected to comply with the additional requirements outlined in Chap. 14. Provided all polar class vessel requirements as specified are met, such vessels may be distinguished in the Record by Ice Class followed by ice class **PC7** through **PC1**, as applicable. Vessels which are assigned a **PC** notation and complying with the relevant requirements of Chaps. 12, 13 and 14 may be given the additional notation **'ICE BREAKER'**. The **ICE BREAKER** notation refers to any vessel having an operational profile that includes escort or ice management functions, having powering and dimensions that allow it to undertake aggressive operations in ice-covered waters. These vessels are to be distinguished in the Record by the notation **ICE BREAKER** followed by an appropriate Ice Class notation (for example, ✠ **A1, ICE BREAKER**, **ICE CLASS PC3**). For vessels which are assigned a **PC** notation, the hull form and propulsion power must be such that the ship can operate independently and at continuous speed in a representative ice condition, as defined in Table 11.2 for the corresponding polar class. For vessels and vessel-shaped units which are intentionally not designed to operate independently in ice, such operational intent or limitations are to be explicitly stated in the Certificate of Classification. For vessels which are assigned a **PC** notation **PC1** through **PC5**, bows with vertical sides, and bulbous bows are generally to be avoided. Bow angles should in general be within the range specified in Chap. 13. For vessels which are assigned a **PC** notation **PC6** and **PC7** and are designed with a bow with vertical sides or bulbous bows, operational limitations (restricted from intentional ramming) in design conditions are to be stated in the Certificate of Classification.

11.1.1 Selection of Polar Classes

The **PC** notation and descriptions are given in Table 11.2. It is the responsibility of the vessel owner and or operator to select the most appropriate polar class for the vessel when applying for the notation. The descriptions in Table 11.2 are intended to guide ship designers, vessel owners and operators and maritime administrations in selecting the most appropriate polar class to match the requirements for the vessel with its intended voyage or service. The **PC** notations are used throughout this chapter to convey the differences between classes with respect to operational capability and strength.

11.2 Strengthening for Navigation in Ice

11.2.1 Structural Requirements for Polar Class Vessels

The structural requirements discussed in this chapter apply to polar class vessels in accordance with Table 11.1. For the benefit of readers, the following definitions are provided as a point of reference. *Ice belt.* The ice belt is that reinforced portion of the shell and hull appendages that overlaps the upper and lower ice waterlines and is subject to the design ice loads. The required ice belt overlap extends from 1.5 m (4.9 ft) below the lower ice waterline to 1.0 m (3.2 ft) or 1.5 m (4.9 ft) above the upper ice waterline, depending upon polar class. In the bow area, the overlap increases linearly to 2.0 m (6.5 ft) above the upper ice waterline at the stem. Refer to Fig. 11.1 for details. *Upper and lower ice waterlines.* The upper and lower ice waterlines upon which the design of the vessel has been based are to be indicated on the Certificate of Classification. The upper ice waterline (UIWL) is to be defined by the maximum drafts fore, amidships and aft. The lower ice waterline (LIWL) is to be defined by the minimum drafts fore, amidships and aft. The lower ice waterline is to be determined with due regard to the vessel's ice-going capability in the ballast loading conditions. The propeller is to be fully submerged at the lower ice waterline. *Displacement.* The displacement, D, is the moulded displacement in metric tons (long tons) at the upper ice waterline.

Length. The vessel's scantling length, L, is the distance in metres (feet) measured on the waterline at the scantling draft from the fore side of the stem to the centreline of the rudder stock. For use with the Class Rules, L is generally not to be less than 96% and need not be greater than 97% of the extreme length on the waterline at the scantling draft. The forward end of L is expected to coincide with the fore side of the stem on the waterline on which L is measured. In ships without a rudder stock (e.g., ships fitted with azimuth thrusters), L is to be taken equal to 97% of the extreme length on the waterline at the scantling draft. In ships with an unusual stern and bow arrangement the length, L, is to be specially considered. For polar class vessels, the L is measured on the upper ice waterline, in m (ft). The length L_{UI} is the distance, in m (ft), measured horizontally from the fore side of the stem at the intersection with the upper ice waterline (UIWL) to the after side of the rudder post, or the centre of the rudder stock if there is no rudder post. L_{UI} is not to be less than 96%, and need not be greater than 97%, of the extreme length of the upper ice waterline (UIWL)

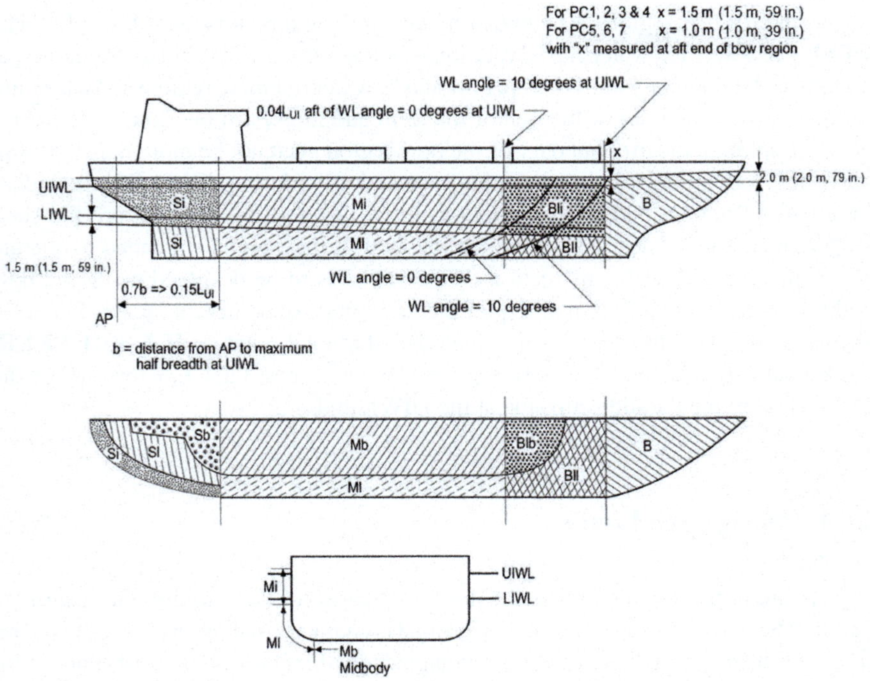

Fig. 11.1 Hull area extents

measured horizontally from the fore side of the stem. In ships with unusual stern and bow arrangement the length L_{UI} will be specially considered. This is not to be confused with the freeboard length L_f, which is defined as the distance in metres (feet) on a waterline at 85% of the least moulded depth measured from the top of the keel from the fore side of the stem to the centreline of the rudder stock or 96% of the length on that waterline, whichever is greater. Where the stem is a fair concave curve above the waterline at 85% of the least moulded depth and where the aftermost point of the stem is above the waterline, the forward end of the length, L_f, is to be taken at the aftermost point of the stem above that waterline. In ships designed with a raked keel, the waterline on which this length is measured is to be parallel to the designed waterline.

11.3 Hull Areas

The hull of all polar class vessels is divided into areas reflecting the magnitude of the loads that are expected to act upon them. In the longitudinal direction, there are four regions: the *bow*, *bow intermediate*, *midbody* and the *stern*. The bow intermediate, midbody and stern regions are further divided in the vertical direction into the *bottom*, *lower* and *ice belt* regions. The extent of each hull area is illustrated Fig. 11.1. The UIWL and lower ice waterline LIWL are as defined in Fig. 11.1 notwithstanding, at no time is the boundary between the bow and bow intermediate regions to be forward of the intersection point of the line of the stem and the vessel baseline. Figure 11.1 notwithstanding, the aft boundary of the bow region need not be more than 0.45 L_{UI} aft of the fore side of the stem at the intersection with the UIWL. The boundary between the bottom and lower regions is to be taken at the point where the shell tangent is inclined 7 degrees from the horizontal. If the vessel is intended to operate astern in ice regions, the aft section of the vessel is to be designed using the bow and bow intermediate hull area requirements as prescribed herein: again, Fig. 11.1 notwithstanding, if the vessel is assigned the additional notation **ICE-BREAKER**, the forward boundary of the stern region is to be at least 0.04 L_{UI} forward of the section where the parallel ship side at the UIWL ends.

11.4 Design Ice Loads

A glancing impact on the bow is the design scenario for determining the scantlings required to resist ice loads. The design ice load is characterised by an average pressure (P_{avg}) uniformly distributed over a rectangular load patch of height (b) and width (w). Within the bow area of all polar class vessels, and within the bow intermediate ice belt area of polar class **PC 6** and **PC 7**, the ice load parameters are functions of the actual bow shape. To determine the ice load parameters (P_{avg}, b and w), it is required to calculate the following ice load characteristics for sub-regions of the bow area; shape coefficient (fa_i), total glancing impact force (F_i), line load (Q_i) and pressure (P_i). In other ice-strengthened areas, the ice load parameters (P_{avg}, b_{NonBow} and w_{NonBow}) are determined independently of the hull shape and based on a fixed load patch aspect ratio, $AR = 3.6$. Design ice forces calculated according to the bow characteristics discussed in *bow area (3)* are applicable for bow forms where the buttock angle γ at the stem is positive and less than 80°, and the normal frame angle β' at the centre of the foremost sub-region, as defined in *bow area (1)*, is greater than 10°. Design ice forces calculated according to *bow area (4)* are applicable for ships which are assigned the polar class **PC6** or **PC7** and have a bow form with vertical sides. This includes bows where the normal frame angles β' at the considered sub-regions, as defined in *bow area (1)*, are between 0° and 10°. For ships which are assigned the polar class **PC6** or **PC7**, and equipped with bulbous bows, the design ice forces on the bow are to be determined according to *bow area (4)*. In addition, the design forces are not to be taken less than those given in *bow area (3)*, assuming $fa = 0.6$ and $AR = 1.3$. For ships with bow forms other than those defined in *bow area (5)* to *bow area (7)*, design forces are to be specially considered. Vessel structures that are not directly subjected to ice loads may still experience inertial loads of stowed cargo and equipment resulting from ship/ice interaction. These inertial loads, based on the maximum accelerations as given in the section *Machinery requirements for polar class vessels*, are to be considered in the design of these structures.

Table 11.3 Class factors to be used in accordance with *bow area (3)*

Polar class	Crushing failure class factor (CFC)	Flexural failure class factor (CFF)	Load patch dimensions class factor (CFD)	Displacement class factor (CFDIS)	Longitudinal strength class factor (CFL)
PC1	17.69 (1804,1794)	68.60 (6995,6885)	2.01 (0.122,0.308)	250 (250,246)	7.46 (753,473)
PC2	9.89 (1009,1003)	46.80 (4772,4697)	1.75 (0.1062,0.268)	210 (210,207)	5.46 (551, 346)
PC3	6.06 (618,614)	21.17 (2159,2125)	1.53 (0.093,0.234)	180 (180,177)	4.17 (421, 264)
PC4	4.50 (459,456)	13.48 (1375,1353)	1.42 (0.086,0.218)	130 (130,128)	3.15 (318, 200)
PC5	3.10 (316,314)	9.00 (918,903)	1.31 (0.080,0.201)	70 (70,69)	2.50 (252, 158)
PC6	2.40 (245,243)	5.49 (560,551)	1.17 (0.071,0.179)	40 (40,39)	2.37 (239, 150)
PC7	1.80 (184,183)	4.06 (414,407)	1.11 (0.0673,0.170)	22 (22,22)	1.81 (183, 115)

Note There are three system of units employed in this chapter. The first is SI, as is used in the IACS Unified Requirement. The second is the MKS system, and the third is the US customary units. In this chapter units and constants are shown as SI (MKS, US), as for example: MPa (kgf/mm^2, psi). In many cases the SI and MKS values are the same, but three values are always given for complete clarity

Table 11.4 Class factors to be used in accordance with *bow area (4)*

Polar class	Crushing failure class factor (CFCV)	Line load class factor (CFQV)	Pressure class factor (CFPV)
PC6	3.43 (350, 347)	2.82 (1.039, 2.608)	0.65 (0.00497, 7.137)
PC7	2.60 (265, 263)	2.33 (0.859, 2.155)	0.65 (0.00497, 7.137)

11.4.1 Glancing Impact Load Characteristics

The parameters defining the glancing impact load characteristics are reflected in the *Class factors* listed in Tables 11.3 and 11.4.

11.4.2 Bow Area

In the Bow area, the force *F*, line load (*Q*), pressure (*P*) and load patch aspect ratio (*AR*) associated with the glancing impact load scenario are functions of the hull angles measured at the UIWL. The influence of the hull angles is captured through calculation of a bow shape coefficient (f_a). The hull angles are defined in Fig. 11.2. The waterline length of the bow region is generally to be divided into four sub-regions of equal length. The force (F), line load (Q), pressure (P) and load patch aspect ratio (AR) are to be calculated with respect to the mid-length position of each sub-region (each maximum of F, Q and P is to be used in the calculation of the ice load parameters P_{avg}, b and w). The Bow area load characteristics for bow forms defined in *Bow area (4)* are determined as follows:

Note

B' = normal frame angle at upper ice waterline, degrees.

α = upper ice waterline angle, degrees.

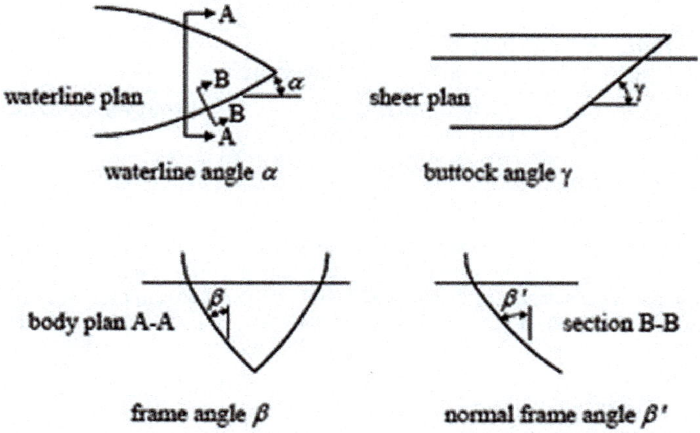

Fig. 11.2 Definition of hull angles

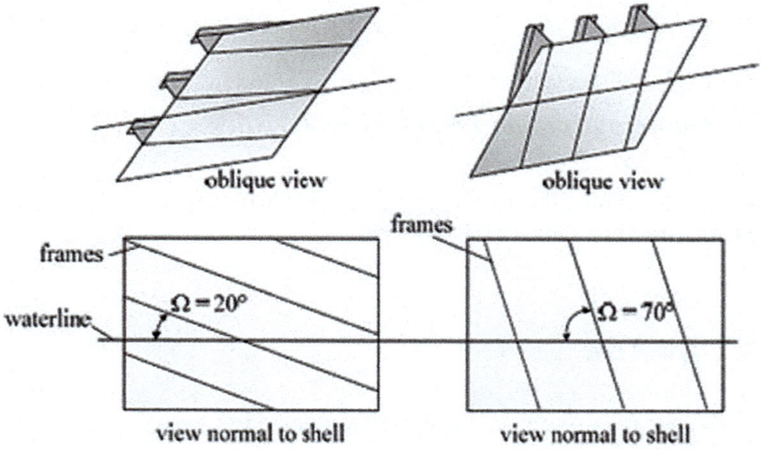

Fig. 11.3 Shell framing angle Ω

γ = buttock angle at upper ice waterline (angle of buttock line measured from horizontal), degrees.

$\tan(\beta) = \tan(\alpha)/\tan(\gamma)$.

$\tan(\beta') = \tan(\beta)\cos(\alpha)$.

(a) *Shape Coefficient*. Shape coefficient, fa_i, is to be taken as:

$$f_{ai} = minimum\left(fa_{i,1};\ fa_{i,2};\ fa_{i,3}\right)$$

where:

$fa_{i,1}$ $= \left[0.097 - 0.68\left(\frac{x}{L_{UI}} - 0.15\right)^2\right] \cdot \alpha_i/(\beta_{i\prime})^{0.5}$.

$fa_{i,2}$ $= 1.2 \cdot CF_F\left(\sin(\beta_{i\prime}) \cdot CF_C \cdot D_{UI}^{0.64}\right)$.

$fa_{i,3}$ $= 0.60$

i = sub-region considered.

L_{UI} = length as defined in *Length*, in m (m, ft).

X = distance from the fore side of the stem at the intersection with the UIWL to station under consideration, in m (m, ft).

α = waterline angle, in degrees, refer to Fig. 11.2.

β' = normal frame angle, in degrees, refer to Fig 11.2.

D_{UI} = displacement as defined in *Displacement*, not to be taken less than 5 kt (5 kt, 4.9 kLt).

CF_C = Crushing Failure Class Factor from Table 11.1.

CF_F = Flexural Failure Class Factor from Table 11.1.

(b) *Force*. Force, F, in MN (tf, Ltf) is to be taken as:

$$F_i = f_{ai} \cdot CF_C \cdot D_{UI}^{0.64}$$

where:

i = sub-region considered.

fa_i = shape coefficient of sub-region *i*.

CF_C = Crushing Failure Class Factor from Table 11.1.

D_{UI} = displacement as defined in *Displacement*, not to be taken less than 5kt (5kt, 4.9kLt).

(c) *Load Patch Aspect Ratio*. Load patch aspect ratio, AR, is to be taken as:

$$AR_i = 7.46.\sin(\beta_{i\prime}) \geq 1.3$$

where:

i = sub-region considered.
β_i' = normal frame angle of sub-region *i*, in degrees.

(d) *Line Load*. Line load, *Q*, in MN/m (tf/cm, Ltf/in) is to be taken as:

$$Q_i = F_i^{0.61} \cdot CF_D / AR_i^{0.35}$$

where:

i = sub-region considered.
F_i = force of sub-region *i*, in MN (tf, Ltf).
CF_D = Load Patch Dimensions Class Factor from Table 11.1.

(e) *Pressure*. Pressure, *P*, in MPa (kgf/mm^2, psi) is to be taken as:

$$P_i = c_1 \cdot F_i^{0.22} \cdot CF_D^2 \cdot AR_i^{0.3}$$

where:

i = sub-region considered.
F_i = force of sub-region *i*, in MN (tf, Ltf).
CF_D = Load Patch Dimensions Class Factor from Table 11.1.
AR_i = load patch aspect ratio of sub-region *i*.
c_1 = 1 (10, 2240).

(1) The bow area load characteristics for bow forms defined in *Design ice loads* are determined as follows:

(a) Shape coefficient

$$fa_i = \frac{\alpha_i}{30}$$

where:

i = sub-region considered.
α_i = waterline angle, in degrees, refer to Fig. 11.2.

(b) *Force*. Force, *F*, in MN (tf, Ltf) is to be taken as:

$$F_i = f_{ai} \cdot CF_{CV} \cdot D_{UI}^{0.47}$$

where:

i = sub-region considered
fa_i = shape coefficient of sun region i
CF_{CV} = Crushing failure class factor
D_{UI} = displacement as defined in *Displacement*, not to be taken less than 5 kt (5 kt, 4.9kLt)

(c) *Line Load.* Line load, Q, in MN/m (tf/cm, Ltf/in) is to be taken as:

$$Q_i = F_i^{0.22} \cdot CF_{QV}$$

where:

i = sub-region considered
F_i = force of sub-region i, in MN (tf, Ltf)
CF_{QV} = Pressure Class Factor from

(d) *Pressure.* Pressure, P, in MPa (kgf/mm^2, psi) is to be taken as:

$$P_i = F_i^{0.56} \cdot CF_{PV}$$

where:

i = sub-region considered.
Fi = force of sub-region i, in MN (tf, Ltf).
CF_{PV} = Pressure Class Factor from Table 11.2.

11.4.3 Hull Areas Other Than the Bow

In the hull areas other than the bow, the force (F_{NonBow}) and line load (Q_{NonBow}) used in the determination of the load patch dimensions (b_{NonBow}, w_{NonBow}) and design pressure ($_{Pavg)}$ are determined as follows:

(1) *Force.* Force, F_{NonBow}, in MN (tf, Ltf) is to be taken as:

$$F_{NonBow} = 0.36 \cdot CF_C \cdot DF$$

where:

CF_C = Crushing Force Class Factor from Table 11.1.
DF = vessel displacement factor.
$= D_{UI}^{0.64}$ if $D_{UI} \leq CF_{DIS}$
$= CF_{DIS}^{0.64} + 0.10(D_{UI} - CF_{DIS})$ if $D_{UI} \geq CF_{DIS}$

D_{UI} = displacement as defined in *Displacement*, not to be taken less than 10 kt (10 kt, 9.8 kLt)
CF_{DIS} = Displacement Class Factor from Table 11.1.

(2) *Line Load.* Line Load, Q_{NonBow}, in MN/m (tf/cm, Ltf/in) is to be taken as:

$$Q_{NonBow} = 0.639 \cdot F_{NonBow}^{0.61} \cdot CF_D$$

where:

F_{NonBow} = force from 6-1-2/5.7.1, in MN (tf, Ltf)
CF_D = Load Patch Dimensions Class Factor from Table 11.1.

11.4.4 Design Load Patch

Bow area. In the bow area, and the bow intermediate ice belt area for vessels with class notation **PC6** and **PC7**, the design load patch has dimensions of width, w_{Bow}, and height, b_{Bow}, expressed in m (cm, in) and defined as follows:

(a) $w_{Bow} = F_{Bow}/Q_{Bow}$
(b) $b_{Bow} = c1 Q_{Bow}/P_{Bow}$

where:

F_{Bow} = maximum F_i in the Bow area, in MN (tf, Ltf)
Q_{Bow} = maximum Q_i in the Bow area, in MN/m (tf/cm, Ltf/in)
P_{Bow} = maximum P_i in the Bow area, in MPa (kgf/mm^2, psi)
c_1 = 1 (10, 2240)

Other hull areas. In hull areas other than those covered in *bow area*, the design load patch has dimensions of width, w_{NonBow}, and height, b_{NonBow}, expressed in m (cm, in.) and defined as follows:

(a) $w_{NonBow} = F_{NonBow}/Q_{NonBow}$
(b) $b_{NonBow} = wnonbow/^{3.6}$

where:

F_{NonBow} = force determined using the data in *Force*, in MN (tf, Ltf)
Q_{NonBow} = line load determined using the data *Line load*, in MN/m (tf/cm, Ltf/in)

11.4.5 Pressure Within the Design Load Patch

Average pressure. The average pressure, P_{avg}, in MPa (kgf/mm^2, psi) within a design load patch is determined as follows:

$$P_{avg} = c_1 F/(b \cdot w)$$

where:

F $= F_{Bow}$ or F_{NonBow} as appropriate for the hull area under consideration, in MN (tf, Ltf)

b $= b_{Bow}$ or b_{NonBow} as appropriate for the hull area under consideration, in m (cm, in)

w $= w_{Bow}$ or w_{NonBow} as appropriate for the hull area under consideration, in m (cm, in)

c_1 $= 1$ (10, 2240)

Areas of higher, concentrated pressure. Areas of higher, concentrated pressure exist within the load patch. In general, smaller areas have higher local pressures. Accordingly, the peak pressure factors listed in Table 11.3 are used to account for the pressure concentration on localised structural members.

11.4.6 Hull Area Factors

Associated with each hull area is an *AF* that reflects the relative magnitude of the load expected in that area. The *AF* for each hull area for polar class vessels are listed in Tables 11.4 and 11.5. For ships assigned the additional notation, **ICE BREAKER**, the AF for each hull area is listed in Tables 11.6 and 11.7. In the event that a structural member spans across the boundary of a hull area, the largest hull area factor is to be used in the scantling determination of the member. Due to their increased manoeuvrability, vessels having propulsion arrangements with azimuth thruster(s) or "podded" propellers are required to have specially considered Stern Ice belt (*Si*) and Stern Lower (*Sℓ*) hull area factors. The adjusted hull area factors are listed in Tables 11.5 and 11.7 for Polar Class vessels and ships assigned with the additional notation **ICE BREAKER**, respectively (Table 11.9).

11.5 Shell Plate Requirements

11.5.1 Required Minimum Shell Plate Thickness

The required minimum shell plate thickness, *t*, expressed in mm (mm, in), is given by:

$$T = \text{tnet} + \text{t}_\text{s}$$

Table 11.5 Peak pressure factors

Structural member		Peak pressure factor (PPF_i)
Plating	Transversely framed	$PPF_p = (1.8\text{-}s/c_2) \geq 1.2$
	Longitudinally framed	$PPF_p = (2.2\text{–}1.2\ s/c_2) \geq 1.5$
Frames in transverse framing systems	With load distributing stringers	$PPF_t = (1.6\text{-}s/c_2) \geq 1.0$
	With no load distributing stringers	$PPF_p = (1.8\text{-}s/c_2) \geq 1.2$
Frames in bottom structures		$PPF_s = 1.0$
Load carrying stringers side and longitudinals web frames		$PPF_s = 1.0$, if $S_w \geq 0.5$ w
		$PPF_s = 2.0\text{–}2.0\ S_w/w$, if $S_w \geq 0.5$ w

where S = frame or longitudinal spacing, in m (m, in.)
$C_2 = 1(1,39.4)$
S_w = web framing spacing, in m (cm, in.)
W = ice load patch width, in m (cm, in.)

Table 11.6 Hull AF for vessels intended to operate ahead only

Hull area		Area	Polar class						
			PC1	PC2	PC3	PC4	PC5	PC6	PC7
Bow (B)	All	B	1.00	1.00	1.00	1.00	1.00	1.00	1.00
Bow intermediate (BI)	Icebelt	BI_i	0.90	0.85	0.85	0.80	0.80	1.00*	1.00*
	Lower	$BI\ell$	0.70	0.65	0.65	0.60	0.55	0.55	0.50
	Bottom	BI_b	0.55	0.50	0.45	0.40	0.35	0.30	0.25
Midbody (M)	Icebelt	M_i	0.70	0.65	0.55	0.55	0.50	0.45	0.45
	Lower	$M\ell$	0.50	0.45	0.40	0.35	0.30	0.25	0.25
	Bottom	M_b	0.30	0.30	0.25	**	**	**	**
Stern (S)	Icebelt	S_i	0.75	0.70	0.65	0.60	0.50	0.40	0.35
	Lower	$S\ell$	0.45	0.40	0.35	0.30	0.25	0.25	0.25
	Bottom	S_b	0.35	0.30	0.30	0.25	0.15	**	**

* *Indicates that strengthening for ice loads is not necessary*

where:

t_{net} = plate thickness required to resist ice loads in accordance with the section *Shell plate thickness to resist ice load.*

t_s = corrosion and abrasion allowance in accordance with the section *Corrosion/abrasion additions and steel renewal.*

Table 11.7 Hull AF for vessels intended to operate ahead and astern

Hull area		Area	Polar class						
			PC1	PC2	PC3	PC4	PC5	PC6	PC7
Bow (B)	All	B	1.00	1.00	1.00	1.00	1.00	1.00	1.00
Bow intermediate (BI)	Icebelt	BI_i	0.90	0.85	0.85	0.80	0.80	1.00*	1.00*
	Lower	$BI\ell$	0.70	0.65	0.65	0.60	0.55	0.55	0.50
	Bottom	BI_b	0.55	0.50	0.45	0.40	0.35	0.30	0.25
Midbody (M)	Icebelt	M_i	0.70	0.65	0.55	0.55	0.50	0.45	0.45
	Lower	$M\ell$	0.50	0.45	0.40	0.35	0.30	0.25	0.25
	Bottom	M_b	0.30	0.30	0.25	**	**	**	**
Stern intermediate (SI)	Icebelt	SI_i	0.90	0.85	0.85	0.80	0.80	1.00*	1.00*
	Lower	$SI\ell$	0.70	0.65	0.65	0.60	0.55	0.55	0.50
	Bottom	SI_b	0.55	0.50	0.45	0.40	0.35	0.30	0.25
Stern (S)	All	S	1.00	1.00	1.00	1.00	1.00	1.00	1.00

* *Indicates that strengthening for ice loads is not necessary*

11.5.2 Shell Plate Thickness to Resist Ice Load

The thickness of shell plating required to resist the design ice load, t_{net}, expressed in mm (mm, in), depends on the orientation of the framing. In the case of transversely-framed plating $\Omega \geq 70$ °, including all bottom plating (i.e., plating in hull areas BI_b, M_b and S_b), the net thickness is given by:

$$t_{net} = n_0 \cdot s \cdot \frac{\left[\frac{(AF \cdot PPF_p \cdot P_{avg})}{\sigma_y}\right]^{0.5}}{\left[1 + \frac{c_3 s}{(2 \cdot b)}\right]}$$

In the case of longitudinally-framed plating ($\Omega \leq 20°$), when $b \geq s$, the net thickness is given by:

$$t_{net} = n_0 \cdot s \cdot \frac{\left[\frac{(AF \cdot PPF_p \cdot P_{avg})}{\sigma_y}\right]^{0.5}}{\left[1 + \frac{s}{(2 \cdot \ell)}\right]}$$

In the case of longitudinally-framed plating ($\Omega \leq 20°$), when b < s, the net thickness is given by:

$$t_{net} = n_0 \cdot s \cdot \left[\frac{(AF \cdot PPF_p \cdot P_{avg})}{\sigma_y}\right]^{0.5} \cdot \frac{\left[2 \cdot \frac{b}{(c_3 s)} - \frac{b}{(c_3 s)}^2\right]^{0.5}}{[1 + s/(2 \cdot \ell]}$$

In the case of obliquely-framed plating ($70° > \Omega > 20°$), linear interpolation is to be used.

c_3 = 1 (100, 1)
n_0 = 500 (500, 0.5)
Ω = smallest angle between the chord of the waterline and the line of the first level framing as illustrated in Fig. 11.3, in degrees.
s = transverse frame spacing in transversely-framed vessels or longitudinal frame spacing in longitudinally-framed vessels, in m (m, in).
AF = Hull Area Factor from
PPF_p = Peak Pressure Factor from Table 11.3
P_{avg} = average patch pressure determined by *Pressure within the design load patch*, in MPa (kgf/mm^2, psi)
σ_y = minimum upper yield stress of the material, in N/mm^2 (kgf/mm^2, psi), but not greater than 690 N/mm^2 (70 kgf/mm^2, 100,000 psi)
b = height of design load patch, in m (cm, in), where b is to be taken not greater than $(\ell - s/4)/c_3$ in the case of determination of the net thickness for transversely framed playing, refer to (1) above.
ℓ = distance between frame supports in m (m, in) (i.e., equal to the frame span as given in the section *Framing span*), but not reduced for any fitted end brackets, in m (m, in). When a load-distributing stringer is fitted, the length ℓ need not be taken greater than the distance from the stringer to the most distant frame support.

11.5.3 Changes in Plating Thickness

Changes in plating thickness in the transverse direction from the ice belt to the bottom and in the longitudinal direction within the ice belt are to be gradually tapered.

11.6 Framing

Framing members of polar class vessels are to be designed to withstand the ice loads defined in the previous section *Design ice loads* for local transverse and longitudinal frames, and the later section *Framing (web frames and load carrying stringers)* for web frames and load-carrying stringers. The term "framing member" refers to transverse and longitudinal local frames, load-carrying stringers and load-carrying web frames in the areas of the hull exposed to ice pressure, refer to Fig. 11.1.

11.6.1 Fixity

The strength of a framing member is dependent upon the fixity that is provided at its supports. Fixity can be assumed where framing members are either continuous through the support or attached to a supporting section with a connection bracket. In other cases, simple support is to be assumed unless the connection can be demonstrated to provide significant rotational restraint. Fixity is to be ensured at the support of any framing which terminates within an ice-strengthened area. The details of framing member intersection with other framing members, including plated structures, as well as the details for securing the ends of framing members at supporting sections, must be prepared and submitted for Class review.

11.6.2 Framing Span

The effective span of a framing member is to be determined on the basis of its moulded length. If brackets are fitted, the effective span may be reduced provided the bracket is in accordance with Table 11.10a and b and rigidity of the supporting member where the bracket being attached is adequate. Brackets are to be configured to ensure stability in the elastic and post-yield response regions.

11.6.3 Scantlings

When calculating the section modulus and shear area of a framing member, net thicknesses of the web, flange (if fitted) and attached shell plating are to be used. The shear area of a framing member may include that material contained over the full depth of the member (i.e., web area including portion of flange, if fitted), but excluding attached shell plating.

11.6.4 Net Effective Shear Area

The actual net effective shear area, A_w, in cm^2 (cm^2, in^2) of a transverse or longitudinal local frame is given by:

$$A_w = h \cdot_{twn} \cdot sin\varphi_w / c_4^2$$

where:

c_4 = 10 (10, 1)

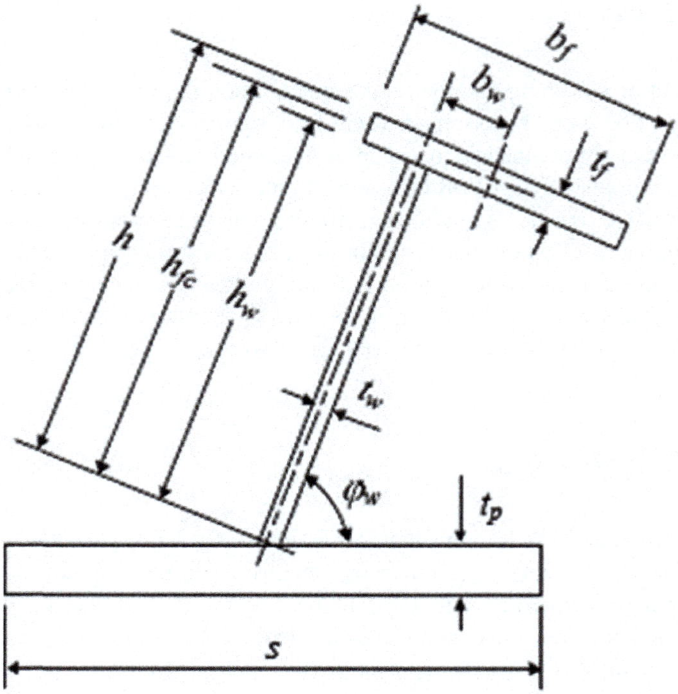

Fig. 11.4 Stiffener geometry

h = height of stiffener, in mm (mm, in), refer to Fig. 11.4.

t_{wn} = net web thickness, in mm (mm, in).
$= t_w - t_c$.

t_w = as-built web thickness, in mm (mm, in), refer to Fig. 11.4.

t_c = corrosion deduction, in mm (mm, in), to be subtracted from the web and flange thickness (but not less than as required by the later section *Corrosion/abrasion additions and steel renewal.*

φ_w = smallest angle between shell plate and stiffener web, measured at the midspan of the stiffener, refer to Fig. 11.4. The angle φ_w may be taken as 90° provided the smallest angle is not less than 75°.

11.6.5 Net Effective Plastic Section Modulus

When the net cross-sectional area of the attached plate flange, A_{pn}, exceeds the net cross-sectional area of the frame, A_{frn}, to which the shell plate flange is attached, the actual net effective plastic section modulus, Z_p, in cm^3 (cm^3, in^3), of a transverse or longitudinal frame is given by:

$$Z_p = \frac{A_{frn} t_{pn}}{2c_4} = \frac{h_w^2 t_{wn} sin\varphi_w}{2 \cdot c_4^3} + \frac{A_{fn}\left(h_{fc} sin\varphi_w - b_w cos\varphi_w\right)}{c_4}$$

where:

A_{pn} = net cross-sectional area of the attached plate flange, in cm^2 (cm^2, in^2)

= $\frac{t_{pn}s}{c_4^2}$.

A_{frn} = net cross-sectional area of the local frame, in cm^2 (cm^2, in^2)

= $\frac{h_w t_{wn} + b_f t f_n}{c_4^2}$

t_{pn} = fitted net shell plate thickness, in mm (mm, in), (is to comply with t_{net} as discussed in the section *Shell plate thickness to resist ice load.*

h_w = height of local frame web, in mm (mm, in.), refer to Fig. 11.4.

b_f = width of local frame flange, in (mm, in.), refer to Fig. 11.4.

t_{fn} = net thickness of local frame flange, in (mm, in.), refer to Fig. 11.4.

A_{fn} = net cross-sectional area of local frame flange, in cm^2 (cm^2, in^2)

= $\frac{b_f t_{fn}}{c_4^2}$

h_{fc} = height of local frame measured to centre of the flange area, mm (mm, in), refer to Fig. 11.4

= $h_w + \frac{t_{fn}}{2}$.

b_w = distance from mid thickness plane of local frame web to the centre of the flange area, in mm (mm, in), refer to Fig. 11.4.

c_4, h, t_{wn}, t_c and φ_w are as given in *Net effective shear area* and s is as given in *Shell plate thickness to resist ice load*. When the net cross-sectional area of the local frame, A_{frn}, exceeds the net cross-sectional area of the attached plate flange, A_{pn}, the plastic neutral axis is located a distance z_{na}, in mm (mm, in), above the attached shell plate, given by:

$$z_{na} = \frac{\left(c_4^2 A_{fn+} h_w t_{wn} - c_4^3 t_{pn} s\right)}{(2t_{wn})}$$

and the net effective plastic section modulus, Z_p, in cm^3 (cm^3, in^3), of a transverse or longitudinal frame is given by:

$$Z_p = t_{pn} s\left(z_{na} + \frac{t_{pn}}{2}\right) sin\varphi_w + \left[\frac{\left[(h_w - z_{na})^2 + z_{na}^2\right] t_{wn} sin\varphi_w}{2 \cdot c_4^3} + \frac{A_{fn}\left[\left(h_{fc-} z_{na}\right) sin\varphi_w - b_w cos\varphi_w\right]}{c_4}\right]$$

11.6.6 Oblique Framing

In the case of oblique framing arrangement (70° > Ω > 20°, where Ω is defined as given in *Shell plate thickness to resist ice load*), linear interpolation is to be used.

11.7 Framing (Local Frames in Bottom Structures and Transverse Local Frames Within Side Structures)

11.7.1 Plastic Strength

The local frames in bottom structures (i.e., hull areas BI_b, M_b and S_b) and transverse local frames within side structures are to be dimensioned such that the combined effects of shear and bending do not exceed the plastic strength of the member. The plastic strength is defined by the magnitude of midspan load that causes the development of a plastic collapse mechanism. For bottom structure the patch load should be applied with the dimension *b* parallel with the frame direction.

11.7.2 Required Shear Area

The actual net effective shear area of the frame, A_w, as defined in *Net effective shear area,* is required to comply with the following condition: $A_w \geq A_t$ in cm^2 (cm^2, in^2) where:

$$A_t = 100^{n1} \cdot 0.5 \cdot LL \cdot s \cdot \frac{(AF \cdot PPF \cdot P_{avg})}{(0.577 \cdot \sigma_y)}$$

where:

n_1 = 2 (1, 0).
LL = length of loaded portion of span, the lesser of *a* and *b*, in m (cm, in).
a = local frame span as defined in *Framing span*, in m (cm, in).
b = height of design ice load patch according to *Design load patch* (*bow area:* $b_{Bow} = c_1 Q_{Bow}/P_{Bow}$) or *Other hull areas:* $b_{NonBow} = w_{Nonbow}/3.6$)
s = spacing of local frame, in m (m, in).
AF = Hull Area Factor from Tables 11.4, 11.5, 11.6 and 11.7.
PPF = Peak Pressure Factor, PPF_t or PPF_s, as appropriate from Table 11.3.
P_{avg} = average pressure within load patch according to *Pressure within the design load patch*, in MPa (kgf/mm^2, psi)
σ_y = minimum upper yield stress of the material, in N/mm^2 (kgf/mm^2, psi), but not greater than 690 N/mm^2 (70 kgf/mm^2, 100,000 psi)

11.7.3 Required Plastic Section Modulus

The actual net effective plastic section modulus of the plate/stiffener combination, *Zp*, is to comply with the following condition: $Z_p \geq Z_{pt}$, in cm^3 (cm^3, in^3) where Z_{pt} is to be the greater calculated on the basis of two load conditions:

(1) Ice load acting at the midspan of the local frame; and
(2) The ice load acting near a support.

The A_1 parameter, in the equation below, reflects these two conditions.

$$Z_{pt} = 100^{n2} \cdot \frac{LL \cdot Y \cdot s \cdot \left(AF \cdot PPF \cdot P_{avg}\right) \cdot a \cdot A_1}{4 \cdot \sigma_y}$$

where:

n_2 = 3 (1, 0).
Y = $1-0.5 \cdot (LL/a)$.
A_1 = maximum of:
A_1A = $\frac{1}{\left(1+\frac{j}{2}+k_w \cdot \frac{j}{2} \cdot \left[\left(1-a_1^2\right)^{0.5}-1\right]\right)}$.
A_1B = $\frac{\left[1-\frac{1}{(2 \cdot a_1 \cdot Y)}\right]}{\left(0.275+1.44 \cdot k_z^{0.7}\right)}$.
J = 1 for a local frame with one simple support outside the ice-strengthened areas
= 2 for a local frame without any simple supports.
a_1 = A_t/A_w.
A_t = minimum shear area of the local frame as given in *Required shear area*, in cm^2 (cm^2, in^2).
A_w = net effective shear area of the local frame (calculated according to *Net effective shear area*), in cm^2 (cm^2, in^2)
k_w = $\frac{1}{\left(1+2 \cdot \frac{A_{fn}}{A_w}\right)}$ with A_{fn} as given in *Net effective plastic section modulus.*
k_z = z_p/Z_p in general
= 0.0 when the frame is arranged with end bracket.
z_p = sum of individual plastic section moduli of flange and shell plate as fitted, in cm^3 (cm^3, in^3)
= $\frac{\left(b_f \cdot \frac{t_{fn}^2}{4}+b_{eff} \cdot \frac{t_{pn}^2}{4}\right)}{c_5}$.
c5 = 1000 (1000, 1).
b_f = flange breadth, in mm (mm, in), refer to Fig. 11.4.
t_{fn} = net flange thickness, in mm (mm, in)
= $t_f - t_c$ (t_c as given in *Net effective shear area*).
t_f = as-built flange thickness, in mm (mm, in), refer to Fig. 11.4.
t_{pn} = fitted net shell plate thickness, in mm (mm, in) (not to be less than t_{net} as given in the section *Shell plate requirements*)

b_{eff} = effective width of shell plate flange, in mm (mm, in)
= 0.5 $c_5 \cdot s$.
Z_p = net effective plastic section modulus of the local frame (calculated according to the section on *Net effective plastic section modulus*), in cm^3 (cm^3, in^3)

AF, *PPF*, P_{avg}, *LL*, *b*, *s*, *a*, and σ_y are as given in *Required shear area.*

11.7.4 Structural Stability

The scantlings of the local frame are expected to meet the structural stability requirements set out in the later section *Framing (structural stability).*

11.8 Framing (Longitudinal Local Frames Within Side Structures)

11.8.1 Plastic Strength

Longitudinal local frames within side structures are to be dimensioned such that the combined effects of shear and bending do not exceed the plastic strength of the member. The plastic strength is defined by the magnitude of midspan load that causes the development of a plastic collapse mechanism.

11.8.2 Required Shear Area

The actual net effective shear area of the frame, A_w, as defined in this section, is to comply with the following condition: $A_w \geq A_L$, in cm^2 (cm^2, in^2) where:

$$A_L = 100^{n3} \cdot \left(AF \cdot PPF_s \cdot P_{avg}\right) \cdot 0.5 \cdot b_1 \cdot \frac{a}{\left(0.577 \cdot \sigma_y\right)}$$

where:

n_3 = 2 (0, 0)
AF = Hull Area Factor from Tables 11.4, 11.5, 11.6 and 11.7
PPF_s = Peak Pressure Factor from Table 11.3
P_{avg} = average pressure within load patch according to *Pressure within the design load patch (average pressure)*, in MPa (kgf/mm^2, psi)
b_1 = $k_o \cdot b_2$, in m (cm, in)
k_o = $1 - 0.3/b'$

b' = b/(s · c_6)
c_6 = 1 (100, 1)
b = height of design ice load patch from *Design load patch (bow area:* $b_{Bow} = c_1 Q_{Bow}/P_{Bow}$ or $b_{NonBow} = w_{Nonbow}/3.6$), in m (cm, in)
s = spacing of longitudinal frames, in m (m, in)
b_2 = $b(1 - 0.25 \cdot b')$, in m (cm, in) if $b' < 2$
= $s \cdot c_6$, in m (cm, in) if $b' \geq 2$
a = effective span of longitudinal local frame as given in *Framing span*, in m (cm, in)
σ_y = minimum upper yield stress of the material, in N/mm^2 (kgf/mm^2, psi), but not greater than 690 N/mm^2 (70 kgf/mm^2, 100,000 psi)

11.8.3 Required Plastic Section Modulus

The actual net effective plastic section modulus of the plate/stiffener combination, Z_p, as defined in *Net effective plastic section modulus*, is to comply with the following condition: $Z_p \geq Z_{pL}$ in cm^3 (cm^3, in^3) where:

$$Z_{pL} = 100^{n4} \cdot \left(AF \cdot PPF_s \cdot P_{avg}\right) \cdot b_1 \cdot a^2 \cdot \frac{A_4}{\left(8 \cdot \sigma_y\right)}$$

where:

n_4 = 3 (0, 0)
A_4 = $1/\left(2 + k_{wl} \cdot \left[\left(1 - a_4^2\right)^{0.5} - 1\right]\right)$.
a_4 = A_L/A_w.
A_L = minimum shear area for longitudinal as given in *Required shear area*, in cm^2 (cm^2, in^2)
A_w = net effective shear area of longitudinal (calculated according to *Required shear area*), in cm^2 (cm^2, in^2)
Kwl = $\frac{1}{\left(1+2\cdot\frac{A_{fn}}{A_w}\right)}$ with A_{fn} as given in *Net effective plastic section modulus.*

A_F, PPF_s, P_{avg}, b_1, a, and σ_y are as given in *Required shear area.*

11.8.4 Structural Stability

The scantlings of the longitudinals are to meet the structural stability requirements of the section *Framing (structural stability).*

11.9 Framing (Web Frames and Load-Carrying Stringers)

The structural performance of web frames and load-carrying stringers is to be evaluated utilising direct calculation methods. The evaluation may be performed based on linear or nonlinear analysis. Recognised structural idealisation and calculation methods are to be applied, with detailed requirements agreed upon with Class. Class recommendations regarding the use of nonlinear analysis methods for the strength evaluation of web frames and load-carrying stringers should be heeded. For guidance on how to conduct a Nonlinear Finite Element Analysis (NLFEA) refer to *ABS Guidance Notes on Nonlinear Finite Element Analysis of Marine Structures* or related Class guidance as appropriate. Although shell plating and local frames will typically be included in a direct calculation structural model, direct calculations are not to be utilised as an alternative to the design equations prescribed for the shell plating and local frame requirements given in sections *Shell plate requirements*, *Framing (local frames in bottom structures and transverse local frames within side structures)*, and *Framing (longitudinal local frames within side structures.* Web frames and load-carrying stringers are required to be designed to withstand the ice load patch as defined in the section *Design ice loads*. The load patch is to be applied, without being combined with any other loads. The load patch is to be applied at locations where the combined effects of bending and shear is maximised, or structural stability is at a minimum, such as the areas around cutouts, changes in geometry, or other stress concentrations. For linear analysis the structural response under the load patch and pressure as specified in *Design ice loads* is to be evaluated.

11.9.1 Structural Stability

Where possible, the scantlings of web frames and load-carrying stringers are to meet the structural stability requirements of *Framing (structural stability)*. If it is not possible to meet the stability requirements, NLFEA (or alternative) should be used to demonstrate that at the three load cases specified in *Load patch* that no structural instability has occurred.

11.9.2 Load Patch

For linear analysis methods, where the structural configuration is such that the members do not form part of a grillage system, the appropriate peak pressure factor (PPF) in Table 11.3 is to be used. For nonlinear analysis methods, the structural response under three load cases is to be evaluated:

(1) *Design:* the load patch and pressure specified in *Design ice loads.*
(2) *Overload:* the design load pressure multiplied by the polar class dependent Overload Capacity Factor (CF_O), specified in Table 11.8 is to be applied to the design load patch.
(3) *Reserve:* the Overload case pressure multiplied by 1.25 is to be applied to the design load patch (Table 11.11).

Table 11.8 Hull AF for vessels with additional notation ICE BREAKER and intended to operate ahead only

Hull area		Area	Polar class						
			PC1	PC2	PC3	PC4	PC5	PC6	PC7
Bow (B)	All	B	1.00	1.00	1.00	1.00	1.00	1.00	1.00
Bow intermediate (BI)	Icebelt	BI_i	0.90	0.85	0.85	0.85	0.85	1.00	1.00
	Lower	$BI\ell$	0.70	0.65	0.65	0.65	0.65	0.65	0.65
	Bottom	BI_b	0.55	0.50	0.45	0.45	0.45	0.45	0.45
Midbody (M)	Icebelt	M_i	0.70	0.65	0.55	0.55	0.55	0.55	0.55
	Lower	$M\ell$	0.50	0.45	0.40	0.40	0.40	0.40	0.40
	Bottom	M_b	0.30	0.30	0.25	0.25	0.25	0.25	0.25
Stern (S)	Icebelt	S_i	0.95	0.90	0.80	0.80	0.80	0.80	0.80
	Lower	$S\ell$	0.55	0.50	0.45	0.45	0.45	0.45	0.45
	Bottom	S_b	0.35	0.30	0.30	0.30	0.30	0.30	0.30

Table 11.9 Hull AF for vessels with additional notation ICE BREAKER and intended to operate ahead and astern

Hull area		Area	Polar class						
			PC1	PC2	PC3	PC4	PC5	PC6	PC7
Bow (B)	All	B	1.00	1.00	1.00	1.00	1.00	1.00	1.00
Bow intermediate (BI)	Icebelt	BI_i	0.90	0.85	0.85	0.85	0.85	1.00	1.00
	Lower	$BI\ell$	0.70	0.65	0.65	0.65	0.65	0.65	0.65
	Bottom	BI_b	0.55	0.50	0.45	0.45	0.45	0.45	0.45
Midbody (M)	Icebelt	M_i	0.70	0.65	0.55	0.55	0.55	0.55	0.55
	Lower	$M\ell$	0.50	0.45	0.40	0.40	0.40	0.40	0.40
	Bottom	M_b	0.30	0.30	0.25	0.25	0.25	0.25	0.25
Stern intermediate (SI)	Icebelt	SI_i	0.90	0.85	0.85	0.85	0.85	1.00	1.00
	Lower	$SI\ell$	0.70	0.65	0.65	0.65	0.65	0.65	0.65
	Bottom	SI_b	0.55	0.50	0.45	0.45	0.45	0.45	0.45
Stern (S)	All	S	1.00	1.00	1.00	1.00	1.00	1.00	1.00

Table 11.10 **a** Thickness and flanges of brackets and knees for vessels ≥ 90 m (295 ft) in length. **b** Thickness and flanges of brackets and knees for vessels ≤ 90 m (295 ft) in length

(a)

Depth of longer arm	Thickness		Width of flange
	Plain	Flanged	
Millimeters			
150	6.5		
175	7.0		
200	7.0	6.5	30
225	7.5	6.5	30
250	8.0	6.5	30
275	8.0	7.0	35
300	8.5	7.0	35
325	9.0	7.0	40
350	9.0	7.5	40
375	9.5	7.5	45
400	10.0	7.5	45
425	10.0	8.0	45
450	10.5	8.0	50
475	11.0	8.0	50
500	11.0	8.5	55
525	11.5	8.5	55
550	12.0	8.5	55
600	12.5	9.0	60
650	13.0	9.5	65
700	14.0	9.5	70
750	14.5	10.0	75
800		10.5	80
850		10.5	85
900		11.0	90
950		11.5	90
1000		11.5	95
1050		12.0	100
1100		12.5	105
1150		12.5	110
1200		13.0	110
Inches			
6.0	0.26		
7.5	0.28		

(continued)

Table 11.10 (continued)

(a)

Depth of longer arm	Thickness		Width of flange
	Plain	Flanged	
9.0	0.30	0.26	1 ¼
10.5	0.32	0.26	1 ¼
12.0	0.32	0.28	1 ½
13.5	0.36	0.28	1 ½
15.0	0.38	0.30	1 ¾
16.5	0.40	0.30	1 ¾
18.0	0.42	0.32	2
19.5	0.44	0.32	2
21.0	0.46	0.34	2 ¼
22.5	0.48	0.34	2 ¼
24.0	0.50	0.36	2 ½
25.5	0.52	0.36	2 ½
27.0	0.54	0.38	2 ¾
28.5	0.56	0.38	2 ¾
30.0	0.58	0.40	3
33.0		0.42	3 ¼
36.0		0.44	3 ½
39.0		0.46	3 ¾
42.0		0.48	4
45.0		0.50	4 ¼

(b)

Length of face f, mm	Thickness, mm		Width of flange, mm
	Plain	Flanged	
Metric			
Not exceeding 305	5.0	–	–
Over 305–455	6.5	5.0	38
Over 455–660	8.0	6.5	50
Over 660–915	9.5	8.0	63
Over 915–1370	11.0	9.5	75
Inch			
Not exceeding 12	$^{3}/_{16}$	–	–
Over 12–18	$^{1}/_{4}$	$^{3}/_{16}$	$1^{1}/_{2}$

(continued)

Table 11.10 (continued)

(b)

Length of face f, mm	Thickness, mm		Width of flange, mm
	Plain	Flanged	
Over 18–26	$^5/_{16}$	$^1/_4$	2
Over 26–36	$^3/_8$	$^3/_{16}$	$2^1/_2$
Over 36–54	$^7/_{16}$	$^3/_8$	3

Table 11.11 Overload capacity factor

Polar class	Overload capacity factor (CFo)	
Structure has both web frames and load-carrying stringers?	No—has only one type of structural member	Yes—has both types of structural members
PC1–PC3	1.20	1.10
PC4–PC5	1.25	1.15
PC6–PC7	1.30	1.20

11.9.3 Acceptance Criteria (Linear Analysis)

If the web frames and load-carrying stringers are evaluated based on linear analysis methods, the following are to be considered:

(1) Nominal shear stresses in member web plates are to be less than $0.577\sigma_y$ at the design load.
(2) Nominal von Mises stresses in member flanges is to be less than $1.15\sigma_y$ at the design load.
(3) Web plate and flange elements in compression and shear do not exhibit signs of elastic buckling at the overload load case.

11.9.4 Acceptance Criteria (Nonlinear Analysis)

NLFEA (or alternative) may be used to develop the maximum load deflection curve for the web frame or load-carrying stringer under consideration. The following criteria must be satisfied for the web frame or load-carrying stringer to be considered adequate:

(1) The maximum permanent set (δ_p) after unloading from the design load case pressure specified in (1) above must be less than 0.3% of the web frame or load-carrying stringer span under consideration.

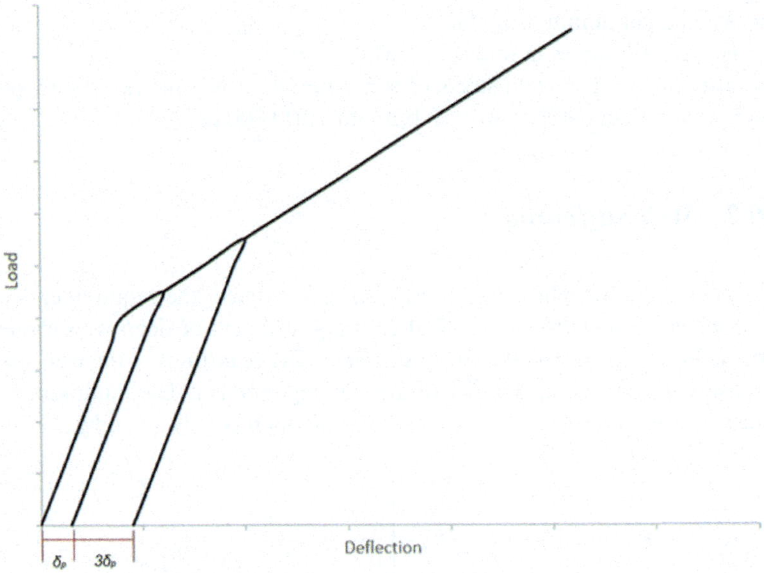

Fig. 11.5 Load deflection curve

(2) The maximum permanent set (δ_p) after unloading from the overload load case pressure specified in (2) above must be less than 0.9% of the web frame or load-carrying stringer span under consideration.

(3) The slope of the maximum load deflection curve must be positive, and no structural instability has occurred at the reserve load case pressure specified in (3) above (Fig. 11.5).

11.10 Framing (Structural Stability)

11.10.1 Framing Members

To prevent local buckling in the web, the ratio of web height (h_w) to net web thickness (t_{wn}) of any framing member is not to exceed:

- For flat bar sections: $\frac{h_w}{t_{wn}} \leq c_7(\sigma_y)^{0.5}$
- For bulb, tee and angle sections: $\frac{h_w}{t_{wn}} \leq c_8(\sigma_y)^{0.5}$

where:

c_7 = 282 (90, 3396).
c_8 = 805 (257, 9695).

h_w = web height in mm (mm, in).
t_{wn} = net web thickness in mm (mm, in).
σ_y = minimum upper yield stress of the material, in N/mm^2 (kgf/mm^2, psi), but not greater than 690 N/mm^2 (70 kgf/mm^2, 100,000 psi)

11.10.2 Web Stiffening

Framing members for which it is not practicable to meet the requirements set out above in *Framing members* (e.g., load-carrying stringers or deep web frames) are required to have their webs effectively stiffened. The scantlings of the web stiffeners are to ensure the structural stability of the framing member. The minimum net web thickness, t_{wn}, in mm (mm, in), for these framing members is given by:

$$t_{wn} = 2.63 \cdot 10^{-3} \cdot h_u \cdot \left[\frac{\sigma_y}{\left(c_{11} + c_{12} \cdot \left(\frac{h_u}{L_w} \right)^2 \right)} \right]^{0.5}$$

where:

h_u = h_w−0. 8 h_f mm (mm, in).
h_w = web height of stringer/web frame, in mm (mm, in) (refer to Fig. 11.6).
h_f = height of framing member penetrating the member under consideration (0 if no such framing member), in mm (mm, in) (refer to Fig. 11.6).
L_w = spacing between supporting structure oriented perpendicular to the member under consideration, in mm (mm, in) (refer to Fig. 11.6).
c_{11} = 5.34 (0.545, 775).
c_{12} = 4 (0.41, 580).
σ_y = minimum upper yield stress of the material, in N/mm^2 (kgf/mm^2, psi), but not greater than 690 N/mm^2 (70 kgf/mm^2, 100,000 psi).

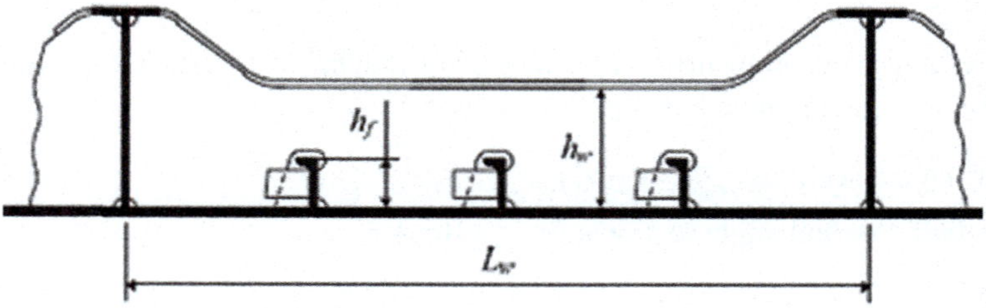

Fig. 11.6 Parameter definition for web stiffening

11.10.3 Web Thickness

In addition, the following is to be satisfied:

$$t_{wn} \geq 0.35 \cdot t_{pn} \cdot \left(\frac{\sigma_y}{c_{13}}\right)^{0.5}$$

where:

σy = minimum upper yield stress of the shell plate in way of the framing member, in N/mm^2 (kgf/mm^2, psi)
c13 = 235 (24, 34,083).
twn = net thickness of the web, in mm (mm, in).
tpn = net thickness of the shell plate in way of the framing member, in mm (mm, in).

11.10.4 Flange Width and Outstand

To prevent local flange buckling of welded profiles, the following are to be satisfied:

(1) The flange width, b_f, is not to be less than five times the net thickness of the web, t_{wn}.
(2) The flange outstand, b_{out}, in mm (mm, in), is to meet the following requirement:

$$\frac{b_{out}}{t_{fn}} \leq \frac{c_{14}}{(\sigma_y)^{0.5}}$$

where:

c_{14} = 155 (49.5, 1867).
t_{fn} = net thickness of flange, in mm (mm, in).
σ_y = minimum upper yield stress of the material, in N/mm^2 (kgf/mm^2, psi), but not greater than 690 N/mm^2 (70 kgf/mm^2, 100,000 psi).

11.11 Plated Structures

Plated structures are those stiffened plate elements in contact with the hull and subject to ice loads. Plated structures are to meet the requirements specified in *Framing (web frames and load carrying stringers)*. These requirements are applicable to an inboard extent which is the lesser of:

(1) Web height of adjacent parallel web frame or stringer; or
(2) 2.5 times the depth of framing that intersects the plated structure.

11.12 End Fixity

The thickness of the plating and the scantlings of attached stiffeners are to be such that the degree of end fixity necessary for the shell framing is ensured.

11.12.1 Stability

The stability of the plated structure is to adequately withstand ice loads, and *Framing (web frames and load carrying stringers)* for plated structures with attached web frames or load-carrying stringers.

11.13 Corrosion/Abrasion Additions and Steel Renewal

Effective protection against corrosion and ice-induced abrasion is recommended for all external surfaces of the shell plating for all Polar Class vessels.

11.13.1 Corrosion/abrasion Additions for Shell Plating

The values of corrosion/abrasion additions, t_s, in mm (mm, in) to be used in determining the shell plate thickness are listed in Tables 11.12 and 11.13.

Table 11.12 Corrosion/abrasion additions for shell plating for vessels intended to operate ahead only

Hull area	*ts*, mm (mm, in)					
	With effective protection			Without effective protection		
	PC1–PC3	PC4 and PC5	PC6 and PC7	PC1–PC3	PC4 and PC5	PC6 and PC7
Bow; bow intermediate icebelt	3.5 (3.5, 0.138)	2.5 (2.5, 0.098)	2.0 (2.0, 0.079)	7.0 (7.0, 0.276)	5.0 (5.0, 0.197)	4.0 (4.0, 0.158)
Bow intermediate lower; midbody and stern icebelt	2.5 (2.5, 0.098)	2.0 (2.0, 0.079)	2.0 (2.0, 0.079)	5.0 (5.0, 0.197)	4.0 (4.0, 0.158)	3.0 (3.0, 0.118)
Midbody and stern lower; bottom	2.0 (2.0, 0.079)	2.0 (2.0, 0.079)	2.0 (2.0, 0.079)	4.0 (4.0, 0.158)	3.0 (3.0, 0.118)	2.5 (2.5, 0.098)

Table 11.13 Corrosion/abrasion additions for shell plating for vessels intended to operate ahead and astern

Hull area	*ts,* mm (mm, in)					
	With effective protection			Without effective protection		
	PC1–PC3	PC4 and PC5	PC6 and PC7	PC1–PC3	PC4 and PC5	PC6 and PC7
Bow; bow intermediate icebelt; stern; stern intermediate icebelt	3.5 (3.5, 0.138)	2.5 (2.5, 0.098)	2.0 (2.0, 0.079)	7.0 (7.0, 0.276)	5.0 (5.0, 0.197)	4.0 (4.0, 0.158)
Bow intermediate lower; midbody icebelt and stern intermediate lower	2.5 (2.5, 0.098)	2.0 (2.0, 0.079)	2.0 (2.0, 0.079)	5.0 (5.0, 0.197)	4.0 (4.0, 0.158)	3.0 (3.0, 0.118)
Midbody lower; bottom	2.0 (2.0, 0.079)	2.0 (2.0, 0.079)	2.0 (2.0, 0.079)	4.0 (4.0, 0.158)	3.0 (3.0, 0.118)	2.5 (2.5, 0.098)

11.13.2 Corrosion/Abrasion Additions for Internal Structures

Polar Class vessels are to have a minimum corrosion/abrasion addition of $t_s = 1.0$ mm (1.0 mm, 0.0394 in) applied to all internal structures within the ice-strengthened hull areas, including plated members adjacent to the shell, as well as stiffener webs and flanges.

11.13.3 Steel Renewal

Steel renewal for ice strengthened structures is required when the gauged thickness is less than $t_{net} + 0.5$ mm (0.5 mm, 0.02 in).

11.14 Materials

All hull structural materials are required to be in accordance with the requirements set by Class and the following paragraphs. Steel grades of plating for hull structures are to be not less than those given in Table 11.12 based on the as-built thickness, the polar class and the material class of structural members in accordance with the guidance in the section on *Material classes* below.

11.14.1 Material Classes

Material classes specified in Table 11.15 are applicable to polar class vessels regardless of the vessel's length. In addition, material classes for weather and sea exposed structural members and for members attached to the weather and sea exposed plating of polar vessels are given in Table 11.15. Where the material classes in Table 11.15 and those in Table 11.6 differ, the higher material class is to be applied (Table 11.14).

Table 11.14 Material classes for structural members of polar class vessels

Structural members	Material class
Shell plating within the bow and bow intermediate icebelt hull areas (B, BI_i)	II
All weather and sea exposed SECONDARY and PRIMARY, structural members outside $0.4L_{UI}$ amidships	I
Plating materials for stem and stern frames, rudder horn, rudder, propeller nozzle, shaft brackets, ice skeg, ice knife and other appendages subject to ice impact loads	II
All inboard framing members attached to the weather and sea-exposed plating including any contiguous inboard member within 600 mm (600 mm, 23.6in) of the plating	I
Weather-exposed plating and attached framing in cargo holds of vessels which by nature of their trade have their cargo hold hatches open during cold weather operations	I
All weather and sea exposed SPECIAL, structural members within $0.2L_{UI}$ from FP	II

Table 11.15 Material grades

Plate thickness t mm (in)	Material class		
	I	**II**	**III**
$t \leq 15$ ($t \leq 0.60$)	A (2), AH	A, AH	A, AH
$15 < t \leq 20$ ($0.60 < t \leq 0.79$)	A, AH	A, AH	B, AH
$20 < t \leq 25$ ($0.79 < t \leq 0.98$)	A, AH	B, AH	D, DH
$25 < t \leq 30$ ($0.98 < t \leq 1.18$)	A, AH	D, DH	D (1), DH
$30 < t \leq 35$ ($1.18 < t \leq 1.38$)	B, AH	D, DH	E, EH
$35 < t \leq 40$ ($1.38 < t \leq 1.57$)	B, AH	D, DH	E, EH
$40 < t \leq 100$ ($1.57 < t \leq 4.00$)	D, DH	E, EH	E, EH
$100 < t \leq 150$ ($4.00 < t \leq 6.00$) (3)	E, EH	E, EH	E, EH

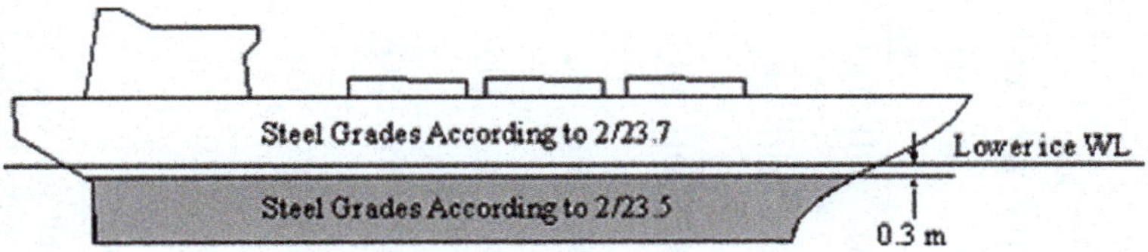

Fig. 11.7 Steel grade requirements for submerged and weather exposed shell plating

11.14.2 Steel Grades

Steel grades for all plating and attached framing of hull structures and appendages situated below the level of 0.3 m (0.3 m, 12 in) below the lower waterline, as shown in Fig. 11.7, are to be obtained based on the *Material class for structural members* in Table 11.15 above, regardless of polar class.

11.14.3 Steel Grades for Weather Exposed Plating

Steel grades for all weather exposed plating of hull structures and appendages situated above the level of 0.3 m (0.3 m, 12in) below the lower ice waterline, as shown in Fig. 11.7, are to be not less than given in Table 11.16.

11.14.4 Castings

Castings are to have specified properties consistent with the expected service temperature for the cast component.

11.15 Longitudinal Strength

A ramming impact on the bow is the design scenario for the evaluation of the longitudinal strength of the hull. Intentional ramming is not considered as a design scenario for ships which are designed with vertical or bulbous bows. Hence, the longitudinal strength requirements given in this section are not to be considered for ships with stem angle γ_{stem} stem equal to or larger than 80°. Ice loads are only to be combined with still water loads. The combined stresses are to be compared against permissible bending and shear stresses at different locations along the vessel's length. In addition, sufficient local buckling strength is also to be maintained.

Table 11.16 Steel grades for weather exposed plating [1,2]

Thickness, t mm (in)	Material class *I*						Material class *II*						Material class *III*								
	PC1–5			PC6 and 7			PC1–5			PC6 and 7			PC1–3			PC4 and 5			PC6 and 7		
	MS	***HT***	***XHT***	***MS***	***HT***	***XHT***	***MS***	***HT***	***XHT***	***MS***	***HT***	***XHT***	***MS***	***HT***	***XHT***	***MS***	***HT***	***XHT***	***MS***	***HT***	***XHT***
$t \leq 10$ $t \leq 0.394$	B	AH	AQ	B	AH	AQ	B	AH	AQ	B	AH	AQ	E	EH	EQ	E	EH	EQ	B	AH	AQ
$10 < t \leq 15$ $0.394 < t \leq 0.591$	B	AH	AQ	B	AH	AQ	D	DH	DQ	B	AH	AQ	E	EH	EQ	E	EH	EQ	D	DH	DQ
$15 < t \leq 20$ $0.591 < t \leq 0.787$	D	DH	DQ	B	AH	AQ	D	DH	DQ	B	AH	AQ	E	EH	EQ	E	EH	EQ	D	DH	DQ
$20 < t \leq 25$ $0.787 < t \leq 0.984$	D	DH	DQ	B	AH	AQ	D	DH	DQ	B	AH	AQ	E	EH	EQ	E	EH	EQ	D	DH	DQ
$25 < t \leq 30$ $0.984 < t \leq 1.18$	D	DH	DQ	B	AH	AQ	E	EH	EQ	D	DH	DQ	E	EH	EQ	E	EH	EQ	E	EH	EQ
$30 < t \leq 35$ $1.18 < t \leq 1.38$	D	DH	DQ	B	AH	AQ	E	EH	EQ	D	DH	DQ	E	EH	EQ	E	EH	EQ	E	EH	EQ
$35 < t \leq 40$ $1.38 < t \leq 1.58$	D	DH	DQ	D	DH	DQ	E	EH	EQ	D	DH	DQ	Ø	FH	FQ	E	EH	EQ	E	EH	EQ
$40 < t \leq 45$ $1.58 < t \leq 1.77$	E	EH	EQ	D	DH	EQ	E	EH	EQ	D	DH	DQ	Ø	FH	FQ	E	EH	EQ	E	EH	EQ
$45 < t \leq 50$ $1.77 < t \leq 1.97$	E	EH	EQ	D	DH	EQ	E	EH	EQ	D	DH	DQ	Ø	FH	FQ	Ø	FH	FQ	E	EH	EQ

Ø Not applicable

Notes (1) **MS**: ordinary strength steel, **HT**: high strength steel, **XHT**: extra high strength steel, (2) Includes weather-exposed plating of hull structures and appendages, as well as their outboard framing members, situated above a level of 0.3 m (0.3 m, 12in) below the lowest ice waterline, (3) Grades **D**, **DH** are allowed for a single strake of side shell playing not more than 1.8, (1.8 m, 70.9 in) wide from 0.3 m (0.3 m, 12in) below the lowest ice waterline

11.15.1 Design Vertical Ice Force at the Bow

The design vertical ice force at the bow, F_{IB}, in MN (tf, Ltf) is to be taken as:

$$F_{IB} = minimum\left(F_{IB,1};\ F_{IB,2}\right)$$

where:

$F_{IB,1}$ $= 0.534 \cdot K_I^{0.15} \cdot sin^{0.2}(\gamma_{stem}) \cdot (D_{UI} \cdot K_h)^{0.5} \cdot CF_L.$
$F_{IB,2}$ $= 1.20 \cdot CF_F.$
K_I $= indentationparameter = K_f / K_h.$

(a) *For the case of a blunt bow form*:

$$K_f = c_{15}\left[2 \cdot C \cdot \left(\frac{B_{UI}}{c_{16}}\right)^{1-e_b} / (1 + e_b)\right]^{0.9} \cdot \frac{\tan(\gamma_{stem})^{-0.9(1+e_b)}\,MN}{m}\left(\frac{tf}{cm}, Ltf/in\right).$$

$c_{15} = 1(1.025, 2.54).$
$c_{16} = 1(1, 3.28).$

(b) *For the case of wedge bow form* ($\alpha_{stem} < 80°$), $e_b = 1$ *and the above simplifies to:*

K_f $= \left[tan(\alpha_{stem})/tan^2(\gamma_{stem})\right]^{0.9}$ MN/m (tf/cm, Ltf/in).
K_h $= c_{17} A_{wp}$ MN/m (tf/cm, Ltf/in).
c_{17} $= 0.01(0.01, 0.00237)$ MN/m^3 [tf/(m^2-cm), Ltf/(ft^2-in)].
CF_L = Longitudinal strength class factor from Table 11.1
e_b = bow shape exponent which best describes the waterplane (Figs. 11.8 and 11.9)
= 1.0 for a simple wedge bow form
= 0.4–0.6 for a spoon bow form
= 0 for a landing craft bow form

An approximate e_b determined by a simple fit is acceptable.

γ_{stem} = stem angle to be measured between the horizontal axis and the stem tangent at the upper ice waterline, in degrees (buttock angle as per Fig. 11.2 measured on the centreline).
α_{stem} = waterline angle measured in way of the stem at the UIWL, in degrees (refer to Fig. 11.8).
C $= 1/\left[2 \cdot (L_B/B_{UI})^{eb}\right].$
B_{UI} = moulded breadth corresponding to the UIWL, in m (m, ft).
L_B = bow length used in the equation $y = \frac{B_{UI}}{2} \cdot \left(\frac{x}{L_B}\right)^{eb}$, in m (m, ft) (refer to Figs. 11.8 and 11.9).
D_{UI} = displacement, not to be taken less than 10kt (10kt, 9.8kLt).
A_{wp} = waterplane area corresponding to the UIWL, in m^2 (m^2, ft^2).
CF_F = Flexural failure class factor from Table 11.1.

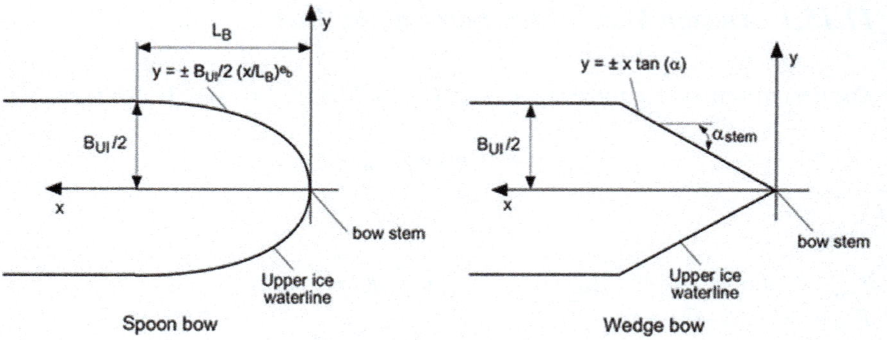

Fig. 11.8 Bow shape definition

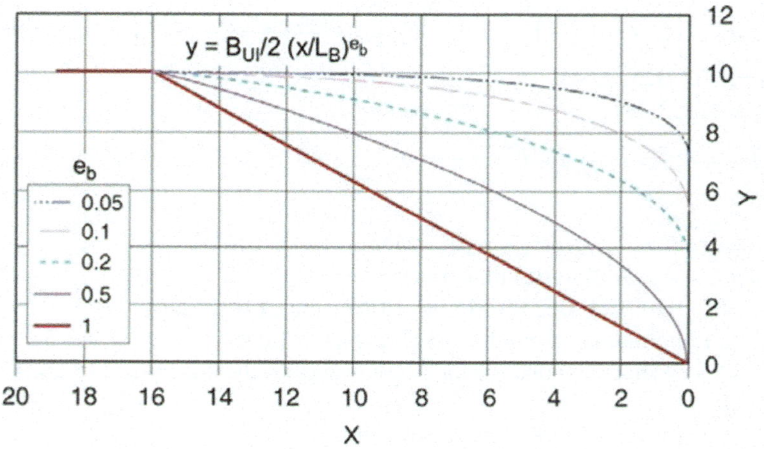

Fig. 11.9 Illustration of e_b effect on the bow shape for $B_{UI} = 20$ and $L_B = 16$

11.15.2 Design Vertical Shear Force

The design vertical ice shear force, F_I, in MN (tf, Ltf) along the hull girder is to be taken as:

$$F_I = C_f \cdot F_{IB}$$

where: C_f = longitudinal distribution factor to be taken as follows:

(1) Positive shear force

$C_f = 0.0$ between the aft end of C_f and 0.6 L_{UI} from aft.
$= 1.0$ between 0.9 L_{UI} from aft and the forward end of L_{UI}.

(2) Negative shear force

$C_f = 0.0$ at the aft end of L_{UI}.
$= -\ 0.5$ between 0.2 L_{UI} and 0.6 L_{UI} from aft.
$= 0.0$ between 0.8 L_{UI} from aft and the forward end of L_{UI}.
Intermediate values are to be determined by linear interpolation.

The applied vertical shear stress, τ_a, is to be determined along the hull girder in a similar manner as the *Shearing strength for vessels of 61 m (200ft) in length or over* by substituting the design vertical ice shear force for the design vertical wave shear force.

11.15.3 Shearing Strength for Vessels of 61 m (200ft) in Length or Over

In calculating the nominal total shear stress, f_s, due to still water and wave induced loads, the maximum algebraic sum of the shearing force in still water F_{sw} and that induced by wave F_w at the station examined is to be used. The thickness of the side shell and the longitudinal bulkhead where fitted, is to be such that the nominal total shear stress f_s, as obtained from *Shearing strength for vessels without effective longitudinal bulkheads* or *Shearing strength for vessels with two or three plane longitudinal bulkheads*, are not greater than 11.0 kN/cm^2 (1.122 tf/cm^2, 7.122 Ltf/in^2).

11.15.4 Shearing Strength for Vessels Without Effective Longitudinal Bulkheads

For vessels without continuous longitudinal bulkheads, the nominal total shear stress f_s in the side shell plating may be obtained from the following equation:

$$f_s = \frac{(F_{sw} + F_w)m}{2t_s I} \text{kN/cm}^2 \left(\text{tf/cm}^2,\ \text{Ltf/in}^2\right)$$

where:

I = moment of inertia of the hull girder at the section under consideration, in cm^4 (in^4).

m = first moment, in cm^3 (in^3), about the neutral axis, of the area of the effective longitudinal material between the horizontal level at which the shear stress is being determined and the vertical extremity of effective longitudinal material, taken at the section under consideration.

t_s = thickness of the side shell plating at the position under consideration, in cm (in).
F_{sw} = hull girder shearing force in still-water, in kN (tf, Ltf).
F_w = F_{wp} or F_{wn}, in kN (tf, Ltf), depending upon loading.

11.15.5 Modification of Hull Girder Shearing Force Peaks

The hull girder shearing force in still water, F_{sw}, to be used for calculating side shell plate shear stress may be modified to account for the loads transmitted through the double bottom structure to the side shell through the transverse bulkhead. For this modification, unless a detailed calculation is performed, the following equation may be used as guidance to determine the shear force carried by the side shell at the transverse bulkhead.

$$F_{sw} = F_{sw} - \text{ FB kN (tf, Ltf)}$$

where:

F_{sw} = hull girder shearing force in still water as obtained by the conventional direct integration method, in kN (tf, Ltf).
F_B = F_{BA} or F_{BF}, whichever is the lesser.
F_{BA} = $\frac{\left(0.45 - \frac{0.2\ell_A}{b_A}\right) W_a b_a}{B}$.
F_{BF} = $\frac{\left(0.45 - \frac{0.2\ell_F}{b_F}\right) W_F b_F}{B}$.
W_A, W_F = total load (net weight or net buoyancy) in the hold immediately abaft or forward of the bulkhead in question, in kN (tf, Ltf).
ℓA, ℓF = length of the adjacent holds respectively, containing W_A and W_F, in m (ft).
F_B, F_B = breadth of the double bottom structure in the holds immediately abaft and forward of the bulkhead in question, respectively, in m (ft). For vessels having double skins with flat inner bottom, it may be measured to the inner skins.
B = breadth of vessel, in m (ft)

11.15.6 Shearing Strength for Vessels with Two or Three Plane Longitudinal Bulkheads

For vessels having continuous longitudinal bulkheads, the total shear stresses in the side shell and the longitudinal bulkheads are to be calculated by an acceptable method. In determining the Stillwater shear force, consideration is to be given to the effects of non-uniform athwartship distribution of loads.

11.15.7 Design Vertical Ice Bending Moment

The design vertical ice bending moment, M_I, in MN-m (tf-m, Ltf-ft) along the hull girder is to be taken as:

$$M_I = 0.1 \cdot C_m \cdot L_{UI} \cdot \sin^{-0.2}(\gamma_{stem}) \cdot F_{IB}$$

where:

L_{UI} = length, in m (m, ft).
γ_{stem} = as given in *Design vertical ice force at the bow.*
F_{IB} = design vertical ice force at the bow, in MN (tf, Ltf).
C_m = longitudinal distribution factor for design vertical ice bending moment to be taken as follows:
= 0.0 at the aft end of L_{UI}.
= 1.0 between 0.5 L_{UI} and 0.7 L_{UI} from aft.
= 0.3 at 0.95 L_{UI} from aft.
= 0.0 at the forward end of L_{UI}.
Intermediate values are to be determined by linear interpolation.

The applied vertical bending stress, σ_4, is to be determined along the hull by substituting the design vertical ice bending moment for the design vertical wave bending moment. The vessel still water bending moment is to be taken as the permissible still water bending moment in sagging condition.

11.15.8 Longitudinal Strength Criteria

The strength criteria provided in Table 11.17 are expected to be satisfied. The design stress is not to exceed the permissible stress.

Table 11.17 Longitudinal strength criteria

Failure mode	Applied stress	Permissible stress when	Permissible stress when
Tension	σ_a	$\eta \cdot \sigma_y$	$\eta \cdot 0.41(\sigma_u + \sigma_y)$
Shear	τ_a	$\eta \cdot \sigma_y/(3)^{0.5}$	$\eta \cdot 0.41(\sigma_u + /\sigma_y)/(3)^{0.5}$
Buckling	σ_a	σ_c	For plating and for web plating of stiffeners
		$\sigma_c/1.1$	for stiffeners
	τ_a	τ_c	

where:

σ_a = applied vertical bending stress, in N/mm^2 (kgf/mm^2, psi).
τ_a = applied vertical shear stress, in N/mm^2 (kgf/mm^2, psi).
σ_y = minimum upper yield stress of the material, in N/mm^2 (kgf/mm^2, psi), but not greater than 690 N/mm^2 (70 kgf/mm^2, 100,000 psi).
σ_u = ultimate tensile strength of material, in N/mm^2 (kgf/mm^2, psi).
σ_c = critical buckling stress in compression, in N/mm^2 (kgf/mm^2, psi).
τ_c = critical buckling stress in shear, in N/mm^2 (kgf/mm^2, psi).
η = 0.6 for ships which are assigned the additional notation **ICE BREAKER**
= 0.8, otherwise

11.16 Stem and Stern Frames

For Polar Class **PC6** and **PC7** vessels requiring Baltic Ice Class **1AA** or **1A** equivalency, the stem and stern requirements of the *Finnish-Swedish Ice Class Rules* may need to be additionally considered.

11.17 Appendages

All appendages are required to be designed to withstand forces appropriate for the location of their attachment to the hull structure or their position within a hull area. Load definition and response criteria are likely to be determined by Class on a case-by-case basis.

11.18 Local Details

For the purpose of transferring ice-induced loads to supporting structure (bending moments and shear forces), local design details must be prepared and submitted for Class review. The loads carried by a member in way of cut-outs must not cause instability. Where necessary, the structure may need to be stiffened.

11.19 Welding

Hull construction welding design must comply with the Class Rules for welding. All welding within ice-strengthened areas must be of the double continuous type. Continuity of strength is to be ensured at all structural connections.

11.19.1 *Filler Metals*

When the ordinary and higher strength hull steels of Table 11.18 or Table 11.19 are applied in accordance with Table 11.12, approved filler metals appropriate to the grades.

Table 11.18 Condition of supply and frequency of impact tests ordinary strength hull structural steel

Grade	Deoxidation	Products	Condition of supply (impact test lot size in tons)					
			Thickness in mm (in)					
			Exceeding:		12.5 (0.5)	25 (1.0)	35 (1.375)	50 (2.0)
			Not exceeding:	12.5 (0.5)	25 (1.0)	35 (1.375)	50 (2.0)	150 (6.0)
A	Rimmed	All		A (-)				
	Semi-killed	All		A (-)				
	Killed	P						N (-) [4] TM (-) CR (50) AR (50)
		S						
B	Semi-killed	ALL		A (-)		A (50)		
	Killed	P						N (50) TM (50) CR (25) AR (25)
		S						
D	Killed and fine grain	P		A (50) N (50)			N (50) TM (50) CR (50)	N (50) TM (50) CR (25)
		S						
E	Killed and fine grain	P		N (P) TM (P)				N (P) TM (P)
		S		N (25) TM (25) CR (15)				

Table 11.19 Condition of supply and frequency of impact tests higher-strength hull structural steel

Grade	Deoxidation	Grain refining element	Products	Condition of supply impact test lot size in tons						
				Thickness in mm (in)						
				Exceeding: → Not exceeding: →	12.5 (0.5)	12.5 (0.5) 20 (0.80)	20 (0.80) 25 (1.0)	25 (1.0) 35 (1.375)	35 (1.375) 50 (2.0)	50 (2.0) 150 (6.0)
AH 32 AH 36	Killed, Fine Grain Practice	Nb V	P		A (50)	N (50*) TM (50) CR (50)				N (50) TM (50) CR (25)
			S		A (50)	N (50*) TM (50) CR (50) AR (25)				N/A
		Al Al + Ti	P		A (50)		AR (25) N (50*) TM (50) CR (50)		N (50*) TM (50) CR (50)	N (50) TM (50) CR (25)
			S		A (50)		N (50*) TM (50) CR (50) AR (25)			N/A
DH 32 DH 36		Nb V	P		A (50)	N (50) TM (50) CR (50)				N (50) TM (50) CR (25)
			S		A (50)	N (50) TM (50) CR (50)				N/A
		Al Al + Ti	P		A (50)		AR (25) N (50) TM (50) CR (50)	N (50) TM (50) CR (50)		N (50) TM (50) CR (25
			S		A (50)		AR (25) N (50) TM (50) CR (50)	N (50) TM (50) CR (50)		N/A
EH 32 EH 36		Any	P		N (P) TM (P)					N (P) TM (P)
			S		N (25) TM (25) CR (15)					N/A
FH 32 FH 36		Any	P		N (P) TM (P) QT (P)					N (P) TM (P)
			S		N (25) TM (25) QT (25)					N/A

11.19.2 Hull Steels Other Than Class Approved Grades

For the welding of hull steels other than the grades given in Table 11.12, weld metal is to exhibit a Charpy V-Notch toughness value at least equivalent to the transverse base metal requirements ($^2/_3$ of longitudinal base metal requirements).

Chapter 12
Machinery Requirements for Polar Class Vessels

12.1 Introduction

This chapter applies to the main propulsion, steering gear, emergency and essential auxiliary systems essential for the safety of polar class vessels and the survivability of the crew. The requirements contained in this chapter are additional to those discussed previously in Chaps. 6, 7, 8, 9, 10 and 11. Pursuant to the specific demands required by Class, the following drawings and technical particulars may be requested for submission by the vessel designer, owner and/or operator:

- *Environmental conditions.* Details of the environmental conditions and the required ice class for the machinery, if different from vessel's ice class.
- *Drawings.* Detailed drawings of the main propulsion machinery, description of the main propulsion, steering, emergency and essential auxiliaries are to include operational limitations. Information on essential main propulsion load control functions.
- *Description detailing.* Description detailing how main, emergency and auxiliary systems are located and protected to prevent problems from freezing, ice and snow and evidence of their capability to operate in intended environmental conditions.
- *Calculations and documentation.* Calculations and documentation indicating compliance with the requirements of this section. The following table shows a sample list of information and calculations required to be submitted:

A. Olsen, *Ship Operations in Extreme Low Temperature Environments*, Springer Series on Naval Architecture, Marine Engineering, Shipbuilding and Shipping 19,
https://doi.org/10.1007/978-3-031-52513-1_12

General	Torsional vibration calculations addressing the effect of ice/propeller interaction (including as a minimum: (a) Shaft speed drop curve(s) due to ice impact; and (b) Shaft response torque curve(s) etc.) Fatigue calculations for the propulsion line components considering ice loads (including S–N curves, Miner's Rule calculations, etc.)
Main engine	Main engine power curve (power supply), geometrical details (i.e., overall dimensions and detailed dimensions of the crank throws) and material properties of the crankshaft, harmonic packs (i.e., excitation tables), torsional damping coefficients between journals
Propeller	Propeller power curve (power demand), main particulars, inertial properties, water entrained factors, torsional and axial damping coefficients
Shaftline (including crankshaft)	Bearing offsets, bearing clearances, bearing radial and axial stiffnesses, bearing torsional and axial damping coefficients

12.2 System Design

12.2.1 General

All machinery installed onboard vessels holding a **PC** notation is to be suitable for operation under the environmental conditions to which it will be exposed in service and is to include all necessary special provisions for that purpose. Attention is directed to the appropriate governmental authorities in the intended regions of operation for additional requirements in consideration of operation in ice such as fuel capacity, refuelling capability, water capacity, radio communications requirements, etc. With reference to damage by freezing, systems subject to damage by freezing are required to be fully drainable.

12.2.2 Propeller Damage

Vessels classed **PC1** to **PC5** inclusive should have some means provided to ensure sufficient vessel operation in the case of propeller damage including the CP mechanism (i.e., pitch control mechanism). Sufficient vessel operation infers that the vessel is able to reach a point of safe harbour where repairs can be undertaken to rectify propeller damage. This may be achieved either through temporary repairs at sea, or by towing the impaired vessel to shore, assuming assistance is available (condition for approval).

12.2.3 Turning Gear

Some means should be provided to free a stuck propeller by turning backwards. This means that any plant intended for unidirectional rotation must be equipped at least with a sufficient turning gear that is capable of turning the propeller in the reverse direction.

12.3 Materials

12.3.1 Materials Exposed to Sea Water

Materials exposed to sea water, such as the propeller blades, propeller hub, and cast thruster body should have an elongation of not less than 15% on a test specimen with a length which is five times the diameter of the test specimen. Charpy V impact tests should be carried out for materials other than bronze and austenitic steel. The average impact energy of 20 J (20 J, 14.75 lbf-ft) taken from three Charpy V tests is to be obtained at – 10 °C (14 °F).

12.3.2 Materials Exposed to Sea Water Temperature

Materials exposed to sea water temperature should be either steel or another approved ductile material. Charpy V impact tests should be carried out for materials other than bronze and austenitic steel. Average impact energy value of 20 J (20 J, 14.75 lbf-ft) taken from three Charpy V tests is to be obtained at – 10 °C (14 °F). This requirement applies to blade bolts, CP-mechanisms, shaft bolts, strut-pod connecting bolts, etc. This does not apply to surface hardened components, such as bearings and gear teeth. For a definition of structural boundaries exposed to sea water temperature refer to Chap. 11, Fig. 11.7.

12.3.3 Materials Exposed to Low Air Temperature

Materials of essential components exposed to low air temperature should be of steel or another approved ductile material. Average impact energy value of 20 J (20 J, 14.75 lbf-ft) taken from three Charpy V tests is to be obtained at 10 °C (50 °F) below the lowest design temperature. This does not apply to surface hardened components, such as bearings and gear teeth. For definition of structural boundaries exposed to air temperature see Chap. 11, Fig. 11.7.

12.4 Ice Interaction Load

12.4.1 Propeller-Ice Interaction

The Class Rules cover open and ducted type propellers situated at the stern of a vessel having CP or FP blades. Ice loads on bow propellers should receive special consideration at the discretion of Class. The given loads are expected, single occurrence, maximum values for the whole ships service life for normal operational conditions. These loads do not cover off-design operational conditions, for example, when a stopped propeller is dragged through ice. The Rules cover loads due to propeller ice interaction also for azimuth and fixed thrusters with geared transmission or integrated electric motor ("geared and podded propulsors"). However, the load models of the regulations do not include propeller/ice interaction loads when ice enters the propeller of a turned azimuthing thruster from the side (radially) or load case when ice block hits on the propeller hub of a pulling propeller. The loads given in the section *Ice interaction loads* are total loads (unless otherwise stated) during ice interaction and are to be applied separately (unless otherwise stated) and are intended for component strength calculations only. F_b is a force bending a propeller blade backwards when the propeller mills an ice block while rotating ahead. F_f is a force bending a propeller blade forwards when a propeller interacts with an ice block while rotating ahead.

12.4.2 Ice Class Factors

Table 12.1 lists the design ice thickness and ice strength index to be used for estimation of the propeller ice loads.

where:

H_{ice} = ice thickness in m (m, ft) for machinery strength design.
S_{ice} = ice strength index for blade ice force.

Table 12.1 Design ice thickness and ice strength index

Ice class	Hice, m (m, ft)	Sice, [-]
PC1	4.0 (4.0, 13.12)	1.2
PC2	3.5 (3.5, 11.48)	1.1
PC3	3.0 (3.0, 9.84)	1.1
PC4	2.5 (2.5, 8.20)	1.1
PC5	2.0 (2.0, 6.56)	1.1
PC6	1.75 (1.75, 5.74)	1
PC7	1.5 (1.5, 4.92)	1

12.4.3 Design Ice Loads for Open Propeller

Maximum backward blade force. The maximum backward blade force, F_b, in kN (tf, Ltf), is to be taken as:

- when $D < D_{limit}$:

$$F_b = c_0 \cdot S_{ice}[n \cdot D]^{0.7} \cdot \left[\frac{EAR}{Z}\right]^{0.3} \cdot D^2$$

- when $D > D_{limit}$:

$$F_b = c_1 \cdot S_{ice}[n \cdot D]^{0.7} \cdot \left[\frac{EAR}{Z}\right]^{0.3} \cdot [H_{ice}]^{1.4} \cdot D$$

where:

D_{limit} $= c_2 \cdot (H_{ice})^{1.4}$ m (m, ft).
C_0 $= 27(2.753, 0.1096)$
C_1 $= 23(2.345, 0.0580)$
C_2 $= 0.85(0.85, 0.528)$
n = nominal rotational speed, in rev/s, (at MCR free running condition) for CP propeller and 85% of the nominal rotational speed (at MCR free running condition) for a FP-propeller (regardless driving engine type).
D = propeller diameter, in m (m, ft).
EAR = expanded blade area ratio.
Z = number of propeller blades.

F_b is to be applied as a uniform pressure distribution to an area on the back (suction) side of the blade for the following load cases:

- *Load case 1:* from 0.6R to the tip and from the blade leading edge to a value of 0.2 chord lengths.
- *Load case 2:* a load equal to 50% of the F_b is to be applied on the propeller tip area outside of 0.9R.
- *Load case 5:* for reversible propellers, a load equal to 60% of the F_b is to be applied from 0.6R to the tip and from the blade trailing edge to a value of 0.2 chord lengths measured from trailing edge.

See load cases 1, 2, and 5 in Table 12.2.

Maximum forward blade force. The maximum forward blade force, F_f, in kN (tf, Ltf) is to be taken as:

- when $D < D_{limit}$:

$$F_f = c_3 \cdot \left[\frac{EAR}{Z}\right] \cdot D^2$$

- when $D \geq D_{limit}$:

$$F_f = 2c_3 \cdot \left[\frac{1}{1-d/D}\right] \cdot H_{ice} \cdot \left[\frac{EAR}{Z}\right] \cdot D$$

where:

D_{limit} $= \left[\frac{2}{1-d/D}\right] \cdot H_{ice}$ m (m, ft).
c_3 = 250(25.493, 2.331)
d = propeller hub diameter, in m (m, ft).
D = propeller diameter, in m (m, ft).
EAR = expanded blade area ratio.
Z = number of propeller blades.

Table 12.2 Load cases for open propeller

	Force	Loaded area	Right-handed propeller blade seen from back
Load case 1	*Fb*	Uniform pressure applied on the back of the blade (suction side) to an area from 0.6*R* to the tip and from the leading edge to 0.2 times the chord length	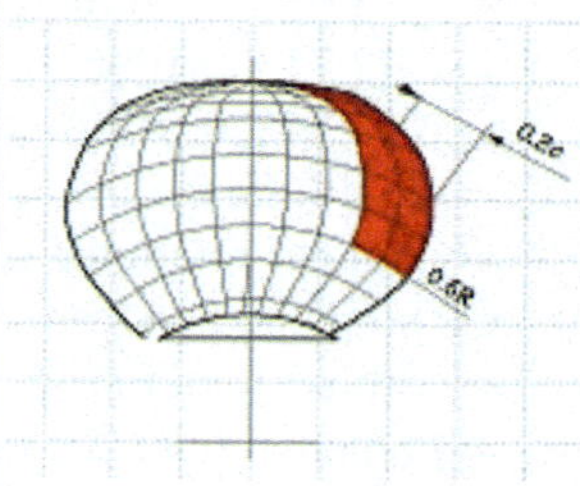
Load case 2	50% of *Fb*	Uniform pressure applied on the back of the blade (suction side) on the propeller tip area outside of 0.9*R* radius	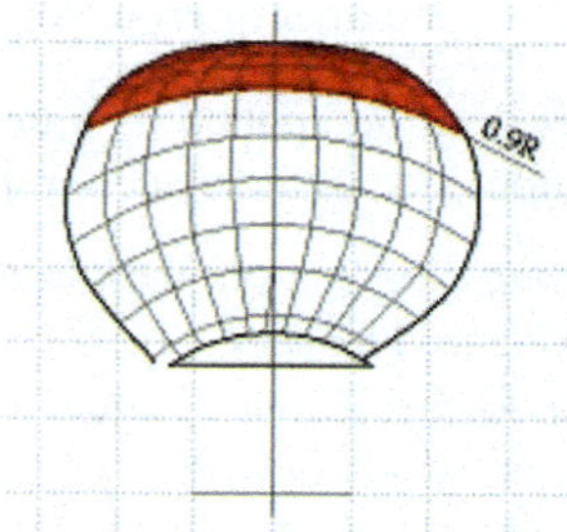
Load case 3	*Ff*	Uniform pressure applied on the blade face (pressure side) to an area from 0.6*R* to the tip and from the leading edge to 0.2 times the chord length	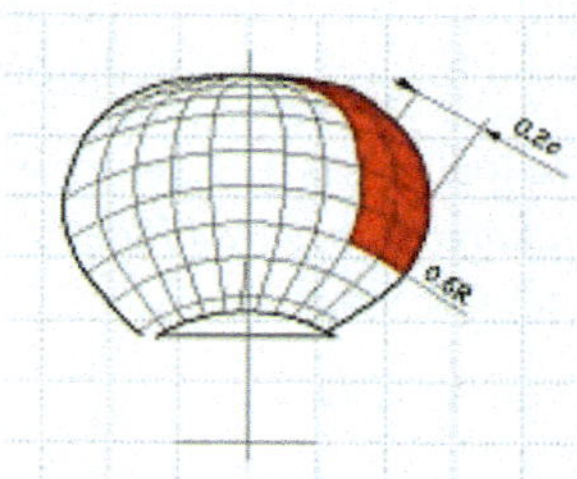

(continued)

Table 12.2 (continued)

	Force	Loaded area	Right-handed propeller blade seen from back
Load case 4	50% of *Ff*	Uniform pressure applied on propeller face (pressure side) on the propeller tip area outside of 0.9*R* radius	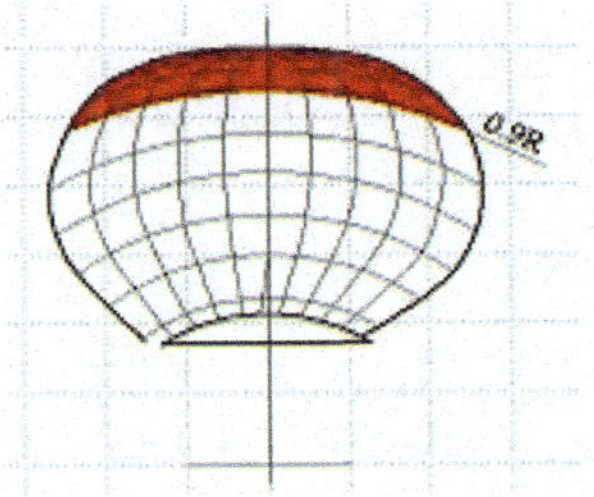
Load case 5	60% of *Ff* or *Fb* whichever is greater	Uniform pressure applied on propeller face (pressure side) to an area from 0.6*R* to the tip and from the trailing edge to 0.2 times the chord length	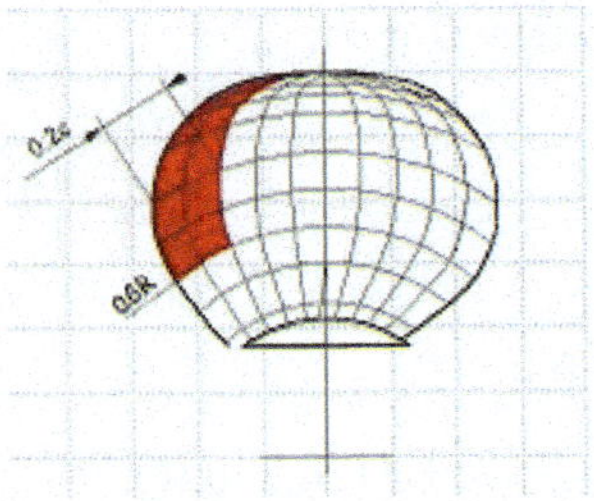

f_f is to be applied as a uniform pressure distribution to an area on the face (pressure) side of the blade for the following loads cases:

- *Load case 3:* from 0.6*R* to the tip and from the blade leading edge to a value of 0.2 chord length.
- *Load case 4:* a load equal to 50% of the f_f is to be applied on the propeller tip area outside of 0.9*R*.
- *Load case 5:* for reversible propellers, a load equal to 60% of f_f is to be applied from 0.6*R* to the tip and from the blade trailing edge to a value of 0.2 chord lengths measured from trailing edge.

See load cases 3, 4, and 5 in Table 12.2.

Maximum blade spindle torque. Spindle torque, Q_{smax}, in kN-m (tf-m, Ltf-ft), around the spindle axis of the blade fitting should be calculated both for the load cases described in *Maximum backward blade force* and *Maximum forward blade force* for f_b and f_f. If these spindle torque values are less than the default value given below, the default minimum value to be used.

Default Value: $Q_{smax} = 0.25 \cdot F \cdot c_{0.7}$

where:

$c_{0.7}$ = length of the blade chord at 0.7R radius, in m (m, ft).
F = either f_b or f_f, in kN (tf, Ltf), whichever has the greater absolute value.

Maximum propeller ice torque applied to the propeller. The maximum propeller ice torque, Q_{max}, in kN-m (tf-m, Ltf-ft) applied to the propeller is to be taken as:

- when $D < D_{limit}$:

$$Q_{max} = k_{open} \cdot \left[1 - \frac{d}{D}\right] \cdot \left[\frac{P_{0.7}}{D}\right]^{0.16} \cdot [n \cdot D]^{0.17} \cdot D^3$$

where:

k_{open} = 14.7(1.50, 0.112) for PC1 – PC5; and
k_{open} = 10.9(1.11, 0.083) for PC6 – PC7.

- when $D \geq D_{limit}$:

$$Q_{max} = 1.9 \cdot k_{open} \cdot \left[1 - \frac{d}{D}\right] \cdot [H_{ice}]^{1.1} \cdot \left[\frac{P_{0.7}}{D}\right]^{0.16} \cdot [n \cdot D]^{0.17} \cdot D^{1.9}$$

where:

D_{limit} = $1.81 H_{ice}$ m (m, ft).
$P_{0.7}$ = propeller pitch at 0.7*R*, in m (m, ft).
n = rotational propeller speed, in rev/s, at bollard condition. If not known, n is to be taken as follows:

Propeller type	n
CP propellers	n_n
FP propellers driven by turbine or electric motor	n_n
FP propellers driven by diesel engine	$0.85 n_n$

Where n_n is the nominal rotational speed at MCR, free running condition. For CP propellers, propeller pitch, $P_{0.7}$ should correspond to MCR in bollard condition. If not known, $P_{0.7}$ should be taken as $0.7 P_{0.7n}$ where $P_{0.7n}$ is propeller pitch at MCR free running condition.

Maximum propeller ice thrust applied to the shaft. The maximum propeller ice thrust, in kN (tf, Ltf), applied to the shaft is to be taken as:

$$T_f = 1.1 \cdot F_f$$

$$T_b = 1.1 \cdot F_{fb}$$

However, the load models of this UR do not include propeller/ice interaction loads when ice block hits on the propeller hub of a pulling propeller.

12.4.4 Design Ice Loads for Ducted Propeller

Maximum backward blade force. The maximum backward blade force, F_b, in kN (tf, Ltf) is to be taken as:

- when $D < D_{limit}$:

$$F_b = c_4 \cdot S_{ice}[n \cdot D]^{0.7} \cdot \left[\frac{EAR}{Z}\right]^{0.3} \cdot D^2$$

- when $D \geq D_{limit}$:

$$F_b = c_5 \cdot S_{ice}[n \cdot D]^{0.7} \cdot \left[\frac{EAR}{Z}\right]^{0.3} \cdot [H_{ice}]^{1.4} \cdot D^{0.6}$$

where:

D_{limit} $= 4H_{ice}$ m (m, ft).
c_4 $= 9.5(0.969, 0.0386)$
c_5 $= 66(6.730, 0.2679)$

n is to be taken as in *Maximum backward blade force.*

F_b is to be applied as a uniform pressure distribution to an area on the back side for the following load cases:

- *Load case 1*: on the back of the blade from 0.6R to the tip and from the blade leading edge to a value of 0.2 chord lengths.
- *Load case 5*: for reversible rotation propellers, a load equal to 60% of F_b is applied on the blade face from 0.6R to the tip and from the blade trailing edge to a value of 0.2 chord lengths measured from trailing edge. See load cases 1 and 5 in Table 12.3.

Maximum forward blade force. The maximum forward blade force, F_f, in kN (tf, Ltf), is to be taken as:

- when $D \leq D_{limit}$:

$$F_f = c_3 \cdot \left[\frac{EAR}{Z}\right] \cdot D^2$$

- when $D \geq D_{limit}$:

$$F_f = 2c_3\left[\frac{1}{1 - d/D}\right] \cdot H_{ice} \cdot \left[\frac{EAR}{Z}\right] \cdot D$$

Table 12.3 Load cases for ducted propeller

	Force	Loaded area	Right-handed propeller blade seen from back
Load case 1	*Fb*	Uniform pressure applied on the back of the blade (suction side) to an area from 0.6*R* to the tip and from the leading edge to 0.2 times the chord length	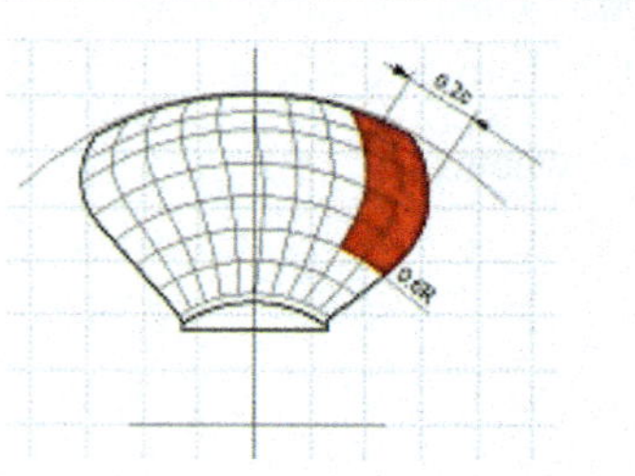
Load case 3	F_f	Uniform pressure applied on the blade face (pressure side) to an area from 0.6*R* to the tip and from the leading edge to 0.5 times the chord length	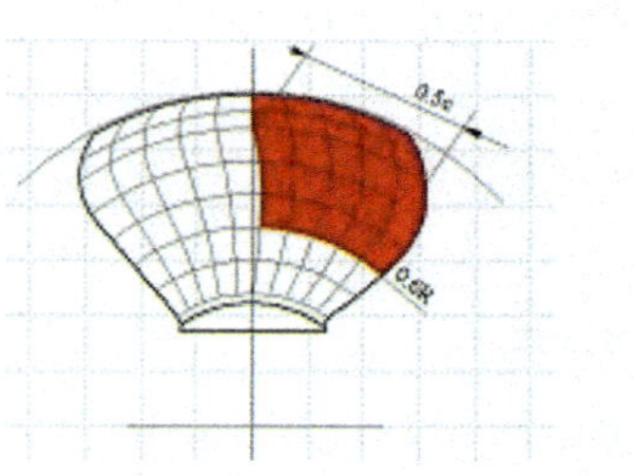
Load case 5	60% of F_f or *Fb* whichever is greater	Uniform pressure applied on propeller face (pressure side) to an area from 0.6*R* to the tip and from the trailing edge to 0.2 times the chord length	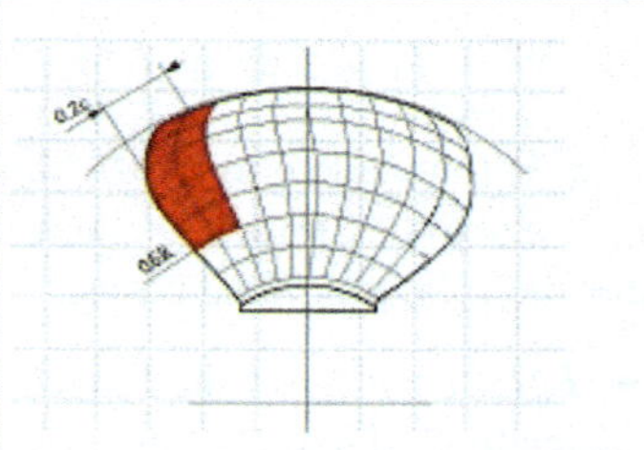

where:

D_{limit} $= \left[\frac{2}{1-d/D}\right] \cdot H_{ice}$ m (m, ft).
c_3 = 250(25.493, 2.331)
d = propeller hub diameter, in m (m, ft).
D = propeller diameter, in m (m, ft).
EAR = expanded blade area ratio.
Z = number of propeller blades.

F_f is to be applied as a uniform pressure distribution to an area on the face (pressure) side for the following load cases:

- *Load case 3:* from 0.6*R* to the tip and from the blade leading edge to a value of 0.2 chord length.
- *Load case 4:* a load equal to 50% of the F_f is to be applied on the propeller tip area outside of 0.9*R*.

- *Load case 5:* for reversible propellers, a load equal to 60% of F_f is to be applied from $0.6R$ to the tip and from the blade trailing edge to a value of 0.2 chord lengths measured from trailing edge. Refer to load cases 3 and 5 in Table 12.3.

Maximum blade spindle torque for CP mechanism design. Spindle torque, Q_{smax}, in kN-m (tf-m, Ltf-ft), around the spindle axis of the blade fitting is to be calculated for the load case described in *Ice interaction load*. If these spindle torque values are less than the default value given below, the default value is to be used.

$$Default\ Value: Q_{smax} = 0.25 \cdot F \cdot c_{0.7}$$

where:

$c_{0.7}$ = length of the blade chord at $0.7R$ radius, in m (m, ft).
F = either F_b or F_f, in kN (tf, Ltf), whichever has the greater absolute value.

Maximum propeller ice torque applied to the propeller. The maximum propeller ice torque, Q_{max}, in kN-m (tf-m, Ltf-ft) applied to the propeller is to be taken as:

- when $D < D_{limit}$:

$$Q_{max} = k_{open} \cdot \left[1 - \frac{d}{D}\right] \cdot \left[\frac{P_{0.7}}{D}\right]^{0.16} \cdot [n \cdot D]^{0.17} \cdot D^3$$

where:

k_{open} = 14.7(1.50, 0.112) for **PC1–PC5**; and.
k_{open} = 10.9(1.11, 0.083) for **PC6–PC7**.

- when $D \geq D_{limit}$:

$$Q_{max} = 1.9 \cdot k_{open} \cdot \left[1 - \frac{d}{D}\right] \cdot [H_{ice}]^{1.1} \cdot \left[\frac{P_{0.7}}{D}\right]^{0.16} \cdot [n \cdot D]^{0.17} \cdot D^{1.9}$$

D_{limit} = $1.81 H_{ice}$ m (m, ft).
$P_{0.7}$ = propeller pitch at $0.7R$, in m (m, ft).
n = rotational propeller speed, in rev/s, at bollard condition. If not known, n is to be taken as follows:

Propeller type	n
CP propellers	n_n
FP propellers driven by turbine or electric motor	n_n
FP propellers driven by diesel engine	$0.85n_n$

Where n_n is the nominal rotational speed at MCR, free running condition. For CP propellers, propeller pitch, $P_{0.7}$ should correspond to MCR in bollard condition. If not known, $P_{0.7}$ should be taken as $0.7P_{0.7n}$ where $P_{0.7n}$ is propeller pitch at MCR free running condition.

Maximum propeller ice thrust (applied to the shaft at the location of the propeller). The maximum propeller ice thrust in kN (tf, Ltf), applied to the shaft at the location of the propeller is:

$$T_f = 1.1 \cdot F_f$$

$$T_b = 1.1 \cdot F_b$$

12.4.5 Propeller Blade Loads and Stresses for Fatigue Analysis

Blade stresses. The blade stresses at various selected load levels for fatigue analysis are to be taken proportional to the stresses calculated for maximum loads given in *Design ice loads for open propeller* and *Design ice loads for ducted propeller*. The peak stresses are those determined due to F_f and F_b. The peak stress range $\Delta\sigma_{max}$ and the maximum load amplitude F_{Amax} are determined on the basis of:

$$\Delta F_{max} = 2.F_{Amax} = F_f + F_b$$

12.4.6 Design Loads on Propulsion Line

Torque excitation. The propeller ice torque excitation for shaft line dynamic analysis should be described by a sequence of blade impacts which are of half sine shape and occur at the blade. The torque due to a single blade ice impact as a function of the propeller rotation angle is then:

$$\begin{aligned} Q(\varphi) &= C_q \cdot Q_{\max} \cdot \sin\left[\varphi\left(\frac{180}{\alpha_i}\right)\right] \text{ when } \varphi = 0\ldots\alpha_i \\ Q(\varphi) &= 0 \text{ when } \varphi = \alpha_i\ldots 0 \end{aligned}$$

Where C_q and α_i are parameters given in Table 12.4.

The total ice torque is obtained by summing the torque of single blades considering the phase shift 360°/Z. The number of propeller revolutions during a milling sequence may be obtained with the formula:

$$N_Q = c_6 \cdot H_{ice}$$

Table 12.4 Parameters C_q and α_i

Torque excitation	Propeller-ice interaction	Cq	αi
Case 1	Single ice block	0.75	90
Case 2	Single ice block	1.0	135
Case 3	Two ice blocks with 45° phase in rotation angle	0.5	45

where:

$$c_6 = 2 \text{ rev/s/m} \\ = 0.6096 \text{ rev/s/ft}$$

The number of impacts during one milling sequence for blade order excitation is $Z \cdot N_Q$. In addition, the impacts are to ramp up over 270° and subsequently ramp down over 270°. The total excitation torque from the three cases will then look like Fig. 12.1. Milling torque sequence duration is not valid for pulling bow propellers, which are subject to special consideration. The response torque at any shaft component is to be analysed considering excitation torque at the propeller, actual engine torque, Q_e, and the mass elastic system.

$Q_e =$ actual maximum engine torque at considered speed.

Response torque in the propulsion system. The response torque ($Q_r(t)$) in all components should be determined by means of transient torsional vibration analysis of the propulsion line. Calculations are to be carried out for all excitation cases given above and the response is to be applied on top of the mean hydrodynamic torque in bollard condition at considered propeller rotational speed. The results of the three cases are to be used in the following way as illustrated in Fig. 12.2:

(1) The highest peak torque (between the various lumped masses in the system) is in the following referred to as peak torque Q_{peak}.
(2) The highest torque amplitude during a sequence of impacts is to be determined as half of the range from max to min torque and is referred to as Q_{Amax}.

***Note*:** for transient torsional vibration analysis (time domain), the model should include the ice excitation at the propeller, the mean torque values provided by the prime mover and the hydrodynamic mean torque produced by the propeller as well as any other relevant excitations. The aim of torsional vibration calculations is to estimate the torsional loads for individual shaft line components in order to determine scantlings for safe operation.

Maximum response thrust. Maximum thrust along the propeller shaft line is to be calculated with the formulae below. The factors 2.2 and 1.5 consider the dynamic magnification due to axial vibration. Alternatively, the propeller thrust magnification factor may be calculated by dynamic analysis.

Maximum shaft thrust forwards: $T_r = T + 2.2 \cdot T_f$

Maximum shaft thrust backwards: $T_r = 1.5 \cdot T_b$

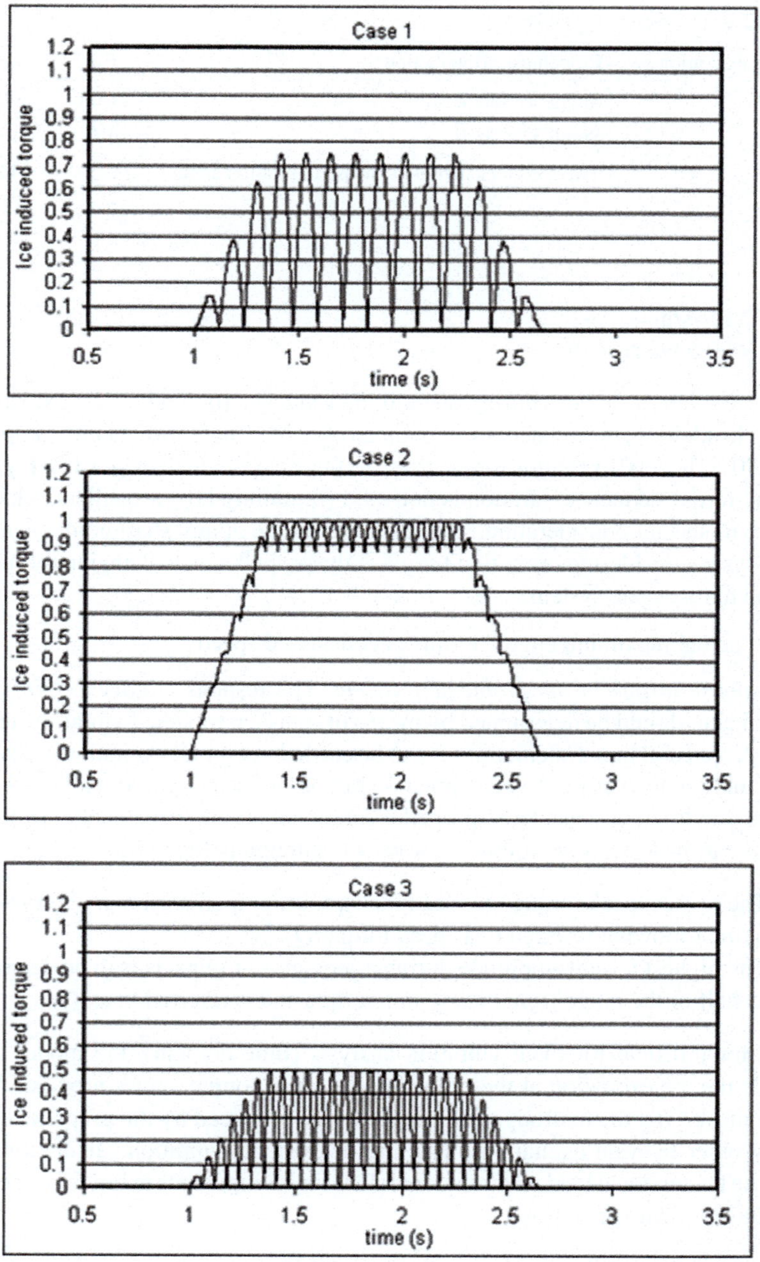

Fig. 12.1 Shape of the propeller ice torque excitation for 90° and 135° single blade impact sequences and 45° double blade impact sequence (figures apply for propellers with four blades)

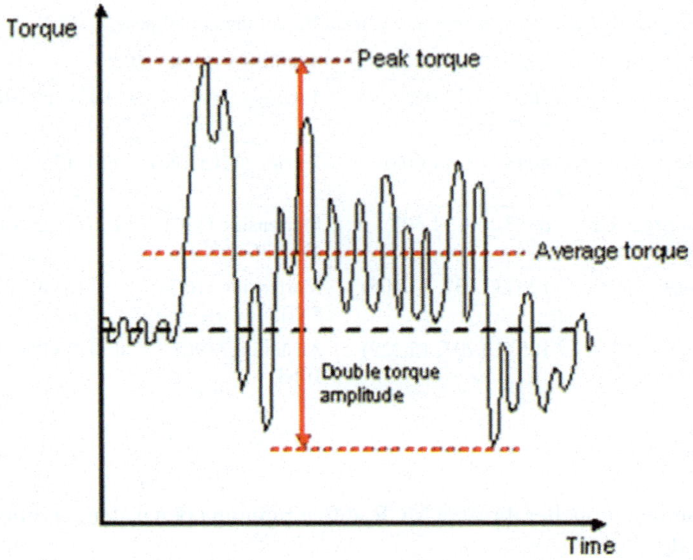

Fig. 12.2 Definitions of peak torque and torque amplitude

where:

T = propeller bollard thrust, in kN (tf, Ltf)
T_f = maximum forward propeller ice thrust, in kN (tf, Ltf)
T_b = maximum backward propeller ice thrust, in kN (tf, Ltf)

If hydrodynamic bollard thrust, T, is not known, T is to be taken as given in Table 12.5:

Table 12.5 Propeller bollard thrust

Propeller type	T
CP propellers (open)	$1.25T_n$
CP propellers (ducted)	$1.1T_n$
FP propellers driven by turbine or electric motor	T_n
FP propellers driven by diesel engine	$0.85T_n$

Table 12.6 Reference number of impacts per propeller rotation speed for each ice class

Ice class	PC1	PC2	PC3	PC4	PC5	PC6	PC7
N_{class}	21×10^6	17×10^6	15×10^6	13×10^6	11×10^6	9×10^6	6×10^6

Table 12.7 Mean fatigue strength, σ_{Fat-E7}, for different material types

Bronze and brass (a = 0.10)		Stainless steel (a = 0.05)	
Type	$^{\sigma}$Fat – E7 MPa (kgf/mm^2, psi)	Type	$^{\sigma}$Fat – E7 MPa (kgf/mm^2, psi)
Mn-Bronze, CU1 (high tensile brass)	80 (8.158, 11,603)	Ferritic (12Cr 1Ni)	120 (12.237, 7405)
Mn-Ni-Bronze, CU2 (high tensile brass)	80 (8.158, 11,603)	Martensitic (13Cr 4Ni/13Cr 6Ni)	150 (15.296,1756)
Ni-Al-Bronze, CU3	120 (12.237, 17,405)	Martensitic (16Cr 5Ni)	165 (16.825, 3931)
Mn-Al-Bronze, CU4	105 (10.707, 15,229)	Austenitic (19Cr 10Ni)	130 (13.256, 8855)

where:

T_n = nominal propeller thrust at MCR at free running open water conditions, in kN (tf, Ltf).

For pulling type propellers ice interaction loads on propeller hub must be considered in addition to the above.

Blade failure load for both open and nozzle propellers. The force is acting at 0.8R in the weakest direction of the blade and at a spindle arm of 1/3 of the distance of axis of blade rotation of leading and trailing edge whichever is the greatest. The blade failure load in kN (tf, Ltf) is:

$$F_{ex} = c_7 \cdot \frac{c \cdot t^2 \cdot \sigma_{ref}}{0.8 \cdot D - 2 \cdot r} \cdot 10^3$$

where:

σ_{ref} = $0.6\sigma_{0.2} + 0.4\sigma_u$ in MPa (kgf/mm^2, psi).
σ_u = specified maximum ultimate tensile strength in MPa (kgf/mm^2, psi).
$\sigma_{0.2}$ = specified maximum yield or 0.2% proof strength in MPa (kgf/mm^2, psi).
c_7 = 0.3 (0.3, 1.9286E-5).
c = actual chord length in m (m, ft).
t = thickness, in m (m, ft), of the cylindrical root section of the blade at the weakest section outside root fillet, typically at the termination of the fillet into the blade profile.
r = radius, in m (m, ft), of the cylindrical root section of the blade at the weakest section outside root fillet, typically at the termination of the fillet into the blade profile.

σ_u and $\sigma_{0.2}$ are representative values for the blade material. Representative in this respect means values for the considered section. These values may either be obtained by means of tests, or commonly accepted "thickness correction factors" approved by the Society. If not available, maximum specified values should be used. Alternatively, the F_{ex} can be determined by means of FEA of the actual blade. Blade bending failure should take place reasonably close to the root fillet end and normally not more 20% of R outside fillet. The blade bending failure is considered to occur when equivalent stress reach σ_{ref1} times 1.5 in elastic model.

12.5 Design

12.5.1 Design Principles

The propulsion line is to be designed according to the pyramid strength principle in terms of its strength. This means that the loss of the propeller blade should not cause any significant damage to other propeller shaft line components. The propulsion line components should withstand maximum and fatigue operational loads with the relevant safety margin. The loads do not need to be considered for shaft alignment or other calculations of normal operational conditions. *Fatigue design in general.* The design loads should be based on the ice excitation and where necessary (shafting) dynamic analysis, as described by a sequence of blade impacts. The shaft response torque should be determined by means of transient torsional vibration analysis of the propulsion line. The components are to be designed so as to prevent accumulated fatigue failure when considering the loads according to *Propeller blade loads and stresses for fatigue analysis* and *Design loads on propulsion line* using the linear elastic Miner's rule.

$$D = \frac{n_1}{N_1} + \frac{n_2}{N_2} + \cdots + \frac{n_k}{N_k} \leq 1 or D = \sum_{J=1}^{j=k} \frac{n_j}{N_j} \leq 1$$

The stress distribution should be divided into a frequency load spectrum having minimum 10 stress blocks (every 10% of the load). Calculation with 5 stress blocks has been found to be too conservative. The maximum allowable load is limited by σ_{ref}. The load distribution (spectrum) is required to be in accordance with the Weibull distribution. *Propeller blades.* The load spectrum for backward loads is normally expected to have a lower number of cycles than the load spectrum for forward loads. Taking this into account in a fatigue analysis introduces complications that are not justified considering all uncertainties involved. The blade stress amplitude distribution is therefore simplified (and at the same time disregarding mean stresses for fatigue purpose) and assumed to be as:

$$\sigma_A(N) = \sigma_{A\max} \cdot \left\{1 - \frac{\log(N)}{\log(N_{ice})}\right\}^{\frac{1}{k}}$$

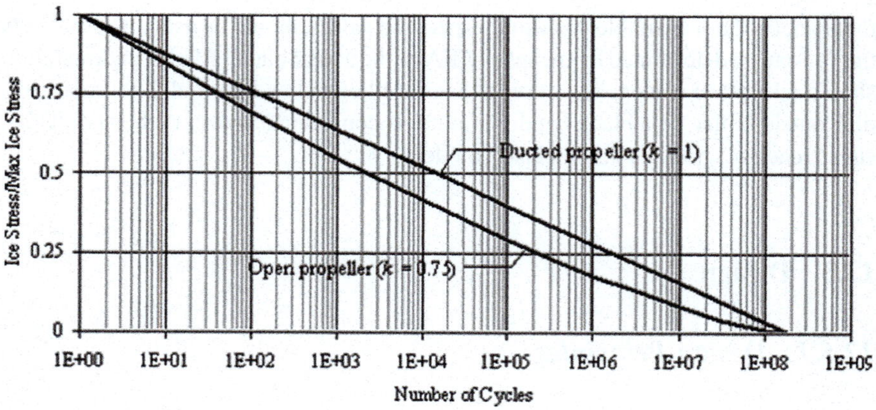

Fig. 12.3 Ice load distribution for ducted and open propeller

where:

k = Weibull exponent
= 0.75 for open propeller
= 1.0 for nozzle propeller

This is illustrated in the cumulative stress spectrum in Fig. 12.3.

Number of load cycles N_{ice} in the load spectrum per blade is to be determined according to the formula:

$$N_{ice} = k_1 \cdot k_2 \cdot N_{class} \cdot n$$

where:

N_{class} = reference number of impacts per propeller rotation speed for each ice class as indicated in Table 12.6

k_1 = 1 for centre propeller
= 2 for wing propeller
= 3 for pulling propeller (wing and centre)
= for pulling bow propellers number of load cycles is expected to increase in range of 10 times

k_2 = $0.8 - f$ when $f < 0$
= $0.8 - 0.4 \cdot f$ when $0 \leq f \leq 1$
= $0.6 - 0.2 \cdot f$ when $1 < f \leq 2.5$
= 0.1 when $f > 2.5$

f = immersion function.
= $\frac{h_0 - H_{ice}}{D/2} - 1$

h_o = depth, in m (m, ft), of the propeller centreline at the minimum ballast waterline in ice LIWL of the ship.

Propulsion line components. The strength of the propulsion line should be designed:

(1) For maximum loads in subsection *Design ice loads for open propeller* and *Design ice loads for ducted propeller* (for open and ducted propellers respectively).
(2) Such that the plastic bending of a propeller blade should not cause damages in other propulsion line components; and
(3) With sufficient fatigue strength as determined by the following criteria:

Cumulative fatigue calculations should be made according to the Miner's rule. The torque and thrust amplitude distribution (spectrum) in the propulsion line is to be taken as (because Weibull exponent k = 1):

$$Q_A(N) = Q_{A\max} \cdot \left\{1 - \frac{\log(N)}{\log(Z \cdot N_{ice})}\right\}$$

This is illustrated by the example in Fig 12.4.

$Q_A max$ is the average response torque shown in Fig. 12.2, calculated by means of transient torsional vibration analysis. The number of load cycles in the load spectrum is determined as $Z \cdot N_{ice}$. The Weibull exponent is $k = 1.0$ both for open propeller torque and for ducted propeller torque (and bending forces). The load distribution is an accumulated load spectrum, and the load spectrum is divided into minimum ten load blocks for the Miner summarising method. The load spectrum used is counting the number cycles for 100% load to be the number of cycles above the next step (e.g., 90% load) which means that the calculation is on the conservative side. Consequently, the fewer stress blocks used the more conservative is the calculated safety margin (Fig. 12.5).

The load spectrum is divided into z-number of load blocks for the Miner summarising method. The following formula can be used for calculation of the number of cycles for each load block.

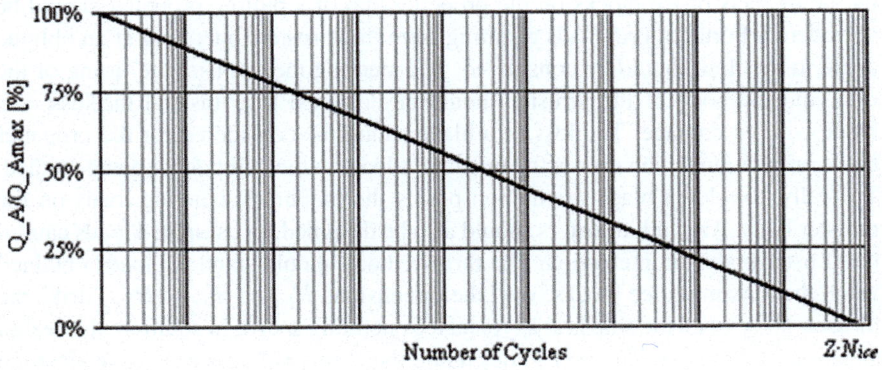

Fig. 12.4 Cumulative torque distribution

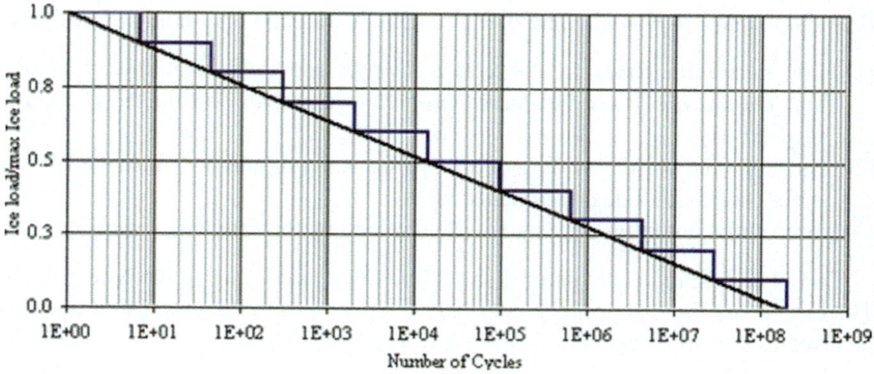

Fig. 12.5 Example of ice load distribution for the shafting ($k = 1$), divided into load blocks

$$n_i = N_{ice}^{1-\left(1-\frac{i}{z}\right)^k} - \sum_{i=1}^{i} n_{i-1}$$

where:

$i =$ single load block
$z =$ number of load blocks.

12.5.2 *Azimuthing Main Propulsors*

In addition to the above requirements, special consideration should be given to those loading cases which are extraordinary for propulsion units when compared with conventional propellers. The estimation of loading cases must reflect the way of operation of the ship and the thrusters. In this respect, for example, the loads caused by the impacts of ice blocks on the propeller hub of a pulling propeller should be considered. Furthermore, loads resulting from the thrusters operating at an oblique angle to the flow should be considered. The steering mechanism, the fitting of the unit, and the body of the thruster should be designed to withstand the loss of a blade without damage. The loss of a blade should be considered for the propeller blade orientation which causes the maximum load on the component being studied. Typically, top-down blade orientation places the maximum bending loads on the thruster body. Azimuth thrusters should also be designed for estimated loads caused by thruster body/ice interaction. The thruster body should stand the loads obtained when the maximum ice blocks, with the dimensions $H_{ice} \cdot 2H_{ice} \cdot 3H_{ice}$, strike the thruster body when the ship is at a typical ice operating speed. In addition, the design situation in which an ice sheet glides along the ship's hull and presses against the thruster body should be considered. The thickness of the sheet should be taken as the thickness of the maximum ice block entering the propeller, as defined in *Ice class factors*.

Design criteria for azimuthing propulsors. Azimuth propulsors should be designed for the following loads:

(1) Ice pressure on strut based on defined location area of the strut/ice interaction.
(2) Ice pressure on pod based on defined location area of thruster body/ice interaction.
(3) Plastic bending of one propeller blade in the worst position (typically top-down) without consequential damages to any other part.
(4) Steering gear design torque, Q_{SG}, in kN-m (tf-m, Ltf-ft), should be minimum 60% of steering torque expected at propeller ice milling condition defined as Q_{max}.

$$Q_{SG} = 0.6 \cdot \left(\frac{Q_{max}}{0.8R}\right) \cdot \ell$$

where:

ℓ = distance from propeller plane to steering (azimuth) axis, in m (m, ft)

(5) Steering gear should be protected by effective means limiting excessive torque caused by:

(a) Ice milling torque exceeding design torque and leading to rotation of unit.
(b) Torque caused by plastic bending of one propeller blade in the worse position (related to steering gear) and leading to rotation of the unit.

(6) Steering gear should be ready for operation after above load, (5)(a) or (5)(b) has gone.

12.5.3 Propeller Blade Design

Maximum blade stresses. Blade stresses (equivalent and principal stresses) are to be calculated using the backward and forward loads given in the section *Design ice loads for open propeller* and *Design ice loads for ducted propeller.* The stresses should be calculated with recognised and well documented FE-analysis or other acceptable alternative method. The stresses on the blade should not exceed the allowable stresses for the blade material given below. Calculated blade equivalent stress for maximum ice load is to comply with the following:

$$\sigma_{calc} < \sigma_{all} = \sigma_{ref}/S$$

where:

S = 1.5
σ_{ref} = reference stress, defined as:
= $0.7\sigma_u$ or.
= $0.6\sigma_{0.2} + 0.4\sigma_u$, whichever is less
$\sigma_u, \sigma_{0.2}$ = minimum specified representative values for the blade material according to the approved maker's specification.

Blade fatigue design. Propeller blades are to be designed so as to prevent accumulated fatigue when considering the loads according to the section *Propeller blades* and using the Miner's rule. For simplification purposes it is permitted to arrange the blade stress distribution into a frequency spectrum having min. ten classes (every 10% load). The S–N curve characteristics are based on two slopes, the first slope 4.5 is from 100 to 10^8 load cycles; the second slope 10 is above 10^8 load cycles.

(1) The maximum allowable stress is limited by σ_{ref}/S
(2) The fatigue strength σ_{Fat-E7} is the fatigue limit at 10 million load cycles.

The geometrical size factor, K_{size}, is:

$$1 - a \cdot \ln\left(\frac{t}{C_8}\right)$$

where:

$a =$ given in Table 12.7.
$t =$ actual blade thickness at considered section, in mm (mm, in).
$c_8 =$ 25 (25, 0.98425).

The mean stress effect (K_{mean}) is:

$$K_{mean} = 1.0 - \left(\frac{1.4 \cdot \sigma_{mean}}{\sigma_u}\right)^{0.75}$$

The fatigue limit for 10 million load cycles is then:

$$\sigma_{E7} = \frac{\sigma_{Fat-E7}}{S} \cdot K_{size} \cdot K_{mean}$$

where:

$S =$ 1.5

The S–N curve is extended by using the first slope (4.5) to 100 million load cycles due to the variable loading effect. σ_{Fat-E7} can be defined from fatigue test results from approved fatigue tests at 50% survival probability and stress ratio $R = -1$.

12.5.4 *Blade Flange, Bolts and Propeller Hub and CP Mechanism*

The blade bolts, the CP mechanism, the propeller boss, and the fitting of the propeller to the propeller shaft should be designed to withstand the maximum and fatigue design loads, as defined in the section *Ice interaction load.* The safety factor against yielding should be greater than 1.3 and that against fatigue greater than 1.5. In addition, the safety factor for loads resulting from loss of the propeller blade through plastic bending, as defined in the section *Blade failure load for both open and nozzle propellers,* should be greater than 1 against yielding. Blade bolts should withstand following bending moment, M_{bolt}, in kN-m (tf-m, Ltf-ft) considered around bolt pitch circle, or another relevant axis for not circular joints, parallel to considered root section:

$$M_{bolt} = SF_{ex}\left(0.8\frac{D}{2} - r_{bolt}\right)$$

where:

r_{bolt} = radius to the bolt plan, in m (m, ft).
S = 1.0

Blade bolt pretension should be sufficient to avoid separation between mating surfaces with maximum forward and backward ice loads in *Design ice loads for open propeller* and *Design ice loads for ducted propeller* (open and ducted respectively). Separate means (e.g., dowel pins) have to be provided in order to withstand a spindle torque resulting from blade failure Q_{sex}(*Blade failure load for both open and nozzle propellers*) or ice interaction Q_{Smax} (refer to *Maximum blade spindle torque for CP mechanism design*), whichever is greater. A safety of $S = 1$ is required. d in mm (mm, in) is:

$$d = c_8\sqrt{\frac{Q_s \cdot 8 \cdot \sqrt{3}}{PCD \cdot i \cdot \pi \cdot \sigma_{0.2}}}$$

where:

c_8 = 1000 (1,000, 163.95)
S = 1.3 for Q_{smax}
= 1.0 for Q_{sex}
PCD = pitch circle diameter, in mm (mm, in)
i = number of pins
Q_s = max $(SQ_{smax}; SQ_{sex}) - Q_{fr1} - Q_{fr2}$ kN-m (tf-m, Ltf-ft)
Q_{sex} = $F_{ex}\frac{1}{3}L_{ex}$ kN-m (tf-m, Ltf-ft)

Q_{fr1} = friction torque in blade bearings caused by the reaction forces due to F_{ex}, in kN-m (tf-m, Ltf-ft)

Q_{fr2} = friction between connected surfaces resulting from blade bolt pretension forces, in kN-m (tf-m, Ltf-ft)

L_{ex} = maximum of distance from spindle axis to the leading, or trailing edge at radius 0.8*R*, in m (m, ft)

Friction coefficient = 0.15 may normally be applied in calculation of Q_{fr}. The blade failure spindle torque Q_{sex} should not lead to any consequential damages. Fatigue strength is to be considered for parts transmitting the spindle torque from blades to a servo system considering ice spindle torque acting on one blade. The maximum amplitude is defined as:

$Q_{samax} = \frac{Q_{sb}+Q_{sf}}{2}$ kN-m(tf-m, Ltf-ft).

Provided that calculated stresses duly considering local stress, concentrations are less than yield strength, or maximum 70% of σ_u of respective materials, detailed fatigue analysis is not required. In opposite case components should be analysed for cumulative fatigue. Similar approach as used for shafting may be applied.

Servo pressure. Design pressure for servo system should be taken as a pressure caused by Q_{smax} or Q_{sex} when not protected by relief valves, reduced by relevant friction losses in bearings caused by the respective ice loads. Design pressure should in any case be less than relief valve set pressure.

12.5.5 Propulsion Line Components

The main propulsion line's components (i.e., propulsion shafts, couplings etc.) are to be reviewed by applying the loads determined in the following sections *Maximum propeller ice thrust applied to the shaft; maximum propeller ice thrust (applied to the shaft at the location of the propeller; torque excitation; response torque in the propulsion system; maximum response thrust;* and *design principles.* The strength evaluation under the applied loads is to verify the loads corresponding to the propeller blade failure load should not cause damage or deformation in the remaining propulsion line components. The fatigue strength evaluation is to be based on the cumulative fatigue analyses according to Miner's Rule, as applicable. The applicable highest peak torque and the corresponding load spectrum are to be determined for each of the components or connections in question, as applicable. The requirements in this section are complementary to those set by Class in reference to propulsion shafting. The loads considered in this section do not need to be considered for shaft alignment or other calculations of normal operational conditions.

Propeller fitting to the shaft: keyless cone mounting. The friction capacity (at 0 °C (32 °F)) should be at least 2.0 times the highest peak torque, as determined in the section *Response torque in the propulsion system*, without exceeding the permissible hub stresses. The necessary surface pressure in MPa (kgf/mm², psi) can be determined as:

$$p_{0^\circ C} = c_9 \frac{2 \cdot 2 \cdot 0 \cdot Q_{peak}}{\pi \cdot \mu \cdot D_S^2 \cdot L}$$

where:

c_9 = 0.001 (0.001, 15.556)
μ = 0.14 for steel-steel
= 0.13 for steel-bronze
D_S = is the shrinkage diameter at mid-length of taper, in m (m, ft).
L = is the effective length of taper, in m (m, ft).

Above friction coefficients may be increased by 0.04 if glycerine is used in wet mounting. *Key mounting.* Key mounting is not permitted. *Flange mounting.* The flange thickness is to be at least 25% of the shaft diameter. Any additional stress raisers such as recesses for bolt heads should not interfere with the flange fillet unless the flange thickness is increased correspondingly. The flange fillet radius is to be at least 10% of the shaft diameter. The diameter of ream fitted (light press fit) bolts should be chosen so that the peak torque does not cause shear stresses beyond 30% of the yield strength of the bolts. The bolts are to be designed so that the blade failure load F_{ex} (refer to *Blade failure load for both open and nozzle propellers*) does not cause yielding.

Propeller shaft. The propeller shaft is to be designed to fulfil the following:

(1) The blade failure load F_{ex} (refer to *Blade failure load for both open and nozzle propellers*) applied parallel to the shaft (forward or backwards) should not cause yielding. Bending moment need not to be combined with any other loads. The diameter d, in mm (mm, in), in way of the aft stern tube bearing should not be less than:

$$d = c_{10} \cdot \sqrt[3]{\frac{F_{ex} \cdot D}{\sigma_{0.2} \cdot \left(1 - \frac{d_i^4}{d^4}\right)}}$$

where:

c_{10} = 160 (160, 48)
$\sigma_{0.2}$ = minimum specified yield or 0.2% proof strength of the propeller shaft material, in MPa (kgf/mm², psi)
d = propeller shaft diameter, in mm (mm, in)
d_i = propeller shaft inner diameter, in mm (mm, in)

Forward from the aft stern tube bearing the diameter may be reduced based on direct calculation of actual bending moments, or by the assumption that the bending moment caused by F_{ex} is linearly reduced to 50% at the next bearing and in front of this linearly to zero at third bearing. Bending due to maximum blade forces F_b and F_f have been disregarded since the resulting stress levels are much below the stresses due to the blade failure load.

(2) The stresses due to the peak torque, Q_{peak}, in kN-m (tf-m, Ltf-ft), should have a minimum safety factor of 1.5 against yielding in plain sections and 1.0 in way of stress concentrations in order to avoid bent shafts. This equates to a minimum diameter of:

$$\text{Plain shaft} : d = c_{11} \cdot \sqrt[3]{\frac{Q_{peak}}{\sigma_{0.2} \cdot \left(1 - \frac{d_i^4}{d^4}\right)}} \text{ mm (mm, in)}$$

$$\text{Notched shaft} : d = c_{12} \cdot \sqrt[3]{\frac{Q_{peak} \cdot \alpha_t}{\sigma_{0.2} \cdot \left(1 - \frac{d_i^4}{d^4}\right)}} \text{ mm (mm, in)}$$

where:

c_{11} = 237 (237, 71).
c_{12} = 207 (207, 62).
α_t = the local stress concentration factor in torsion. Notched shaft diameter should in any case not be less than the required plain shaft diameter.

(3) The torque amplitudes with the foreseen number of cycles should be used in an accumulated fatigue evaluation where the safety factors are as defined in *Design principles*. If the plant also has high engine excited torsional vibrations (e.g., direct coupled 2-stroke engines), this has also to be considered.

(4) For plants with reversing direction of rotation the stress range $\Delta\tau \cdot \alpha_t$ resulting from forward $\Delta\tau \cdot Q_{peakf}$ to astern Q_{peakb} should not exceed twice the yield strength (in order to avoid stress–strain hysteresis loop) with a safety factor of 1.5, i.e.:

$$\Delta\tau \cdot \alpha_t \leq \frac{2 \cdot \sigma_y}{\sqrt{3} \cdot 1 \cdot 5} \text{ MPA(kgf /mm}^2\text{, psi)}$$

The fatigue strengths σ_f and τ_F (3 million cycles) of shaft materials may be assessed on the basis of the material's yield or 0.2% proof strength as:

$$\sigma_F = 0.436 \cdot \sigma_{0.2} + 77 = \tau_F \cdot \sqrt{3} \text{ MPa(kgf /mm}^2\text{, psi)}$$

This is valid for small, polished specimens (no notch) and reversed stresses, see "VDEH 1983 Bericht Nr. ABF11 Berechnung von Wöhlerlinien für Bauteile aus Stahl". The high cycle fatigue (HCF) is to be assessed based on the above fatigue strengths, notch factors (i.e., geometrical stress concentration factors and notch sensitivity), size factors, mean stress influence and the required safety factor of 1.5. The low cycle fatigue (LCF) representing 10^3 cycles is to be based on the lower value of either half of the stress range criterion [see (4)] or the smaller value of yield or 0.7 of tensile strength/$\sqrt{3}$. Both criteria utilise a safety factor of 1.5. The LCF and HCF as given above represent the upper and lower knees in a stress-cycle diagram. Since the required safety factors are included in these values, a Miner sum of unity is acceptable.

Intermediate shafts. The intermediate shafts are to be designed to fulfil the following:

(1) The stresses due to the peak torque Q_{peak}, in kN-m (tf-m, Ltf-ft), should have a minimum safety factor of 1.5 against yielding in plain sections and 1.0 in way of stress concentrations in order to avoid bent shafts. This equates to a minimum diameter of:

$$\text{Plain shaft}: d = c_{11} \cdot \sqrt[3]{\frac{Q_{peak}}{\sigma_{0.2} \cdot \left(1 - \frac{d_i^4}{d^4}\right)}} \text{ mm (mm, in)}$$

$$\text{Notched shaft}: d = c_{12} \cdot \sqrt[3]{\frac{Q_{peak} \cdot \alpha_t}{\sigma_{0.2} \cdot \left(1 - \frac{d_i^4}{d^4}\right)}} \text{ mm (mm, in)}$$

where:

c_{11} = 237 (237, 71)
c_{12} = 207 (207, 62)
α_t = the local stress concentration factor in torsion.
$\sigma_{0.2}$ = minimum specified yield or 0.2% proof strength of the shaft material, in MPa (kgf/mm^2, psi)
d = shaft diameter, in mm (mm, in)
d_i = shaft inner diameter, in mm (mm, in)

(2) The torque amplitudes with the foreseen number of cycles should be used in an accumulated fatigue evaluation where a minimum safety factor of 1.5 is required. If the plant also has high engine excited torsional vibrations (e.g., direct coupled 2-stroke engines), this has also to be considered.

(3) For plants with reversing direction of rotation the stress range $\Delta\tau \cdot \alpha_t$ resulting from forward Q_{peakf} to astern Q_{peakb} should not exceed twice the yield strength (in order to avoid stress–strain hysteresis loop) with a safety factor of 1.5, i.e.:

$$\Delta\tau \cdot \alpha_t \leq \frac{2 \cdot \sigma_y}{\sqrt{3} \cdot 1 \cdot 5} \text{ MPA(kgf /mm}^2\text{, psi)}$$

The fatigue strengths σ_F and τ_F (3 million cycles) of shaft materials may be assessed on the basis of the material's yield or 0.2% proof strength as:

$$\sigma_F = 0.436 \cdot \sigma_{0.2} + 77 = \tau_F \cdot \sqrt{3}\,\text{MPa}(\text{kgf}/\text{mm}^2, \text{psi})$$

This is valid for small, polished specimens (no notch) and reversed stresses, see "VDEH 1983 Bericht Nr. ABF11 Berechnung von Wöhlerlinien für Bauteile aus Stahl". The high cycle fatigue (HCF) is to be assessed based on the above fatigue strengths, notch factors (i.e., geometrical stress concentration factors and notch sensitivity), size factors and the required safety factor of 1.5. The low cycle fatigue (LCF) representing 103 cycles is to be based on the lower value of either half of the stress range criterion [see (3)] or the smaller value of yield or 0.7 of tensile strength/$\sqrt{3}$. Both criteria utilise a safety factor of 1.5. The LCF and HCF as given above represent the upper and lower knees in a stress-cycle diagram. Since the required safety factors are included in these values, a Miner sum of unity is acceptable.

Shaft connections: shrink fit couplings (keyless). The friction capacity should be at least 1.8 times the highest peak torque, Q_{peak}, in kN-m (tf-m, Ltfft), as determined in the section *Response torque in the propulsion system*, without exceeding the permissible hub stresses. The necessary surface pressure can be determined as:

$$p = c_9 \frac{2 \cdot 1 \cdot 8 \cdot Q_{peak}}{\pi \cdot \mu \cdot D_S^2 \cdot L}\,\text{MPa}(\text{kgf}/\text{mm}^2, \text{psi})$$

where:

c_9 = 0.001 (0.001, 15.556)
μ = 0.14 for steel to steel with oil injection (0.18 if glycerine injection)
D_S = is the shrinkage diameter at mid-length of taper, in m (m, ft).
L = is the effective length of taper, in m (m, ft).

Key mounting. Key mounting is not permitted. *Flange mounting.* The flange thickness is to be at least 20% of the shaft diameter (see IACS UR M34). Any additional stress raisers such as recesses for bolt heads should not interfere with the flange fillet unless the flange thickness is increased correspondingly. The flange fillet radius is to be at least 8% of the shaft diameter (see IACS UR M34). The diameter of ream fitted (light press fit) bolts or pins should be chosen so that the peak torque does not cause shear stresses beyond 30% of the yield strength of the bolts or pins. The bolts are to be designed so that the blade failure load (refer to *Blade failure load for both open and nozzle propellers*) in backwards direction does not cause yielding.

Gear transmissions: shafts. Shafts in gear transmissions should meet the same safety level as intermediate shafts, but where relevant, bending stresses and torsional stresses should be combined (e.g., by von Mises). Maximum permissible deflection in order to maintain sufficient tooth contact pattern is to be considered for the relevant parts of the gear shafts. *Gearing.* The gearing should fulfil following three acceptance criteria:

(1) Tooth root stresses
(2) Pitting of flanks
(3) Scuffing.

In addition to the above three criteria, subsurface fatigue may need to be considered. Common for all criteria is the influence of load distribution over the face width. All relevant parameters are to be considered, such as elastic deflections (of mesh, shafts and gear bodies), accuracy tolerances, helix modifications, and working positions in bearings (especially for twin input single output gears). The load spectrum (refer to the section *Design principles*) may be applied in such a way that the numbers of load cycles for the output wheel are multiplied by a factor of (number of pinions on the wheel / number of propeller blades Z). For pinions and wheels with higher speed the numbers of load cycles are found by multiplication with the gear ratios. The peak torque $\left(Q_{peak}\right)$ is also to be considered. Cylindrical gears can be assessed on the basis of the international standard ISO 6336 Pt.1–6, provided that "methods B" are used. Other acceptable alternative methods may also be considered on a case-by-case basis, provided that they are reasonably equivalent. Bevel gears should be assessed on the basis of standards within the classification societies. Tooth root safety should be assessed against the peak torque, torque amplitudes (with the pertinent average torque) as well as the ordinary loads (free water running) by means of accumulated fatigue analyses. The resulting safety factor is to be at least 1.5. (Ref ISO 6336 Pt 1, 3 and 6). The safety against pitting should be assessed in the same way as tooth root stresses, but with a minimum resulting safety factor of 1.2. (Ref ISO 6336 Pt 1, 2 and 6). The scuffing safety (flash temperature method—Ref. ISO-TR 13,989) based on the peak torque should be at least 1.2 when the FZG class of the oil is assumed one stage below specification. The safety against subsurface fatigue of flanks for surface hardened gears (oblique fracture from active flank to opposite root) is to be assessed at the discretion of each society.

Clutches. Clutches should have a static friction torque of at least 1.3 times the peak torque and dynamic friction torque $^2/_3$ of the static. Emergency operation of clutch after failure of (e.g., operating pressure) should be made possible within reasonably short time. If this is arranged by bolts, it should be on the engine side of the clutch in order to ensure access to all bolts by turning the engine.

Elastic couplings. There should be a separation margin of at least 20% between the peak torque and the torque where any twist limitation is reached. The torque amplitude (or range Δ) should not lead to fatigue cracking, i.e., exceeding the permissible vibratory torque. The permissible torque may be determined by interpolation in a log–log torque cycle diagram where T_{Kmax1} respectively ΔT_{Kmax} refer to 50,000 cycles and T_{KV} TKV refer to 10^6 cycles. See illustration in Figs. 12.6, 12.7 and 12.8.

Crankshafts. Special considerations apply for plants with large inertia (e.g., flywheel, tuning wheel or PTO) in the non-driving end of the engine.

Bearings. All shaft bearings are to be designed to withstand the propeller blade ice interaction loads according to *Design ice loads for open propeller* and *Design ice*

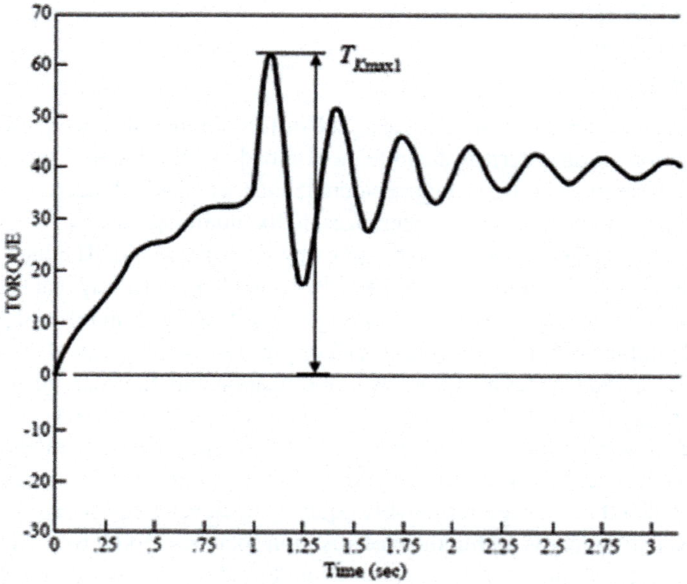

Fig. 12.6 Log–log torque-cycle diagram defining $\boldsymbol{T}_{\boldsymbol{Kmax1}}$

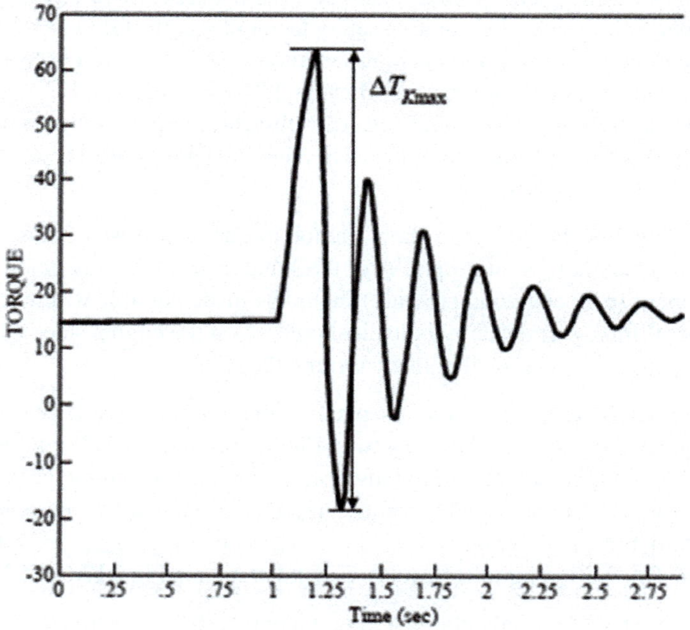

Fig. 12.7 Log–log torque-cycle diagram defining $\Delta \boldsymbol{T}_{\boldsymbol{Kmax}}$

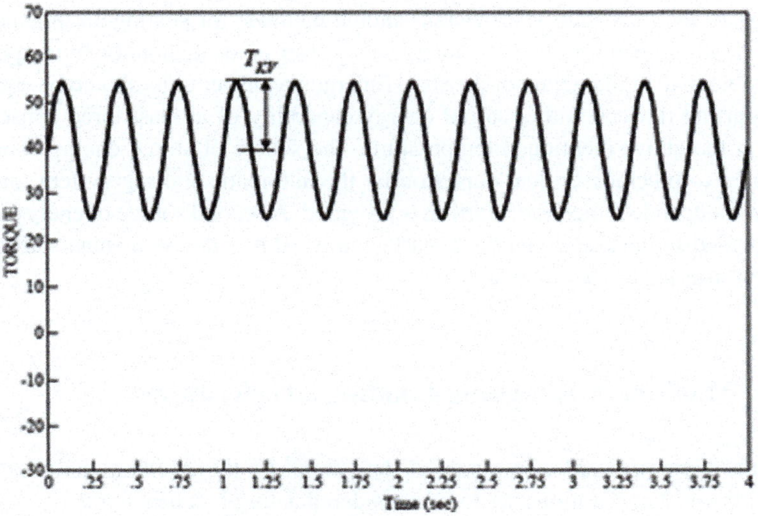

Fig. 12.8 Log–log torque-cycle diagram defining T_{KV}

loads for ducted propeller. For the purpose of calculation, the shafts are assumed to rotate at rated speed. Reaction forces due to the response torque (e.g., in gear transmissions) are to be considered. Additionally, the aft stern tube bearing as well as the next shaft line bearing are to withstand F_{ex} as given in *Blade failure load for both open and nozzle propellers*, in such a way that the ship can maintain operational capability. Rolling bearings are to have a L_{10a} lifetime of at least 40,000 h according to ISO-281. Thrust bearings and their housings are to be designed to withstand maximum response thrust (refer to *Maximum response thrust*) and the force resulting from the blade failure force F_{ex} in *Blade failure load for both open and nozzle propellers*. For the purpose of calculation except for F_{ex} the shafts are assumed to rotate at rated speed. For pulling propellers special consideration is to be given to loads from ice interaction on propeller hub.

Seals. Seals are to prevent egress of pollutants and be suitable for the operating temperatures. Contingency plans for preventing the egress of pollutants under failure conditions are to be documented. Seals are to be of proven design.

12.5.6 Prime Movers

Propulsion engines. Engines are to be capable of being started and running the propeller in bollard condition. Propulsion plants with CP propeller are to be capable being operated even in case with the CP system in full pitch as limited by mechanical stoppers.

Emergency power units. Provisions should be made for heating arrangements to ensure ready starting of the cold emergency power units at an ambient temperature applicable to the Polar Class of the ship. Emergency power units should be equipped with starting devices with a stored energy capability of at least three consecutive starts at the above-mentioned temperature. The source of stored energy should be protected to preclude critical depletion by the automatic starting system, unless a second independent means of starting is provided. A second source of energy should be provided for an additional three starts within 30 min unless manual starting can be demonstrated to be effective.

12.6 Machinery Fastening Loading Accelerations

Essential equipment and supports should be suitable for the accelerations as indicated in as follows. Accelerations are to be considered acting independently.

12.6.1 Longitudinal Impact Accelerations

Maximum longitudinal impact acceleration, $a\ell$, at any point along the hull girder:

$$a_\ell = g_c \cdot (F_{IB}/\Delta) \cdot \left\{[1.1 \cdot \tan(\gamma + \varphi)] + \left[7 \cdot \left(\frac{H}{L}\right)\right]\right\} \text{ m/s}^2(\text{m/s}^2, \text{ ft/s}^2)$$

where:

g_c = 1 (9.80665, 32.174)

12.6.2 Vertical Acceleration

Combined vertical impact acceleration, a_v, at any point along the hull girder:

$$a_v = g_c \cdot 2.5 \cdot (F_{IB}/\Delta) \cdot F_X \text{ m/s}^2(\text{m/s}^2, \text{ ft/s}^2)$$

where:

g_c = 1 (9.80665, 32.174)
F_X = 1.3 at FP
= 0.2 at midships
= 0.4 at AP
= 1.3 at AP for vessels conducting ice breaking astern.

intermediate values to be interpolated linearly.

12.6.3 Transverse Impact Acceleration

Combined transverse impact acceleration, at, at any point along hull girder:

$$a_t = g_c \cdot 3 \cdot F_i \cdot \frac{F_X}{\Delta} \text{ m/s}^2(\text{m/s}^2, \text{ft/s}^2)$$

where:

g_c = 1 (9.80665, 32.174)
F_X = 1.5 at FP
= 0.25 at midships
= 0.5 at AP
= 1.5 at AP for vessels conducting ice breaking astern.
intermediate values to be interpolated linearly
ϕ = maximum friction angle between steel and ice, normally taken as 10, in degrees
γ = bow stem angle at waterline, in degrees
Δ = displacement in (tonnes, tonnes, Lton)
L = length between perpendiculars, in m (m, ft).
H = distance from the water line to the point being considered, in m (m, ft).
F_{ib} = vertical impact force in kN (tf, Ltf).
F_i = total force in kN (tf, Ltf) normal to shell plating in the bow area due to oblique ice impact.

12.7 Auxiliary Systems

Machinery protection. Machinery should be protected from the harmful effects of ingestion or accumulation of ice or snow. Where continuous operation is necessary, means should be provided to purge the system of accumulated ice or snow. Some means should be provided to prevent damage due to freezing, to tanks containing liquids. *Vent and discharge pipes.* Vent pipes, intake and discharge pipes and associated systems should be designed to prevent blockage due to freezing or ice and snow accumulation.

12.8 Sea Inlets and Cooling Water Systems

Cooling water systems for machinery. Cooling water systems for machinery that are essential for the propulsion and safety of the vessel, including sea chests inlets, are to be designed for the environmental conditions applicable to the ice class. *Sea chests.* At least two sea chests are to be arranged as ice boxes for Polar Class **PC1** to **PC5** inclusive where the calculated volume for each of the ice boxes should be at least 1 m^3 (35.314ft^3) for every 750 kW (750 kW, 1005 HP) of the total installed power. For Polar Classes **PC6** and **PC7**, at least one ice box for supplying water for cooling and firefighting purposes is to be connected to the cooling-water discharge by a branch pipe having the same cross-sectional area as the main pipe-line, in order to stay free from ice and slush ice. As far as practicable, the sea inlet chest is to be situated well aft, adjacent to the keel, located preferably near the centreline. *Ice boxes.* Ice boxes are to be designed for an effective separation of ice and the venting of air. *Sea inlet valves.* Sea inlet valves are to be secured directly to the ice boxes. The valves are to be a full-bore type. *Vent pipes.* Ice boxes and sea bays are to have vent pipes and fitted with shut off valves which are connected directly to the shell. *Sea bays freezing prevention.* Some means is to be provided to prevent the freezing of sea bays, ice boxes, ship side valves and fittings positioned above the load water line. *Cooling seawater recirculation.* Efficient means are to be provided to recirculate cooling seawater to the ice box. The total sectional area of the circulating pipes is not to be less than the area of the cooling water discharge pipe. *Ice box access.* Detachable gratings or manholes must be provided for access and egress to the ice box. These manholes are to be located above the deepest load line with access provided to the ice box from above. *Openings in vessel sides.* Openings in vessel sides for ice boxes are to be fitted with gratings, or holes or slots in shell plates. The net area through these openings is to be not less than five times the area of the inlet pipe. The diameter of holes and width of slot in shell plating is to be not less than 20 mm (0.787 in). Gratings of the ice boxes are to be provided with a means of clearing. Clearing pipes are to be provided with screw-down type non return valves.

12.9 Ballast Tanks

Efficient means are to be provided to prevent freezing in fore and after peak tanks and wing tanks located above the water line and where otherwise found necessary.

12.10 Ventilation System

Air intakes location. The air intakes for machinery and accommodation ventilation are to be located on both sides of the vessel. *Air intakes heating.* Accommodation and ventilation air intakes are to be provided with some appropriate means of heating. *Machinery air intakes.* The temperature of inlets air provided to machinery from the air intakes is to be suitable for the safe operation of the machinery.

Table 12.8 Rudder actuator holding torque multipliers

Ice class	PC1	PC2	PC3	PC4	PC5	PC6	PC7
Factor	5	5	3	3	3	2	1.5

Table 12.9 Assumed turning speeds for torque relief arrangements

Ice class	PC1–2	PC3–5	PC6–7
Turning speeds [deg/s]	8	6	4

12.11 Steering Systems

Rudder stops are to be provided. The design ice force on rudder should be transmitted to the rudder stops without damage to the steering system. An ice knife should in general be fitted to protect the rudder in centre position. The ice knife should extend below BWL.

12.11.1 Rudder Actuator Holding Torque

The effective holding torque of the rudder actuator, at safety valve set pressure, is obtained by multiplying the open water requirement at design speed [maximum 9.26 m/s (18kn)] by the factors given in Table 12.8.

12.11.2 Torque Relief Arrangements

The rudder actuator is to be protected by torque relief arrangements, assuming the turning speeds given in Table 12.9 without undue pressure rise (ref. UR M42 for undue pressure rise):

12.11.3 Fast Acting Torque Relief Arrangements

Additional fast acting torque relief arrangements (acting at 15% higher pressure than set pressure of safety valves in the section *Steel grades*) are to provide effective protection of the rudder actuator in case of the rudder is pushed rapidly hard over against the stops assuming turning speeds given in Table 12.10.

The arrangement is to be such that steering capacity can be speedily regained.

Table 12.10 Rudder actuator holding torque multipliers

Ice class	PC1–2	PC3–5	PC6–7
Turning speeds [deg/s]	40	20	10

12.12 Alternative Designs

As an alternative to this section, a comprehensive design study may be submitted and may be requested to be validated by an agreed test programme.

Chapter 13
Requirements for Enhanced Polar Class Notation

13.1 Introduction

Vessels that comply with the requirements outlined and discussed in this chapter, in addition to those previously discussed in Chaps. 11 and 12 may be considered for a **PC** notion as described in Chap. 11, Table 11.1. Where issued, the **PC** notation is followed by the annotation **'ENHANCED'**. (e.g., **ICE CLASS PC3, ENHANCED**).

13.2 Transverse Framing

13.2.1 Main and Intermediate Frames

Upper ends of frames. Main and intermediate frames are required to extend up to the first deck or platform above the ice belt. They are to be welded and bracketed to the deck beams or to the deck longitudinals, as shown in Figs. 13.1 and 13.2. For ice classes **PC4** through **PC7**, where the lowest or only deck, or the lowest platform, is situated above the ice belt so that the distance between the deck, or platform, and the upper boundary of the ice belt exceeds "*d*" metres (feet), given in Table 13.1, the upper ends of intermediate frames in the midbody and stern areas may terminate at a deep stringer situated at least 0.6 m (2 ft) above the ice belt. For ice classes **PC6** and **PC7** in tween deck spaces, where the tween deck is 0.5 m (1.6 ft) or more above the upper ice waterline but within the ice belt, the upper ends of intermediate frames may terminate for ice class **PC6** and **PC7** at a stringer situated at least 0.5 m (1.6 ft) above the ice belt. The upper ends of the frames terminated at a deep stringer are to be welded and bracketed to it as shown in Fig. 13.3. The intermediate frames terminated at an intercostal stringer or longitudinal are to be welded to it as shown in Fig. 13.4.

A. Olsen, *Ship Operations in Extreme Low Temperature Environments*, Springer Series on Naval Architecture, Marine Engineering, Shipbuilding and Shipping 19,
https://doi.org/10.1007/978-3-031-52513-1_13

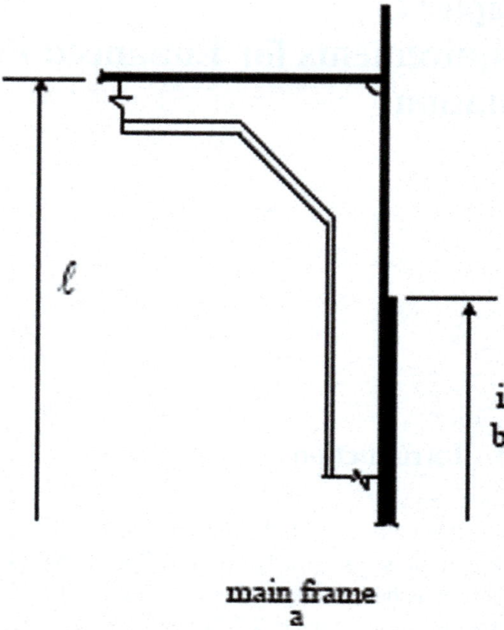

Fig. 13.1 Upper end terminations of frames

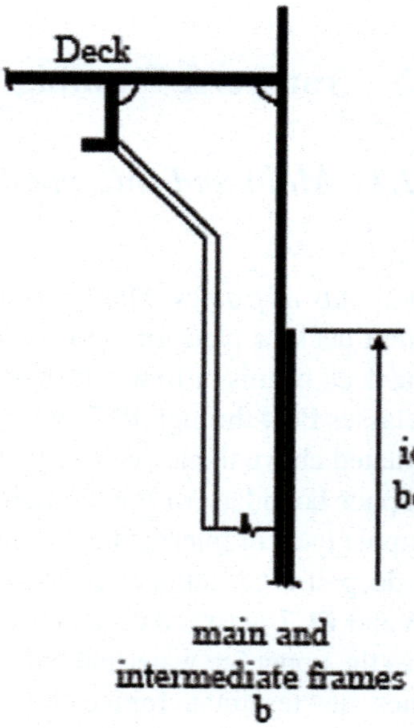

Fig. 13.2 Upper end terminations of frames

Table 13.1 Distance *d*, m (ft)

Ice class	Where web frames are fitted
PC4	5.2 (17)
PC5	4.0 (13)
PC6	3.0 (10)
PC7	3.0 (10)

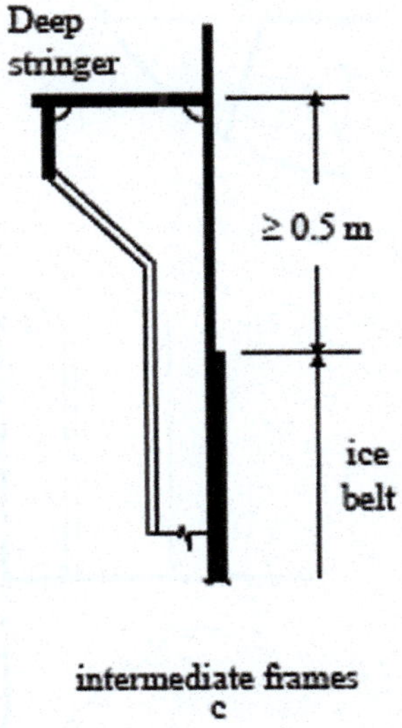

Fig. 13.3 Upper end terminations of frames

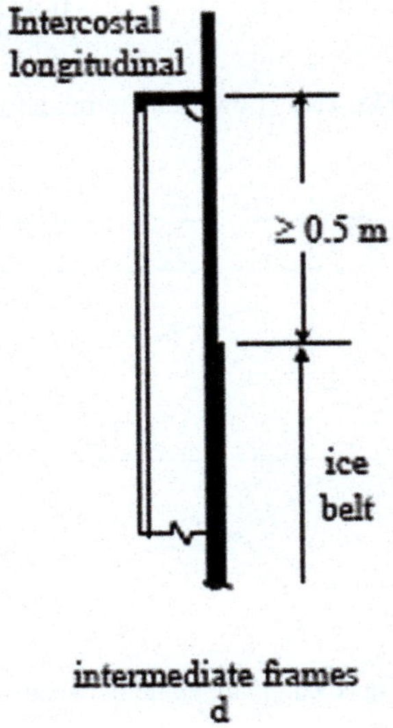

Fig. 13.4 Upper end terminations of frames

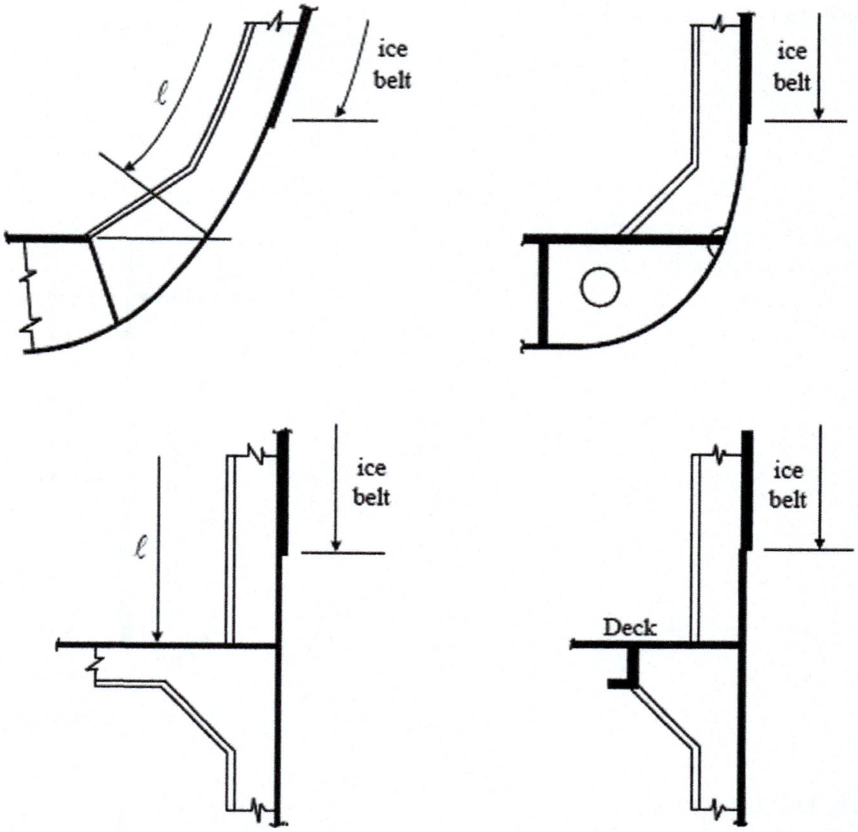

Fig. 13.5 Lower end terminations of frames

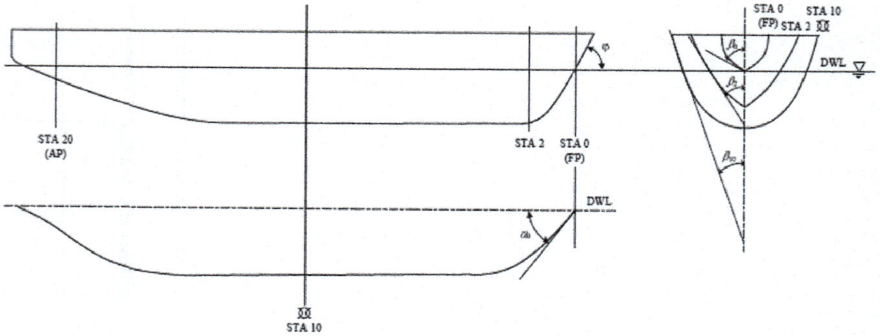

Fig. 13.6 Flare angle between side shell line and CP at DWL

Lower ends of frames. Main and intermediate frames are to extend down to the inner bottom or to the double bottom margin plate. For ice classes **PC4** through **PC7**, the intermediate frames may terminate at a deck 1.0 m (3.3 ft) below the ice belt. The main and intermediate frames are to be attached and bracketed either to the inner bottom or to the double bottom margin plate or to the deck beams, or deck or to the stringer as shown in Fig. 13.3. For vessels not having a double bottom, the intermediate frames are to extend down to a point below the top of the bottom transverses and are to terminate at an intercostal longitudinal. For ice classes **PC6** and **PC7**, the intermediate frames need not extend below the top of the floors, provided they terminate on an intercostal longitudinal not less than 0.8 m (2.6 ft) below the ice belt. The intermediate frames are to be attached to the bottom intercostal longitudinals.

Connection to stringers and decks. Main and intermediate frames are to be attached and bracketed to each supporting (deep) stringer, deck and deck beam within the ice belt (Fig. 13.5).

13.2.2 Web Frames

The web frames are to be attached and bracketed to the solid floors and the beams at each ice deck.

13.2.3 Ice Stringers

Arrangements. Deep continuous or intercostal stringers are to be fitted within the ice belt throughout the length of the vessel. The spacing between adjacent stringers, or between the stringer and a deck or the inner bottom, measured along the shell is to be not more than indicated in Table 13.2. One of the ice stringers is to be fitted about 200–400 mm (8–16 in) below the upper ice waterline, if there is no deck in this area. For ice classes **PC1** through **PC7**, another stringer is to be fitted about 100–300 mm (4–12 in) below the lower ice waterline, if there is no deck or similar support in this area.

Table 13.2 Maximum stringer spacing, m (ft)

Ice class	For framing without web frames	System with web frames
PC1 through **PC4**	1.5 (5)	2.1 (7)
PC5 through **PC7**	1.5 (5)	2.7 (9)

Scantlings and connections. Where ice stringers are intercostal, the following criteria should be met: (1) the intercostal stringers should be fitted between frames and their scantlings are to be not less than those for main frames; (2) the intercostal stringers are to be welded to the main and intermediate frames; (3) the web plate and the flange, or face, of intercostal ice stringers are to be attached to those of the main and intermediate frames; and (4) the intercostal stringers are to be bracketed to the bulkheads, side transverses, or web frames. Alternatively, where deep ice stringers are fitted, the following criteria should be met: (1) the shear area of the deep ice stringer within one frame space from the web frame is to be not less than that of the web frames; (2) the depth of the ice stringer at the midspan between the web frames is to be not less than twice the depth of the main frame; (3) the face, or flange, area of the deep stringer is to be not less than that of the web frame; (4) the web plate and the face, or flange, of deep ice stringers are to be attached to those of the web frames; (5) the deep stringer referred to in *Main and intermediate frames* at which the upper ends of frames are terminated, is to have the scantlings as required in *Ice stringers*; and (6) the deep stringers are to be bracketed to the bulkheads or side transverses, so that the shear area at the bulkhead is twice that of the ice stringer web. Stiffeners or tripping brackets may be fitted as required by Class.

13.3 Longitudinal Framing

13.3.1 Struts

Where one or more struts are fitted as an effective supporting system for the ice belt structure, they are to be located within the ice belt and spaced so to divide the supported web into spans of approximately equal length. Inboard ends of the struts are to be supported sufficiently by longitudinal bulkhead transverses having a section modulus not less than 0.9 of that discussed in Chap. 11, *Framing (web frames and load carrying stringers)*. The sectional area of the strut is to be obtained from the following equation:

$$A = \left(\frac{bs_1}{K}\right)\left(\frac{P}{\sigma_y}\right)K_o\text{cm}^2\left(in^2\right)$$

where:

b = as defined in *Design load patch* for particular area of the ice belt, in m (ft).
s_1 = distance between web frames in mm (in) measured along lower ice waterline in way of compartment being considered.
K = 0.04–0.0175 (ℓ/r) for SI and MKS units
= 0.0333–0.00175 (ℓ/r) for US units.
ℓ = unsupported span of the strut, m (ft).
r = least radius of gyration, cm (in).

P $= C_1 P_{ave} AF$
C_1 = 0.60 for bow area as defined in *Hull areas.*
= 0.50 for all other areas
P_{ave} = as defined in *Pressure within the design load patch* for a particular area of the ice belt.
AF = hull area factor
σ_y = minimum upper yield stress of the material, in N/mm^2, but not greater than 690 N/mm^2(70 kgf/mm^2, 100,000 psi).
K_o $= (2.44/\ell)^{1/2}$ (ℓ in m)
$= (8/\ell)^{1/2}$ (ℓ in ft), but not less than 0.4.

13.4 Peak Frames

Main and intermediate frames in forepeaks are to extend down to the floors or the bottom transverses or the stem. The section modulus of each peak frame is to be as given in Chap. 14, *Required plastic section modulus.* The spacing between the deep ice stringers or platforms measured along the shell is to be not more than 1.5 m (5ft) for forepeaks of ice classes **PC1** through **PC4**. For the forepeaks of ice classes **PC5** through **PC7**, the distance is to be not more than 2.1 m (7 ft). For ice classes **PC1** through **PC4**, transverse peak frames are to be fitted so that the angle between the web of the transverse frame and the shell plating, φ_w, is not less than 40° at any waterline within the ice belt. If this angle is less than 60°, the section modulus of the transverse peak frames is to be increased by the factor.

$$K = 2\cos\varphi w$$

where $40° \leq \varphi w \leq 60°$.

For all ice classes, the intermediate frames are to extend down to the bottom structure and up to the first deck above the ice belt.

13.5 Double Bottom Hulls

13.5.1 Inner Bottom

An inner bottom is to be fitted between the peaks in all vessels of ice classes **PC1** to **PC3** and in **PC4** ice class vessels of lengths of 61 m (200 ft) and over.

13.5.2 Transversely Framed Bottom

For ice classes **PC1** through **PC5**, solid floors are to be fitted at each web frame along the length of the vessel, and, in addition, at each main frame within the bow, lower intermediate and lower stern areas of the ice belt. Spacing of the solid floors is to be not more than required by Class, as applicable. Open floors or bilge brackets extending to longitudinals, or side girders are to be fitted at each intermediate frame that extends to the inner bottom. The distance between bottom side girders is to be not more than 2.4 m (8 ft) for the bow area of ice classes **PC1** through **PC3** and 3.0 m (10 ft) elsewhere for ice classes **PC1** through **PC5**. Spacing of the side girders is to be not more than required by Class.

13.5.3 Longitudinally Framed Bottom

For ice classes **PC1** through **PC5**, solid bottom transverses or solid floors are to be fitted at each web frame along the length of the vessel, but at not more than 1.8 m (6 ft) within the bow, lower intermediate and lower stern areas of the ice belt. Spacing of the solid floors is to be not more than required by Class, as applicable. Special consideration will be given to wider spacings. Open floors or bilge brackets extending to the outboard longitudinals are to be fitted throughout at each frame that extends to the inner bottom. The spacing of the bottom longitudinals within the bow, lower intermediate and lower stern areas of the ice belt is to be not more than 0.6 m (2 ft) for ice classes **PC1** through **PC3** and 0.7 m (2.3 ft) for ice classes **PC4** through **PC7**.

13.6 Ice Decks

The following requirements apply to decks or parts of decks situated within the ice belt as defined in this chapter, *Definitions*. For vessels not having decks in the ice belt and for vessels of ice classes **PC1** through **PC4** having only one deck in the ice belt, the following requirements apply also to decks or parts of decks above and below the ice belt to which the main and intermediate frames extend.

13.6.1 Deck Plating

The thickness of the stringer plate is to be not less than:

$$t = k(s^2bp)^{\frac{1}{3}}\,\text{mm(in)}$$

where:

k = 0.12 (0.257, 0.00523).
s = distance between the deck beams, in mm (in).
b = as defined in Chap. 14, *Design load patch*, in m (ft), for the particular area of the ice belt.
P = $C_1 P_{ave} AF$.
C_1 = 0.60 for bow area as defined in Chap. 14, *Hull areas*
= 0.50 for all other areas.
P_{ave} = as defined in Chap. 14, *Pressure within the design load patch* for the particular area of the ice belt.
AF = hull area factor for the particular area of the ice belt.

The width of the stringer plate is to be not less than five times the depth of the main frame for ice classes **PC1** and **PC2** and four times the main frame depth for **PC3** to **PC7** ice classes. For ice classes **PC1** through **PC7**, the thickness of the deck plating is to be not less than 0.75 times the required thickness of the stringer plate.

13.6.2 Deck Transverses and Deck Beams

Transversely framed decks. Partial beams or brackets are to be fitted at every intermediate frame for ice classes **PC1** to **PC5.** These partial beams or brackets are to be extended from the frames to a deck longitudinal or deck girder. The length of these partial beams or brackets is to be not less than the width of the stringer plate.

Longitudinally framed decks. Deck transverses are to be fitted at every web frame and, in addition, not less than at every second main frame for ice classes **PC1** to **PC4,** at every third main frame for ice classes **PC5** to **PC7.** Partial beams or brackets are to be fitted at all other main frames and at every intermediate frame for ice classes **PC1** to **PC7.**The partial beams or brackets are to be extended from the frames to a deck longitudinal or deck girder situated not less than 1.5 s from the inboard edge of the frames, where s is as defined in *Deck plating.*

Scantlings. The sectional area of the beams and deck transverses is to be not less than:

$$A = K_1 sb\left(\frac{P}{\sigma_y}\right)\cos\beta \text{cm}^2$$

$$A = 1.2K_1 sb\left(\frac{P}{\sigma_y}\right)\cos\beta \text{in}^2$$

The moment of inertia of the beams is to be not less than:

$$MI = kK_2 s\ell^2 bP \cos\beta \text{cm}^4 \left(\text{in}^4\right)$$

where:

k = 1.0 (9.81, 0.1191).
P = $C_1 P_{ave} AF$
C_1 = 0.60 for bow area as defined in Chap. 14, *Hull areas*
= 0.50 for all other areas
P_{ave} = as defined in Chap. 14, *Pressure within the design load patch*, in N/mm^2 (kgf/mm^2, ksi), for the particular area of the ice belt
A_F = hull area factor
b = as defined in Chap. 14, *Design load patch*, in m (ft), for the particular area of the ice belt
s = distance between the beams, in mm (in)
ℓ = the span of the beam, measured in m (ft), between the inboard edge of the frame and the deck longitudinal or deck girder supporting the beam
σ_y = minimum upper yield stress of the material, in N/mm^2, but not greater than 690 N/mm^2 (70 kgf/mm^2, 100,000 psi)
β = as defined in Chap. 14, *Bow area*, in degrees, for the particular area of the ice belt
K_1 = 8.5 for ice classes **PC1** to **PC5.**
= 6.6 for ice classes **PC6** and **PC7**
K_2 = 0.24 for ice classes **PC1** to **PC5**
= 0.13 for ice classes **PC6** and **PC7**

The sectional area and the moment of inertia of the partial beams and of the brackets are to be not less than required above. The beams and the partial beams are to be bracketed to the deck longitudinals or deck girders. Beams or partial beams or brackets fitted at the web frames are to be reinforced so that their section modulus, S_M is to be not less than:

$$SM = \frac{K_3 SM_{wf\ell} wf}{\ell} \text{cm}^2 \left(\text{in}^3\right)$$

where:

SM_{wf} = section modulus of the web frame in cm^3 (in^3)
ℓ_{wf} = span of the web frame, measured in m (ft), between supports, with no reduction for fitted end brackets, if any
K_3 = 0.8 for ice classes **PC1** through **PC5**
= 0.5 for ice classes **PC6** and **PC7**

When calculating the section modulus and the moment of inertia of a framing member, net thicknesses of the web, flange (if fitted) and attached shell plating are to be used.

13.6.3 *Decks with Wide Openings*

Within the bow intermediate and midbody areas of the ice belt, the cross-sectional area of the deck outside the line of openings is to be not less than:

$$A = Kb\ell\left(\frac{P}{\sigma_y}\right) \cdot 103\ \text{cm}^2$$

$$A = 14.4Kb\ell\left(\frac{P}{\sigma_y}\right)\text{in}^2$$

where:

K = 8.2 for ice classes **PC1** to **PC5**
= 6.2 for ice classes **PC6** and **PC7**
b = as defined in Chap. 14, *Design load patch*, in m (ft), for the particular area of the ice belt
ℓ = the length of the opening, in m (ft), but need not be taken as more than 0.1 L
P = $C_1 P_{ave} AF$
C_1 = 0.60 for bow area as defined in Chap. 14, *Hull areas*
= 0.50 for all other areas
P_{ave} = as defined in Chap. 14, *Pressure within the design load patch*, for the particular area of the ice belt
AF = Hull area factor
σ_y = as defined in Chap. 16, *Deck transverses and deck beams*
L = as defined in this chapter, *Length*, in m (ft)

13.7 Bulkheads

For ice classes **PC1** to **PC5**, those parts of transverse bulkheads situated within the ice belt are not to be vertically corrugated.

13.7.1 *Scantlings*

For ice classes **PC1** to **PC7**, the thickness of that part of the bulkhead adjacent to the side shell and within the ice belt is to be not less than the thickness of the adjacent frames or of the stringers connected to the bulkhead, whichever is greater. The width of these parts of the bulkhead is to be not less than shown in Table 13.3. These parts of the bulkhead adjacent to the shell within the ice belt are to be fitted with stiffeners normal to the shell plating. The stiffeners are to be welded to a vertical

Table 13.3 Minimum width of reinforced bulkhead plating

Ice class	Area of the ice belt			
	Peak bulkheads m (ft)	Bow and bow intermediate areas m (ft)	Midbody area m (ft)	Stern area m (ft)
PC1 through **PC4**	1.6 (5.2)	1.4 (4.6)	1.2 (4.0)	1.4 (4.6)
PC5 through **PC7**	1.2 (4.0)	1.2 (4.0)	1.0 (3.3)	1.0 (3.3)

Table 13.4 Solid stem bar coefficients

Ice class	A_0 cm^2 (in^2)	K_1
PC1	750 (116.2)	0.28
PC2	750 (116.2)	0.28
PC3	700 (108.5)	0.27
PC4	500 (77.5)	0.24
PC5	200 (31.0)	0.18
PC6	62 (9.6)	0.13
PC7	62 (9.6)	0.13

bulkhead stiffener and welded and bracketed to the side longitudinals. Where the shell is transversely framed, brackets are to be welded to the shell and extended and attached to adjacent frames.

If a vessel is intended to operate astern in ice regions, the width of the reinforced parts of the bulkhead adjacent to the *Stern and stern intermediate ice belt areas* is to be not less than that required for *Bow and bow intermediate areas* shown in Table 13.3.

13.8 Stem and Stern Frames

The relevant Class requirements for hull structures and arrangements, with particular reference to longitudinal strength, must be complied with. The stem and stern frame for ice class **PC1** through **PC5**, and for ice class **PC6** and **PC7** vessels of displacements more than 50,000 tonnes (49,200 Lt), are to be constructed of rolled bar, cast or forged steel. Shaped plate stem may be used for **PC6** and **PC7** vessels of and less than 50,000 tonnes (49,200 Lt). The shaped plate stem used in other cases is to be specially considered. All joints and connections are to fully develop the strength of the stem and stern frame. All rudders are to be protected against ice impacts for going astern.

13.8.1 Stem

Solid stem. The cross-sectional area of a stem made of rolled bar, cast or forged steel from the centre vertical keel to 0.01 L above the ice belt is to be not less than:

$$A = K_1 D^{\frac{1}{3}}(L - 61) + A_o \text{cm}^2$$

$$A = 0.0473 K_1 D^{\frac{1}{3}}(L - 200) + A_o \text{in}^2$$

where:

K_1 and A_o	= as given in Table 13.4
D	= as defined in this chapter, *Definitions*
L	= as defined in this chapter, *Definitions*, in m (ft), but is not to be taken less than 61 m (200 ft)

For vessels of displacements less than 2500 tonnes (2460 Lt) the cross-sectional area given by the above equation may be reduced 10%. The cross-sectional area of the stem above the ice belt may be reduced gradually to the value permitted by Class.

Shaped plate stem. Thickness of shaped plate stems within the bow area of the ice belt is to be not less than.

$t = 0.8s(P/P\sigma_y)^{\frac{1}{2}} + t_s$ but not less than 0.04R.where:

t	= required thickness of plate stem, in mm (in)
s	= distance between frames, brackets (breast hooks) or stiffeners, in mm (in)
P	= 0.75 Pbow , as defined in Chap. 14, *Structural requirements for Polar Class vessels*
σ_y	= minimum upper yield stress of the material, in N/mm^2, but not greater than 690 N/mm^2 (70 kgf/mm^2, 100 psi)
t_s	= corrosion/abrasion addition for the bow area, as defined in Chap. 14, *Corrosion/abrasion additions and steel renewal*, in mm
R	= the inside radius of the stem at the given section, in mm (in). Need not be taken greater than 625 mm (24.6 in) for ice classes **PC6** and **PC7**

At any section, the fore and aft length of the stem plate is to be not less than 15t.

Arrangement. The outer surface of connections of the shell plating to the stem is to be flush. The stem is to be supported by floors, webs, frames, breasthooks or brackets spaced not more than 610 mm (24 in). In addition, shaped plate stems are to be supported on the centreline by a plate, web or bulkhead having the same thickness as the centre vertical keel and a width not less than 610 mm (24 in).

Table 13.5 Stern post coefficient

Ice class	K
PC1	2.0
PC2	1.9
PC3	1.8
PC4	1.6
PC5	1.4
PC6	1.2
PC7	1.2

13.8.2 Stern Frame

The stern post is to be of a size acceptable to Class, with all thicknesses increased by coefficient K, as given in Table 13.5.

13.9 Towing Arrangements

13.9.1 Bow

Polar Class vessels intended to be escorted by a higher ice class leading vessel, are to be fitted with a tow chock pipe and a tow bitt on the bow. The chock and the bitt are to be properly connected to the stem frame. The portions of the decks at which the chock and the bitt are attached are to meet requirements of Chap. 16, *Ice decks.* The shell plating and framing below and 1.5 m (5 ft) around the chock are to be as required by Chap. 14, *Shell plate requirements*; Chap. 16, *Framing (General)*; and Chap. 16, *Framing (local frames in bottom structures and transverse local frames in side structures*) for the bow area of the ice belt for ice classes **PC6** and **PC7** and for the intermediate area of the ice belt for ice classes **PC2** through **PC5** and where the corrosion and abrasion allowance, t_s, is as given in Table 13.6. The stem frame below the connections with the chock is to be as required by Chap. 16, *Stern and stem frames* for the portion of the stem within the ice belt.

Where a bulbous bow is fitted, the bulb is not to extend beyond the fore end of the lower ice waterline specified in the section *Upper and lower ice waterlines.*

13.9.2 Stern

Vessels of ice classes **PC1** through **PC6** intended to be used as leading vessels assisting passage of a lower ice class vessel as listed in Table 13.2 are to be equipped

Table 13.6 Corrosion/abrasion additions for shell plating around the chock

Hull area	ts, mm					
	With effective protection			Without effective protection		
	PC4 and PC5	PC6 and PC7	PC1–PC3	PC4 and PC5	PC6 and PC7	PC1–PC3
Shell plating below and 1.5 m (5ft) around the chock	**1.0**	**1.0**	**3.0**	**2.0**	**1.5**	**1.0**

with a towing system. Both the arrangement of the towing system and the shape of the stern are to be suitable for towing the assisted vessel in immediate contact. The portion of the upper deck which the towed vessel may contact is to be as required by *Ice decks*. The shell plating and framing adjacent to this portion of the upper deck are to be as required by this chapter, *Framing (general)* and this chapter, *Framing (local frames in bottom structures and transverse local frames inside structures)* for the stern area of the ice belt.

13.10 Machinery Arangements

13.10.1 Propulsion Arrangements

In addition to the regular governor, all propulsion engines and turbines are to be fitted with a separate overspeed device so adjusted that the speed cannot exceed the maximum rated speed by more than 20%.

13.10.2 Electric Propulsion

Propulsion motors are to be fitted with automatic protection against excessive torque, overloading and temperature. This protection is to automatically limit these parameters but is not to cause loss of propulsion power.

13.10.3 Boilers

Vessels propelled by steam machinery are to be fitted with at least two boilers of equal capacity.

13.10.4 Protection Against Excessive Torques

For vessels of all classes, if torsionally flexible couplings or torque-limiting devices are fitted in the propulsion system, positive means are to be provided for transmitting full torque to the propeller in the event of failure of the flexible element. In addition, for vessels of classes **PC1** through **PC4**, couplings of the elastomer-in-shear type are not to be fitted in those portions of the propulsion system which are subject to shock loading from the propeller.

13.10.5 Propeller Arrangements

Propeller arrangements, the shape of the stern and the propeller protecting structures are to be adequate for the intended service. Special consideration is to be given to the propeller protection when moving astern. For **PC1** through **PC5** ice class vessels, the following condition is to be complied with.

$$0.5b_x - b_x \geq kd$$

where:

B_x = breadth of the lower ice waterline, at the hull section in way of the propeller tips, in m (ft)
b_x = distance from the vessel centreline to the outermost propeller blade tip, in m (ft)
k = 0.25 for open propellers
= 0.10 for ducted propellers
d = propeller diameter, in m (ft)

13.10.6 Tunnel Thrusters

The mechanical components of a tunnel thruster (i.e., propellers, gears, shafts, couplings, etc.) are to meet the applicable requirements of Class, for a theoretical input torque of twice the prime mover output torque in order to simulate, in a conservative way, the effect of ice on all the torque transmitting components. Alternatively, a comprehensive study to determine the effect of ice—propeller interaction and the resultant ice torque may be considered. In this way the mechanical components of the tunnel thruster are to meet the applicable requirements using the so determined ice-torque.

13.10.6.1 Cooling Water Arrangements

The following apply to vessels of ice classes **PC1** through **PC5**.

Sea bay or tank. The suctions for cooling water for all machinery essential to the propulsion of the vessel and for firefighting purposes are to be taken from a sea bay or tank located as close as practicable to the keel. The sea bay or tank is to be supplied with water from at least two independent sea suctions with at least one on each side of the hull. The area of each sea suction opening is to be not less than six times the total cross-sectional area of all pump suctions connected to the sea bay.

Sea suctions. Suitable strainers are to be provided between the sea suctions and the sea bay. Valves are to be provided to permit isolation of the strainers, both from the sea suctions and from the sea bay. The cross-sectional area of such valves and strainers and associated piping for each sea suction is not to be less than the total cross-sectional area of all pump suctions connected to the sea bay.

Sea water pumps. Each sea water pump serving machinery essential to the propulsion of the vessel is to draw sea water directly from the sea bay. Design flow velocity in any suction line is not to exceed 2 m (6.6 ft) per second.

Cooling water recirculation. The discharge line from the cooling system is to be provided with suitable piping, valves and fittings to permit the discharge flow to be recirculated. The recirculation piping is to connect with the suction piping at a point on the seaward side of the strainer sea shut-off valves. Piping, valves and fittings for the recirculation line are to be of at least the same cross-sectional area as the overboard discharge line.

13.10.7 Starting Air System

For vessels of Ice Class **PC1** through **PC5**, in addition to the applicable requirements of Class, starting air systems are to comply with the following: (1) at least two independently driven starting-air compressors are to be provided. The total capacity of the compressors is to be sufficient to charge the air receiver from empty to maximum pressure in not more than 30 min; and (2) the smallest of the starting air compressors is to have not less than two-thirds the capacity of the largest.

13.11 Power of Propulsion Machinery

For polar classes **PC1** through **PC7**, the total ahead power delivered to the propellers, is to be sufficient for the vessel to maintain a design service speed under the ice conditions described in Table 13.2, as related to the appropriate vessel notation. An appropriate analytical approach or ice model testing results are to be submitted for review. Where the design is in an early stage or ice model testing is not planned, the requirement for minimum power/astern power may be used for an assessment of power of propulsion machinery, unless otherwise any specific methodology is provided by the cognisant authorities having jurisdiction over the water in which the vessel is intended to operate. The requirements of the cognisant authorities or administrations may also need to be recognised or complied with.

13.11.1 Minimum Powering Criteria

The total propulsion power delivered to the propellers is recommended to satisfy either of two criteria, namely (1) the thickness of consolidated level ice passable by a vessel of ice classes **PC1** though **PC7** in stable continuous icebreaking is to be as defined in Chap. 16, *Maximum thickness of consolidated level ice;* and (2) the total power delivered to the propellers at the maximum continuous rate has to be as defined in this chapter, *Total power delivered to the propellers.*

13.11.2 Maximum Thickness of Consolidated Level Ice

The maximum thickness, h_{max}, of consolidated level ice (in the absence of wind/current driven ice compressions) passable in stable continuous icebreaking is not to be less than the value of h_o given in Table 13.7, i.e.:

$$h_{max} \geq h_0$$

The value of h_{max} can be determined at design stages by the following formula:

$$h_{max} = f_u f_s f_p \frac{(N_p d_{pr})^{1/3} \Delta^{1/6}}{B^{0.5}} \geq h_0 m$$

Table 13.7 Nominal values of powering criteria

Ice class	Nominal ice thickness h_0, m	
	Ice breakers	Ice class vessels
PC1	2.8	2.5
PC2	2.2	1.9
PC3	1.6	1.3
PC4	1.2	1.0
PC5	0.8	0.6
PC6	0.6	0.5
PC7	–	0.5

where:

f_u = 0.615

f_s = factor of hull shape

$= \frac{cos^{1.5}\varphi \cdot sin^{0.5}\left(\frac{\alpha_0+\beta_0+\beta_2}{3}\right)}{sin^{1.5}(90^\circ-\beta_{10})} \cdot \left(\frac{B}{L}\right)^{0.2}$

f_p = factor of propellers arrangement, as follows:
= 0.88 for single screw ships
= 0.99 for twin screw ships
= 1.06 for triple screw ships with all three propellers of the same diameter
For vessels with azimuthing propellers only
= 0.9 for single azimuthing pod
= 1.0 for two azimuthing pods

Np = total power delivered to the propellers, MW

B = maximum breadth of ship at DWL, m

L = LBP, m

Δ = displacement of ship at DWL or at the deepest WL for ice conditions, whichever is greater, tons

d_{pr} = diameter of the propellers, m

φ = stem inclination angle to the waterplane at DWL, degrees

α_0 = angle between DWL and CL at FP, degrees

β_0 = flare angle between side shell line and CP at DWL at STA 0 (FP), degrees

β_2 = flare angle between side shell line and CP at DWL at STA 2, degrees

β_{10} = flare angle between side shell line and CP at DWL at the midship (STA 10), degrees

Special consideration will be given to vessels with other arrangements of propellers including the use of both bow and stern shaft-line propellers operating jointly, or a mix of shaft-line and azimuthing propellers, or other (Fig. 13.6).

13.11.3 Total Power Delivered to Propellers

The total ahead maximum continuous rated power, *N*, delivered to the propellers, is to be not less than the values obtained as follows:

(1) For Ice Class **PC1** through **PC4** and vessels assigned the **ICE BREAKER** notation.

$$N = kA(B)^{0.8}(L)^{0.4}\left[1 + me^{-5\Delta x10-6}\right]\text{kW}$$

(2) For Ice Class **PC5** through **PC7**

The smaller of the values obtained from the following two equations:

$$N = kA(B)^{0.8}(L)^{0.4}\left[1 + me^{-5\Delta x10-6}\right]\text{kW}$$

Table 13.8 Power coefficients

Ice class	A in SI units	M	C	K
PC1	360	1.3	–	–
PC2	270	1.0	–	–
PC3	200	0.8	–	–
PC4	136	0.6	–	–
PC5	107	0.6	1500	400
PC6	93	0.6	1000	350
PC7	93	0.6	1000	350

Table 13.9 Design speed for rudders, couplings and pintles

Ice class	Minimum design speed, knots
PC1	29
PC2	29
PC3	28
PC4	26
PC5	23
PC6	20
PC7	20

$$N = k\left(C + K\frac{\Delta}{1000}\right)\text{kW}$$

where:

N = total propulsion power delivered to propellers
e = base of natural logarithms
k = unit system factor, = 0.735

B, L, and Δ are as defined above.
A, m, C, and K are coefficients given in Table 13.8.

13.11.4 Powering Criteria Obtained from Ice Model Tests

At later design and construction stages, when results of ice model tests are available, the value of $h_{\max}$ calculated in Chap. 16, *Maximum thickness of consolidated level ice*, or/and the value of N_p calculated in Chap. 16, *Total power delivered to propellers* can be superseded by the results of self-propelled model tests in an ice model testing basin. The model tests have to be conducted according to a standard procedure approved by International Association of Ice Model Testing Basins. Using standard series model propellers will be approved in the self-propelled model tests used to produce the required values of h_{max} and N_p provided that the standard series model propellers are most similar to the actual full-scale propellers approved for the vessel.

13.11.5 Astern Power

The following requirements apply to all main propulsion systems fitted to a vessel with an **ENHANCED PC** notation. **PC1, ENHANCED** through **PC6, ENHANCED** with **ICE BREAKER** notation: total astern power delivered to the propellers is to be not less than that required in Chap. 16, *Power of propulsion machinery.* All vessels with **ENHANCED POLAR CLASS** notation intended to operate astern in accordance with Chap. 16, Hull areas: total astern power delivered to the propellers is to be not less than that required in Chap. 16, *Power of propulsion machinery.* **PC1, ENHANCED** through **PC3, ENHANCED**: total astern power delivered to the propellers is not to be less than 90% of that required in Chap. 16, *Power of propulsion machinery.* **PC4, ENHANCED**: total astern power delivered to the propellers is not to be less than 85% of that required in Chap. 16, *Power of propulsion machinery.* **PC5, ENHANCED** through **PC7, ENHANCED**: total astern power delivered to the propellers is not to be less than 70% of that required in Chap. 16, *Power of propulsion machinery.*

13.11.6 Flexible Couplings

Flexible couplings which may be subject to damage from overheating are to be provided with temperature monitoring devices or equivalent means of overload protection with alarms at each engine control station.

13.12 Bossings

The bossings are to be designed to withstand the design ice forces F_n, as specified by Chap. 16, *Design ice forces* where d_1 is the diameter of the bossing. The bossing plating thickness is to be not less than required by Chap. 16, *Shell plate thickness to resist ice load* for the stern ice belt area, where s is the distance between stiffeners.

13.13 Rudder and Steering Arrangements

13.13.1 Rudder Arrangements

Multiple rudders. Where two or more rudders are provided, they are to be mechanically independent. *Pintles.* Rudders must have at least two pintles. *Rudder stops.* Rudders are to be protected by strong and effective external rudder stops and provided with mechanical means of locking the rudder parallel to the centreline for use in the astern condition. *Ice knife.* Rudders are to be protected by ice knives or other similar structures located abaft the rudder. Clearance between the ice knife and the rudder is not to exceed 100 mm (4 in).

13.13.2 Rudder Stocks, Couplings and Pintles

Ice classes **PC1** *through* **PC5.** In addition to the requirements established by Class, the rudder stocks, couplings and pintles are to meet the ice strengthening requirements. *Ice classes* **PC6** *and* **PC7.** For ice classes **PC6** and **PC7**, rudder stocks, pintles, gudgeons and other bolting arrangements to the stern frames are to meet the requirements set by Class (Table 13.9).

Ice classes **PC1** *through* **PC7.** The stresses in these members with the load F applied as follows are not to exceed the shear yielding strength which may be taken as 0.577 times the specified yield point of the material.

$$F = 2K_3(Dt)^{1/2}kN(tf, Ltf)$$

where:

K_3 = as given in Table 13.10
D = ship displacement, in tonnes (long tons), as specified in Chap. 14, *Hull areas other than the bow*
t = thickness of the rudder, in m (ft), measured at the level of F and at 10% of the rudder length from the trailing edge.

F is to be applied to the after edge of the rudder in a direction parallel to the centreline of the vessel at all locations below the ice waterline within the middle 40% of the rudder height to determine the most severe requirements. Alternatively, F may be spread over any 60% of the rudder height as a uniform load. No other force need be considered simultaneously with F.

Table 13.10 Design ice force coefficient

Ice class	$K1$	$K3$
	SI units (MKS, US)	SI units (MKS, US)
PC1	55 (5.6, 3.1)	294 (30.0, 16.4)
PC2	53 (5.4, 3.0)	286 (29.2, 16.0)
PC3	49 (5.0, 2.7)	243 (24.8, 13.6)
PC4	43 (4.4, 2.4)	188 (19.2, 10.0)
PC5	32 (3.3, 1.8)	110 (11.2, 6.1)
PC6	20 (2.1, 1.1)	59 (6.0, 3.3)
PC7	20 (2.1, 1.1)	59 (6.0, 3.3)

13.13.3 Double Plate Rudder

For double plate rudders, the minimum thickness of plates is to be not less than required by this chapter, *Appendages.*

13.14 Propeller Nozzles

13.14.1 General

This section applies to fixed nozzles. Special consideration will be given to steering nozzles. The nozzles are to be supported at least at the upper and lower ends. The strength, rigidity and resistance to buckling of the nozzle are to be adequate for the design ice forces given in Chap. 16, *Appendages.* All the critical loading cases are to be considered. In no case under the design ice forces are the normal and axial displacements of the inside ring to exceed 10% of the clearance between the inside plating of the nozzle and the propeller blade tips, or 0.5% of the inside ring diameter, whichever is less. Nozzles are to be protected by stern structures as much as possible against direct impacts with large ice features.

13.14.2 Design Ice Forces

The design ice forces are to be not less than those obtained from the following equations:

$$F_n = K_1 K_2 (D d_1)^{1/2} \text{kN (tf, Ltf)}$$

$$F_f = K_3 K_4 [(d_1 - d_2)]^{1/2} \text{kN (tf, Ltf)}$$

where:

F_n = the design ice force applied normal to the outside surface of the nozzle in the most critical location

K_2 = 1 for the external sides of a single nozzle of a single screw vessel, = 1.1 for the outboard external sides of the outermost nozzles of vessels with two or more screws, = 0.25 for the external sides of nozzles situated between the outermost ones and for the internal sides of any nozzles, = 0.8 for bottoms of the nozzles

D = ship displacement, in tonnes (long tons), as specified in Chap. 14, *Hull areas other than the bow*

d_1 = maximum outer diameter of the nozzle, in m (ft)

d_2 = minimum internal diameter of the nozzle, in m (ft)

F_f = the design ice force applied to the ends of the nozzle, parallel to the propeller axis, in the most critical locations

K_4 = 1 for aft end face of the nozzle having no rudder behind, = 0.7 for the aft end face of the nozzle with a rudder behind, = 0.6 for the fore end face of the nozzle

K_1 and K_3 are as given in Table 13.10.

Values of K_2 and K_4 less than above will be approved, provided the stern and bottom hull structures effectively protect the nozzle against large ice fragments.

13.14.3 Plate Thickness

The plate thickness of both inner and outer surfaces of the nozzle is to be not less than required by Chap. 14, *Shell plate requirements* for the stern ice belt area where the corrosion and abrasion allowance, t_s, is as given in Table 13.11.

Table 13.11 Corrosion/abrasion additions for nozzle surface plating

Hull area	*ts*, mm					
	With effective protection			Without effective protection		
	PC4 and PC5	PC6 and PC7	PC1–PC3	PC4 and PC5	PC6 and PC7	PC1–PC3
Nozzle surface plating	0	0	0	2.0	1.8	1.0

13.15 Hull Structural Materials

13.15.1 Inspection

In addition to the non-destructive inspection requirements contained on the other Class mandated rules, all intersections of full penetration welds within the ice belt structure of ice class vessels **PC1 ENHANCED** to **PC4 ENHANCED** are to be inspected by radiographic or ultrasonic methods and are to meet the Class A requirements set by Class for non-destructive Inspection. Additional inspections may also be required by the Class approved surveyor for other locations including block connection joints.

Chapter 14
Requirements for Vessels Intended for Navigation in First-Year Ice

14.1 Introduction

Vessels which are to be distinguished in the Record by Ice Class followed by ice class **A0**, **B0**, **C0**, **D0** or **E0** are expected to meet the applicable requirements set by Class, as summarised in this chapter. Non-self-propelled vessels are expected to comply with separate requirements as applicable. Moreover, those vessels requiring ice breaker assistance are required to comply with the additional requirements set by Class.

14.1.1 Selection of Ice Class

The requirements in this section are intended primarily for vessels intended for navigation in first-year ice. The ice classes are as follows:

- **ICE CLASS A0**
- **ICE CLASS B0**
- **ICE CLASS C0**
- **ICE CLASS D0**
- **ICE CLASS E0**

A. Olsen, *Ship Operations in Extreme Low Temperature Environments*, Springer Series on Naval Architecture, Marine Engineering, Shipbuilding and Shipping 19,
https://doi.org/10.1007/978-3-031-52513-1_14

For the guidance of the vessel owner in selecting the most suitable ice class, ice conditions suitable for respective ice classes are shown in Table 14.1. The conditions of first-year ice are shown in Table 14.2.

Table 14.1 Regions and periods for navigation in ice for selecting ice class

Ice class	Navigating independently or when escorted by an ice breaker of the following ice classes	Year around navigation in water with first year ice with the ice conditions
A 0	Escorted by **PC4** or higher ice class vessel	Extreme
B 0	Escorted by **PC3** or higher ice class vessel	Extreme
A0, B 0, C 0	Escorted by **PC5** or higher ice class vessel	Very severe
A 0	Independently	Severe
B 0	Independently	Medium
C0	Independently	Light
D 0	Independently	Very light
E 0	Independently	Very light drift ice [in coastal areas]

Table 14.2 Ice conditions of first-year ice versus concentration and thickness of ice cover

Thickness of first-year ice cover in m (ft)	Concentration of ice[a]			
	Very close and consolidated ice, fast ice (from 10/10 to 9/10 or from 8/8 to 7/8)	Close ice (from 9/10 to 6/10 or from 7/8 to 5/8)	Open ice (from 6/10 to 3/10 or from 5/8 to 2/8) and fresh channel[b] in fast ice (more than 6/10 or 5/8)	Very open ice (less than 3/10 or 2/8), fresh channel[b] in fast ice (6/10 or 5/8 and less) and brash ice
1.0 (3.3) and above	Extreme	Extreme	Very severe	Severe
From 0.6 (2) to 1.0 (3.3)	Extreme	Very severe	Severe	Medium
From 0.3 (1) to 0.6 (2)	Very severe	Severe		Light
Less than 0.3 (1)	Severe	Medium	Light	Very light

Notes

[a] These ratios of mean area density of Ice in any given area are from the "World Meteorological Organisation Sea Ice Nomenclature" and give the ratio of area of Ice concentration to the total area of sea surface within some large geographic locales

[b] Provided the channel is wider than the ship

Table 14.3 Dimensions of ice belt areas m (ft)

Ice class	A	B	C	D	F*	S	Q
A0	0.8 (2.6)	0.6 (2.0)	$0.5D$	$0.2 + 0.004L$ $(0.7 + 0.004L)$	$0.3L$	$0.10L$	10.0 (33.0)
B0	0.6 (2.0)	0.5 (1.6)	0	$0.1 + 0.003L$ $(0.3 + 0.003L)$	$0.3L$	$0.10L$	09.0 (30.0)
C0	0.6 (2.0)	0.5 (1.6)	0	$0.00025L$	$0.3L$	$0.10L$	6.6 (22.0)
D0, E0	0.5 (1.6)	0.5 (1.6)	0	$0.002L$	$0.3L$	0	4.5 (15.0)

14.1.2 Extent and Length of Ice Belt Areas

The ice belt for self-propelled vessels is subdivided into the following areas:

- *For ice class* **A0** *through* **C0**:

Bow, midbody and stern areas.

- *For ice class* **D0** *and* **E0**:

Bow area.

For all first-year ice classes, the lowest extent of the bow area need not extend below a line drawn between Qm (ft) below the lower ice waterline at the stem and Bm (ft) below the lower ice waterline at the stern (refer to Table 14.3 for the values of Q and B). The extent and length of each area is shown in Fig. 14.1a, b, c and Table 14.3.

For ships with upper ice waterline parallel to the centreline, F is to be as shown in Fig. 14.1c. In any case, the bow area is to extend aft not less than to a section at:

$M = 0.2L$ abaft the fore-end of the lower ice waterline, or
$N = 0.05L$ abaft point where the moulded stem line crosses the baseline, whichever is located aft.

14.2 Design Ice Loads

14.2.1 Design Ice Pressure on the Bow Area

The design ice pressure on the bow area is to be not less than that obtained from the following equations:

$$P_b = P_o F_b$$

P_b design ice pressure on the bow area, in N/mm^2 (kgf/mm^2, ksi).

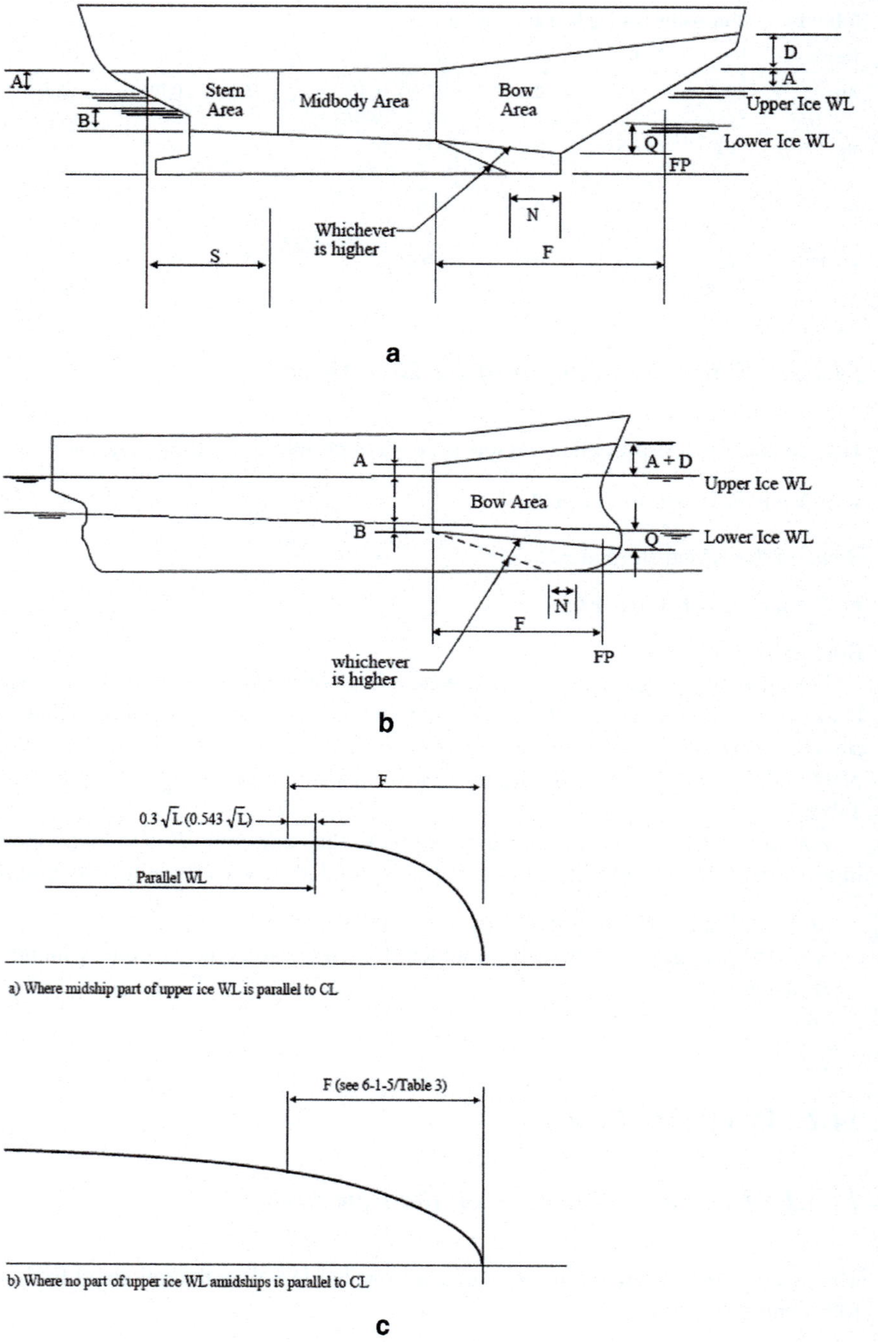

Fig. 14.1 **a** Ice belt areas (ice class A0 through C0). **b** Ice belt areas (ice class D0 and E0). **c** Ice belt areas (definition of *F*)

For all first-year ice classes:

$$P_o = \mathrm{B(D/n)}^{0.2}$$

where

B coefficient, as given in Table 11.4
D displacement
N 1000 (1,000,984).

$$F_b = (F_{b1})(F_{b2})$$

F_{b1} coefficient is given in Fig. 14.2. It is to be determined for each bow section at the upper and lower ice waterlines depending on αb and βb and the maximum value obtained is to be used; if the values of coefficient F_{b1} obtained for the different sections and at different ice waterlines vary by more than 15%, different coefficients F_{b1} and, correspondingly, different design ice pressures may be used for the appropriate parts of the bow area. F_{b1} is not to be taken less than 0.80 but need not be taken as more than 1.25 for vessels with conventional bows; for vessels fitted with bulbous bows, the F_{b1} coefficient within the bulb area is to be as given in Table 14.4.

$$F_{b2} = 1 + i(1.3 + 0.001D)^{-2}$$

i coefficient given in Table 11.4
ab angle between the centreline and a tangent to the ice waterline being considered at the bow section being considered
βb angle between the vertical and tangent to the bow section at the level of the ice waterline being considered.

14.2.2 Design Ice Pressures on Other Ice Belt Areas

Design ice pressures on other parts of the ice belt are to be obtained from the following equations:

For the midbody:

$$P_m = K_m P_o \quad or \quad P_m = K_m P_b, \text{ whichever is less}$$

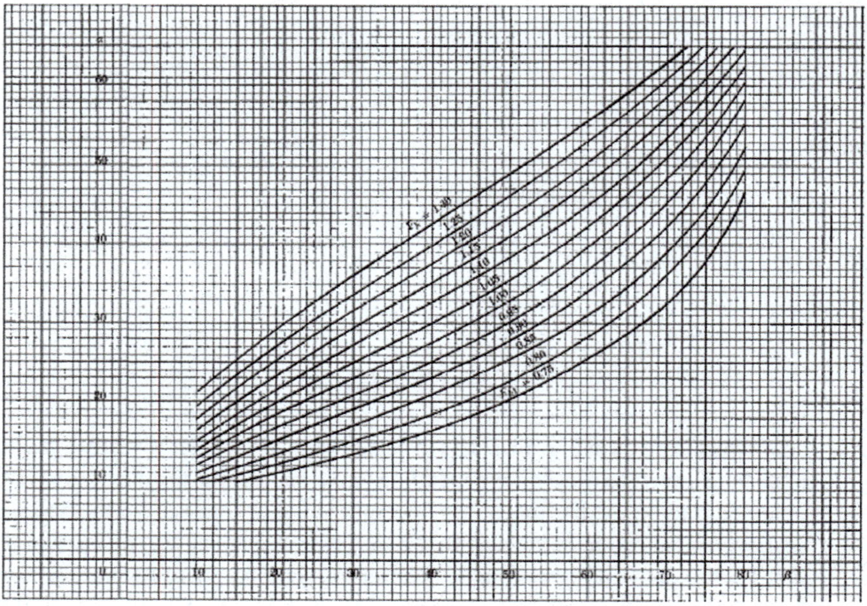

Fig. 14.2 Coefficients F_{b1} versus angles α_b and β_b

Table 14.4 Bow area ice pressure coefficients

Ice class	B N/mm^2 (kgf/mm^2, ksi)	I	F_{b1}
A0	0.997 (0.102, 0.142)	2	1.35
B0	0.750 (0.076, 0.109)	0	1.25
C0	0.60 (0.061, 0.086)	0	1.25
D0	0.50 (0.051,0.071)	0	1.25
E0	0.30 (0.031,0.043)	0	1.25

* Within the bulbous bow area

For the stern:

$$P_s = K_s P_b$$

P_m and P_s design ice pressures on corresponding area, in N/mm^2 (kgf/mm^2, ksi)
K_s coefficient, as given in Table 14.5
K_m coefficient, as given in Table 14.5 or by $2(3 + 4\sin\beta_m)^{-1}$, whichever is less
β_m as defined for β_b, but for the section at amidships.

Table 14.5 Ice pressure coefficients in other areas

Ice class	K_s	K_m
A0	0.35	20.45
B0	0.22	0.35
C0	0.11	0.22

14.2.3 *Extent of Design Ice Load*

In a vertical direction, the design ice pressure is considered to be uniformly distributed on the side structure. The vertical extent of the design ice pressure is to be obtained from the following equations:

For the bow:

$$b_b = 0.61 + b_o F_{b1} \text{m}$$
$$b_b = 2 + b_o F_{b1} \text{ft}$$

For the midbody:

$$b_m = 0.65 + 0.5 b_o \text{m}$$
$$b_m = 2.13 + 0.5 b_o \text{ft}$$

For the stern:

$$b_s = 0.61 + 0.7 b_o \text{m}$$
$$b_s = 2 + 0.7 b_o \text{ft}$$

where b_b, b_m and b_s are the vertical extent of the design ice pressure, in m (ft)

b_o $R\,(N/k)^{0.\,25}\,(D/n)^{\,0.2}$
R coefficient, as given in Table 14.6
F_{b1} coefficient
k 746 (100, 986)
N total maximum continuous power delivered to the propellers, in kW (mhp, hp).

For **A0** to **C0** ice class vessels fitted with bulbous bows, the extent b_b within the bulbous area of the ice belt is to be 30% more. D and n are as (Table 14.6).

Table 14.6 Extent of ice load coefficients

Ice class	*R*, *m* (ft)
A0	0.020 (0.066)
B0	0
C0	0
D0	0
E0	0

14.3 Shell Plating

14.3.1 Ice Belt with Transverse Framing

The thickness of the ice belt shell plating is to be not less than that obtained from the following equation:

$$t = 0.60\ \mathrm{s}\left(\frac{P}{Y}\right)^{\frac{1}{2}} + Ct_o\ \mathrm{mm(in.)}$$

where

- t thickness of the shell plating, in mm (in.)
- s distance measured along the shell between adjacent frames, in mm (in.)
- P design ice pressure in appropriate region, as given above, in N/mm^2 (kgf/mm^2, ksi)
- Y minimum yield strength of the material, in N/mm^2 (kgf/mm^2, ksi)
- C 1—for the bow area

  0.80—for the midbody area

  0.65—for the stern area
- *To* as given in Table 14.7.

14.3.2 Ice Belt with Longitudinal Framing

The thickness of ice belt shell plating is to be not less than that obtained from the following equation:

$$t = 0.7\ \mathrm{s}\left(\frac{P}{Y}\right)^{\frac{1}{2}} + Ct_o\ \mathrm{mm(in.)}$$

Table 14.7 Minimum thickness and abrasion allowance of ice belt plating

Ice class	Minimum thickness	t_o mm (in.)
A0	12 (0.47)	3 (0.118)
B0	10 (0.39)	3 (0.118)
C0	8 (0.315)	3 (0.118)
D0	8 (0.315)	1 (0.04)
E0	8 (0.315)	1 (0.04)

* Values of t_o may be reduced down to $0.3t_o$ if an abrasive-resistant coating is used for the ice belt plating. Special approval of this will be based on necessary evidence including submission of results of operational experience in ice

where

S distance between longitudinal frames, in mm (in.)

t, P, Y, t_o, C are as defined above. The thickness of ice belt plating is also to be not less than the thickness given in Table 14.7, plus 1 mm (0.04 in.).

14.3.3 Changes in Plating Thickness

Plating thickness in the transverse direction from the ice belt to the bottom and in the longitudinal direction within the ice belt is to be gradually tapered.

14.4 Transverse Framing[1]

14.4.1 Ice Belt Frame Spacing

Except for the midbody and stern areas of ice class **C0** and the bow area of ice class **E0**, spacing between any adjacent frames measured along the centreline is in general not to exceed one half of the standard frame spacing. A larger spacing between any adjacent frames may be approved if the intermediate frames have end fixity similar to that of the main frames. In no case is the spacing between any adjacent frames measured alongside plating to exceed 0.75 of the standard frame spacing allowance.

[1] Main frames: are the hold, tween deck and peak frames; intermediate frames are the additional frames fitted within the ice belt between the main frames, to comply with ice belt frame spacing; standard frame spacing is the frame spacing measured along the centreline.

14.4.2 Main and Intermediate Frames

Section modulus. The section modulus, *SM*, of each transverse main and intermediate frame in association with the width of plating, *s*, to which it is attached is to be not less than that obtained from the following equation

$$SM = K s \ell b (P/Y)\ \text{cm}^3$$

$$SM = 0.144 K s \ell b (P/Y)\ \text{in.}^3$$

where

K 160–100 b/ℓ $K^1 K^2$
s distance between adjacent frames, in mm (in), measured along the lowest ice waterline in way of the compartment being considered.
ℓ span of the main frame, in m (ft), measured along the frame between decks or between deck and inner bottom.
b vertical extent of the design ice pressure, in m (ft).
P the design ice pressure.
Y minimum yield strength of the material, in N/mm^2 (kgf/mm^2, ksi).

For framing system with supporting stringers, coefficient K_1 is to be obtained from the equation:

$$K_1 = 2/(3 + j)$$

where j = number of the supporting stringers.

For framing system without supporting stringers, coefficient K_1 is to be as given in Table 14.8.

K_2 1.1 for the midship area of the ice belt for ice classes **A0** through C0
1 elsewhere

The web thickness, *t*, of the main and intermediate frames is to be not less than:

$$t = 0.013h + 6\ mm$$

$$t = 0.013h + 0.24\text{in.}$$

Table 14.8 Coefficient K_1 for the framing system without supporting stringers

Termination of the upper and lower ends of the main and intermediate frames	At the upper deck (or platform) of the adjacent upward spaces	Other
At bottom structures or at the lower deck of the adjacent downward spaces (hold, tween deck, tank, etc.)	0.9	1
Other	1	1.15

where h is the depth of the main and intermediate frame, in mm (in.). In no case is the web thickness t to be less than the following:

Ice class **A0**	9 mm (0.35 in.)
Ice class **B0**	8.5 mm (0.34 in.)
Ice class **C0**, **D0** and **E0**	8.0 mm (0.31 in.)

Upper end of frames. Main and intermediate frames are required to extend up to the first deck or platform above the ice belt. They are to be welded and bracketed to the deck beams or to the deck longitudinals, as shown in Fig. 14.3a, b. For ice classes **A0** through **E0**, where the lowest or only deck, or the lowest platform, is situated above the ice belt so that the distance between the deck, or platform, and the upper boundary of the ice belt exceeds "d" metres (feet), given in Table 14.9, the upper ends of intermediate frames in the midbody and stern areas (**A0** through **C0**) or bow area (**D0** and **E0**) may terminate at a deep stringer situated at least 0.6 m (2 ft) above the ice belt. For ice classes **A0, B0, C0, D0** and **E0** in tween deck spaces, where the tween deck is 0.5 m (1.6 ft) or more above the upper ice waterline but within the ice belt, the upper ends of intermediate frames may terminate for ice class **A0** at a stringer situated at least 0.5 m (1.6 ft) above the ice belt, and for ice classes **B0, C0, D0** and **E0** at an intercostal longitudinal at least 0.5 m (1.6 ft) above the ice belt. The upper ends of the frames terminated at a deep stringer are to be welded and bracketed to it as shown in Fig. 14.3c. The intermediate frames terminated at an intercostal stringer or longitudinal are to be welded to it as shown in Fig. 14.3d.

Lower end of frames. Main and intermediate frames are to extend down to the inner bottom or to the double bottom margin plate. For ice class **A0**, the intermediate frames may terminate at a deck 1.0 m (3.3 ft) below the ice belt. For ice classes **B0, C0, D0** and **E0**, the intermediate frames may terminate at a stringer or intercostal longitudinal situated at least 1.0 m (3.3 ft) below the ice belt. The main and intermediate frames are to be attached and bracketed either to the inner bottom or to the double bottom margin plate or to the deck beams, or deck or to the stringer as shown in Fig. 14.4. For vessels without a double bottom, the intermediate frames are to extend down to a point below the top of the bottom transverses and are to terminate at an intercostal longitudinal. For ice classes **A0, B0, C0, D0** and **E0**, the intermediate frames need not extend below the top of the floors, provided they terminate on an intercostal

Table 14.9 Coefficient K_1 for the framing system without supporting stringers

Ice class	Where web frames are fitted	No web frames are fitted
A0	3.0 (10)	–
B0	2.1 (7)	3 (10)
C0	1.2 (4)	1.8 (6)
D0	1.2 (4)	1.8 (6)
E0	1.2 (4)	1.8 (6)

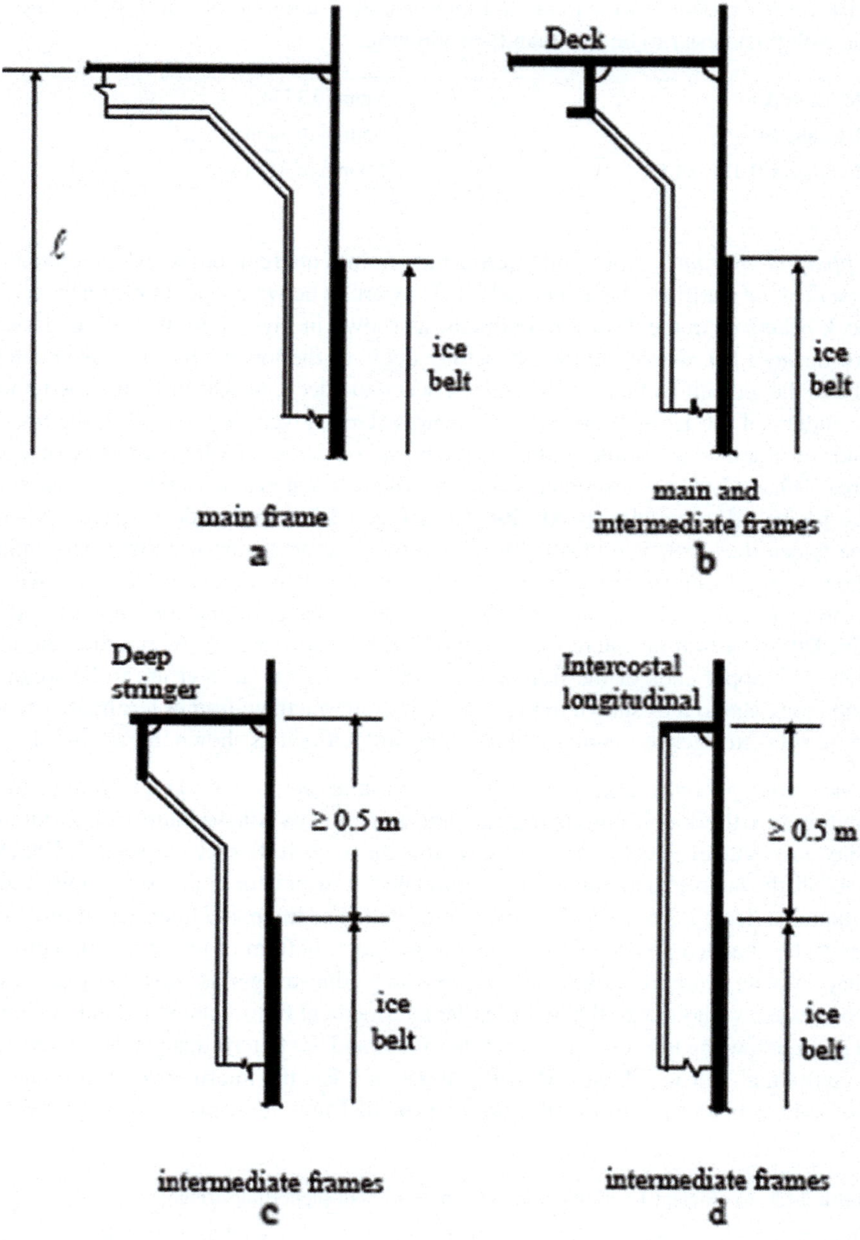

Fig. 14.3 **a** Upper end terminations of frames. **b** Upper end terminations of frames. **c** Upper end terminations of frames. **d** Upper end terminations of frames

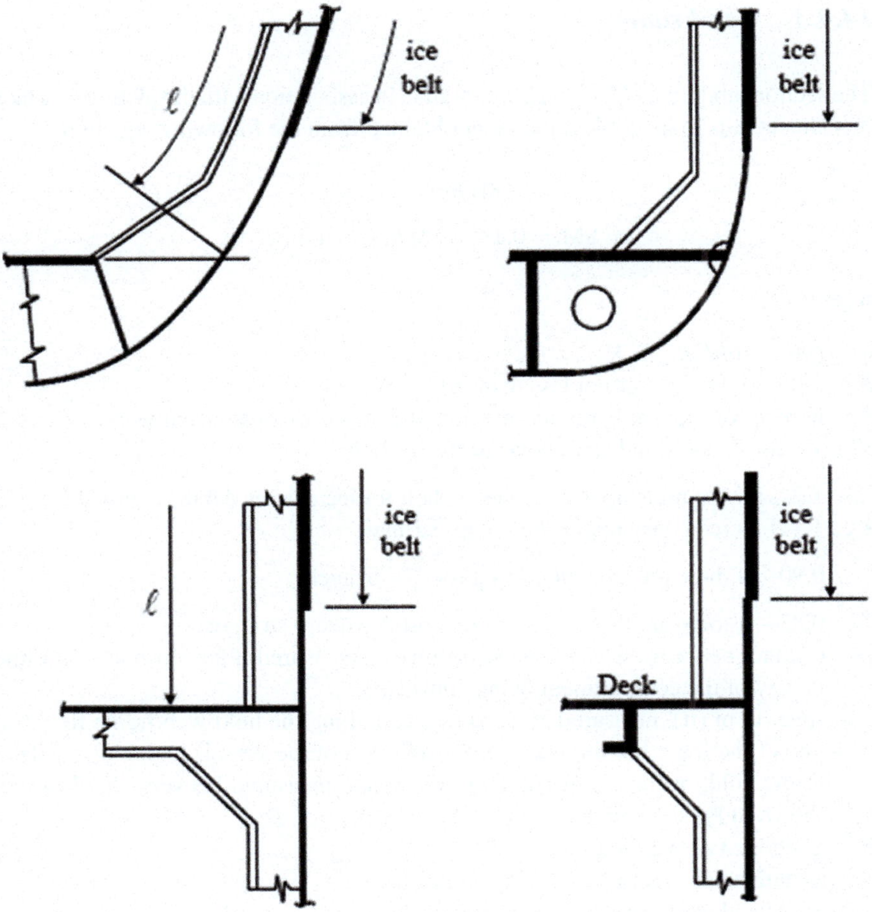

Fig. 14.4 Lower end terminations of frames

longitudinal not less than 0.8 m (2.6 ft) below the ice belt. The intermediate frames are to be attached to the bottom intercostal longitudinals.

Connection to stringers and decks. Main and intermediate frames are to be attached and bracketed to each supporting (deep) stringer, deck and deck beam within the ice belt.

14.4.3 Web Frames

The section modulus, *SM*, of each web frame, in association with the plating to which it is attached, is to be not less than that obtained from the following equation:

$$\mathrm{SM} = Ks1\ell b(P/Y)\ \mathrm{cm}^3.$$

$$\mathrm{SM} = 0.144Ks1\ell b(P/Y)\ \mathrm{in.}^3.$$

where

K $(96 - 36b/\ell)K_1K_2K_3$
K_1 $1.06 - 0.0024i^2$, but not less than 0.4
I number of the main and intermediate frames between adjacent web frames
K_2 1—for the bow and stern areas of the ice belt

1.2—for the midship area of the ice belt for ice classes **A0** through **C0**
K_3 1—if there is one supporting (deep) stringer

0.90—if there are two supporting (deep) stringers

0.85—if there are three or more supporting (deep) stringers
s_1 distance between the web frames, in mm (in), measured along lower ice waterline in way of the compartment being considered
ℓ span, in m (ft), measured in a straight line along the hold web frame from the line of the inner bottom (extended to the side of the vessel) to the lowest deck of the hold, or for the tween deck web frame measured between the decks as shown in Figs. 14.3a, b and 14.4
B as defined above, in m (ft)
P as defined above, in N/mm^2 (kgf/mm^2, ksi)
Y as defined above, in N/mm^2 (kgf/mm^2, ksi).

In determining the section modulus, the effective width of the plating is to be the distance between the webs or 0.125ℓ, whichever is less.

Thickness of the web plate, *t*, is to be not less than that obtained from the following equation:

$$t = 0.01h + 8\ \mathrm{mm}$$

$$t = 0.01h + 0.32\ \mathrm{in.}$$

where h is the depth of the web frame; t need not exceed 15 mm (0.59 in.). The web frames are to be attached and bracketed to the solid floors and the beams at each ice deck.

14.4.4 Ice Stringers

Arrangements. Deep continuous or intercostal stringers are to be fitted in the bow area of the ice belt for ice class **C0**, **D0** and **E0** vessels and within the ice belt throughout the entire length of the vessel for ice class **A0** and **B0** vessels. The spacing between adjacent stringers or between the stringer and a deck or the double bottom measured along the shell is to be not more than indicated in Table 14.10. One of the ice stringers is to be fitted about 200–400 mm (8–16 in.) below the upper ice waterline if there is no deck in this area. For ice class **A0**, another stringer is to be fitted about 100–300 mm (4–12 in.) below the lower ice waterline if there is no deck or similar support in this area.

Scantlings and connections. Where ice stringers are intercostal, the following criteria should be met: (1) the intercostal stringers should be fitted between frames and their scantlings are to be not less than those for main frames; (2) the intercostal stringers are to be welded to the main and intermediate frames; (3) The web plate and the flange, or face, of intercostal ice stringers are to be attached to those of the main and intermediate frames; and (4) the intercostal stringers are to be bracketed to the bulkheads, side transverses, or web frames. Alternatively, where deep ice stringers are fitted, the following criteria should be met: (1) the shear area of the deep ice stringer within one frame space from the web frame is to be not less than that of the web frames; (2) the depth of the ice stringer at the midspan between the web frames is to be not less than twice the depth of the main frame; (3) the face, or flange, area of the deep stringer is to be not less than that of the web frame; (4) the web plate and the face, or flange, of deep ice stringers are to be attached to those of the web frames; (5) the deep stringer, at which the upper ends of frames are terminated, is to have the scantlings as required by the provisions for ice stringers; and (6) The deep stringers are to be bracketed to the bulkheads or side transverses, so that the shear area at the bulkhead is twice that of the ice stringer web. Stiffeners or tripping brackets are to be fitted as required by Class.

Table 14.10 Maximum stringer spacing, m (ft)

Ice class	For framing without web frames	System with web frames
A0 through **E0**	1.5 (5)	2.7 (9)

14.5 Longitudinal Framing

14.5.1 Spacing of Longitudinals

The spacing measured along the shell between adjacent longitudinals and between the longitudinal and the double bottom or a deck within the ice belt is not to exceed one half of the spacing required.

14.5.2 Section Modulus

The section modulus, *SM*, of each longitudinal, in association with the width of plating, *s*, to which it is attached, is to be not less than that obtained from the following equation:

$$SM = 70s\ell 2Ko(P/Y)\ \text{cm}^3$$
$$SM = 10s\ell 2Ko(P/Y)\ \text{in.}^3$$

where

S spacing of longitudinals, in mm (in.)
ℓ span, in m (ft), of the longitudinals measured at the lower ice waterline
K_o $(2.44/\ell)^{1/2}$ (ℓ in m)

$(8/\ell)^{1/2}$ (ℓ in ft), but not less than 0.4
P design ice pressure
Y minimum yield stress of the material, in N/mm^2 (kgf/mm^2, ksi).

The longitudinals are to be attached and bracketed to the webs and to the bulkheads. The total net shear area of the brackets and longitudinals is to be at least twice the net shear area of the longitudinal.

14.5.3 Web Frames

The section modulus, *SM*, of the web frame, in association with the plating to which it is attached, is to be not less than that obtained from the following equation:

$$SM = KoKs1\ell b(P/Y)\ \text{cm}^3$$
$$SM = 0.144KoKs1\ell b(P/Y)\ \text{in.}^3$$

where

SM required section modulus of the web frame, in cm^3 ($in.^3$)
K_o as defined above
K 165 without struts

100 with one horizontal strut

80 with two struts

70 with three struts
s_1 in mm (in.)
ℓ in m (ft)
b for particular area of the ice belt, in m (ft)
P as defined for particular area of the ice belt
Y as defined above.

In determining the section modulus, the effective width of plating is to be the distance between the webs or 0.125ℓ, whichever is less. The net sectional area of the web plate including effective end brackets, where applicable, is to be not less than that obtained from the following equation:

$$\mathrm{A} = \mathrm{K1SM}/\ell\ \mathrm{cm}^2$$

$$\mathrm{A} = 8.33\mathrm{K1SM}/\ell\ \mathrm{in.}^2$$

where

K1 0.009 without struts.

0.015 with one or more struts.

Plate thickness is to be not less than those stated in the section on *Struts*.

14.5.4 Struts

Where one or more struts are fitted as an effective supporting system for the ice belt structure, they are to be located within the ice belt and spaced so as to divide the supported web into spans of equal length. Inboard ends of the struts are to be supported sufficiently by longitudinal bulkhead transverses having a section modulus not less than 0.9 of those stated in the section on *Web frames*. The sectional area of the strut is to be obtained from the following equation:

$$A = (bs1/K)(P/Y)K_o\ \mathrm{cm}^2(\mathrm{in.}^2)$$

where

b for particular area of the ice belt, in m (ft)
s_1 in mm (in.)
K $0.04 - 0.0175(\ell/\text{r})$for SI and MKS units
$0.0333 - 0.00175(\ell/\text{r})$ for US units
ℓ unsupported span of the strut, m (ft)
r least radius of gyration, cm (in.)
P for particular area of the ice belt
Y as discussed in the section on *Changes in plating thickness*
K_o as defined in the section on *Web frames.*

14.6 Alternative Framing Arrangements

Where framing arrangements differing from those given in the section on *Transverse framing* and *Longitudinal framing* are used for the ice belt structures, special approval of the framing members will be based on submitted stress analysis of the structure.

14.7 Peak Frames

Main and intermediate frames in forepeaks are to extend down to the floors or the bottom transverses or the stem. The section modulus of each peak frame is to be as given in *Section modulus* where ℓ, in m (ft), is measured between deep ice stringers and $K_1 = 1$. For the forepeaks of ice classes **A0** and **B0**, the distance is not to be more than 2.1 m (7 ft). For ice classes **A0** and **B0**, the intermediate frames are to extend down to the bottom structure and up to the first deck above the ice belt. Intermediate frames in the forepeak for ice class **C0**, **D0** and **E0** may terminate at the first stringer above the ice belt.

14.8 Double Bottomed Hulls

14.8.1 Longitudinally Framed Bottom

Open floors or bilge brackets extending to the outboard longitudinals are to be fitted throughout at each frame that extends to the inner bottom, except ice classes **B0**, **C0**, **D0** and **E0**, where only the bow area is to comply with this requirement. The spacing of the bottom longitudinals within the bow, lower intermediate and lower stern areas of the ice belt is to be not more than 0.7 m (2.3 ft) for ice class **A0**.

14.9 Ice Decks

14.9.1 General

The following requirements apply to decks or parts of decks situated within the ice belt. For vessels not having decks in the ice belt, the following requirements apply also to decks or parts of decks above and below the ice belt to which the main and intermediate frames extend. Where there are three or more decks within the ice belt, the deck or parts of the deck situated within the upper area of the ice belt need not comply with these requirements.

14.9.2 Deck Plating

The thickness of the stringer plate is to be not less than:

$$t = k\left(s^2 bP\right)^{\frac{1}{3}}$$

where

k 0.12 (0.257, 0.0523)
s distance between the deck beams, in mm (in.)
b in m (ft), for the particular area of the ice belt
P for the particular area of the ice belt.

The width of the stringer plate is to be not less than four times the main frame depth for **A0** ice class. For ice class **A0**, the thickness of the deck plating is to be not less than 0.75 times the required thickness of the stringer plate.

14.9.3 Deck Transverses and Deck Beams

Longitudinally framed decks. Deck transverses are to be fitted at every third main frame for ice class **A0** and at every fourth main frame for ice class **B0**. Partial beams or brackets are to be fitted at all other main frames and at every intermediate frame for ice class **A0**, and at all other main frames for ice classes **B0**, **C0**, **D0** and **E0**. The partial beams or brackets are to be extended from the frames to a deck longitudinal or deck girder situated not less than 1.5 s from the inboard edge of the frames, where s is as defined in the section *Deck plating*.

Scantlings. The sectional area of the beams and deck transverses is to be not less than:

$$A = K_1 sb\left(\frac{P}{Y}\right)\cos\beta \text{ cm}^2$$

$$A = 1.2K_1 sb\left(\frac{P}{Y}\right)\cos\beta \text{ in.}^2$$

The moment of inertia of the beams is to be not less than:

$$MI = kK_2 s\ell bP\cos\beta \text{ cm}^4(\text{in.}^4)$$

where

- k 1.0 (9.81, 0.1191)
- P as defined in the section *Design ice pressure on the bow area* or *Design ice pressure on the other ice belt areas*, in N/mm^2 (kgf/mm^2, ksi), for the particular area of the ice belt
- b as defined in the section *Extent of design ice load*, in m (ft), for the particular area of the ice belt
- s distance between the beams, in mm (in.)
- ℓ span of the beam, measured in m (ft), between the inboard edge of the frame and the deck longitudinal or deck girder supporting the beam
- Y as defined in *Ice belt with transverse framing*
- β as defined in the section *Design ice pressure on the bow area* or *Design ice pressure on the other ice belt areas*, in degrees, for the particular area of the ice belt
- K_1 6.6
- K_2 0.13.

The sectional area and the moment of inertia of the partial beams and of the brackets are to be not less than required above. The beams and the partial beams are to be bracketed to the deck longitudinals or deck girders. Beams or partial beams or brackets fitted at the web frames are to be reinforced so that their section modulus, *SM* is to be not less than:

$$SM = K_3 SM_{wf}\ell_{wf}/_{\ell} \text{ cm}^3(\text{in.}^3)$$

where SM_{wf} and ℓ_{wf} are the section modulus and the span of the web frame, as defined in the section *Web frames*, respectively.

$$K_3 = 0.5$$

14.9.4 Decks with Wide Openings

Within the midbody area of the ice belt, the cross-sectional area of the deck outside the line of openings is to be not less than:

$$A = Kb\ell\left(\frac{P}{Y}\right) \cdot 10^3 \text{ cm}^2$$

$$A = 14.4Kb\ell\left(\frac{P}{Y}\right) \text{ in.}^2$$

where

K 6.2 for ice classes **A0** and **B0**
b as defined in the section *Extent of design ice load*, in m (ft), for the particular area of the ice belt
ℓ length of the opening, in m (ft), but need not be taken as more than 0.1 L
P as defined in the section *Design ice pressures on other ice belt areas*, for the particular area of the ice belt
Y as defined in the section *Ice belt with transverse framing.*
L in m (ft).

14.10 Bulkheads

14.10.1 Scantlings

For ice class **A0**, the thickness of that part of the bulkhead adjacent to the side shell and within the ice belt is to be not less than the thickness of the adjacent frames or of the stringers connected to the bulkhead, whichever is greater. The width of these parts of the bulkhead is to be not less than shown in Table 14.11. These parts of the bulkhead adjacent to the shell within the ice belt are to be fitted with stiffeners normal to the shell plating. The stiffeners are to be welded to a vertical bulkhead stiffener and welded and bracketed to the side longitudinals. Where the shell is transversely framed, brackets are to be welded to the shell and extended and attached to adjacent frames.

Table 14.11 Minimum width of reinforced bulkhead plating

Ice class	Area of the ice belt			
	Peak bulkheads m (ft)	Bow and intermediate areas m (ft)	Midbody area m (ft)	Stern area m (ft)
A0	1.2 (4.0)	1.2 (4.0)	1.0 (3.3)	1.0 (3.3)

14.11 Stem and Stern Frame

The stem and stern frame for ice class **A0** vessels of displacements more than 50,000 tonnes (49,200 Lt) are to be constructed of rolled bar, cast or forged steel. Shaped plate stem may be used elsewhere. All joints and connections are to fully develop the strength of the stem and stern frame. All rudders are to be protected against ice impacts for going astern.

14.11.1 Stem

Solid stem. The cross-sectional area of a stem made of rolled bar, cast or forged steel from the centre vertical keel to 0.01L above the ice belt is to be not less than:

$$A = K_1 D^{\frac{1}{3}} (L - 61) + A_o \text{ cm}^2$$

$$A = 0.0473 K_1 D^{\frac{1}{3}} (L - 200) + A_o \text{ in.}^2$$

where (Table 14.12)

K_1 and A_o as given in Table 14.12
D displacement.
L length, in m (ft), but is not to be taken less than 61 m (200 ft).

For vessels of displacements less than 2500 tonnes (2460 Lt) the cross-sectional area given by the above equation may be reduced 10%. The cross-sectional area of the stem above the ice belt may be reduced gradually.

Shaped plate stem. Thickness of shaped plate stems within the bow area of the ice belt is to be not less than

$$t = 0.8s \left(\frac{P}{Y} \right)^{\frac{1}{2}} + t_o \quad \text{but not less than } 0.04R$$

Table 14.12 Solid stem bar coefficients

Ice class	Ao cm^2 (in.^{2})	K1
A0	62 (9.6)	0.13
B0	50 (7.8)	0.705
C0	45 (7.0)	0.095
D0	45 (7.0)	0.095
E0	45 (7.0)	0.095

Table 14.13 Stern post coefficient

Ice class	K
A0	1.2
B0	1.12
C0	1.07
D0	1.05
E0	1.05

where

t	required thickness of plate stem, in mm (in.)
s	distance between frames, brackets (breast hooks) or stiffeners, in mm (in.)
P, Y and t_o	are as defined in the section *Shell plating*
R	inside radius of the stem at the given section, in mm (in). Need not be taken greater than 625 mm (24.6 in.) for ice classes **A0** through **E0.**

At any section, the fore and aft length of the stem plate is to be not less than $15t$.

Arrangement. The outer surface of connections of the shell plating to the stem is to be flush. The stem is to be supported by floors, webs, frames, breasthooks or brackets spaced not more than 610 mm (24 in.). In addition, shaped plate stems are to be supported in the centreline by a plate, web or bulkhead having the same thickness as the centre vertical keel and a width not less than 610 mm (24 in.).

14.11.2 Stern Frame

The stern post is to be of size acceptable to Class, with all thicknesses increased by coefficient K, as given in Table 14.13.

14.12 Power of Propulsion Machinery

14.12.1 Minimum Power

For ice classes **A0** through **C0**, the total ahead horsepower delivered to the propellers, N, is to be not less than the lesser of the values obtained from the following two equations:

$$(1)\ N = kA(B)^{0.8}(L)^{0.4}\left(1 + me^{-5D\times10^{-6}}\right)\ \text{kW (mhp, hp)}$$

$$(2)\ N = k\left(C + KD \times \frac{n}{1000}\right)$$

Table 14.14 Power coefficients

Ice class	*A* SI and MKS (US units)	*m*	*C*	*K*	*No* kW (mhp, hp)
A0	93 (22.4)	0.6	1000	350	1490 (2030, 2000)
B0	79 (19.0)	0.6	500	300	746 (1040, 1000)
C0	64 (15.4)	0.6	0	250	373 (507, 500)

where

B maximum breadth of the vessel, in m (ft), at the upper ice waterline
L length of the vessel, in m (ft)
e base of natural logarithms
D as defined in 6-1-5/5.7
n 1 (1.016)
k 0.735 (1, 0.986)

A, m, C and K are as given in Table 14.14.

For vessels with unconventional features, the power delivered to the propellers may also be less than given in equation (i), if the particular vessel is able to progress continuously in any ice condition corresponding to its ice class. Special approval of this will be based on necessary evidence including the submission of results of full-scale and model tests. Special consideration will be given when the value of N determined from the equations in the section *Power of propulsion machinery* is less than N_o in Table 14.14.

14.12.2 Astern Power

The astern power delivered to the propellers for ice classes **A0** to **C0** is to be not less than 70% of that required in the section *Power of propulsion machinery*. For ice class **D0** and **E0**, sufficient power for going astern must be provided to secure proper control of the vessel in all normal circumstances. The astern power of the main propelling machinery must be capable of maintaining in free route astern at least 70% of the ahead rpm corresponding to the maximum continuous ahead power. For main propulsion systems with reversing gears, controllable pitch propellers or electric propulsion drive, running astern is not to lead to overload of the propulsion machinery. Main propulsion systems are to undergo tests to demonstrate the astern response characteristics. The tests are to be conducted at least over the manoeuvring range of the propulsion system and from all control positions. A test plan is to be provided by the yard and accepted by the surveyor. If specific operational characteristics have been defined by the manufacturer these shall be included in the test plan. The ability of the machinery, including the blade pitch control system of controllable pitch propellers, to reverse the direction of thrust of the propeller in sufficient time, and so to bring

Table 14.15 Ice conditions for towing or pushing barges

Ice class	Towed/pushed	Towed by ice class PC5 vessel	Towed by ice class PC4 vessel
A0	Severe	Very severe	Extreme
B0	Medium	Severe	
C0	Light	Medium	
D0	Very light	Light	

[a] Breadth of towed barge not to exceed the breadth of towing vessels

the vessel to rest within a reasonable distance from maximum ahead service speed, is to be demonstrated and recorded during trials. Where steam turbines are used for main propulsion, they are to be capable of maintaining in free route astern at least 70% of the ahead revolutions for a period of at least 15 min. The astern trial is to be limited to 30 min or is to be in accordance with manufacturer's recommendation to avoid overheating of the turbine due to the effects of "windage" and friction.

14.13 Non-self-Propelled Vessels

Barges designed for being towed and/or pushed in broken ice and built to the requirements of this section and related sections of the *ABS Rules for Building and Classing of Steel Barges* (or equivalent) will be designated by ice classes **A0**, **B0**, **C0** and **D0**. Non-self-propelled vessels other than barges covered by the Class Rules will be subject to special consideration.

14.13.1 Ice Classes

For the guidance of the owner, the ice conditions considered appropriate for towing or pushing barges are shown below (Table 14.15):

Barges intended to be pushed in "very severe" or "extreme" ice conditions will be subject to special consideration.

14.13.2 Ice Belt

The ice belt is divided into three parts: bow, midbody and aft areas, except that for class **D0**, the ice belt applies to bow area only. For barges designed for tow by either end, bow area requirements apply to both ends. For such barges, the midbody and two bow areas of the ice belt are to be used. The bow area of the ice belt is to extend forward from the section 0.025L aft of either the point where the rake reaches the bottom or where the lightest ice waterline reaches its greatest breadth, whichever is greater. The aft area of the ice belt is to extend aft of the section 0.025L forward of the point where the lightest ice waterline reaches its greatest breadth. The midbody area of the ice belt extends between the bow and aft areas. Upper boundary of the ice belt throughout the length of the barge is to be not less than 0.75 m (30 in.) above the deepest ice waterline for ice class **A0** and not less than 0.6 m (24 in.) above the deepest ice waterline for ice classes **B0** and **C0** and not less than 0.5 m (20 in.) above the deepest ice waterline for ice class **D0**. The lower boundary of the ice belt is to be not less than 0.6 m (24 in.) below the lightest ice waterline for the midbody and aft areas of ice class **A0**. In the bow area of ice class **A0**, the ice belt is to extend to the bottom of the side shell and is to include the bottom shell in way of the rake. For ice classes **B0**, **C0** and **D0**, the lower boundary of the ice belt is to be not less than 0.5 m (20 in.) below the lightest ice waterline throughout the length of the barge.

14.13.3 Design Ice Loads

The design ice pressure on the bow area, P_{bow}, is to be as given for P_b in the section *Ice interaction load,* where $F_{b1} = 1.25$ for vertical structures and $F_{b1} = 1$ for the rakes. The design ice pressure on the midship and aft areas, P_{mid} and P_{aft} are to be:

$$P_{mid} = K_m P_{bow}$$
$$P_{bow} = K_s P_{bow}$$

where K_m and K_s are as given in Table 14.5.

The vertical extent of the design ice pressure for all of the ice belt areas is to be:

0.61 m (24 in.)	for ice class **A0**
0.51 m (20 in.)	for ice class **B0**
0.45 m (18 in.)	for ice class **C0**
0.40 m (16 in.)	for ice class **D0**

14.13.4 Structural Arrangements

The thickness of the shell plating within the ice belt areas is to be as required by *Ice belt with transverse framing* or *Ice belt with longitudinal framing.* Structural arrangements and scantlings of the ice belt framing members are to be as required by the sections *Transverse framing, Longitudinal framing* and *Peak frames.* Decks and bulkheads situated within the ice belt and, where there are no decks within the ice belt, the deck above and below the ice belt to which the main and intermediate frames are extended are to comply with the requirements set out in the sections *Ice decks* and *Bulkheads*.

14.14 Hull Structural Materials

All hull structural materials are to be in accordance with the requirements of Part 2, Chap. 6. In addition, material grades for ice belt structures and exposed shell and main strength deck structures are to be selected based on the design service temperature and material class, as defined as follows.

14.14.1 Design Service Temperature

The design service temperature is to be taken in accordance with Table 14.16. Design service temperature for insulated members will be specially considered upon submission of substantiating data.

Table 14.16 t_D, °C (°F)

Zone	Ice class		
	A0	B0 and C0	D0 and E0
Ice belt structures (other than Area c)			
1. External plating	– 30 (– 22)	– 20 (– 4)	– 10 (14)
2. Framing[a] for all items above	–20 (– 4)	– 10 (14)	0 (32)
Above ice belt[c]			
1. External plating	– 30 (– 22)	– 20 (– 4)	– 10 (14)
2. Framing[a] for external plating	– 20 (– 4)	– 10 (14)	0 (32)
3. Plating[b] and framing in enclosed spaces			
(i) Heated space	0 (32)	0 (32)	0 (32)
(ii) Unheated space	– 10 (14)	0 (32)	0 (32)
More than 0.3 m (1 ft) below the lower ice waterline	0 (32)	0 (32)	0 (32)

Notes
[a] Includes bulkheads and decks attached to the external plating within 600 mm (23.5 in.) from the plating
[b] Excludes those portions covered by Note 1 above
[c] Above Area c for class **D0** and **E0** excluding the bow area

Table 14.17 Material class of structural members

Structural members	Ice classes	
	A0	B0 and below
Within ice belt (other than Area c)		
1. Bottom and side shell plating-bow, intermediate and lower intermediate areas	III	I
2. Bottom and side shell plating-other ice belt areas	II	I
3. Framing[a]—bow and intermediate areas	II	I
4. Framing[a]—other ice belt areas	I	I
5. Stem, ice knife, propeller nozzle, shaft bracket, rudder, stern frame and rudder horn	III	I
6. Other structures	I	I
Above Ice Belt		
1. Sheer strake and deck stringer		
(i) within 0.4L amidships	III	III
(ii) outside 0.4L amidships	II	II
2. Side shell[d] and strength deck plating[b,c,e]	I	I
3. Other structures[b,c]	I	I
More than 0.3 m (1 ft) below the lower ice waterline	No additional requirements for ice class	

Material class given in this table refers to the classes in Table 14.18, 14.19 and 14.20 or in the section *Criteria for other steels*, as applicable.

Notes

[a] Includes bulkheads and decks attached to the external plating within 600 mm (23.5 in.) from the plating.

[b] Excludes those portions covered by Note 1 above.

[c] Above Area c for class **D0** and **E0** excluding the bow area.

[d] Single side strakes for ships exceeding 150 m (492 ft) without inner continuous longitudinal bulkheads between bottom and the single strength deck are not to be less than grade B/AH within cargo region in ships.

[e] Not to be less than grade B/AH within 0.4L amidships in ships with length exceeding 150 m (492 ft) and single strength deck.

14.14.2 Material Class of Structural Members

The material class of hull structural members is to be in accordance with Table 14.17.

14.14.3 Criteria for ABS Grade Steels

For those rolled steel products in the *ABS Rules for Materials and Welding (Part 2)*, the appropriate grade to be used for respective material class and thickness is shown

Table 14.18 Material grades (class I)

Design service temperature					
Thickness in mm (in)	0 °C (32 °F)	– 10 °C (14 °F)	– 20 °C (–4 °F)	– 30 °C (–22 °F)	– 40 °C (–40 °F)
$t < 12.5$ ($t < 0.50$)	A, AH	A, AH	A, AH	A, AH	B[b], AH
$12.5 < t \leq 20$ ($0.50 < t \leq 0.79$)	A, AH	A, AH	A, AH	B, AH	D, DH
$20 < t \leq 25$ ($0.79 < t \leq 0.98$)	A, AH	A, AH	B, AH	D, DH	D[a], DH[a]
$25 < t \leq 30$ ($0.98 < t \leq 1.18$)	A, AH	A, AH	D, DH	D, DH	E, EH
$30 < t \leq 35$ ($1.18 < t \leq 1.38$)	A, AH	B, AH	D, DH	D, DH	E, EH
$35 < t \leq 40$ ($1.38 < t \leq 1.57$)	A, AH	D, DH	D, DH	D, DH	E, EH
$40 < t \leq 51$ ($1.57 < t \leq 2.00$)	B, AH	D, DH	D, DH	D, DH	E, EH

Notes
[a] To be normalised
[b] May be "A" if fully killed.

Table 14.19 Material grades (class II)

Design service temperature					
Thickness in mm (in.)	0 °C (32 °F)	– 10 °C (14° F)	– 20 °C (–4 °F)	– 30 °C (–22 °F)	– 40 °C (–40 °F)
$t \leq 12.5$ ($t \leq 0.50$)	A, AH	A, AH	A, AH	B^2, AH	D, DH
$12.5 < t \leq 20$ ($0.50 < t \leq 0.79$)	A, AH	A, AH	B, AH	D, DH	D[a], DH[a]
$20 < t \leq 25$ ($0.79 < t \leq 0.98$)	A, AH	B, AH	D, DH	D[a], DH[a]	E, EH
$25 < t \leq 30$ ($0.98 < t \leq 1.18$)	A, AH	B, AH	D, DH	E, EH	E, EH
$30 < t \leq 35$ ($1.18 < t \leq 1.38$)	B, AH	D, DH	D, DH	E, EH	E, EH
$35 < t \leq 40$ ($1.38 < t \leq 1.57$)	B, AH	D, DH	D, DH	E, EH	E, EH
$40 < t \leq 51$ ($1.57 < t \leq 2.00$)	D, DH	D, DH	D, DH	E, EH	E, EH

Notes
[a] To be normalised.
[b] May be "A" if fully killed.

Table 14.20 Material grade (class III)

Design service temperature					
Thickness in mm (in)	0 °C (32 °F)	−10 °C (14 °F)	− 20 °C (− 4 °F)	− 30 °C (− 22 °F)	− 40 °C (− 40 °F)
$t < 12.5$ ($t < 0.50$)	A, AH	A, AH	B[b], AH	D, DH	D[a], DH[a]
$12.5 < t \leq 20$ ($0.50 < t \leq 0.79$)	A, AH	B, AH	D, DH[a]	D[a], DH[a]	E, EH
$20 < t \leq 25$ ($0.79 < t \leq 0.98$)	B, AH	D, DH	D[a], DH[a]	E, EH	E, EH
$25 < t \leq 30$ ($0.98 < t \leq 1.18$)	B, AH	D, DH	E, EH	E. EH	E, EH
$30 < t \leq 35$ ($1.18 < t \leq 1.38$)	D, DH	D, DH	E, EH	E, EH	–
$35 < t \leq 40$ ($1.38 < t \leq 1.57$)	D, DH	D, DH	E, EH	E, EH	

Notes
[a] To be normalised.
[b] May be "A" if fully killed.

in Tables 14.18, 14.19 and 14.20. Where the selection of steel results in a higher grade, such higher grade is to be used.

14.14.4 Criteria for Other Steels

Yield strength below 410 N/mm², (42 kgf/mm², 60 ksi). Where steels other than those in the *ABS Rules for Materials and Welding (Part 2)* are intended, their specifications are to be submitted for approval. These steels are to comply with the following impact test requirements:

Yield strength			CVN (longitudinal)		
N/mm²	(kgf/mm²)	(ksi)	J	(kgf-m)	(ft-lbf)
235–305	(24–31)	(34–44)	27	(2.8)	(20)
315–400	(32–41)	(45.5–58)	34	(3.5)	(25)

At the following temperatures:

Class I—t_D.
Class II—10 °C (18 °F) below t_D.
Class III—20 °C (36 °F) below t_D.

Yield strength 410–690 N/mm² (42–70 kgf/mm², 60–100 ksi). Where steels of this strength level are intended, their specifications are to be submitted to Class for approval. These steels are to comply with the impact test requirements of 34 J (3.5 kgf-m, 25 ft-1bf) at the following temperatures:

t_D	Test temperature
0 °C (32 °F)	– 30 °C (– 22 °F)
– 10 °C (14 °F)	– 40 °C (– 40 °F)
– 20 °C (– 4 °F)	– 40 °C (– 40 °F)
– 30 °C (– 22 °F)	– 50 °C (– 58 °F)
– 40 °C (– 40 °F)	– 60 °C (– 76 °F)

Alternative requirements. As an alternative to the requirements in *Criteria for other steels (yield strength below 410 N/mm², (42 kgf/mm², 60 ksi)* and *Criteria for other steels (yield strength 410–690 N/mm² (42–70 kgf/mm², 60–100 ksi),* higher strength steels may comply with the following:

(1) For transverse specimens, $^2/_3$ of energy values shown in *Criteria for other steels (yield strength below 410 N/mm², (42 kgf/mm², 60 ksi)* and *Criteria for other steels (yield strength 410–690 N/mm² (42–70 kgf/mm², 60–100 ksi).*
(2) For longitudinal specimens, lateral expansion is not to be less than 0.5 mm (0.02 in.). For transverse specimens, lateral expansion is not to be less than 0.38 mm (0.015 in.).
(3) Nil-ductility temperature (NDT), as determined by drop weight tests, is to be 5 °C (9 °F) below the temperature specified in *Criteria for other steels (yield strength below 410 N/mm², (42 kgf/mm², 60 ksi)* and *Criteria for other steels (yield strength 410–690 N/mm² (42–70 kgf/mm², 60–100 ksi).*

14.14.5 Weld Metal

Hull steels. When ordinary and higher strength hull steels are applied in accordance with Tables 14.18, 14.19 and 14.20, approved filler metals appropriate to the grades may be used.

Criteria for other steels. For the welding of hull steels other than the grades in Tables 14.18, 14.19 and 14.20, weld metal is to exhibit a Charpy V-Notch toughness value at least equivalent to the transverse base metal requirements ($^2/_3$ of longitudinal base metal requirements).

14.15 Weld Design

Weld design of hull construction must comply with the Class Rules for welding quality. Special attention is to be paid to welds in structures attached to side shell, such as transverse bulkheads, decks, frames, web frames and side shell stringers, within the ice belt, which are to be of double continuous weld.

14.16 Towing Arrangements

14.16.1 Bow

Every ice class vessel intended to be escorted by a higher ice class leading vessel, as given in Table 14.1, is to be fitted with a tow chock pipe and a tow bitt on the bow. The chock and the bitt are to be properly connected to the stem frame. The portions of the decks at which the chock and the bitt are attached are to meet requirements of the section *Ice decks*. The shell plating and framing below and 1.5 m (5 ft) around the chock are to be as required in the section *Shell plating* and *Transverse framing* for the bow area of the ice belt for ice classes **A0**, **B0**, **C0**, **D0** and **E0**. The stem frame below the connections with the chock is to be as required in the section *Stem and stern frame* for the portion of the stem within the ice belt. Where a bulbous bow is fitted, the bulb is not to extend beyond the fore end of the lower ice waterline.

14.17 Propeller Nozzles

This section applies to fixed nozzles. Special consideration will be given to steering nozzles for ice class **A0**. For ice class **A0**, the nozzles are to be supported at least at the upper and lower ends. For ice classes **B0**, **C0**, **D0** and **E0**, the nozzles supported only at the upper ends are to be attached to the hull for a width of not less than $^1/_6$ of the outer circumference of the nozzle. The strength, rigidity and resistance to buckling of the nozzle are to be adequate for the design ice forces given in the section *Design ice forces*. All the critical loading cases are to be considered. In no case under the design ice forces are the normal and axial displacements of the inside ring to exceed 10% of the clearance between the inside plating of the nozzle and the propeller blade tips, or 0.5% of the inside ring diameter, whichever is less. Nozzles are to be protected by stern structures as much as possible against direct impacts with large ice features.

14.17.1 Design Ice Forces

The design ice forces are to be not less than those obtained from the following equations:

$$F_n = K_1 K_2 (Dd_1)^{\frac{1}{2}} \text{ kN (tf, Ltf)}$$

$$F_f = K_3 K_4 D(d_1 - d_2)^{\frac{1}{2}} \text{ kN (tf, Ltf)}$$

where

F_n design ice force applied normal to the outside surface of the nozzle in the most critical location

K_2 1 for the external sides of a single nozzle of a single screw vessel

1.1 for the outboard external sides of the outermost nozzles of vessels with two or more screws

0.25 for the external sides of nozzles situated between the outermost ones and for the internal sides of any nozzles

0.8 for bottoms of the nozzles

D ship displacement, in tonnes (long tons).

d_1 maximum outer diameter of the nozzle, in m (ft).

d_2 minimum internal diameter of the nozzle, in m (ft).

F_f design ice force applied to the ends of the nozzle, parallel to the propeller axis, in the most critical locations.

K_4 1 for aft end face of the nozzle having no rudder behind.

0.7 for the aft end face of the nozzle with a rudder behind

0.6 for the fore end face of the nozzle

K_1 and K_3 are as given in Table 14.21.

Values of K_2 and K_4 less than above will be approved, provided the stern and bottom hull structures effectively protect the nozzle against large ice fragments.

Table 14.21 Design ice force coefficient

Ice class	K_1	K_3
	SI units (MKS, US)	SI units (MKS, US)
A0	20 (2.1, 1.1)	59 (6.0, 3.3)
B0	13 (1.3, 0.7)	35 (3.6, 2.0)
C0	9 (0.9, 0.5)	22 (2.2, 1.2)
D0	7 (0.7, 0.4)	18 (1.8, 1.0)

14.17.2 Plate Thickness

The plate thickness of both inner and outer surfaces of the nozzle is to be not less than required in the section *Transverse framing* for the stern ice belt area with coefficient $C = 0.3$. A value of $C = 0$ will be considered for a high abrasion-resistant coating of the nozzle. In this case, the results of operational experience information, required in the note to Table 14.7, are to be submitted.

14.18 Rudder and Steering Arrangements

All ice classes, multiple rudders. Where two or more rudders are provided, they are to be mechanically independent.

Ice class **A0.**

Pintles. Rudders are to have at least two pintles.

Locking. Rudders are to be protected by strong and effective external rudder stops and provided with mechanical means of locking the rudder parallel to the centreline for use in the astern condition.

Ice classes A0 through B0

Ice knife. Rudders are to be protected by ice knives or other similar structures located abaft the rudder. The clearance between the ice knife and the rudder must not exceed 100 mm (4 in.).

14.18.1 Rudder Stocks, Couplings and Pintles

14.18.1.1 *Ice Classes* A0 *Through* E0

For ice classes **A0** through **E0**, rudder stocks, pintles, gudgeons and other bolting arrangements to the stern frames are to meet the Class requirements in association with V_i as defined below, in lieu of V.

V_i = the greater of V, as defined by Class, or the minimum design speed in Table 14.22.

14.18.1.2 Ice Class A0

The stresses in these members with the load F applied as follows are not to exceed the shear yielding strength which may be taken as 0.577 times the specified yield point of the material.

Table 14.22 Design speed for rudders, couplings and pintles

Ice class	Minimum design speed, knots
A0	18
B0	16
C0	14
D0	12

$$F = 2K_3(Dt)^{\frac{1}{2}}\ \text{kN (tf, Ltf)}$$

where

K_3 as given in Table 14.21
D ship displacement, in tonnes (long tons)
t thickness of the rudder, in m (ft), measured at the level of F and at 10% of the rudder length from the trailing edge.

F is to be applied to the after edge of the rudder in a direction parallel to the centreline of the vessel at all locations below the ice waterline within the middle 40% of the rudder height to determine the most severe requirements. Alternatively, F may be spread over any 60% of the rudder height as a uniform load. No other force need be considered simultaneously with F.

14.18.2 Double Plate Rudder

For double plate rudders, the minimum thickness of plates is to be not less than required in the section *Plate thickness.*

14.19 Bossings

The bossings are to be designed to withstand the design ice forces F_n, as specified by the section *Propeller nozzles (design ice forces)* where d_1 is the diameter of the bossing. The bossing plating thickness is to be not less than required by the section *Shell plating (ice belt with transverse framing)* for the stern ice belt area, where s is the distance between stiffeners.

14.20 Machinery Arrangements

All machinery is to be suitable for operation under the environmental conditions to which it will be exposed in service and is to include all necessary special provisions for that purpose.

14.20.1 Governmental Authority

Attention is directed to the appropriate governmental authorities in the intended regions of operation for additional requirements in consideration of operation in ice such as fuel capacity, refuelling capability, water capacity, radio communications requirements, etc.

14.20.2 Propulsion Arrangements

In addition to the regular governor, all propulsion engines and turbines are to be fitted with a separate overspeed device so adjusted that the speed cannot exceed the maximum rated speed by more than 20%.

14.20.3 Electric Propulsion

Propulsion motors are to be fitted with automatic protection against excessive torque, overloading and temperature. This protection is to automatically limit these parameters but is not to cause loss of propulsion power.

14.20.4 Boilers

Vessels propelled by steam machinery are to be fitted with at least two boilers of equal capacity.

14.20.5 Protection Against Excessive Torques

For vessels of all classes, if torsionally flexible couplings or torque-limiting devices are fitted in the propulsion system, positive means are to be provided for transmitting full torque to the propeller in the event of failure of the flexible element. Ratings for flexible couplings are to be in accordance with the guidance in the section *Flexible couplings*.

14.20.6 Sea Chests

For vessels of Ice Class **A0**, **B0, C0, D0** and **E0**, at least one sea chest for supplying water for cooling and fire-fighting purposes is to be connected to the cooling-water discharge by a branch pipe having the same cross-sectional area as the main pipeline, in order to stay free from ice and slush ice. As far as practicable, the sea inlet chest is to be situated well aft, adjacent to the keel.

14.21 Materials for Propellers and Propulsion Shafting (2022)

Propeller materials are to be in accordance with the applicable requirements set by Class. In addition to the applicable requirements set by Class, the material, for propeller shafts and other shafting that are exposed to sea water, is to have a Charpy V-notch impact value of not less than 20 J (2.1 kgf-m, 15 ft-lbf) at a temperature of – 10 °C (14 °F) for all ice classes, except ice class **D0** and **E0**. The propulsion shafts and couplings are to be made of steel.

14.22 Determination of Ice Torque for Propulsion Systems

The ice torque M for determining the dimensions of propellers and gears is to be in accordance with Table 14.23 and associated notes.

14.23 Propellers

14.23.1 Propeller Arrangements

Propeller arrangements, the shape of the stern and the propeller protecting structures are to be adequate for the intended service. Special consideration is to be given to the propeller protection when moving astern (Table 14.24).

Table 14.23 Value of ice torque M

Location of propeller	Centreline	Off centreline
Propellers protected by nozzle		
Nozzle protected (see Note [a])		
Class **A0-B0**	$0.75M_1$ (see Note [b])	$0.85M_1$ (see Note [b])
Class **C0-E0**	$0.85M_1$	$0.9M_1$
Nozzle unprotected		
Class **A0-C0**	$0.9M_1$ (see Note [c])	$0.9M_1$ (see Note [c])
Class **D0** and **E0**	$0.9M_1$	$0.9M_1$
Open propellers		
Class **A0-E0**	M_1	M_1

$M_1 = mD^2$, in kN-m (tf-m, Ltf-ft)
m = value from 6-1-5/51.1
Notes
[a] These requirements apply where the nozzle is well protected by ice knives, fins or other adequate stern arrangement from large ice fragments entering into nozzle from forward or backward motion of the vessel. These reductions are subject to special consideration
[b] To be not less than required for the second lower ice class
[c] To be not less than required for the next lower ice class
[d] Need not be greater than required for next higher ice class

Table 14.24 Values of m

Ice class	SI units	MKS units	US units
A0	15.7	1.60	0.48
B0	13.0	1.33	0.40
C0	12.1	1.23	0.37
D0	11.1	1.13	0.34
E0	8.8	0.90	0.27

Table 14.25 Propulsion shaft diameter factor $\boldsymbol{k}_1$

	Solid propellers with hubs	
	Not Larger than 0.25D	Larger than 0.25D and CPP's
Tail shaft	1.08	1.15
Tube shaft	1.03	1.10
Intermediate shaft(s)	0.87	0.95
Thrust shaft	0.95	1.01

14.23.2 Propeller Section

Width and thickness. The thickness T and width W of propeller blade sections are to be determined so that the WT^2 calculated by the actual designed W and T is not less than that required by the following equations:

- At the 0.25 radius for solid propellers

$$WT^2 = \left[\frac{a_1}{U(0.65 + 0.7P_{0.25}}\right][a_2CN/nR) + a_3M]\ \text{cm}^3(\text{in.}^3)$$

- At the 0.35 radius for solid propellers with hubs larger than 0.25 propeller diameter

$$WT^2 = \left[\frac{a_4}{U(0.65 + 0.7P_{0.35}}\right][a_2CN/nR) + a_5M]\ \text{cm}^3(\text{in.}^3)$$

- At the 0.35 radius for controllable-pitch propellers

$$WT^2 = \left[\frac{a_4}{U(0.65 + 0.49P_{nominal}}\right][a_2CN/nR) + a_5M]\ \text{cm}^3(\text{in.}^3)$$

- At the 0.6 radius for solid propellers

$$WT^2 = \left[\frac{a_6}{U(0.65 + 0.7P_{0.6}}\right][a_2CN/nR) + a_7M]\ \text{cm}^3(\text{in.}^3)$$

- At the 0.6 radius for controllable-pitch propellers

$$WT^2 = \left[\frac{a_6}{U(0.65 + 0.49P_{nominal}}\right][a_2CN/nR) + a_7M]\ \text{cm}^3(\text{in.}^3)$$

where

a_1 2650 (270, 27,000)
a_2 272 (200, 176)
a_3 22.4 (220, 59.134)
a_4 2108 (215, 21,500)
a_5 23.5 (230, 61.822)
a_6 932 (95, 9500)
a_7 28.6 (280, 75.261)
W expanded width of a cylindrical section at the appropriate radius, cm (in.)
T maximum thickness at the appropriate radius from propeller drawing, cm (in.)

U tensile strength of propeller material, N/mm^2 (kgf/mm^2, psi)
P pitch at the appropriate radius divided by the propeller diameter (for controllable-pitch propellers, the nominal value of pitch is to be used)
C 1 for $N \leq 7{,}460$ kW (10,140 mhp, 10,000 hp)

$0.667 + \frac{N}{22480}$ for 7,460 kW < N < 29,840 kW

$0.667 + \frac{N}{30420}$ or 10,140 mhp < N < 40,560 mhp

$0.667 + \frac{N}{30000}$ for 10,000 hp < N < 40,000 hp

2 for $N \geq 29{,}840$ kW (40,560 mhp, 40,000 hp)
N per propeller
n number of blades
R rpm at the maximum continuous rating
M ice torque, as defined in the section *Determination of ice torque for propulsion systems.*

Blade tip thickness. The minimum blade thickness t_a, in mm (in), at the tip of the blade (*D*/2) is to be determined from the following equations:

For Classes **A0, B0, C0, D0** and **E0**

$$t_a = (a_4 + a_2 D\sqrt{a_3/U} \text{ mm (in)}$$

where

a_2 2 (2, 0.024)
a_3 490 (50, 71,000)
a_4 15 (15, 0.591)
D propeller diameter, m (ft)
U tensile strength of the propeller material, N/mm^2 (kgf/mm^2, psi).

Blade bolts. For built-up or controllable-pitch propellers, the cross-sectional area of the bolts at the root of the thread is to be determined by the following equation:

$$\alpha = 0.082UWT^2/U_b nr$$

where

α area of each bolt at root of thread, in mm^2 (in.^{2})
U tensile strength of the propeller material, N/mm^2 (kgf/mm^2, psi)
U_b tensile strength of the bolt material, N/mm^2 (kgf/mm^2, psi)
n number of bolts on one side of blade (if n is not the same on both sides of the blade, the smaller number is to be used)
r radius of bolt pitch circle, in mm (in.).

W and T are as defined in the section *Propulsion shafting diameters*, in mm (in).

14.23.3 Additional Requirements

Rule required thickness. Where the blade thickness derived from the equations in the section *Propellers (Propeller arrangements)* is less than the required thickness set by Class, the latter is to be used.

Other sections. The thicknesses of propeller sections at radii intermediate to those specified are to be determined from fair curves connecting the required section thicknesses.

Blade edges. The thickness of blade edges is not to be less than 50% of the required tip thickness t_a, measured at a point 1.25 t_a from the leading edge for controllable-pitch propellers, and from each edge for solid propellers.

Controllable-pitch propellers. The strength of the internal mechanisms of controllable-pitch propellers is to be at least 1.5 times that of the blade in the weakest direction of the blade for a load applied on the blade at the 0.9 radius and at an offset from the blade spindle axis equal to two-thirds the distance from the spindle axis to the leading or trailing edge (whichever is greater, as measured at the 0.9 radius).

Highly skewed propellers. Where highly skewed propellers are utilised, stress calculations considering both the ahead and astern operating conditions as well as the above ice loads are to be submitted for review.

14.23.4 Friction Fitting of Propeller Hubs and Shaft Couplings

Friction fitting of propeller hubs, shaft couplings or other torque transmitting components in those portions of the shaft line subject to shock loading from the propeller, is to have a factor of safety against slip considering both propulsion torque and ice torque of at least 2.4. Detailed stress and fitting calculations for all friction-fitted components are to be submitted for review.

14.24 Propulsion Shafting Diameters

The diameters of the propulsion shafts are to be not less than that obtained from the following equation:

$$d = k_o k_1 (W_a T_a^2 U / Y)^{\frac{1}{3}} \text{ cm (in.)}$$

where

d diameter of the shaft being considered, measured at its aft bearing, cm (in.)
k_o 1.00.
k_1 as given in Table 14.25
w_a, T_a actual values of the propeller blade expanded width and maximum thickness measured at the blade section at the 0.25 radius for solid propellers with the propeller hub not larger than 0.25D and at the 0.35 radius otherwise; in cm (in).
U tensile strength of the propeller material, N/mm^2 (kgf/mm^2, psi).
Y yield strength of the shaft steel, N/mm^2 (kgf/mm^2, psi).

14.25 Reduction Gears

Pinions, gears and gear shafts are to be designed to withstand an increase in torque over that normally required for ice-free service. The following corrected ice torque (T_i) is to be used as determined by Class:

$$T_i = T + C\left[\frac{M I_H R^2}{(I_L + I_H R^2}\right]$$

where

T_i ice corrected torque, N-m (kgf-cm, lbf-in.)
T torque corresponding to maximum continuous power, N-m (kgf-cm, lbf-in.)
M ice torque, as defined in the section *Determination of ice torque for propulsion systems*, kN-m (tf-m, Ltf-ft)
t_a sum of mass moment of inertia of machinery components rotating at higher rpm (drive side)
R gear ratio (pinion rpm/gear wheel rpm)
t_a sum of mass moment of inertia of machinery components rotating at lower rpm (driven side) including propeller with an addition of 30% for water
C 1,000 (100,000, 26,800).

I_H and I_L are to be expressed in the same units.

14.26 Flexible Couplings

Torsionally flexible couplings are to be selected so that the ice-corrected torque, as determined section *Reduction gears*, does not exceed the coupling manufacturer's recommended rating for continuous operation. When the rotating speed of the coupling differs from that of the propeller, the ice-corrected torque is to be suitably adjusted for the gear ratio. If a torque-limiting device is installed between the propeller and the flexible coupling, the maximum input torque to the torque-limiting device may be taken as the basis for selecting the coupling, in lieu of the ice-corrected torque. Flexible couplings which may be subject to damage from overheating are to be provided with temperature-monitoring devices or equivalent means of overload protection with alarms at each engine control station.

14.27 Tunnel Thrusters

Where **APS**, **PAS**, or dynamic positioning systems **DPS** notations are assigned, the mechanical components of a tunnel thruster (i.e., propellers, gears, shafts, couplings, etc.) are to meet the applicable requirements of propulsion systems in this section. Alternatively, the specific Class Rules may be applied to the mechanical components of a tunnel thruster when a comprehensive study to determine the effect of ice is submitted for consideration.

Chapter 15
Baltic Ice Class Notation

15.1 Introduction

Vessels to be distinguished in the vessel's record by **ICE CLASS** followed by ice class **I AA** through **I C** (or equivalent, as specified by the vessel's Class notation protocols) are expected to comply the applicable guidance contained in this chapter. All vessels designated such are to be self-propelled and equipped with a radio telephone capable of VHF communications. The ice strengthening requirements outlined in this chapter are based in part on the *Finnish-Swedish Ice Class Rules 2017*, which were developed for vessels sailing in the Baltic Sea area in winter or in other sea areas in similar ice conditions. For additional guidance, refer to the latest revision of the *Guidelines for the Application of the Finnish-Swedish Ice Class Rules.*

15.2 Assignment of Baltic Ice Class

15.2.1 Ice Class

The requirements outlined in this chapter are intended primarily for vessels sailing in the Baltic Sea region in winter or in other sea areas in similar ice conditions and are assigned to ice classes in accordance with established Class notations. For context, this book will refer to the ABS Ice Class notations:

- **Ice class I AA**: vessels with such structure, engine output and other properties that they are normally capable of navigating in difficult ice conditions without the assistance of ice breakers
- **Ice class I A**: vessels with such structure, engine output and other properties that they are capable of navigating in difficult ice conditions, with the assistance of ice breakers when necessary

A. Olsen, *Ship Operations in Extreme Low Temperature Environments*, Springer Series on Naval Architecture, Marine Engineering, Shipbuilding and Shipping 19,
https://doi.org/10.1007/978-3-031-52513-1_15

- **Ice class I B**: vessels with such structure, engine output and other properties that they are capable of navigating in moderate ice conditions, with the assistance of ice breakers when necessary
- **Ice class I C**: vessels with such structure, engine output and other properties that they are capable of navigating in light ice conditions, with the assistance of ice breakers when necessary

The minimum rule required engine output power in either kW or hp (**Minimum engine output power XX kW/hp**) must be specified in the notation; for instance:

Ice Class 1A (**Minimum Engine Output Power 1520 kW**)

The maritime administrations of Sweden and Finland provide icebreaker assistance to vessels bound for their ports in winter. Depending on the ice conditions, restrictions by the Swedish and Finnish administrations may apply to the size and ice class of the vessel. Where no specific requirements are given, vessels are assumed to be normal seagoing cargo vessels of conventional proportions, hull form and propulsion arrangement. Any vessel having very unconventional proportions, hull form or propulsion arrangement, or any other characteristics, may be assigned a lower ice class by the relevant maritime administration.

15.2.2 General Suitability for Winter Conditions

In the northern Baltic Sea area, the air temperature is below 0 °C (32 °F) for much of the winter and may occasionally fall to around – 30 °C (– 22 °F), and for short periods of time temperatures as low as – 40 °C (– 40 °F) can be encountered. This should be considered when designing structures, equipment and arrangements essential to the safety and operation of the ship. Matters to be borne in mind include (e.g., the functioning of hydraulic systems, the danger of water piping and tanks freezing, the start-up of emergency diesel engines, the strength of materials at low temperature, etc.). The following temperatures are given for reference in the Baltic Sea region:

- Ambient temperature: – 30 °C (– 22 °F)
- Sea water temperature: –2 °C (28.4 °F)

Equipment and material exposed to the weather should be capable of withstanding and remaining operable at the design temperature for extended periods.

***Note**: there have been no reported cases of brittle fracturing when material grades designed for normal worldwide service are used for winter navigation in Baltic Sea region. The propulsion and auxiliary machinery should be capable of full operation in ambient conditions, as required in winter conditions. For example, the engine suction air should be sufficiently heated before entering the engine, or other alternative solutions, such as a specially adapted waste-gate, should be considered.*

15.3 Maximum and Minimum Draft Fore and Aft

The maximum and minimum ice class drafts at fore and aft perpendiculars are to be determined in accordance with the upper and lower ice waterlines and the drafts of the ship at fore and aft perpendiculars, when ice conditions require the ship to be ice-strengthened, should always be between the upper and lower ice waterlines. Restrictions on drafts when operating in ice should be documented and kept onboard readily available to the master. The maximum and minimum ice class drafts fore, amidships and aft are to be indicated in the classification certificate. For vessels built on or after 1 July 2007, if the summer load line in fresh water is anywhere located at a higher level than the UIWL, the vessel's sides are to be provided with a warning triangle and with an ice class draft mark at the maximum permissible ice class draft amidships. Vessels built before 1 July 2007 are to be provided with such a marking, if the UIWL is below the summer load line, not later than the first scheduled dry docking after 1 July 2007. The draft and trim, limited by the UIWL, must not be exceeded when the vessel is navigating in ice. The salinity of the sea water along the intended route should be considered when loading the vessel. The vessel is to always be loaded down at least to the draft of LIWL amidships when navigating in ice. Any ballast tank, situated above the LIWL and needed to load down the vessel to this waterline, is to be equipped with devices to prevent the water from freezing. In determining the LIWL, regard is to be paid to the need for ensuring a reasonable degree of ice-going capability in ballast. The highest point of the propeller is to be fully submerged, and if possible, at a depth of at least h_0 below the water surface in all loading conditions. The forward draft is to be at least:

$$d_f = (2 + 0.00025\Delta)h_o \text{ m}$$
$$d_f = (2 + 0.000254\Delta)h_o \text{ ft}$$

but need not exceed $4h_o$ where

- Δ displacement of the vessel, in metric tons (long tons), at the upper ice waterline (UIWL) amidships, as defined in *Upper and lower ice waterlines*
- $4h_o$ level ice thickness, in m (ft), as defined in *Hull structural design (vertical extent of design ice pressure).*

15.4 Power of Propulsion Machinery

The minimum required engine output power P is to be determined in accordance with *Propulsion machinery output, Ice Classes* **I AA, 1 A, I B** *and* **I C** and stated in the Classification certificate.

15.4.1 Propulsion Machinery Output, Ice Classes I AA, 1 A, I B and I C

Note: *for reference purposes, the propulsion machinery output requirements for* **I AA, I A, I B** *and* **I C** *in the1985 Finnish-Swedish Ice Class Rules were amended as follows for vessels with the keel laid or which are at a similar stage of construction on or after 1 September 2003.*

15.4.1.1 Definitions

The dimensions of the vessel are defined below in Fig. 15.1.

L length of the vessel between perpendiculars at the UIWL, m (m, ft)
L_{BOW} length of the bow, m (m, ft)
L_{PAR} length of the parallel midship body, m (m, ft)
B maximum breadth of the vessel at the UIWL, m (m, ft)
T actual ice class draughts of the vessel in accordance with *Propulsion machinery output, Ice Classes* **I AA, 1 A, I B** *and* **I C**. Draughts to be used are the maximum draught amidships corresponding to UIWL and the minimum draught corresponding to LIWL, m (m, ft)
A_{Wf} area of waterline of the bow, m^2 (m^2, ft^2)
H_F thickness of the brash ice layer displaced by the bow, m (m, ft)
H_M thickness of the brash ice in mid channel, m (m, ft)
α the angle of the waterline at *B/4*, degrees
φ_1 the rake of the stem at the centreline, degrees
φ_2 the rake at the bow, at *B/4*, degrees
Ψ flare angle calculated as $\Psi = \arctan(\tan\varphi/\sin\alpha)$ using angles α and φ at each location. For *Power of propulsion machinery*, flare angle is calculated using $\varphi = \varphi_2$
D_P diameter of the propeller, m (m, ft).

For a vessel with a bulbous bow, $\varphi 1$ is to be taken as 90°.

15.4.2 Power Calculation

To be entitled to ice class **I AA, I A, I B** or **I C** (or equivalent), a vessel the keel of which is laid, or which is at a similar stage of construction on or after 1 September 2003 is to comply with the following requirements regarding its engine output. The engine output requirement should be calculated for two drafts. Draughts to be used are the maximum draft amidships referred to as UIWL and the minimum draft referred to as LIWL, as defined in *Maximum and minimum draught fore and aft*. In the calculations, the vessel's parameters which depend on the draft are to be determined

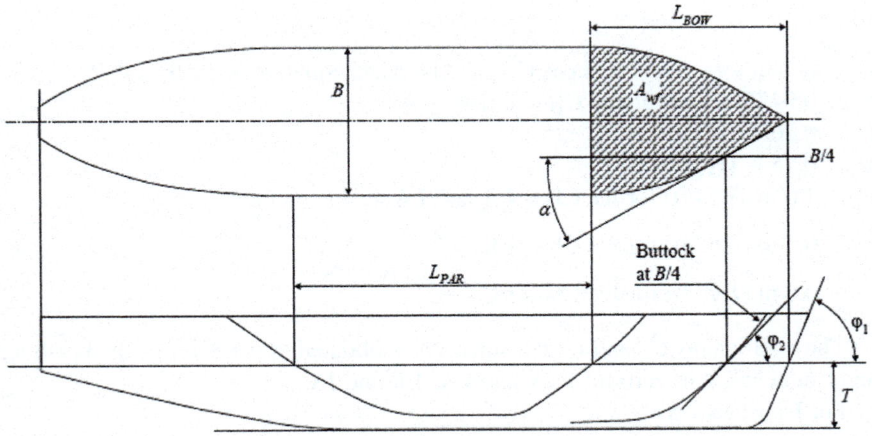

Fig. 15.1 Vessels' dimensions

at the appropriate draught, but L and B are to be determined only at the UIWL. The engine output should not be less than the greater of these two outputs. The engine output is to be not less than determined by the formula below and in no case less than 1000 kW (1360 mhp; 1341 hp) for Ice Class **I A, I B** and **I C**, and not less than 2800 kW (3807 mhp; 3754 hp) for Ice Class **I AA**.

$$p = K_c \frac{R_{CH}/1000)^{3/2}}{D_p} \text{ kW(mhp, hp)}$$

where K_c is to be taken as follows:

Propeller type or propulsion machinery	CP propeller or electric or hydraulic propulsion machinery			FP propeller		
	SI units	MKS units	US units	SI units	MKS units	US units
1 propeller	2.03	84.76	83.79	2.26	94.37	93.29
2 propellers	1.44	60.13	59.44	1.6	66.81	66.04
3 propellers	1.18	49.27	48.71	1.31	54.70	54.07

These K_c values apply for conventional propulsion systems. Other methods may be used for determining the required power for advanced propulsion systems (refer to *Propulsion machinery output, Ice Classes* **I AA, 1 A, I B** *and* **I C**). R_{CH} is the ice resistance of the vessel in a channel with brash ice and a consolidated layer.

$$R_{CH} = C_1 + C_2 + C_3 C_\mu (H_F + H_M)^2 (B + C_\varnothing H_F) + C_F L_{PAR} H_F^2 + C_5 \left[\frac{LT}{B^2}\right]^{3A} \frac{Wf}{L} \text{ N (kgf, lbf)}$$

where

C_μ 0.15cos φ_2 + sin Ψ sin α, C_μ is to be taken equal or larger than 0.45.
$C_\emptyset$ $0.047\Psi - 2.115$, and $C_\emptyset = 0$ if $\Psi \leq 45°$
H_F $0.26 + (\text{H B})^{0.5}$ m.
H_F $0.85 + (\text{H B})^{0.5}$ ft.
H_M 1.0 m (3.28 ft) for Ice Class **I A** and **I AA**

0.8 m (2.62 ft) for Ice Class **I B**

0.6 m (1.97 ft) for Ice Class **I C**.

The coefficients C_1 and C_2 consider a consolidated upper layer of the brash ice and can be taken as zero for Ice Class **I A, I B** and **I C**.

For Ice Class **I AA**:

$$C_1 = f_1 \frac{BL_{PAR}}{\left(\frac{2LT}{N}\right) + 1} + (1 + 0.021\varphi_1)(f_2 B + f_3 L_{BOW} + f_4 BL_{BOW})\ \text{N (kgf, lbf)}$$

$$C_2 = (1 + 0.063\varphi_1)(g_1 + g_2 B) + g_3(1 + 1.2T/B)\frac{B^2}{\sqrt{L}}\ \text{N (kgf, lbf)}$$

	SI units	MKS units	US units
f_1	23 N/m^2	2.35 kgf/m^2	0.48 lbf/ft^2
f_2	45.8 N/m	4.67 kgf/m	3.138 lbf/ft
f_3	14.7 N/m	1.50 kgf/m	1.007 lbf/ft
f_4	29 N/m^2	2.96 kgf/m^2	0.61 lbf/ft^2
g_1	1530 N	156.02 kgf	343.96 lbf
g_2	170 N/m	17.34 kgf/m	11.649 lbf/ft
g_3	400 N/m$^{1.5}$	40.79 kgf/m$^{1.5}$	15.132 lbf/ft$^{1.5}$
C_3	845 N/m^3	86.2 kgf/m^3	5.38 lbf/ft^3
C_4	42 N/m^3	4.28 kgf/m^3	0.267 lbf/ft^3
C_5	825 N/m	84.1 kgf/m	56.5 lbf/ft

$$\Psi = \arctan[\tan\varphi_2/\sin\alpha]\ \text{deg}.$$

If the value of the term $\left(\frac{LT}{B^2}\right)^3$ is less than 5, the value 5 should be used and if the value of the term is more than 20, the value 20 should be used.

15.4.2.1 Other Methods of Determining KC and RCH

The Administration may for an individual vessel, in lieu of the K_C or R_{CH} values defined in *Power of propulsion machinery*, approve the use of K_C and R_{CH} values based on more exact calculations or values based on model test. Such an approval will be given on the understanding that it can be revoked if experience with the vessel's performance in practice motivates this. The design requirement for ice classes is a minimum speed of 5kn in the following brash ice channels:

I AA $H_M = 1.0$ m (3.28 ft) and a 0.1 m (0.328 ft) thick consolidated layer of ice
I A $H_M = 1.0$ m (3.28 ft)
I B $H_M = 0.8$ m (2.62 ft)
I C $H_M = 0.6$ m (1.97 ft).

15.5 Hull Structural Design

The requirements for the hull scantlings are based on certain assumptions concerning the nature of the ice load on the structure. These assumptions are from full scale observations made in the Northern Baltic. The local ice pressure on small areas can reach high values. This pressure may be well more than the normal uniaxial crushing strength of sea ice since the stress field is multi-axial. It has also been observed that the ice pressure on a frame can be greater than on the shell plating at mid- spacing between frames. This is due to the different flexural stiffness of the frames and shell plating. The load distribution on the side structure is assumed to be as shown in Fig. 15.2.

The formulae and values given in this section may be substituted by direct analysis if the Administration or Class deems them to be invalid or inapplicable for a given structural arrangement or detail. Otherwise, direct analysis is not to be used as an alternative to the analytical procedures prescribed by explicit requirements in *Shell plating, framing* and *ice stringers*. Direct analyses are to be conducted using the load patch defined in *Hull structural design (vertical extent of design ice pressure)* and *Hull structural design (design ice pressure)* (p, h and ℓ_a). The pressure to be used is

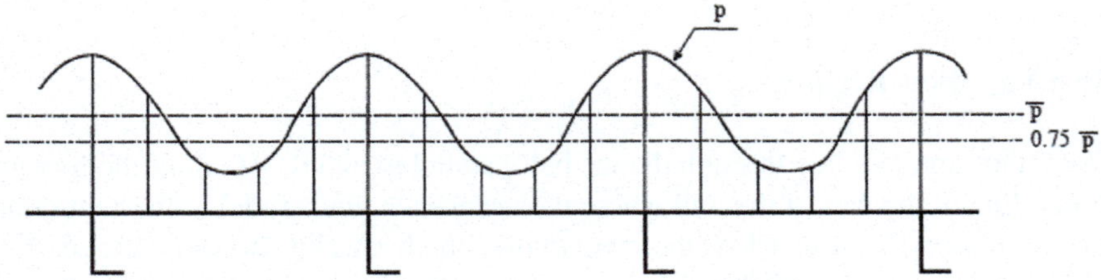

Fig. 15.2 Ice load distribution on ship's side

$1.8p$ where p is determined according to *Design ice pressure*. The load patch is to be applied at locations where the capacity of the structure under the combined effects of bending and shear are minimised. In particular, the structure is to be checked with load centred at UIWL, $0.5h_0$ below the LIWL, and positioned several vertical locations in between. Several horizontal locations should also be checked, especially the locations centred at the mid-span—or spacing. Further, if the load length ℓ_a cannot be determined directly from the arrangement of the structure, several values of ℓ_a should be checked using corresponding values for c_a. Acceptance criterion for designs is that the combined stresses from bending and shear, using the von Mises yield criterion, are lower than the yield point σ_y. When the direct calculation is using beam theory, the allowable shear stress is not to be larger than $0.9\ \tau_y$, where $\tau_y = \sigma_y/\sqrt{3}$. Where the scantlings given by these requirements are less than those required by the Class Rules for a non-ice-strengthened vessel, the greater requirements are to apply. The frame spacings and spans defined in the following text are normally (in accordance with the Class Rules) assumed to be measured along the plate and perpendicular to the axis of the stiffener for plates, along the flange for members with a flange, and along the free edge for flat bar stiffeners. For curved members, the span (or spacing) is defined as the chord length between span (or spacing) points. The span points are defined by the intersection between the flange or upper edge of the member and the supporting structural element (stringer, web frame, deck or bulkhead). Figure 15.3 illustrates the determination of span and spacing for curved members. The effective breadth of the attached plate to be used for calculating the combined section modulus of the stiffener, stringer and web frame and attached plate is to be taken as the Class Rules require. The requirements for the section modulus and shear area of the frames, stringers and web frames in *Framing* and *Ice stringers*, and *Web frames* are with respect to effective member cross section. For such cases where the member is not normal to the plating, the section properties are to be adjusted in accordance with the Class Rules.

15.5.1 Hull Regions

For the application of this section the vessel's ice belt is divided forward and aft into the following regions, refer also to Fig. 15.4.

15.5.1.1 Bow Region

From the stem to a line through the ice belt parallel to and $0.04L$ aft of the forward borderline of the part of the hull where the waterlines run parallel to the centreline. For ice classes **I AA** and **I A**, the overlap over the borderline need not exceed 6 m (19.7 ft); for ice classes **I B** and **I C**, this overlap need not exceed 5 m (16.4 ft).

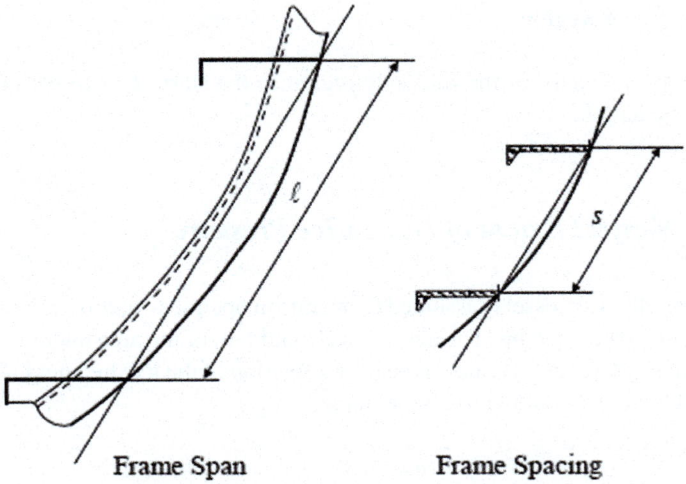

Fig. 15.3 Definition of the frame span and frame spacing for curved members

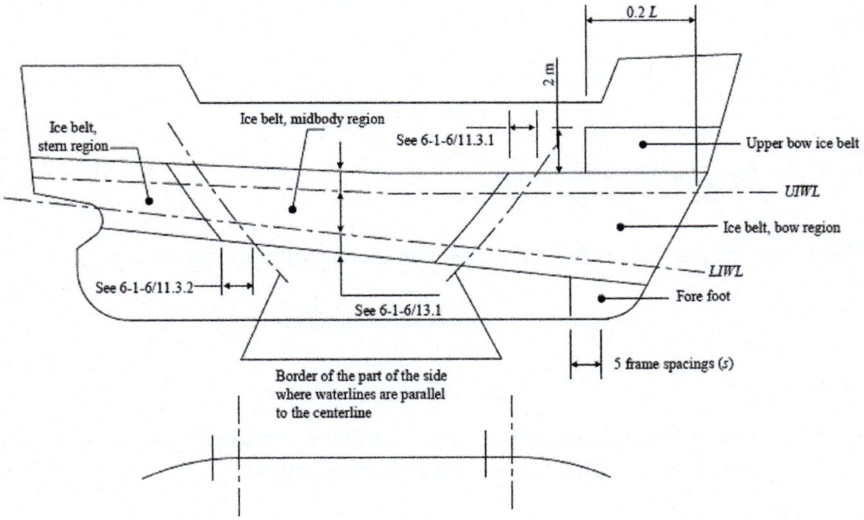

Fig. 15.4 Ice strengthened regions of the hull

15.5.1.2 Midbody Region

From the aft boundary of the bow region to a line parallel to and $0.04L$ aft of the aft borderline of the part of the hull where the waterlines run parallel to the centreline. For ice classes **I AA** and **I A**, the overlap over the borderline need not exceed 6 m (19.7 ft); for ice classes **I B** and **I C**, this overlap need not exceed 5 m (16.4 ft).

15.5.1.3 Stern Region

From the aft boundary of the midbody region to the stern. *L* is to be taken as the vessel's rule length.

15.5.2 Vertical Extent of Design Ice Pressure

An ice strengthened vessel is assumed to operate in open sea conditions with level ice thickness not exceeding ho. The design load height, h, of the area under ice pressure at any time is, however, assumed to be only a fraction of the ice thickness. The values for h_0 and h are given in the following table:

Ice Class	h_0 m (ft)	h m (ft)
I AA	1.0 (3.28)	0.35 (1.15)
I A	0.8 (2.62)	0.30 (0.98)
I B	0.6 (1.97)	0.25 (0.82)
I C	0.4 (1.31)	0.22 (0.72)

15.5.3 Design Ice Pressure

The design ice pressure is to be not less than given by the following equation:

$$p = c_d \cdot c_1 \cdot c_a \cdot p_o \ \text{N/mm}^2 (\text{kgf/mm}^2, \text{psi})$$

where

c_d a factor which considers the influence of the size and propulsion machinery output of the vessel. This factor is taken as maximum $c_d = 1$

k $\frac{\left(ak + \frac{b}{1000}\right)}{\sqrt{n \cdot P/1000}}$.

a and *b* are given in the following table:

	Region			
	Bow	Midbody and stern		
	k ≤ 12	k > 12	k ≤ 12	k > 12
a	30	6	8	2
b	230	518	214	286

n 1.0 (1.0, 1.016)

Δ displacement of the vessel, in metric tons (long tons), at the upper ice waterline (UIWL) amidships, as defined in *Upper and lower ice waterlines*

P the actual continuous propulsion machinery output, in kW, as defined in *Propulsion machinery output*

c_1 factor which considers the probability that the design ice pressure occurs in a certain region of the hull for the ice class.

The value of c_1 is given in the following table:

	Region		
	Bow	Midbody	Stern
I AA	1.0	1.0	0.75
I A	1.0	0.85	0.65
I B	1.0	0.70	0.45
I C	1.0	0.50	0.25

c_a a factor which considers the probability that the full length of the area under consideration will be under pressure at the same time

$\sqrt{\frac{\ell_0}{\ell_a}}$, maximum 1.0, minimum 0.35.

ℓ_0 0.6 m (2 ft)

ℓa is as given in the following table:

Structure	Type of framing	ℓ_a m (ft)
Shell	Transverse	Frame spacing
	Longitudinal	1.7 times spacing of frame
Frames	Transverse	Frame spacing
	Longitudinal	Span of frame
Ice stringer		Span of stringer
Web frame		2 times spacing of web frames

p_o the nominal ice pressure; the value 5.6 N/mm^2 (0.571 kgf/mm^2, 812 psi) is to be used

15.6 Shell Plating

15.6.1 Vertical Extent of Ice Strengthening for Plating (Ice Belt)

The vertical extension of the ice belt is given in the following table (refer to Fig. 15.4):

Ice Class	Hull region	Above UIWL m (ft)	Below LIWL m (ft)
I AA	Bow	0.60 (1.97)	1.20 (3.94 ft)
	Midbody		
	Stern		1.0 (3.28 ft)
I A	Bow	0.50 (1.64)	0.90 (2.95)
	Midbody		0.75 (2.46)
	Stern		
I B and I C	Bow	0.40 (1.31)	0.70 (2.30)
	Midbody		0.6 (1.97)
	Stern		

In addition, the following areas are to be strengthened:

15.6.1.1 Fore Foot

For ice class **I AA**, the shell plating below the ice belt from the stem to a position five main frame spaces abaft the point where the bow profile departs from the keel line should be ice-strengthened in the same way as the bow region.

15.6.1.2 Upper Bow Ice Belt

For ice class **I AA** and **I A**, on vessels with an open water service speed equal to or exceeding 18kn, the shell plating from the upper limit of the ice belt to 2 m (6.56 ft) above it and from the stem to a position at least 0.2L abaft the forward perpendicular is to be at least the thickness required for the ice belt in the Midbody region. A similar strengthening of the bow region is also advisable for a ship with a lower service speed when, based on the model tests, for example, it is evident that the ship will have a high bow wave. Side lights, side scuttles, etc., are not to be situated in the ice belt. If the weather deck in any part of the vessel is situated below the upper limit of the ice belt (e.g., in way of the well of a raised quarter decker), the bulwark is to be given at least the same strength as is required for the shell in the ice belt. The strength of the construction of the freeing ports is to meet the requirements for the bulwark.

15.6.2 *Ice Belt Plating Thickness*

With transverse framing, the thickness of the shell plating is to be not less than given by the following equation:

$$t = as\sqrt{f_1 P_{PL}/\sigma_y} + t_c \text{ mm in.}$$

With longitudinal framing, the thickness of the shell plating is to be not less than given by the following equation:

$$t = as\sqrt{p/f_2\sigma_y + t_c} \text{ mm in.}$$

where

s frame spacing, in m (ft)
P_{PL} 0.75 p, in N/mm^2 (kgf/mm^2, psi)
p as given in *Hull structure design*
$f1$ $1.3 - 4.2/[(h/s) + 1.8]^2$; maximum 1.0
$f2$ $0.6 + 0.4/(h/s)$; when $h/s \leq 1$
$1.4 - 0.4\ (h/s)$; when $1 \leq h/s < 1.8$
h as given in *Vertical Extent of Design Ice Pressure*, in m (ft)
σ yield strength of the material, in N/mm^2 (kgf/mm^2, psi)
a 667 (8)

Use of steels with yield strengths greater than 390 N/mm^2 (40 kgf/mm^2, 56,565 psi) are subject to special consideration.

t_c increment for abrasion and corrosion, in mm (in); normally, t_c is to be 2 mm (0.08 in.); however, if a special surface coating by experience is shown capable to withstand the abrasion of ice and is applied and maintained effective, lower values may be approved.

15.7 Framing

15.7.1 End Attachments

Within the ice strengthened area, all frames are to be effectively attached to all supporting structures. A longitudinal frame should be attached to all the supporting web frames and bulkheads by brackets. When a transversal frame terminates at a stringer or deck, a bracket or similar construction is to be fitted. When a frame is running through the supporting structure, both sides of the web plate of the frame are to be connected to the structure (by direct welding, collar plate or lug). When a bracket is installed, it is to have at least the same thickness as the web plate of the frame and the edge is to be appropriately stiffened against buckling.

Frames. (1) *Welding.* Frames are to be attached to the shell by double continuous welding. Scallops are to be avoided, except where frames cross shell plate butts; (2) *web thickness.* The web thickness of the frames is to be at least the maximum of the following:

$$\frac{hw\sqrt{\sigma_y}}{C}$$

where

h_w web height
C 805 for profiles

282 for flat bars.

- Half of the net thickness of the shell plating, $t - t_c$. For the purpose of calculating the web thickness of frames, the required thickness of the shell plating is to be calculated according to *Ice Belt Plating Thickness* using the yield strength σ_y of the frames
- 9 mm (0.35 in.)

Where there is a deck, top or bottom plating of a tank, tank top or bulkhead in lieu of a frame, the plate thickness of it should be calculated as above, to a depth corresponding to the height of the adjacent frames. In such a case, the material properties of the deck, top or bottom plating of the tank, tank top or bulkhead and the frame height h_w of the adjacent frames should be used in the calculations, and the constant C should be 805; and (3) *slanted frames*. Frames that are not normal to the plating or the profile is unsymmetrical, and the span exceeds 4.0 m (13.1 ft) are to be supported against tripping by brackets, intercostals, stringers or similar at a distance preferably not exceeding 1.3 m (4.25 ft). If the span is less than 4.0 m (13.1 ft), the supports against tripping are required for unsymmetrical profiles and stiffeners the web of which is not normal to plating in the following regions:

- **I AA**All hull regions
- **I A**Bow and Midbody regions
- **I B** and **I C**Bow region.

15.7.2 Vertical Extent of Ice Strengthening for Framing

The vertical extent of the ice strengthening of framing is to be at least as given in the following table:

Ice class	Hull region	Above UIWL	Below UIWL
I AA	Bow	1.2 (3.94)	Down to double bottom or below top of floors
	Midbody		2.0 (6.560)
	Stern		1.6 (5.25)
I A, I B, I C	Bow	1.0 (3.28)	1.6 (5.250)
	Midbody		1.3 (4.270)
	Stern		1.0 (3.28)

Where an upper bow ice belt is required, refer to *Vertical extent of ice strengthening for plating (ice belt)*, the ice strengthening of the framing is to be extended at least to the top of this ice belt. Where the ice strengthening would go beyond a deck or a tank top by not more than 250 mm (9.8 in.), it may be terminated at that deck or tank top.

15.7.3 Transverse Framing

Section modulus and shear area. The section modulus, *SM*, of a main or intermediate frame is to be not less than that obtained from the equation:

$$SM = n\left(\frac{p \cdot h \cdot s \cdot \ell}{m_t \cdot \sigma_y}\right) \text{ cm}^3\left(\text{in.}^3\right)$$

and the effective shear area is calculated from

$$A = k\left(\frac{\sqrt{3} \cdot f_3 \cdot p \cdot h \cdot s}{2\sigma_y}\right) \text{ cm}^2\left(\text{in.}^2\right)$$

where

n 10^6 (1728)
k 10^4 (144)
p ice pressure, as given in *Design ice pressure*, in N/mm^2 (kgf/mm^2, psi)
s frame spacing, in m (ft)
h height of load area, as given in *Vertical extent of design ice pressure*, in m (ft)
ℓ span of the frame, in m (ft)
$m\ t$ $7m_o/[7 - 5\ (h/\ell)]$
f_3 is a factor which considers the maximum shear force versus the load location and the shear stress distribution

1.2
σ_y yield strength, as defined in *Ice belt plating thickness*, in N/mm^2 (kgf/mm^2, psi).

m_o values are given in Fig. 15.5.

The boundary conditions shown are for the main and intermediate frames. Possible different conditions for the main frames are assumed to have been taken care of by interaction between the frames and are reflected in the m_o values. The load is considered applied at mid span. Where less than 15% of the span, ℓ, of the frame is situated within the ice-strengthening zone for frames as defined in *Vertical extent of ice strengthening for framing*, ordinary frame scantlings may be used.

Upper end of transverse frames. The upper end of an ice-strengthened part of a main frame and of an intermediate ice frame is to be attached to a deck or ice stringer, refer to *Ice stringers*. Where an intermediate ice frame terminates above a deck or ice stringer that is situated at or above the upper limit of the ice belt, refer to *Vertical extent of ice strengthening for plating (ice belt)*, the part above the deck or stringer

Fig. 15.5 Web frame model

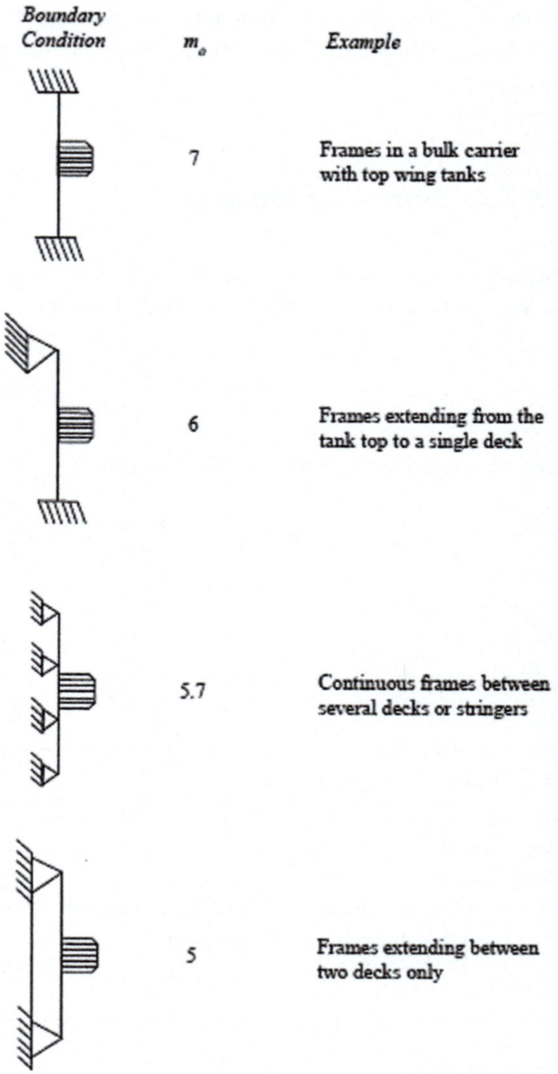

may have scantlings as required for a non-ice-strengthened vessel and the upper end of the intermediate frame may be connected to the adjacent main frames by a header of the same scantlings as the main frame.

Lower end of transverse framing. The lower end of an ice-strengthened part of a main frame and of an intermediate ice frame is to be attached to a deck, tank top or ice stringer, refer to *Ice stringers*. Where an intermediate ice frame terminates below a deck, tank top or ice stringer which is situated at or below the lower limit of the ice belt, refer to *Vertical extent of ice strengthening for plating (ice belt)*, the lower end of the frame may be connected to the adjacent main frames by a header of the same scantlings as the main frame. Note that the main frames below the lower edge of the ice belt must be ice strengthened, refer to *Vertical extent of ice strengthening for framing.*

15.7.4 Longitudinal Framing

The following requirements are intended for longitudinal frames with all end conditions.

Frames with and without brackets. The section modulus, *SM*, of a longitudinal frame is to be not less than that obtained from the equation:

$$SM = n(f_4 ph\ell^2/m_1\sigma_y)\ \text{cm}^3\left(\text{in.}^3\right)$$

The effective shear area, *A*, is to be not less than that obtained from the equation:

$$A = k(\sqrt{3} f_4 f_5 ph\ell/\sigma_y)\ \text{cm}^2\left(\text{in.}^2\right)$$

In calculating the actual shear area of the frames, the area of the brackets is not to be considered.

f_4 factor which considers the load distribution to adjacent frames
(1 – 0.2*h/s*)

f_5 factor which considers the pressure definition and maximum shear force versus load location and the shear stress distribution

2.16

p ice pressure, as given in *Design ice pressure*, in N/mm^2 (kgf/mm^2, psi)

h height of load area, as given in *Vertical extent of design ice pressure*, in m (ft)

s frame spacing, in m (ft)

n 106 (1728)

k 5 × 103 (72)

ℓ total span of frame, in m (ft)

m_1 boundary condition factor; $m_1 = 13.3$ for a continuous beam. Where the boundary conditions deviate significantly from those of a continuous beam (e.g., in an end field), a smaller boundary factor may be required. For frames without brackets a value $m_1 = 11.0$ is to be used

σ_y yield strength, as defined in *Vertical extent of ice strengthening for plating (ice belt)*, in N/mm^2 (kgf/mm^2, psi).

15.8 Ice Stringers

15.8.1 Stringers Within the Ice Belt

The section modulus, *SM*, of a stringer within the ice belt (refer to *Vertical extent of ice strengthening for plating (ice belt)*) is to be not less than that obtained from the equation:

$$SM = \left(\frac{f_6 \cdot f_7 \cdot p \cdot h \cdot \ell^2}{m \cdot \sigma_y}\right) \text{cm}^3 \left(\text{in.}^3\right)$$

The effective shear area, A, is to be not less than that obtained from the equation:

$$A = k\left(\frac{\sqrt{3} \cdot f_6 \cdot f_7 \cdot f_8 \cdot p \cdot h \cdot \ell}{2\sigma_y}\right) \text{cm}^2 \left(\text{in.}^2\right)$$

where

p ice pressure, as given in *Design ice pressure*, in N/mm^2 (kgf/mm^2, psi)
h height of load area, as given in *Vertical extent of design ice pressure*, in m (ft).

The product ($p \times h$) is not to be taken as less than 0.15 SI units (0.0153 MKS units, 71.4 US units)

n 10^6 (1728)
k 10^4 (144)
ℓ span of stringer, in m (ft)
m_s boundary condition factor as defined in *Longitudinal framing*
f_6 factor which considers the distribution of the load on the transverse frames
0.9
f_7 factor that considers the design point of stringers

1.8
f_8 factor that considers the maximum shear force versus load location and the shear stress distribution
1.2
σ_y yield strength, as defined in *Ice belt plating thickness*, in N/mm^2 (kgf/mm^2, psi).

15.8.2 Stringers Outside the Ice Belt

The section modulus, *SM*, of a stringer outside the ice belt that supports ice strengthened frames is to be not less than that obtained from the equation:

$$SM = n\left(\frac{f_9 \cdot f_{10} \cdot p \cdot h \cdot \ell^2}{m_s \cdot \sigma_y}\right)\left(1 - \frac{h_s}{\ell_s}\right) \text{ cm}^2(\text{in.}^2)$$

The effective shear area, *A*, is to be not less than that obtained from the equation:

$$A = k\left(\frac{\sqrt{3} f_9 \cdot f_{10} \cdot f_{11} \cdot p \cdot h \cdot \ell}{2\sigma_y}\right)\left(1 - \frac{h_s}{\ell_s}\right) \text{ cm}^2(\text{in.}^2)$$

where

p ice pressure, as given in *Design ice pressure*, in N/mm^2 (kgf/mm^2, psi)
h height of load area, as given in *Vertical extent of design ice pressure*, in m (ft).

The product ($p \times h$) is to be not taken as less than 0.15 SI units (0.0153 MKS units, 71.4 US units).

n 10^6 (1728)
k 10^4 (144)
ℓ span of stringer, in m (ft)
m_s boundary condition factor; ms = 13.3 for a continuous beam
ℓ_s the distance to the adjacent ice stringer, in m (ft)
h_s the distance to the ice belt, in m (ft)
f_9 factor which considers the distribution of load on transverse frames.
0.80
f_{10} factor that considers the design point of stringers

1.8
f_{11} factor that considers the maximum shear force versus load location and the shear stress distribution

1.2
σ yield strength, as defined in *Ice belt plating thickness*, in N/mm^2 (kgf/mm^2, psi).

15.8.3 Deck Strips

The deck strips abreast of hatches serving as ice stringers are to comply with the section modulus and shear area requirements in *Stringers within the ice belt* and *stringers outside the ice belt*, respectively. In the case of exceptionally long hatches, the product ($p \times h$) may be taken as less than 0.15 SI units (0.0153 MKS units, 71.4 US units), but in no caseless than 0.10 SI units (0.0102 MKS units, 47.6 US units). In designing weather deck hatch covers and their fittings, special attention is to be paid to the deflection of the vessel's sides due to ice pressure in way of exceptionally long (more than *B*/2) hatch openings.

15.9 Web Frames

15.9.1 Design Ice Load

The design load, F, on a web frame from an ice stringer or from longitudinal framing may be obtained from the following equation:

$$F = nf_{12}phs \text{ kN (tf, Ltf)}$$

where

n 10^3 (0.0643)
f12 a factor that considers the design point of web frames
1.8
p ice pressure in N/mm^2 (kgf/mm^2, psi); in calculating c_a however, ℓa is to be taken as $2S$
h height of ice load area, in m (ft).

The product ($p \times h$) is not to be taken as less than 0.15 SI units (0.0153 MKS units, 71.4 US units).

S distance between web frames, in m (ft).

In case the supported stringer is outside the ice belt, the force F should be multiplied by $(1 - h_s/\ell_s)$, where h_s and ℓ_s should be taken as defined in *Stringers outside the ice belt*.

15.9.2 Section Modulus and Shear Area

The section modulus and shear area may be obtained from the following equations:

- *Effective Shear Area*

$$A = k\left(\frac{\sqrt{3} \cdot f_{13 \cdot \alpha \cdot Q}}{\sigma_y}\right) \text{cm}^2\left(\text{in.}^2\right)$$

where

Q maximum calculated shear force under the load F, as given in Design ice load
k 10 (2240)
f_{13} 1.1

factor that considers the shear force distribution
α as given in the table below

σ_y yield strength, as defined in *Ice belt plating thickness*, in N/mm^2 (kgf/mm^2, psi)
F as in *Design ice load*.

- *Section Modulus*

$$SM = n\left(\frac{M}{\sigma_y}\right)\sqrt{\frac{1}{1-(\frac{\gamma A}{A_a})^2}}\ \text{cm}^3(\text{in.}^3)$$

where

n 1000 (26,880)
M maximum calculated bending moment under the load F; this is to be taken as

$0.193F\ell$

γ as given in the table below
A required shear area
A_a actual cross-sectional area of the web frame, in cm^2 (in.^{2})

$A_f + A_{w.}$

- *Factors α and γ*

A_f actual cross section area of free flange, in cm^2 (in.^{2}).
A actual effective cross section area of web plate, in cm^2 (in.^{2}).

A_f/A_w	0	0.2	0.4	0.6	0.8	1.0	1.2	1.4	1.6	1.8	2.0
α	1.5	1.23	1.16	1.11	1.09	1.07	1.06	1.05	1.05	1.04	1.04
γ	0	0.44	0.62	0.71	0.76	0.80	0.83	0.85	0.87	0.88	0.89

15.10 Bow

15.10.1 Stem

The stem may be made of rolled, cast or forged steel or of shaped steel plates as shown in Fig. 15.6. The thickness of a shaped plate stem and, in the case of a blunt bow, any part of the shell where $\alpha \geq 30°$ and $\Psi \geq 75°$ (refer to *Propulsion machinery output, ice classes* **I AA, 1 A, I B** *and* **I C** for angle definitions), is to be obtained from the equation in *Ice belt plating thickness* where:

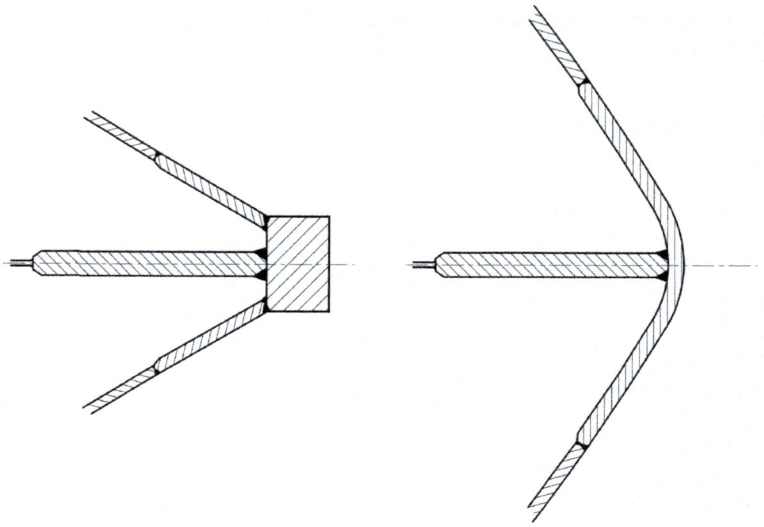

Fig. 15.6 Examples of suitable ice stems

s spacing of elements supporting the plate, in m (ft).
P_{PL} p, in N/mm^2 (kgf/mm^2, psi), refer to *Design Ice Pressure.*
ℓ_a spacing of vertical supporting elements, in m (ft).

The stem and that part of a blunt bow defined above is to be supported by floors, breasthooks or brackets spaced not more than 0.6 m (1.97 ft) apart and of a thickness at least half the shell plate thickness. This reinforcement of the stem is to extend from the keel to a point 0.75 m (2.46 ft) above *UIWL*, or where an upper Bow ice belt is required, refer to *Vertical extent of ice strengthening for plating (ice belt)*, to the upper limit of this upper Bow ice belt.

15.11 Stern

The introduction of new propulsion arrangements with azimuthing thrusters, which provide improved manoeuvrability, will result in increased ice loading of the Stern region and the stern area. This fact should be considered in the design of the aft/ stern structure. To avoid extremely high loads on propeller blade tips, the minimum distance between propeller(s) and hull (including stern frame) should not be less than h_0 (refer to *Hull structural design (vertical extent of design ice pressure)*). On twin and triple screw vessels, the ice strengthening of the shell and framing is to extend to the double bottom for 1.5 m (4.92 ft) forward and aft of the side propellers. Shafting and stern tubes of side propellers are to be normally enclosed within plated bossing. If detached struts are used, their design, strength and attachment to the hull is to be duly considered for ice loading.

15.12 Rudder and Steering

15.12.1 Minimum Design Speed

The scantlings of rudder post, rudder stock, pintles, steering gear, etc., as well as the capacity of the steering gear are to comply with the Class Rules. Where the design ahead speed of the vessel, as defined in the Class Rules is less than the minimum speed indicated in the table below, the latter speed is to be used:

Class	Minimum speed (knots)
I AA	20
I A	18
I B	16
I C	14

For use with the minimum ahead speeds in the above table, k_c (coefficient depending on rudder cross section (profile type)) may be taken as 80% of that specified in the rudder design. Also, k_1 for rudders situated behind nozzles need not be taken as greater than 1.0. The local scantlings of rudders are to be determined assuming that the whole rudder belongs to the ice belt. Further, the rudder plating and frames are to be designed using the ice pressure p for the plating and frames in the midbody region.

15.12.2 Rudder and Rudder Stock Protection

For the ice classes **I AA** and **I A**, the rudder (rudder stock and the upper part of the rudder) is to be protected from direct contact with intact ice by an ice knife that extends below the *LIWL*, if practicable (or equivalent means). Special consideration should be given to the design of the rudder and the ice knife for ships with flap-type rudders.

15.12.3 Overload Design

For ice classes **I AA** and **I A**, due regard is to be given to the excessive loads caused by the rudder being forced out of the midship position when going astern in ice or backing into an ice ridge. Suitable arrangements such as rudder stops are to be installed to absorb these loads. Relief valves for the hydraulic pressure in rudder turning mechanism(s) are to be installed. The components of the steering gear (e.g., rudder stock, rudder coupling, rudder horn, etc.) are to be dimensioned to withstand loads causing yield stresses within the required diameter of rudder stock.

15.13 Propulsion Machinery

The requirements discussed in the section *Propulsion machinery* apply to propulsion machinery covering open- and ducted-type propellers with controllable pitch or fixed pitch design for the ice classes **I AA**, **I A**, **I B** and **I C**. The given propeller loads are the expected ice loads for the whole ship's service life under normal operational conditions, including loads resulting from the changing rotational direction of FP propellers. However, these loads do not cover off-design operational conditions, for example when a stopped propeller is dragged through ice. The requirements also apply to azimuthing and fixed thrusters for main propulsion, considering loads resulting from propeller-ice interaction and loads on the thruster body-ice interaction. However, the load models do not include propeller/ice interaction loads when ice enters the propeller of a turned azimuthing thruster from the side (radially). The given azimuthing thruster body loads are the expected ice loads for the ship's service life under normal operational conditions. The local strength of the thruster body should be sufficient to withstand local ice pressure when the thruster body is designed for extreme loads. The thruster global vibrations caused by blade order excitation on the propeller may cause significant vibratory loads (Table 15.1).

15.13.1 Symbols

c	chord length of blade section, m (ft)
$c_{0.7}$	chord length of blade section at $0.7R$ propeller radius, m (ft)
CP	controllable pitch
D	propeller diameter, m (ft)
d	external diameter of propeller hub (at propeller plane), m (ft)
D_{limit}	limit value for propeller diameter, m (ft)
EAR	expanded blade area ratio
F_b	maximum backward blade force for the ship's service life, kN (kgf, lbf)
F_{ex}	ultimate blade load resulting from blade loss through plastic bending, kN (kgf, lbf)
F_f	maximum forward blade force for the ship's service life, kN (kgf, lbf)
F_{ice}	ice load, kN (kgf, lbf)
$(F_{ice})_{\max}$	maximum ice load for the ship's service life, kN (kgf, lbf)
FP	fixed pitch
h_0	depth of the propeller centreline from the lower ice waterline, m (ft)
H_{ice}	thickness of maximum design ice block entering to propeller, m (ft)

Table 15.1 Definition of loads

	Definition	Use of the load in design process
F_b	The maximum lifetime backward force on a propeller blade resulting from propeller/ice interaction, including hydrodynamic loads on that blade. The direction of the force is perpendicular to 0.7*R* chord line. Refer to Fig. 15.7	Design force for strength calculation of the propeller blade
F_f	The maximum lifetime forward force on a propeller blade resulting from propeller/ice interaction, including hydrodynamic loads on that blade. The direction of the force is perpendicular to 0.7*R* chord line	Design force for calculation of strength of the propeller blade
$Q_{s\,\max}$	The maximum lifetime spindle torque on a propeller blade resulting from propeller/ice interaction, including hydrodynamic loads on that blade	In designing the propeller strength, the spindle torque is automatically considered because the propeller load is acting on the blade as distributed pressure on the leading edge or tip area
T_b	The maximum lifetime thrust on a propeller (all blades) resulting from propeller/ice interaction. The direction of the thrust is the propeller shaft direction, and the force is opposite to the hydrodynamic thrust	Is used for estimation of the response thrust T_r. T_b can be used as an estimate of excitation for axial vibration calculations. However, axial vibration calculations are not required in the Class Rules
T_f	The maximum lifetime thrust on a propeller (all blades) resulting from propeller/ice interaction. The direction of the thrust is the propeller shaft direction acting in the direction of hydrodynamic thrust	Is used for estimation of the response thrust T_r. T_f can be used as an estimate of excitation for axial vibration calculations. However, axial vibration calculations are not required in the Class Rules
Q_{max}	The maximum ice-induced torque resulting from propeller/ice interaction on one propeller blade, including hydrodynamic loads on that blade	Is used for estimation of the response torque (Q_r) along the propulsion shaft line and as excitation for torsional vibration calculations
Fex	Ultimate blade load resulting from blade loss through plastic bending. The force that is needed to cause total failure of the blade so that plastic hinge appears in the root area. The force is acting on 0.8*R*. Spindle arm is to be taken as 2/ of the distance between the axis of blade rotation and leading/trailing edge (whichever is the greater) at the 0.8*R* radius	Blade failure load is used to dimension the blade bolts, pitch control mechanism, propeller shaft, propeller shaft bearing and trust bearing. The objective is to guarantee that total propeller blade failure does not lead to damage to other components

(continued)

Table 15.1 (continued)

	Definition	Use of the load in design process
Q_r	Maximum response torque along the propeller shaft line, considering the dynamic behaviour of the shaft line for ice excitation (torsional vibration) and the hydrodynamic mean torque on the propeller	Design torque for propeller shaft line components
T_r	Maximum response thrust along shaft line, considering the dynamic behaviour of the shaft line for ice excitation (axial vibration) and the hydrodynamic mean thrust on the propeller	Design thrust for propeller shaft line components
F_{ti}	Maximum response force caused by ice block impacts on the thruster body or the propeller hub	Design load for thruster body and slewing bearings
F_{tr}	Maximum response force on the thruster body caused by ice ridge-thruster body interaction	Design load for thruster body and slewing bearings

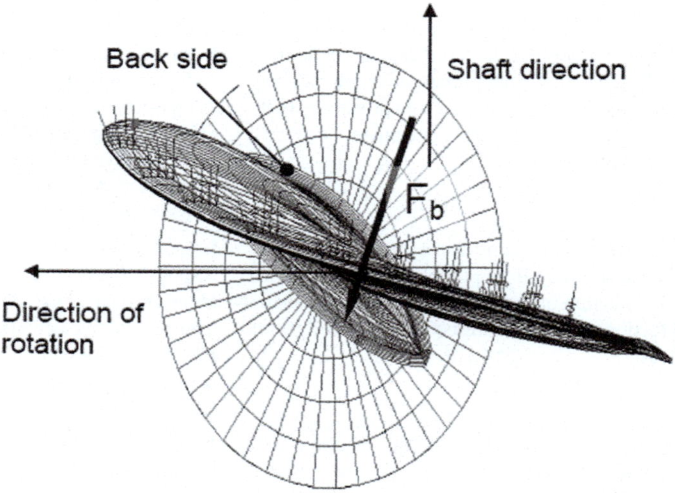

Fig. 15.7 Direction of the backward blade force resultant taken perpendicular to chord line at radius 0.7*R*

Table 15.2 Types of ice operation

Ice class	Operation of the ship
I AA	Operation in ice channels and in level ice The ship may proceed by ramming
I A, I B, I C	Operation in ice channels

I_e	equivalent mass moment of inertia of all parts on engine side of component under consideration, kg-m^2 (lb-ft^2)
I_t	equivalent mass moment of inertia of the whole propulsion system, kg-m^2 (lb-ft^2)
k	shape parameter for Weibull distribution
$LIWL$	lower ice waterline, m (ft)
m	slope for SN curve in log/log scale
M_{BL}	blade bending moment, kN-m (kgf-m, lbf-ft)
MCR	maximum continuous rating
n	propeller rotational speed, rev/s
n_n	nominal propeller rotational speed at MCR in free running condition, rev/s
N_{class}	reference number of impacts per propeller rotational speed per ice class
N_{ice}	total number of ice loads on propeller blade for the ship's service life
N_R	reference number of load for equivalent fatigue stress (10^8 cycles)
N_Q	number of propeller revolutions during a milling sequence
$P_{0.7}$	propeller pitch at 0.7R radius, m (ft)
$P_{0.7n}$	propeller pitch at 0.7R radius at MCR in free running condition, m (ft)
$P_{0.7b}$	propeller pitch at 0.7R radius at MCR in bollard condition, m (ft)
Q	torque, kN-m (kgf-m, lbf-ft)
Q_{emax}	maximum engine torque, kN-m (kgf-m, lbf-ft)
$Q_{\max}$	maximum torque on the propeller resulting from propeller-ice interaction, kN-m (kgf-m, lbf-ft)
$Q_{\max n}$	maximum torque on the propeller resulting from propeller-ice interaction reduced to the rotational speed in question, kN-m (kgf-m, lbf-ft)
Q_{motor}	electric motor peak torque, kN-m (kgf-m, lbf-ft)
Q_n	nominal torque at MCR in free running condition, kN-m (kgf-m, lbf-ft)
Q_r	maximum response torque along the propeller shaft line, kN-m (kgf-m, lbf-ft)
Q_{peak}	maximum of response torque Q_r, kN-m (kgf-m, lbf-ft)
$Q_{s\,max}$	maximum spindle torque of the blade for the ship's service life, kN-m (kgf-m, lbf-ft)
Q_{sex}	maximum spindle torque due to blade failure caused by plastic bending, kN-m (kgf-m, lbf-ft)
Q_{vib}	vibratory torque at considered component, taken from frequency domain open water torque vibration calculation (TVC), kN-m (kgf-m, lbf-ft)
R	propeller radius, m (ft)
r	blade section radius, m (ft)
T	propeller thrust, kN (kgf, lbf)
T_b	maximum backward propeller ice thrust for the ship's service life, kN (kgf, lbf)
T_f	maximum forward propeller ice thrust for the ship's service life, kN (kgf, lbf)
T_n	propeller thrust at MCR in free running condition, kN (kgf, lbf)
T_r	maximum response thrust along the shaft line, kN (kgf, lbf)

t maximum blade section thickness, m (ft)
Z number of propeller blades
α_i duration of propeller blade/ice interaction expressed in rotation angle, degrees
α_1 phase angle of propeller ice torque for blade order excitation component, degrees
α_2 phase angle of propeller ice torque for twice the blade order excitation component, degrees
$\gamma_{\varepsilon 1}$ reduction factor for fatigue; scatter effect
$\gamma_{\varepsilon 2}$ reduction factor for fatigue; test specimen size effect
γ_v reduction factor for fatigue; variable amplitude loading effect
γm reduction factor for fatigue; mean stress effect
ρ reduction factor for fatigue correlating the maximum stress amplitude to the equivalent fatigue stress for 10^8 stress cycles
σ proof yield strength (at 0.2% offset) of blade material, MPa (kgf/cm^2, psi)
σ_{exp} mean fatigue strength of blade material at 10^8 cycles to failure in sea water, MPa (kgf/ cm^2, psi)
σ_{fat} equivalent fatigue ice load stress amplitude for 10^8 stress cycles, MPa (kgf/cm^2, psi)
σ_{fl} characteristic fatigue strength for blade material, MPa (kgf/cm^2, psi)
σ_{ref1} reference stress $\sigma_{ref1} = 0.6\sigma_{0.2} + 0.4\sigma_u$, MPa (kgf/cm^2, psi)
σ_{ref2} reference stress $\sigma_{ref2} = 0.7\sigma_u$ or $\sigma_{ref2} = 0.6\sigma_{0.2} + 0.4\sigma_u$, whichever is less, MPa (kgf/cm^2, psi)
σ_{st} maximum stress resulting from F_b or F_f, MPa (kgf/cm^2, psi)
σ_u ultimate tensile strength of blade material, MPa (kgf/cm^2, psi)
$(\sigma_{ice})_{bmax}$ principal stress caused by the maximum backward propeller ice load, MPa (kgf/cm^2, psi)
$(\sigma_{ice})_{fmax}$ principal stress caused by the maximum forward propeller ice load, MPa (kgf/cm^2, psi)
$(\sigma_{ice})_{max}$ maximum ice load stress amplitude, MPa (kgf/cm^2, psi).

Ice contact pressure at leading edge is shown with small arrows.

15.13.2 Design Ice Conditions

In estimating the ice loads of the propeller for ice classes, several types of operation as given in Fig. 15.2 were considered. For the estimation of design ice loads, a maximum ice block size is determined. The maximum design ice block entering the propeller is a rectangular ice block with the dimensions $H_{ice} \times 2H_{ice} \times 3H_{ice}$. The thickness of the ice block (H_{ice}) is given in Table 15.3.

Table 15.3 Thickness of the ice block (H_{ice})

Ice class	I AA	I A	I B	I C
Thickness of the design maximum ice block entering the propeller (H_{ice})	1.75 m (5.74ft)	1.5 m (4.92ft)	1.2 m (3.94ft)	1.0 m (3.28ft)

15.13.3 Materials

Materials exposed to sea water. Materials of components exposed to sea water, such as propeller blades, propeller hubs, and thruster body, are to have an elongation of not less than 15% on a test specimen, the gauge length of which is five times the diameter. A Charpy V impact test is to be conducted for materials other than bronze and austenitic steel. An average impact energy value of 20 J (2.04 kgf-m, 14.75 lbf-ft) taken from three tests is to be obtained at minus 10 °C (14 °F). For nodular cast iron, average impact energy of 10 J at minus 10 °C (14 °F) is required accordingly.

Materials exposed to sea water temperature. Materials exposed to sea water temperature are to be of ductile material. An average impact energy value of 20 J (2.04 kgf-m, 14.75 lbf-ft) taken from three tests is to be obtained at minus 10 °C (14 °F). This requirement applies to the propeller shaft, blade bolts, CP mechanisms, shaft bolts, strut-pod connecting bolts, etc. This does not apply to stoppers and surface hardened components, such as bearings and gear teeth. The nodular cast iron of a ferrite structure type may be used for relevant parts other than bolts. The average impact energy for nodular cast iron should be a minimum of 10 J at minus 10 °C (14 °F).

15.13.4 Design Loads

The given loads are intended for component strength calculations only and are total loads including ice- induced loads and hydrodynamic loads during propeller/ice interaction. The presented maximum loads are based on a worst-case scenario that occurs once during the service life of the vessel. Thus, the load level for a higher number of loads is lower. The values of the parameters in the formulae in this section are to be given in the units shown in the symbol list in *Propulsion machinery (symbols)*. If the highest point of the propeller is not at a depth of at least h_0 below the water surface when the ship is in ballast condition, the propulsion system should be designed according to ice class **I A** for ice classes **I B** and **I C**.

Design loads on propeller blades. F_b is the maximum force experienced during the lifetime of the ship that bends a propeller blade backwards when the propeller mills an ice block while rotating ahead. F_f is the maximum force experienced during the lifetime of the ship that bends a propeller blade forwards when the propeller mills an ice block while rotating ahead. These forces originate from different propeller/

ice interaction phenomena, not acting simultaneously. Hence, they are to be applied to one blade separately.

- Maximum backward blade force, F_b, for open propellers:

$$F_b = k \cdot [n \cdot D]^{0.7}\left[\frac{EAR}{Z}\right]^{0.3} \cdot D^2 \text{ kN (kgf, lbf) when } D \le D_{limit}$$
$$k = 27(2753.23,\ 245.48)$$

$$F_b = k \cdot [n \cdot D]^{0.7}\left[\frac{EAR}{Z}\right]^{0.3} \cdot D \cdot H_{ice}^{1.4} \text{ kN (kgf, lbf) when } D > D_{limit}$$
$$k = 23(2345.35,\ 130.01)$$

where

D_{limit}	$0.85 \cdot H_{ice}^{1.4}$ m
D_{limit}	$0.622 \times 0.85 \cdot H_{ice}^{1.4}\ (ft)$
n	nominal rotational speed (at MCR in free running condition) for a CP propeller and 85% of the nominal rotational speed (at MCR in free running condition) for an FP propeller.

- Maximum forward blade force, F_f, for open propellers:

$$F_f = k \cdot \left[\frac{EAR}{Z}\right] \cdot D^2 \text{ kN (kgf, lbf) when } D \le D_{limit}$$
$$k = 250(25492.9,\ 5221.36)$$

$$F_f = k \cdot \left[\frac{EAR}{Z}\right] \cdot D \cdot \frac{1}{\left(1 - \frac{d}{D}\right)} \cdot H_{ice} \text{ kN (kgf, lbf) when } D > D_{limit}$$
$$k = 500(50985.81,\ 10442.72)$$

where
$D_{limit} = \frac{2}{\left(1-\frac{d}{D}\right)} \cdot H_{ice}$ m (ft)

- Loaded area on the blade for open propellers. Load cases 1–4 are to be covered, as given in Table 15.4, for CP and FP propellers. To obtain blade ice loads for a reversing propeller, load case 5 also is to be covered for FP propellers.

- Maximum backward blade ice force, F_b, for ducted propellers:

$$F_b = k \cdot [n \cdot D]^{0.7} \cdot \left[\frac{EAR}{Z}\right]^{0.3} \cdot D^2 \text{ kN (kgf, lbf) when } D > D_{limit}$$
$$k = 9.5(968.73,\ 86.37)$$

Table 15.4 Load cases for open propellers

	Force	Loaded area	Right-handed propeller blade seen from behind
Load case 1	F_b	Uniform pressure applied on the back of the blade (suction side) to an area from 0.6*R* to the tip and from the leading edge to 0.2 times the chord length	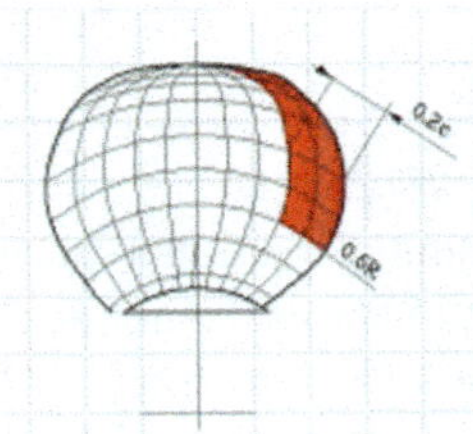
Load case 2	50% of F_b	Uniform pressure applied on the back of the blade (suction side) on the propeller tip area outside 0.9*R* radius	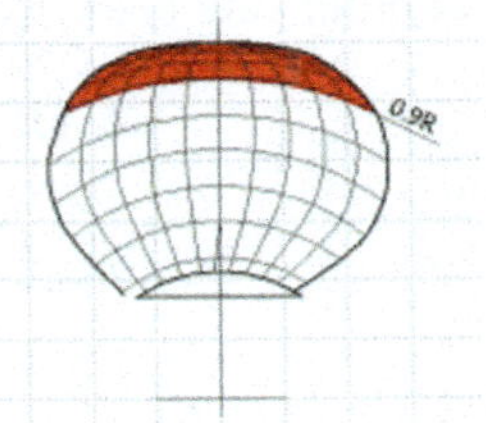
Load case 3	F_f	Uniform pressure applied on the blade face (pressure side) to an area from 0.6*R* to the tip and from the leading edge to 0.2 times the chord length	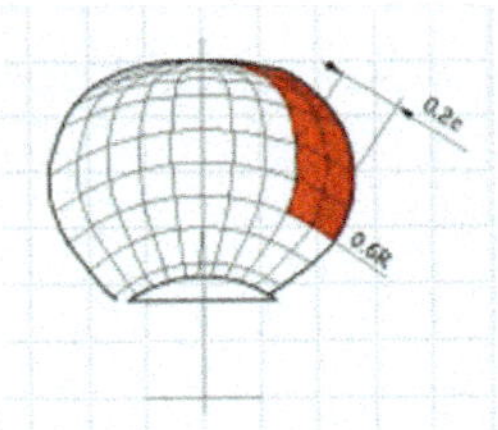
Load case 4	50% of F_f	Uniform pressure applied on propeller face (pressure side) on the propeller tip area outside 0.9*R* radius	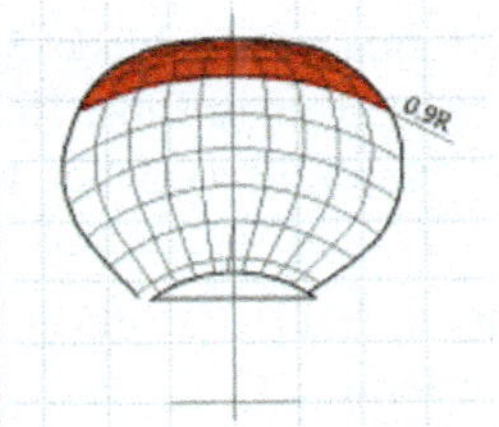
Load case 5	60% of F_f or F_b, whichever is greater	Uniform pressure applied on propeller face (pressure side) to an area from 0.6*R* to the tip and from the trailing edge to 0.2 times the chord length	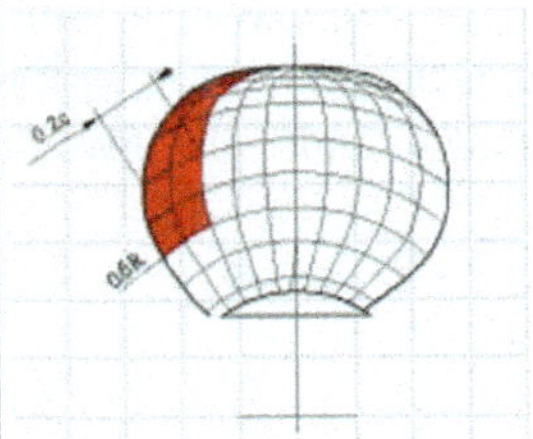

$$F_b = k \cdot [n \cdot D]^{0.7} \cdot \left[\frac{EAR}{Z}\right]^{0.3} \cdot D^{0.6} \cdot H_{ice}^{1.4} \text{ kN (kgf, lbf) when } D > D_{limit}$$

$$k = 66(6730.13, 600.06)$$

where

D_{limit} $4 \cdot H_{ice}$ m (ft)

n nominal rotational speed (at MCR in free running condition) for a CP propeller and 85% of the nominal rotational speed (at MCR in free running condition) for an FP propeller.

- Maximum forward blade ice force, F_f, for ducted propellers:

$$F_f = k \cdot \left[\frac{EAR}{Z}\right] \cdot D^2 \text{ kN (kgf, lbf) when } D \leq D_{limit}$$

$$k = 250(25492.91, 5221.35)$$

$$F_f = k \cdot \left[\frac{EAR}{Z}\right] \cdot D \cdot \frac{1}{\left(1 - \frac{d}{D}\right)} \cdot H_{ice} \text{ kN (kgf, lbf) when } D > D_{limit}$$

$$k = 500(50985.91, 10442.72)$$

where

$$D_{limit} = \frac{2}{\left(1 - \frac{d}{D}\right) \cdot H_{ice}} \text{ m (ft)}$$

- Loaded area on the blade for ducted propellers. Load cases 1 and 3 are to be covered as given in Table 15.5 for all propellers, and an additional load case (load case 5) for an FP propeller, to cover ice loads when the propeller is reversed.

- Maximum blade spindle torque, $Q_{s\,max}$, for open and ducted propellers. The spindle torque, $Q_{s\,max}$, around the axis of the blade fitting is to be determined both for the maximum backward blade force, F_b, and forward blade force, F_f which are applied as in Tables 15.4 and 15.5. The larger of the obtained torques is used as the dimensioning torque. If the above method gives a value which is less than the default value given by the formula below, the default value is to be used.

$$Q_{smax} = 0.25 \cdot F \cdot c_{0.7} \text{ kN-m (kgf-m, lbf-ft)}$$

where

c length of the blade section at $0.7R$ radius.

F either F_b or F_f, whichever has the greater absolute value.

Table 15.5 Load cases for ducted propellers

	Force	Loaded area	Right-handed propeller blade seen from behind
Load case 1	F_b	Uniform pressure applied on the back of the blade (suction side) to an area from 0.6*R* to the tip and from the leading edge to 0.2 times the chord length	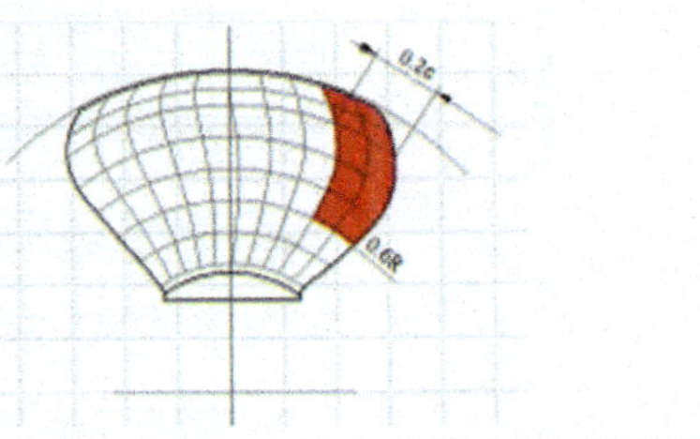
Load case 3	F_f	Uniform pressure applied on the blade face (pressure side) to an area from 0.6*R* to the tip and from the leading edge to 0.5 times the chord length	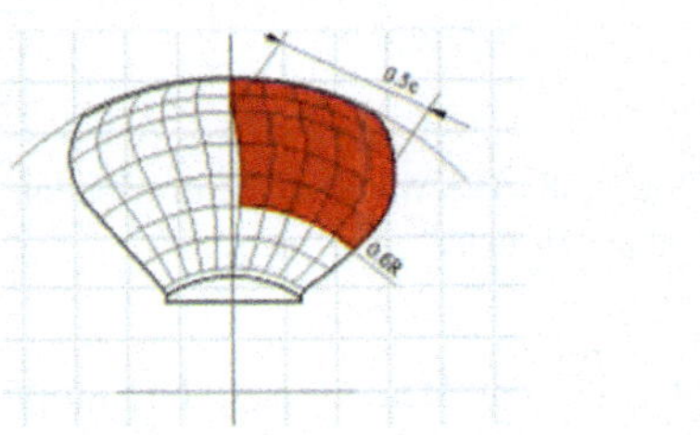
Load case 5	60% of F_f or F_b, whichever is greater	Uniform pressure applied on propeller face (pressure side) to an area from 0.6*R* to the tip and from the trailing edge to 0.2 times the chord length	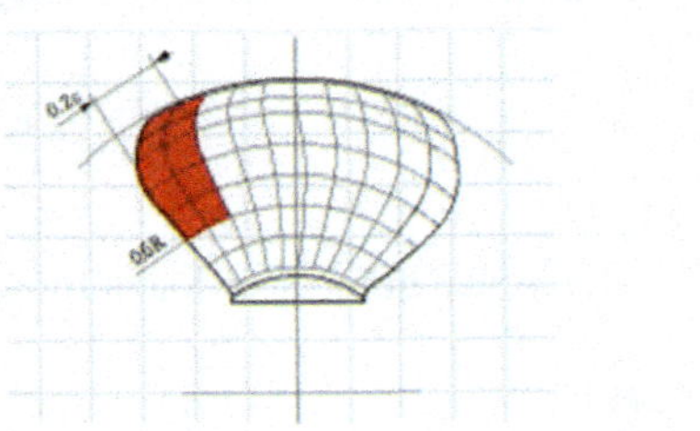

- Load distributions for blade loads. The Weibull-type distribution (probability that F_{ice} exceeds $(F_{ice})_{max}$), as given in Fig. 15.8, is used for the fatigue design of the blade:

$$P\left(\frac{F_{ice}}{(F_{ice})_{ice}} \geq \frac{F}{(F_{ice})_{max}}\right) = e\left(-\left(\frac{F}{F_{ice})_{max}}\right)^k \cdot 1n(N_{ice})\right)$$

where

k shape parameter of the spectrum
N_{ice} number of load cycles in the spectrum
F_{ice} random variable for ice loads on the blade, $0 \leq F_{ice} \leq (F_{ice})_{max}$.

The shape parameter $k = 0.75$ is to be used for the ice force distribution of an open propeller and the shape parameter $k = 1.0$ for that of a ducted propeller blade.

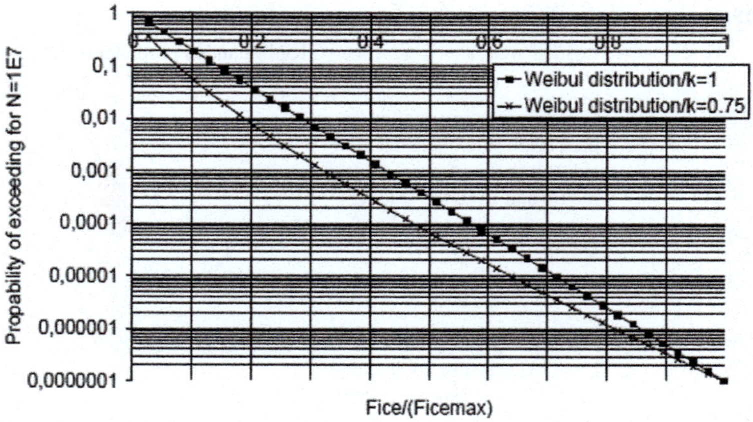

Fig. 15.8 The Weibull-type distribution (probability that F_{ice} exceeds $(F_{ice})_{max}$) that is used for fatigue design

- Number of ice loads. The number of load cycles per propeller blade in the load spectrum is to be determined according to the formula:

$$N_{ice} = k_1 k_2 k_3 N_{class} n_n$$

where

Reference number of loads for ice classes N_{class}:

Class	I AA	I A	I B	I C
Impacts in life/n_n	9 × 106	6 × 106	3.4 × 106	2.1 × 106

Propeller location factor k_1:

Centreline propeller bow first operation	Wing propeller bow first operation	Pulling propeller (wing and centreline) bow propeller or stern first operation
1	2	3

The submersion factor, k_2, is determined from the equation:

k_2 $0.8 - f$ when $f < 0$

$0.8 - 0.4f$ when $0 \leq f \leq 1$

$0.6 - 0.2f$ when $1 \leq f \leq 2.5$

0.1 when $f > 2.5$.

Where the immersion function f is:

$$f = \frac{h_0 - h_{ice}}{D/2} - 1$$
$$f = h_0 - H_{ice} - 1$$

where h_0 is the depth of the propeller centreline at the lower ice waterline (LIWL) of the vessel. Propulsion type factor k_3:

Type	Fixed	Azimuthing
k_3	1	1.2

For components that are subject to loads resulting from propeller/ice interaction with all the propeller blades, the number of load cycles (N_{ice}) is to be multiplied by the number of propeller blades (Z).

Axial design loads for open and ducted propellers. (1) *Maximum ice thrust on propeller.* The maximum forward and backward ice thrusts are:

$$T_f = 1.1 F_f \text{kN (kgf, lbf)}$$
$$T_b = 1.1 F_b \text{kN (kgf, lbf)}$$

(2) *Design thrust along the propulsion shaft line for open and ducted propellers.* The design thrust along the propeller shaft line is to be calculated with the formulae below. The greater value of the forward and backward direction loads is to be taken as the design load for both directions. The factors 2.2 and 1.5 consider the dynamic magnification resulting from axial vibration. In a forward direction:

$$T_r = T + 2.2 T_f \text{kN (kgf, lbf)}$$

In a backward direction:

$$T_r = 1.5 T_b \text{kN (kgf, lbf)}$$

If the hydrodynamic bollard thrust, T, is not known, T is to be taken as follows:

Propeller type	T
CP propellers (open)	$1.25T_n$
CP propellers (ducted)	$1.1T_n$
FP propellers driven by turbine or electric motor	T_n
FP propellers driven by diesel engine (open)	$0.85T_n$
FP propellers driven by diesel engine (ducted)	$0.75T_n$

Where T_n is the nominal propeller thrust at MCR in free running open water condition.

Torsional design loads

- ***Design ice torque on propeller*** Q $_{max}$ ***for open propellers.*** Q_{max} is the maximum torque on a propeller resulting from ice/propeller interaction during the service life of the vessel:

$$Q_{max} = k \cdot \left[1 - \frac{d}{D}\right] \cdot \left[\frac{P_{0.7}}{D}\right]^{0.16} \cdot (nD)^{0.17} \cdot D^3 \text{ kN-m (kgf-m, lbf-ft)}$$

When $D \leq D_{limit}$

$$k = 10.9(1111.49, 186.02)$$

$$Q_{max} = k \cdot \left[1 - \frac{d}{D}\right] \cdot \left[\frac{P_{0.7}}{D}\right]^{0.16} \cdot (nD)^{0.17} \cdot D^{1.9} \cdot H_{ice}^{1.1} \text{ kN-m(kgf-m, lbf-ft)}$$

When $D > D_{limit} D > D_{limit}$

$$k = 20.7(2110.81, 353.26)$$

where $D_{limit} = 1.8 \cdot H_{ice}$ m (ft).

n is the rotational propeller speed at MCR in bollard condition. If unknown, n is to be attributed a value in accordance with the following table.

Propeller type	Rotational speed, n
CP propellers	n_n
FP propellers driven by turbine or electric motor	n_n
FP propellers driven by diesel engine	$0.85n_n$

Where n_n is the nominal rotational speed at MCR in free running open water condition. For CP propellers, the propeller pitch, $P_{0.7}$ should correspond to MCR in bollard condition. If not known, $P_{0.7}$ is to be taken as $0.7P_{0.7n}$, where P0.7n is the propeller pitch at MCR in free running condition.

- ***Design ice torque on propeller*** Q_{max} ***for ducted propellers.*** Q_{max} is the maximum torque on a propeller during the service life of the ship resulting from ice/ propeller interaction.

$$Q_{max} = k \cdot \left[1 - \frac{d}{D}\right] \cdot \left[\frac{P_{0.7}}{D}\right]^{0.16} \cdot (nD)^{0.17} \cdot D^3 \text{ kN-m(kgf-m, lbf-ft)}$$

When $D \leq D_{limit}$

$$k = 7.7(785.18, 131.41)$$

$$Q_{max} = k \cdot \left[1 - \frac{d}{D}\right] \cdot \left[\frac{P_{0.7}}{D}\right]^{0.16} \cdot (nD)^{0.17} \cdot D^{1.9} \cdot H_{ice}^{1.1} \text{ kN-m (kgf-m, lbf-ft)}$$

When $D > D_{limit}$

$$k = 16.6(1488.78, 2749.16)$$

n rotational propeller at MCR speed in bollard condition. If not known, n is to have a value according to the table in *Propulsion machinery (design loads).*

For CP propellers, the propeller pitch, $P_{0.7}$ should correspond to MCR in bollard condition. If not known, $P_{0.7}$ is to be taken as $0.7P_{0.7n}$, where $P_{0.7n}$ is the propeller pitch at MCR in free running condition.

- ***Design torque for non-resonant shaft lines.*** If there is no relevant first blade order torsional resonance in the operational speed range or in the range 20% above and 20% below the maximum operating speeds (bollard condition), the following estimation of the maximum torque can be used. Directly coupled two stroke diesel engines without flexible coupling:

$$Q_{peak} = Q_{peak} + Q_{vib} + Q_{max} \cdot \frac{I_e}{I_t} \text{ kN-m(kgf-m, lbf-ft)}$$

And other plants
$Q_{peak} = Q_{emax} + Q_{max} \cdot \frac{I_e}{I_t}$ kN-m (kgf-m, lbf-ft)
Where

I_e equivalent mass moment of inertia of all parts on the engine side of the component under consideration
I_t equivalent mass moment of inertia of the whole propulsion system.

All the torques and the inertia moments should be reduced to the rotation speed of the component being examined. If the maximum torque, Q_{emax} is unknown, it is to be taken as follows:

- ***Design torque for shaft lines having resonances.*** If there is a first blade order torsional resonance in the operational speed range or in the range 20% above and 20% below the maximum operating speed (bollard condition), the design torque (Q_{peak}) of the shaft component is to be determined by means of torsional vibration analysis of the propulsion line. There are two alternative ways of performing the dynamic analysis.

1. Time domain calculation for estimated milling sequence excitation
2. Frequency domain calculation for blade orders sinusoidal excitation

The frequency domain analysis is considered conservative compared to the time domain simulation, if there is a first blade order resonance in the considered speed range.

- ***Time domain calculation of torsional response.*** Time domain calculations are to be performed for the MCR condition, MCR bollard conditions and for blade order resonant rotational speeds so that the resonant vibration responses can be obtained. The load sequence given herein, for a case where a propeller is milling an ice block, should be used for the strength evaluation of the propulsion line. The given load sequence is not intended for propulsion system stalling analyses. The following load cases are intended to reflect the operational loads on the propulsion system, when the propeller interacts with ice, and the respective reaction of the complete system. The ice impact and system response causes loads in the individual shaft line components. The ice torque Q_{max} may be taken as a constant value in the complete speed range. When considerations at specific shaft speeds are performed, a relevant Q_{max} may be calculated using the relevant speed according to section *Propulsion machinery (design loads)*. Diesel engine plants without an elastic coupling should be calculated at the least favourable phase angle for ice versus engine excitation, when calculated in the time domain. The engine firing pulses should be included in the calculations and their standard steady state harmonics can be used. If there is a blade order resonance just above the MCR speed, calculations are to cover rotational speeds up to 105% of the MCR speed. The propeller ice torque excitation for shaft line transient dynamic analysis in the time domain is defined as a sequence of blade impacts which are of half sine shape. The excitation frequency should follow the propeller rotational speed during the ice interaction sequence. The torque due to a single blade ice impact as a function of the propeller rotation angle is then defined using the formula:

$$Q(\varphi) = C_q \cdot Q_{max} \cdot \sin[\varphi(\frac{180}{\alpha \;\; i})] \text{ when } \varphi = 0 \ldots a_i$$

$$Q(\varphi) = 0 \text{ when } \varphi = a_i \ldots 360$$

where

φ the rotation angle from when the first impact occurs

$C\,q$ and α_i are given in the table below.

Torque excitation	Propeller/ice interaction	Cq	αi [deg]			
			Z = 3	Z = 4	Z = 5	Z = 6
Case 1	Single ice block	0.75	90	90	72	60
Case 2	Single ice block	1.0	135	135	135	135
Case 3	Two ice blocks (phase shift 360/(2·Z) deg.)	0.5	45	45	36	30
Case 4	Single ice block	0.5	45	45	36	30

αi is duration of propeller blade/ice interaction expressed in terms of the propeller rotation angle (refer to Fig. 15.9).

The total ice torque is obtained by summing the torque of single blades, while taking account of the phase shift 360 degrees/Z, see Fig. 15.10 or Fig. 15.11. At the beginning and end of the milling sequence (within the calculated duration) linear ramp functions should be used to increase C_q to its maximum value within one

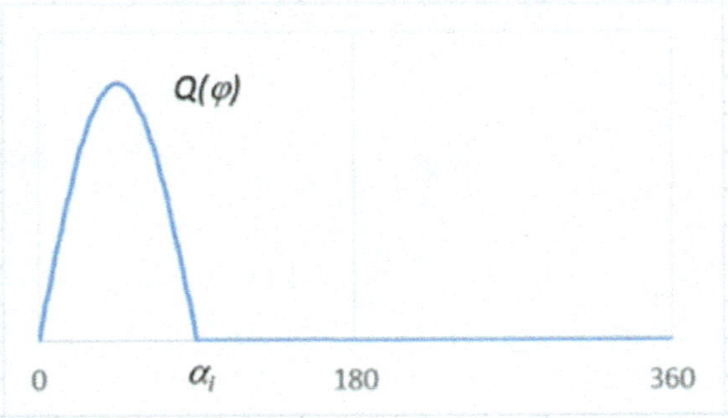

Fig. 15.9 Schematic ice torque due to a single blade ice impact as a function of the propeller rotation angle

propeller revolution and vice versa to decrease it to zero (see the examples of different Z numbers in Fig. 15.10 or Fig. 15.11). The number of propeller revolutions during a milling sequence is to be obtained from the formula:

$$N_Q = 2 \times H_{ice} \text{ where } H_{ice} \text{ in m}$$

$$N_Q = 0.3048 \times 2 \times H_{ice}\ H_{ice} \text{ in ft}$$

The number of impacts is $Z \cdot N_Q$ for blade order excitation. An illustration of all excitation cases for different numbers of blades is given in Fig. 15.10 or Fig. 15.11. A dynamic simulation is to be performed for all excitation cases at the operational rotational speed range. For a fixed pitch propeller propulsion plant, a dynamic simulation should also cover the bollard pull condition with a corresponding rotational speed assuming the maximum possible output of the engine. If a speed drop occurs until the main engine is at a standstill, this indicates that the engine may not be sufficiently powered for the intended service task. For the consideration of loads, the maximum occurring torque during the speed drop process is to be used.

For the time domain calculation, the simulated response torque typically includes the engine mean torque and the propeller mean torque. If this is not the case, the response torques must be obtained using the formula:

$$Q_{peak} = Q_{emax} + Q_{rtd} \text{ kN-m(kgf-m, lbf-ft)}$$

where Q_{rtd} is the maximum simulated torque obtained from the time domain analysis.

- ***Frequency domain calculation of torsional response***. For frequency domain calculations, blade order and twice-the-blade-order excitation may be used. The amplitudes for the blade order and twice-the-blade-order sinusoidal excitation have been derived based on the assumption that the time domain half sine impact sequences were continuous, and the Fourier series components for blade order and twice-the-blade-order components have been derived. The propeller ice torque is then:

$$Q_F((\varphi) = Q_{max} C_{q0} + C_{q1} \sin(Z E_{0\varphi} + \alpha_1) + C_{q2} \sin(2Z E_{0\varphi} + \alpha 2)) \text{ kN-m(kgf-m, lbf-ft)}$$

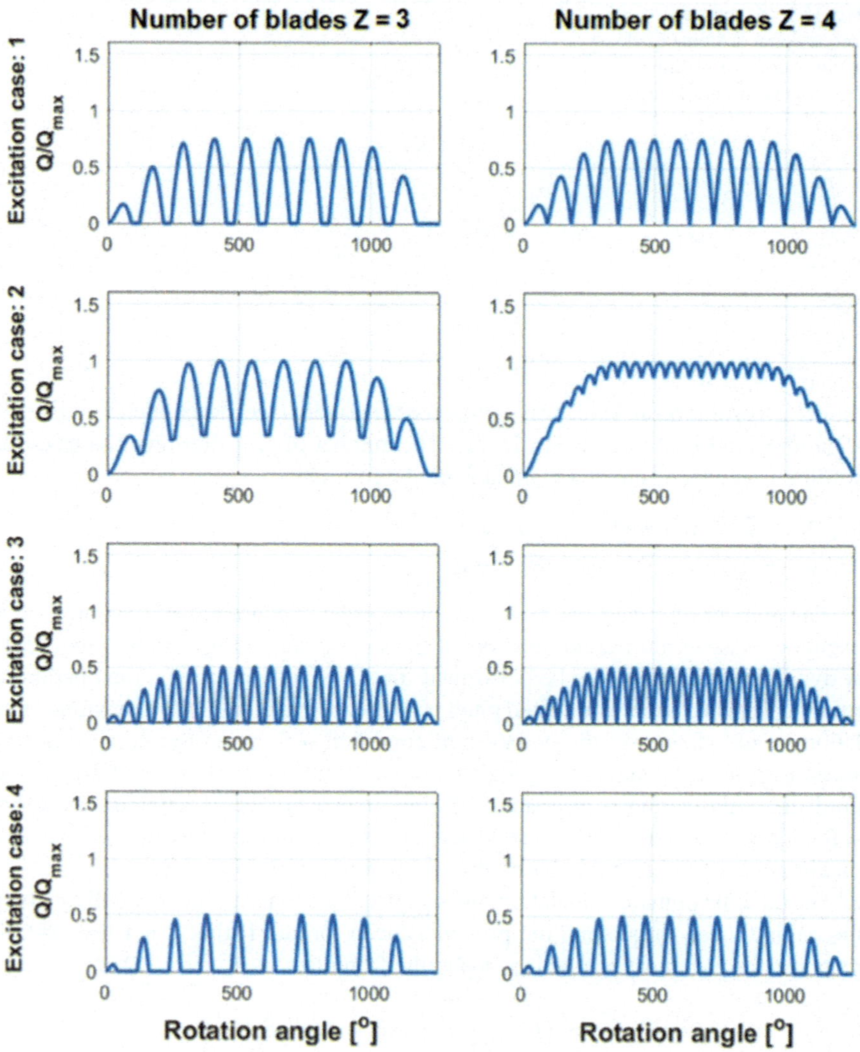

Fig. 15.10 The shape of the propeller ice torque excitation sequences for propellers with 3 or 4 blades

where

C_{q0}	mean torque parameter
C_{q1}	first blade order excitation parameter
C_{q2}	second blade order excitation parameter
α_1, α_2	phase angles of the excitation component
φ	angle of rotation
E_0	number of ice blocks in contact.

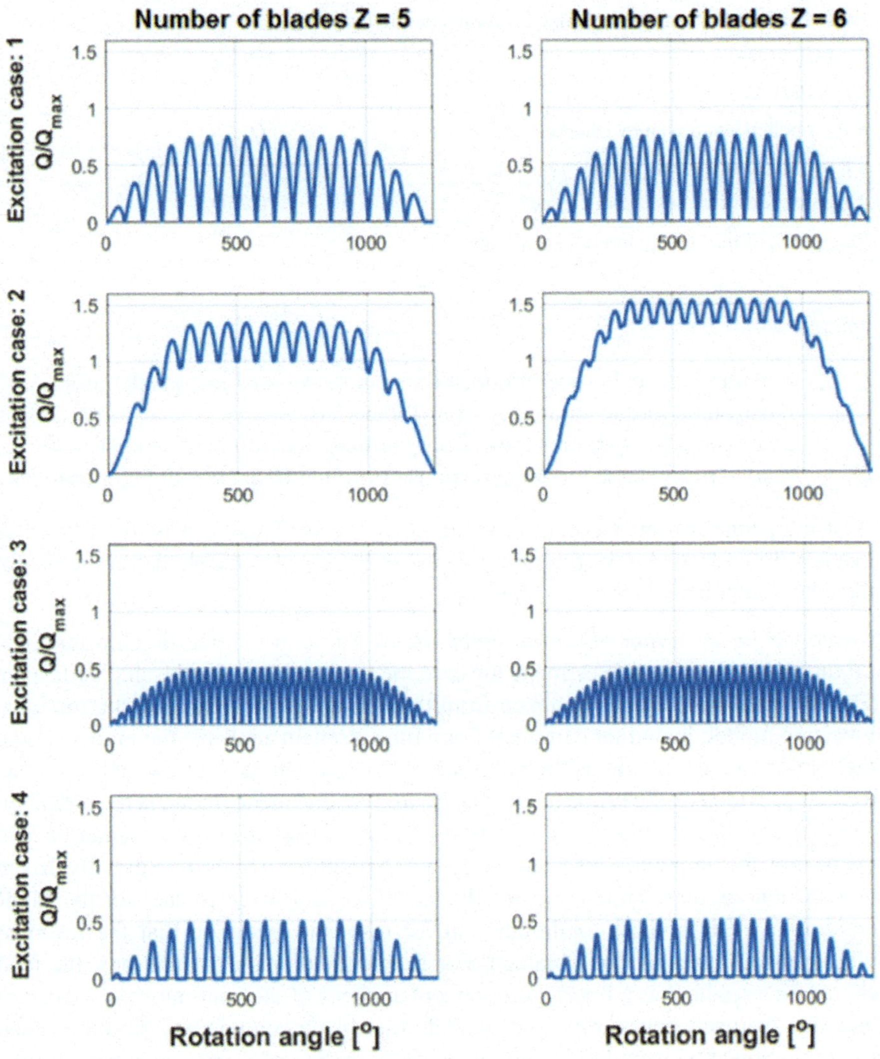

Fig. 15.11 The shape of the propeller ice torque excitation sequences for propellers with 5 or 6 blades

The values of the parameters are given in the following table:

Table 15.6 coefficient values for frequency domain excitation calculation.

The design torque for the frequency domain excitation case is to be obtained using the formula:

$$Q_{peak} = Q_{emax} + Q_{vib} + \left(Q_{max}^{n} C_{q0}\right)^{I_e}/_{I_t} + Q_{rf1} + Q_{rf2} \text{ kN-m(kgf-m, lbf-ft)}$$

Table 15.6 Default values for prime mover maximum torque Q_{emax}

Propeller type	Q_{emax}
Propellers driven by electric motor	Q_{motor}
CP propellers not driven by electric motor	Q_n
FP propellers driven by turbine	Q_n
FP propellers driven by diesel engine	$0.75Q_n$

Note Q_{motor} is the electric motor peak torque

where

Q^n_{max} maximum propeller ice torque at the operation speed in consideration.
C_{q0} mean static torque parameter from Table 15.6
Q_{rf1} blade order torsional response from the frequency domain analysis.
Q_{rf2} second order blade torsional response from the frequency domain analysis.

If the prime mover maximum torque, Q_{emax}, is not known, it is to be taken as given in Table 15.6. All the torque values are to be scaled to the shaft revolutions for the component in question.

Guidance for torsional vibration calculation. The aim of time domain torsional vibration simulations is to estimate the extreme torsional load for the ship's lifespan. The simulation model can be taken from the normal lumped mass elastic torsional vibration model, including damping. For a time, domain analysis, the model should include the ice excitation at the propeller, other relevant excitations and the mean torques provided by the prime mover and hydrodynamic mean torque in the propeller. The calculations should cover variation of phase between the ice excitation and prime mover excitation. This is most relevant to propulsion lines with directly driven combustion engines. Time domain calculations should be calculated for the MCR condition, MCR bollard conditions and for resonant speed, so that the resonant vibration responses can be obtained. For frequency domain calculations, the load should be estimated as a Fourier component analysis of the continuous sequence of half sine load sequences. First and second order blade components should be used for excitation. The calculation should cover the entire relevant rpm range and the simulation of responses at torsional vibration resonances.

Blade failure load

Bending force, F_{ex}. The ultimate load resulting from blade failure because of plastic bending around the blade root is to be calculated with the formula below, or alternatively by means of an appropriate stress analysis, reflecting the non-linear plastic material behaviour of the actual blade. In such a case, the blade failure area may be outside the root section. The ultimate load is acting on the blade at the $0.8R$ radius in the weakest direction of the blade. A blade is regarded as having failed if the tip is bent into an offset position by more than 10% of propeller diameter D.

$$F_{ex} = \frac{k \cdot c \cdot t^2 \cdot \sigma_{ref1}}{0.8 \cdot D - 2 \cdot r} \text{ kN kgf, lbf}$$
$$k = 300(300000, 43.20)$$

where

σ_{ref1} $0.6 \cdot \sigma_{0.2} + 0.4 \cdot \sigma_u$ MPa (kgf/cm^2, psi).
σ_u minimum ultimate tensile strength to be specified on the drawing.
$\sigma_{0.2}$ minimum yield or 0.2% proof strength to be specified on the drawing.

c, *t*, and *r* are, respectively, the actual chord length, maximum thickness, and radius of the cylindrical root section of the blade, which is the weakest section outside the root fillet typically located at the point where the fillet terminates at the blade profile.

Spindle torque, Q_{sex}. The maximum spindle torque due to a blade failure load acting at 0.8*R* should be determined. The force that causes blade failure typically reduces when moving from the propeller centre towards the leading and trailing edges. At a certain distance from the blade centre of rotation, the maximum spindle torque will occur. This maximum spindle torque should be defined by an appropriate stress analysis or using the equation given below.

$$Q_{sex} = \max\left[C_{LE0.8;} 0.8 C_{TE0.8}\right] C_{spex} F_{ex} \text{ kN-m(kgf-m, lbf-ft)}$$

where

C_{spex} $C_{sp} C_{fex} = 0.7\left(1 - \left(\frac{4EAR}{Z}\right)^3\right)$.

If C_{spex} is below 0.3, a value of 0.3 should be used for C_{spex}.

C_{sp} non-dimensional parameter taking account of the spindle arm
C_{fex} non-dimensional parameter taking account of the reduction of the blade failure force at the location of the maximum spindle torque
$C_{LE0.8}$ is the leading-edge portion of the chord length at 0.8*R*
$C_{TE0.8}$ is the trailing edge portion of the chord length at 0.8*R*.

Figure 15.12 illustrates the spindle torque values due to blade failure loads across the entire chord length.

15.13.5 Propeller Design

The strength of the propulsion line is to be designed according to the pyramid strength principle. This means that the loss of the propeller blade is not to cause any considerable damage to other propeller shaft line components.

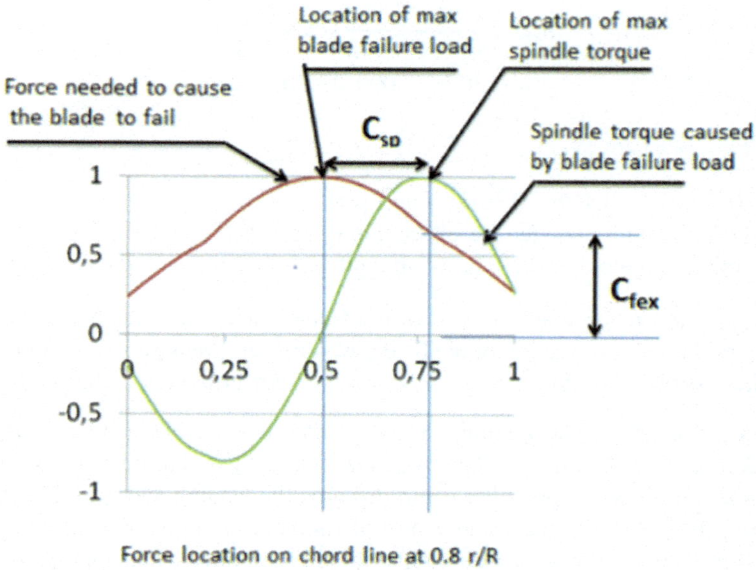

Fig. 15.12 Blade failure load and the related spindle torque when the force acts at a different location on the chord line at radius 0.8*R*

15.13.6 Calculation of Propeller Blade Stresses

The blade stresses are to be calculated for the design loads given in *Propulsion machinery (design loads)*. Finite element analyses are to be used for stress analysis for final approval for all propellers. The following simplified formulae can be used in estimating the blade stresses for all propellers at the root area ($r/R < 0.5$). The root area dimensions will be accepted even if the FEM analysis would show greater stresses at the root area.

$$\sigma_{st} = C_1 \frac{M_{BL}}{k \cdot ct^2} \text{ Mpa}\left(\text{kgf/cm}^2, \text{ psi}\right)$$
$$k = 10^2\left(10^3, 14.4\right)$$

where constant C_1 is the "actual stress"/ "stress obtained with beam equation". If the actual value is not available, C_1 should be taken as 1.6.

M_{BL} $(0.75 - r/R) \cdot R \cdot F$ for relative radius $r/R < 0.5$.

F is the maximum of F_b and F_f, whichever is greater.

Acceptability criterion. The following criterion for calculated blade stresses is to be fulfilled.

$$\frac{\sigma_{ref2}}{\sigma_{st}} \geq 1.3$$

where

σ_{st} calculated stress for the design loads. If FE analysis is used in estimating the stresses, von Mises stresses should be used.

σ_{ref2} reference stress, defined as:

$0.7 \cdot \sigma_u$ or

$0.6 \cdot \sigma_{0.2} + 0.4 \cdot \sigma_u$, whichever is less.

Fatigue design of propeller blade (Note: SI units). The fatigue design of the propeller blade is based on an estimated load distribution for the service life of the ship and the S–N curve for the blade material. An equivalent stress that produces the same fatigue damage as the expected load distribution should be calculated and the acceptability criterion for fatigue should be fulfilled as given in this section. The equivalent stress is normalised for 10^8 (100 million) cycles. For materials with a two-slope SN curve (Fig. 15.13), if the following criterion is fulfilled, fatigue calculations according to this section are not required.

$$\sigma_{exp} \geq B1 \cdot \sigma_{ref2}^{B2} \cdot \log(N_{ice})^{B3}$$

where $B1$, $B2$, and $B3$ are coefficients for open and nozzle propellers are given in the table below.

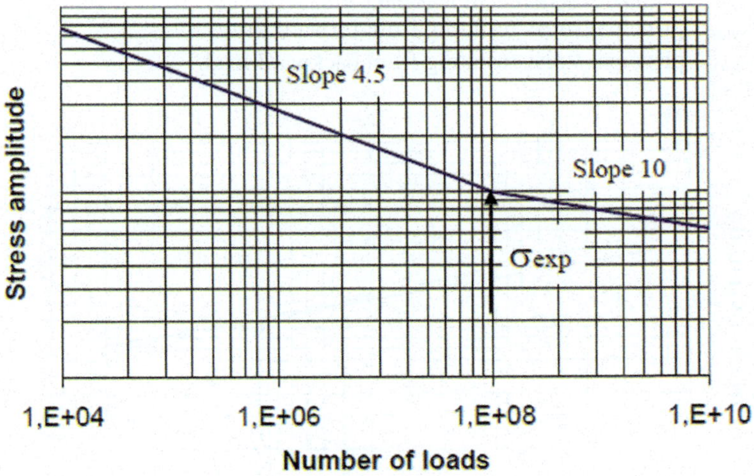

Fig. 15.13 Two-slope S–N curve

	Open propeller	Nozzle propeller
*B*1	0.00328	0.00223
*B*2	1.0076	1.0071
*B*3	2.101	2.471

For calculation of equivalent stress, two types of S–N curves are available:

1. Two slope S–N curve (slopes 4.5 and 10), see Fig. 15.13.
2. One slope S–N curve (the slope can be chosen), see Fig. 15.14.

The type of the S–N curve should be selected to correspond to the material properties of the blade. If the S–N curve is not known the two slope S–N curve is to be used.

Equivalent fatigue stress. The equivalent fatigue stress for 10^8 (100 million) stress cycles which produces the same fatigue damage as the load distribution for the service life of the ship is:

$$\sigma_{fat} = \rho \cdot (\sigma_{ice})_{max}$$

where

$(\sigma_{ice})_{max}$ mean value of the principal stress amplitudes resulting from design forward and backward blade forces at the location being studied.

$0.5 \cdot \left[(\sigma_{ice})_{fmax} - (\sigma_{ice})_{bmax}\right]$.

$(\sigma_{ice})_{fmax}$ principal stress resulting from forward load.

$(\sigma_{ice})_{bmax}$ principal stress resulting from backward load.

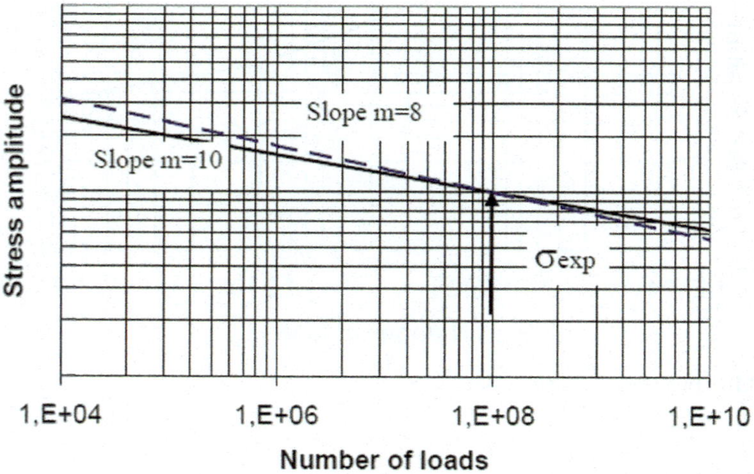

Fig. 15.14 Constant-slope S–N curve

In calculation of $(\sigma_{ice})max$, case 1 and case 3 (or case 2 and case 4) are considered as a pair for $(\sigma_{ice})fmax$, and $(\sigma_{ice})bmax$ calculations. Case 5 is excluded from the fatigue analysis.

Calculation of ρ parameter for two-slope S–N curve. The parameter ρ relates the maximum ice load to the distribution of ice loads according to the regression formula:

$$\rho = C_1 \cdot (\sigma_{ice})_{max}^{C_2} \cdot \sigma_{ft}^{C_3} \cdot \log(N_{ice})^{C_4}$$

where

σ_{fl} $\gamma_{\varepsilon 1} \cdot \gamma_{\varepsilon 2} \cdot \gamma_v \cdot \gamma_m \cdot \sigma_{exp}$.
$\gamma_{\varepsilon 1}$ reduction factor due to scatter (equal to one standard deviation).
$\gamma\varepsilon 2$ reduction factor for test specimen size effect.
γ_v reduction factor for variable amplitude loading.
γm reduction factor for mean stress.
σ exp mean fatigue strength of the blade material at 10^8 cycles to failure in seawater.

The following values are to be used for the reduction factors if actual values are not available:

$$\gamma_\varepsilon = \gamma_{\varepsilon 1} \cdot \gamma_{\varepsilon 2} = 0.67, \gamma_v = 0.75, \text{ and } \gamma_m = 0.75.$$

The coefficients C_1, C_2, C_3, and C_4 are given in Table 15.7. The applicable range of N_{ice} for calculating ρ is $5 \times 10^6 \leq N_{ice} \leq 10^8$.

Calculation of ρ parameter for constant-slope S–N curve. For materials with a constant-slope S–N curve, see Fig. 15.14, the ρ- factor is to be calculated with the following formula:

$$\rho = \left(G\frac{N_{ice}}{N_R}\right)^{1/m}[In(N_{ice})]^{-1/k}$$

where

k shape parameter of the Weibull distribution

1.0 for ducted propellers

0.75 for open propellers

N_R reference number of load cycles (= 108).

Table 15.7 Coefficients C

	Open propeller	Ducted propeller
C1	0.000747	0.000534
C2	0.0645	0.0533
C3	– 0.0565	– 0.0459
C4	2.22	2.584

The applicable range of N for calculating ρ is $5 \times 106 \leq N \leq 10^8$.

Values for the G parameter are given in Table 15.8. Linear interpolation may be used to calculate the G value for other *m/k* ratios than given in the Table 15.8.

Acceptability criterion for fatigue. The equivalent fatigue stress at all locations on the blade is to fulfil the following acceptability criterion:

$$\frac{\sigma_{fl}}{\sigma_{fat}} \geq 1.5$$

where

σ_{fl} $\gamma_\varepsilon \cdot \gamma_{\varepsilon 2} \cdot \gamma_v \cdot \gamma_m \cdot \sigma_{exp}$
$\gamma\varepsilon_1$ reduction factor due to scatter (equal to one standard deviation)
$\gamma_{\varepsilon 2}$ reduction factor for test specimen size effect
γ_v reduction factor for variable amplitude loading
γ_m reduction factor for mean stress
σ_{exp} mean fatigue strength of the blade material at 10^8 cycles to failure in seawater.

The following values are to be used for the reduction factors if actual values are not available: $\gamma_\varepsilon = \gamma_{\varepsilon 1} \cdot \gamma_{\varepsilon 2} = 0.67$, $\gamma_v = 0.75$, and $\gamma_m = 0.75$.

Propeller bossing and CP mechanism. The blade bolts, the CP mechanism, the propeller boss, and the fitting of the propeller to the propeller shaft should be designed to withstand the maximum and fatigue design loads, as defined in *Propulsion machinery (design loads).* the safety factor against yielding should be greater than 1.3 and that against fatigue greater than 1.5. In addition, the safety factor for loads resulting from loss of the propeller blade through plastic bending as defined in *Propulsion machinery (design loads)*is to be greater than 1.0 against yielding.

Propulsion shaft line. The shafts and shafting components, such as the thrust and stern tube bearings, couplings, flanges and sealings, are to be designed to withstand the propeller/ice interaction loads as given in *Propulsion machinery (design loads).* the safety factor is to be at least 1.3 against yielding for extreme operational loads, 1.5 for fatigue loads and 1.0 against yielding for the blade failure load.

Shafts and shafting components. The ultimate load resulting from total blade failure as defined in *Propulsion machinery (design loads)* is not to cause yielding in shafts

Table 15.8 Value for the G parameter for different *m/k* ratios

m/k	*G*	*m/k*	*G*	*m/k*	*G*	*m/k*	*G*
3	6	5.5	287.9	8	40,320	10.5	11.899×10^6
3.5	11.6	6	720	8.5	119,292	11	39.917×10^6
4	24	6.5	1871	9	362,880	11.5	136.843×10^6
4.5	52.3	7	5040	9.5	1.133×10^6	12	479.002×10^6
5	120	7.5	14,034	10	3.629×10^6	–	–

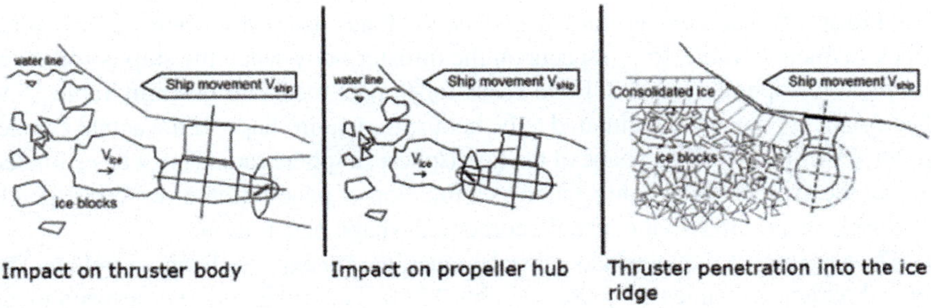

Fig. 15.15 Examples of load scenario types

and shaft components. The loading should consist of the combined axial, bending, and torsion loads, wherever this is significant. The minimum safety factor against yielding is to be 1.0 for bending and torsional stresses. ***Note:*** *the requirements in this section are complementary to those described in Section 4-3-2 of the ABS Rules for Building and Classing Marine Vessels (2023). For fatigue evaluation, cumulative fatigue analyses are to be performed (see Area of operation for recommended method/ practice).The applicable* Q_{peak} *and the corresponding load spectrum should be determined for the component or connection in question, as described in Propulsion machinery (design loads).*

Azimuthing main propulsors. In addition to the above requirements for propeller blade dimensioning, azimuthing thrusters must be designed for thruster body/ice interaction loads. Load formulae are given for estimating once- in-a-lifetime extreme loads on the thruster body, based on the estimated ice condition and ship operational parameters. Two main ice load scenarios have been selected for defining the extreme ice loads. Examples of loads are illustrated in Fig. 15.15. In addition, blade order thruster body vibration responses may be estimated for propeller excitation. The following load scenario types are considered:

1. Ice block impact on the thruster body or propeller hub
2. Thruster penetration into an ice ridge that has a thick consolidated layer
3. Vibratory response of the thruster at blade order frequency.

The steering mechanism, the fitting of the unit, and the body of the thruster should be designed to withstand the plastic bending of a blade without damage. The loss of a blade must be considered for the propeller blade orientation causing the maximum load on the component being studied. Top-down blade orientation typically places the maximum bending loads on the thruster body.

Extreme ice impact loads. When the ship is operated in ice conditions, ice blocks formed in channel side walls or from the ridge consolidated layer may impact on the thruster body and the propeller hub. Exposure to ice impact is very much dependent on the ship size and ship hull design, as well as the location of the thruster. The contact force will grow in terms of thruster/ice contact until the ice block reaches the ship speed. The thruster must withstand the loads occurring when the design ice block defined in Table 15.3 impacts on the thruster body when the ship is sailing at a typical ice operating speed. Load cases for impact loads are given in Table 15.9. The contact geometry is estimated to be hemispherical in shape. If the actual contact geometry differs from the shape of the hemisphere, a sphere radius must be estimated so that the growth of the contact area as a function of penetration of ice corresponds as closely as possible to the actual geometrical shape penetration.

The ice impact contact load must be calculated using the formula below. The related parameter values are given in Table 15.10. The design operation speed in ice can be derived from Tables 15.11 and 15.12, or the ship in question's actual design operation speed in ice can be used. The longitudinal impact speed in Tables 15.11 and 15.12 refers to the impact in the thruster's main operational direction. For the pulling propeller configuration, the longitudinal impact speed is used for load case T2, impact on hub; and for the pushing propeller unit, the longitudinal impact speed is used for load case T1, impact on thruster end cap. For the opposite direction, the impact speed for transversal impact is applied (Fig. 15.16).

$$F_{ti} = C_{DMI} 34.5 R_C^{0.5} (m_{ice} v_s^2)^{0.333} \text{ KN}$$

where

R_c impacting part sphere radius, in m (ft), refer to Fig. 15.16
m_{ice} ice block mass, in kg (lbs)
v_s ship speed at the time of contact, in m/s (knots)
C_{DMI} dynamic magnification factor for impact loads. If unknown, it should be taken from Table 15.11.

For impacts on non-hemispherical areas, such as the impact on the nozzle, the equivalent impact sphere radius must be estimated using the equation below.

$$R_{ceq} = \sqrt{\frac{A}{\pi}} \text{ m (ft)}$$

If the $2R_{ceq}$ is greater than the ice block thickness, the radius is set to half of the ice block thickness. For the impact on the thruster side, the pod body diameter can be used as a basis for determining the radius. For the impact on the propeller hub, the hub diameter can be used as a basis for the radius.

Extreme ice loads on thruster hull when penetrating an ice ridge. In icy conditions, ships typically operate in ice channels. When passing other ships, ships may be subject to loads caused by their thrusters penetrating ice channel walls. There is usually a consolidated layer at the ice surface, below which the ice blocks are loose. In addition, the thruster may penetrate ice ridges when backing. Such a situation is in the case of **I AA** ships, because they may operate independently in difficult ice conditions. However, the thrusters in ships with lower ice classes may also have to withstand such a situation, but at a remarkably lower ship speed. In this load

Table 15.9 Load cases for azimuthing thruster ice impact loads

	Force	Loaded area	
Load case T1a Symmetric longitudinal ice impact on thruster	F_{ti}	Uniform distributed load or uniform pressure, which are applied symmetrically on the impact area	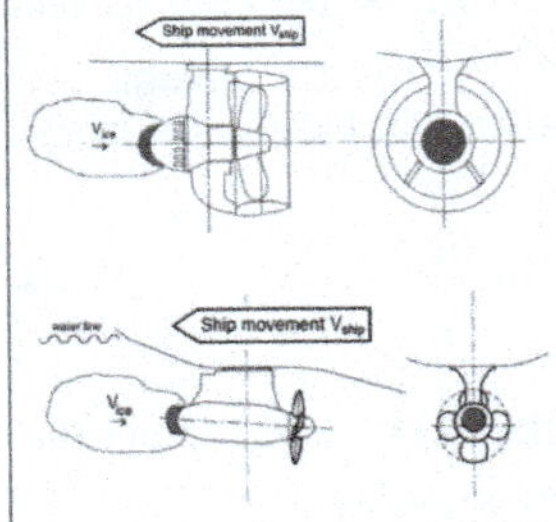
Load case T1b Non-symmetric longitudinal ice impact on thruster	50% of F_{ti}	Uniform distributed load or uniform pressure, which are applied on the other half of the impact area	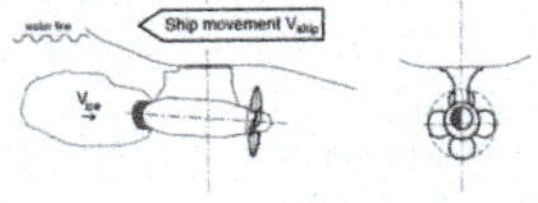
Load case T1c Non-symmetric longitudinal ice impact on nozzle	F_{ti}	Uniform distributed load or uniform pressure, which are applied on the impact area. Contact area is equal to the nozzle thickness (H_{nz}) the contact height (H_{ice})	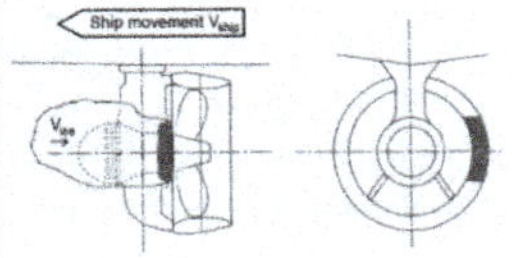
Load case T2a Symmetric longitudinal ice impact on propeller hub	F_{ti}	Uniform distributed load or uniform pressure, which are applied symmetrically on the impact area	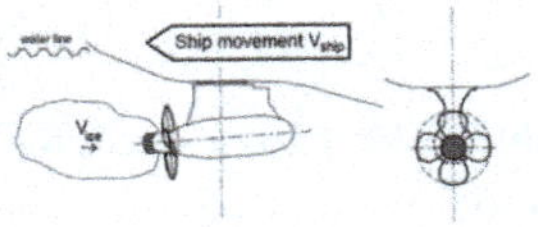
Load case T2b Non-symmetric longitudinal ice impact on propeller hub	50% of F_{ti}	Uniform distributed load or uniform pressure, which are applied on the other half of the impact area	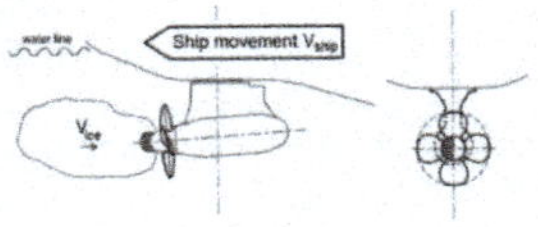
Load case T3a Symmetric lateral ice impact on thruster body	F_{ti}	Uniform distributed load or uniform pressure, which are applied symmetrically on the impact area	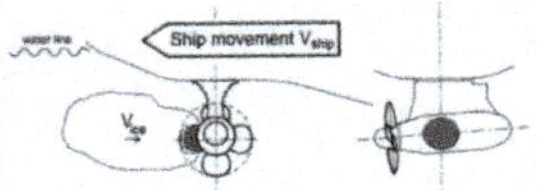
Load case T3b Non-symmetric lateral ice impact on thruster body or nozzle	F_{ti}	Uniform distributed load or uniform pressure, which are applied on the impact area. Nozzle contact radius R to be taken from the nozzle length (L_{nz})	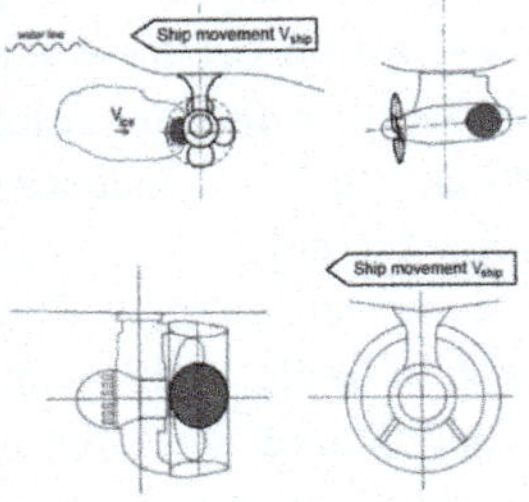

Table 15.10 Parameter values for ice dimensions and dynamic magnification

	I AA	I A	I B	I C
Thickness of the design ice block impacting thruster (2/3 of H_{ice})	1.17 m 3.84 ft	1.0 m 3.28 ft	0.8 m 2.62 ft	0.67 m 2.2 ft
Extreme ice block mass (m_{ice})	8670 kg 19,114 lb	5460 kg 12,037 lb	2800 kg 6173 lb	1600 kg 3527 lb
C_{DMI} (if not known)	1.3	1.2	1.1	1

Table 15.11 Impact speeds for aft centreline thruster

Aft centreline thruster	I AA	I A	I B	I C
Longitudinal impact in main operational direction	6 m/s 11.67 kt	5 m/s 9.72 kt	5 m/s 9.72 kt	5 m/s 9.72 kt
Longitudinal impact in reversing direction (pushing unit propeller hub or pulling unit cover end cap impact)	4 m/s 7.78 kt	3 m/s 5.83 kt	3 m/s 5.83 kt	3 m/s 5.83 kt
Transversal impact in bow first operation	3 m/s 5.83 kt	2 m/s 3.89 kt	2 m/s 3.89 kt	2 m/s 3.89 kt
Transversal impact in stern first operation (double acting ship)	4 m/s 7.78 kt	3 m/s 5.83 kt	3 m/s 5.83 kt	3 m/s 5.83 kt

Table 15.12 Impact speeds for aft wing, bow centreline and bow wing thrusters

Aft wing, bow centreline and bow wing thruster	I AA	I A	I B	I C
Longitudinal impact in main operational direction	6 m/s 11.67 kt	5 m/s 9.72 kt	5 m/s 9.72 kt	5 m/s 9.72 kt
Longitudinal impact in reversing direction (pushing unit propeller hub or pulling unit cover end cap impact)	4 m/s 7.78 kt	3 m/s 5.83 kt	3 m/s 5.83 kt	3 m/s 5.83 kt
Transversal impact	4 m/s 7.78 kt	3 m/s 5.83 kt	3 m/s 5.83 kt	3 m/s 5.83 kt

scenario, the ship is penetrating a ridge in thruster first mode with an initial speed. This situation occurs when a ship with a thruster at the bow moves forward, or a ship with a thruster astern move in backing mode. The maximum load during such an event is considered the extreme load. An event of this kind typically lasts several seconds, due to which the dynamic magnification is considered negligible and is not considered. The load magnitude must be estimated for the load cases shown in Table 15.13, using the equation after Table 15.13. The parameter values for calculations are given in Tables 15.14 and 15.15. The loads must be applied as uniform distributed load or uniform pressure over the thruster surface. The design

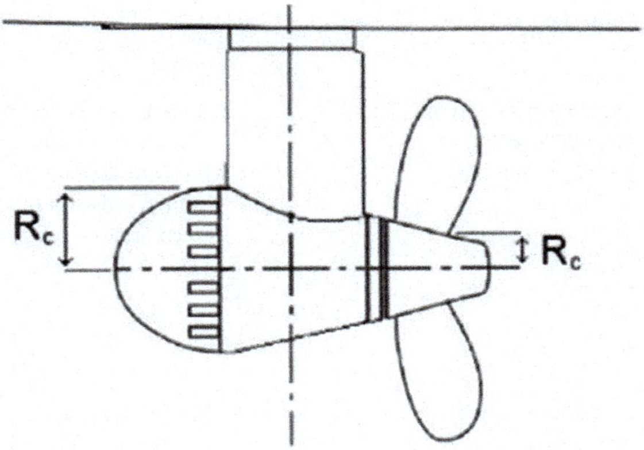

Fig. 15.16 Dimensions used for R_c

operation speed in ice can be derived from Tables 15.14 or Table 15.15. Alternatively, the actual design operation speed in ice of the ship in question can be used.

$$F_{tr} = 32 v_s{}^{0.66} H_r{}^{0.9} A_t^{0.74} \text{ kN}$$

$$F_{tr} = 274.432 v_s{}^{0.66} H_r{}^{0.9} A_t^{0.74} \text{ lbf}$$

where

v_s ship speed, in m/s (knots)
H_r design ridge thickness (the thickness of the consolidated layer is 18% of the total ridge thickness), in m (ft)
A_t projected area of the thruster, in m^2 (ft^2).

When calculating the contact area for thruster-ridge interaction, the loaded area in the vertical direction is limited to the ice ridge thickness, as shown in Fig. 15.17.

Acceptability criterion for static loads. The stresses on the thruster must be calculated for the extreme once-in-a-lifetime loads described in *Propulsion machinery (design).* the nominal von Mises stresses on the thruster body must have a safety margin of 1.3 against the yielding strength of the material. At areas of local stress concentrations, stresses must have a safety margin of 1.0 against yielding. The slewing bearing, bolt connections and other components must be able to maintain operability without incurring damage that requires repair when subject to the loads given in *Propulsion machinery (design)* multiplied by a safety factor of 1.3.

Thruster body global vibration. Evaluating the global vibratory behaviour of the thruster body is important, if the first blade order excitations are in the same frequency range with the thruster global modes of vibration, which occur when the propeller rotational speeds are in the high-power range of the propulsion line. This evaluation is mandatory, and it must be shown that there is either no global first blade order

Table 15.13 Load cases for ridge ice loads

	Force	Loaded area	
Load case T4a Symmetric longitudinal ridge penetration loads	F_{tr}	Uniform distributed load or uniform pressure, which are applied symmetrically on the impact area	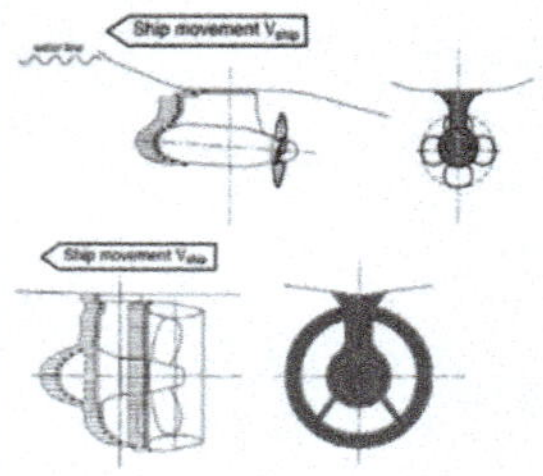
Load case T1b Non-symmetric longitudinal ridge penetration loads	50% of F_{tr}	Uniform distributed load or uniform pressure, which are applied on the other half of the contact area	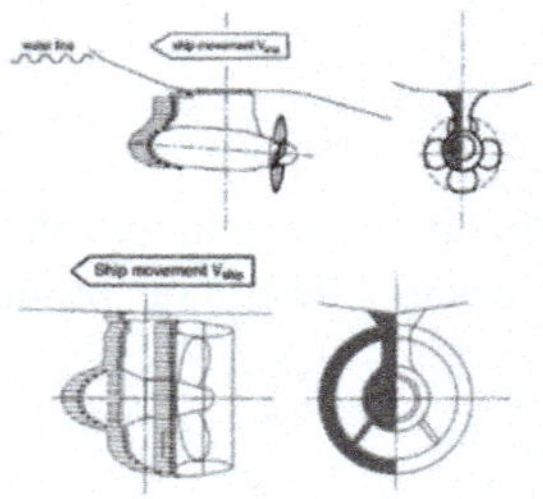
Load case T5a Symmetric lateral ridge penetration loads for ducted azimuthing unit and pushing open propeller unit	F_{tr}	Uniform distributed load or uniform pressure, which are applied symmetrically on the contact area	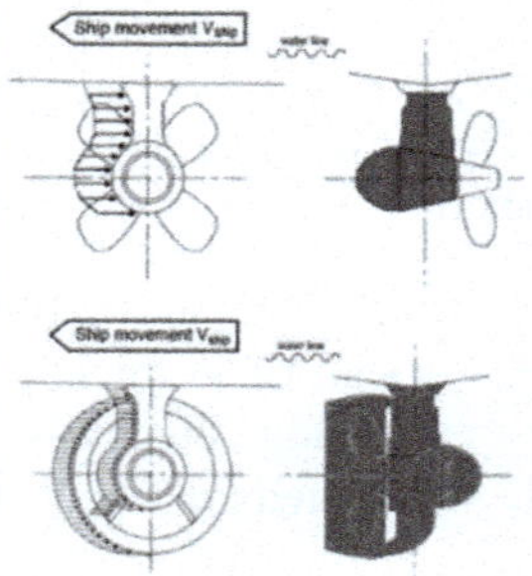
Load case T5b Nonsymmetric lateral ridge penetration loads for all azimuthing units	50% of F_{tr}	Uniform distributed load or uniform pressure, which are applied on the other half of the contact area	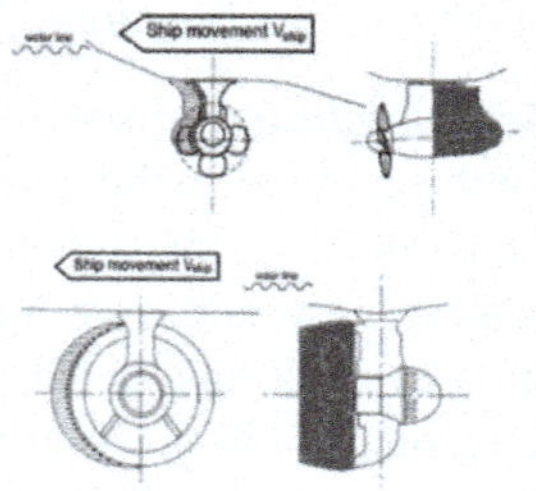

Table 15.14 Parameters for calculating maximum loads when the thruster penetrates an ice ridge aft thruster (bow first operation)

	I AA	I A	I B	I C
Thickness of the design ridge consolidated layer	1.5 m 4.92 ft	1.5 m 4.92 ft	1.2 m 3.94 ft	1.0 m 3.28 ft
Total thickness of the design ridge, H_r	8 m 26.25 ft	8 m 26.25 ft	6.5 m/s 21.33 ft	5 m 16.40 ft
Initial ridge penetration speed (longitudinal loads)	6 m/s 11.66 kt	4 m/s 7.78 kt	4 m/s 7.78 kt	4 m/s 7.78 kt
Initial ridge penetration speed (transversal loads)	3 m/s 5.83 kt	2 m/s 3.89 kt	2 m/s 3.989 kt	2 m/s 3.89 kt

Table 15.15 Parameters for calculating maximum loads when the thruster

	I AA	I A	I B	I C
Thickness of the design ridge consolidated layer	1.5 m 4.92 ft	1.5 m 4.92 ft	1.2 m 3.94 ft	1.0 m 3.28 ft
Total thickness of the design ridge, H_r	8 m 26.25 ft	8 m 26.25 ft	6.5 m/s 21.33 ft	5 m 16.40 ft
Initial ridge penetration speed (longitudinal loads)	4 m/s 7.78 kt	2 m/s 3.89 kt	2 m/s 3.89 kt	2 m/s 3.89 kt
Initial ridge penetration speed (transversal loads)	2 m/s 3.89 kt	1 m/s 1.94 kt	1 m/s 1.94 kt	1 m/s 1.94 kt

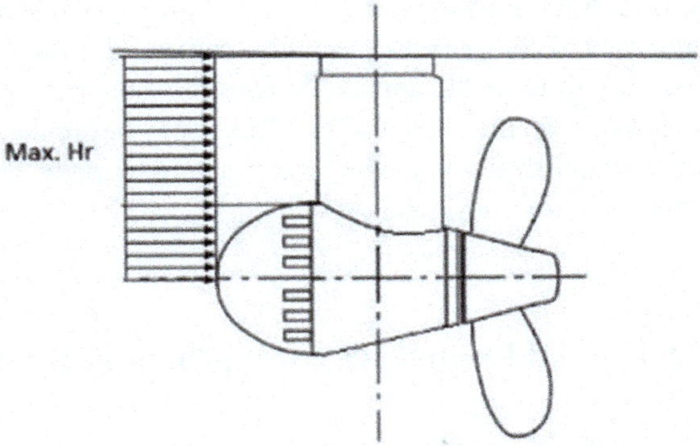

Fig. 15.17 Schematic showing the reduction of the contact area by the maximum ridge thickness

resonance at high operational propeller speeds (above 50% of maximum power) or that the structure is designed to withstand vibratory loads during resonance above 50% of maximum power. When estimating thruster global natural frequencies in the longitudinal and transverse direction, the damping and added mass due to water must be considered. In addition to this, the effect of ship attachment stiffness must be modelled.

15.13.7 Alternative Design Procedure

As an alternative to *Propulsion machinery (design loads)* and *Propulsion machinery (design)*, a comprehensive design study may be conducted to the satisfaction of the Administration. The study is to be based on ice conditions given for different ice classes in *Propulsion machinery (design ice conditions)*. it is to include both fatigue and maximum load design calculations and fulfil the pyramid strength principle, as given in *Propulsion machinery (design)*.

Loading. Loads on the propeller blade and propulsion system should be based on an acceptable estimation of hydrodynamic and ice loads.

Design levels. The analysis is to indicate that all components transmitting random (occasional) forces, excluding propeller blade, are not subjected to stress levels more than the yield stress of the component material, with a reasonable safety margin. Cumulative fatigue damage calculations are to indicate a reasonable safety factor. Due account is to be taken of material properties, stress raisers, and fatigue enhancements. Vibration analysis is to be conducted and is to demonstrate that the overall dynamic system is free from harmful torsional resonances resulting from propeller/ice interaction.

15.14 Tunnel Thrusters

Where **APS**, **PAS** or dynamic positioning systems **DPS** notations are assigned, the mechanical components of a tunnel thruster (i.e., propellers, gears, shafts, couplings, etc.) are to meet the applicable requirements of *Propulsion systems* in this section. Alternatively, the relevant Class Rules may be applied to the mechanical components of a tunnel thruster when a comprehensive study to determine the effect of ice is submitted for consideration.

15.15 Additional Ice Strengthening Requirements

15.15.1 Starting Arrangements

The capacity of the air receivers required for reversible propulsion engines is to be sufficient for at least twelve consecutive starts and that for non-reversible propulsion engines is to be sufficient for six consecutive starts of each engine. If the air receivers supply systems other than starting the propulsion engines, the additional capacity of the receivers is to be sufficient for continued operations of these systems after the capacity for the required number of consecutive engines starts has been used. The capacity of the air compressors is to be sufficient for charging the air receivers from atmospheric to full pressure in one hour. For a vessel with Ice Class **I AA** that requires its propulsion engines to be reversed for astern operations, the compressors are to be able to charge the air receivers in half an hour.

15.15.2 Sea Inlet, Cooling Water Systems and Fire Main

The sea water system is to be designed to ensure a supply of water for the cooling water system and for at least one of the fire pumps when navigating in ice. For this purpose, at least one sea water inlet chest is to be arranged as follows.

(1) The sea inlet should be situated near the centreline of the ship and well aft, if possible
(2) Guidance for designing the volume of the chest should be around one cubic metre (35.3 cubic feet) for every 750 kW (1,033 mhp; 1,019 hp) in engine output of the ship, including the output of auxiliary engines necessary for the operation of the ship
(3) The sea chest should be sufficiently high to allow ice to accumulate above the inlet pipe
(4) A pipe for discharge cooling water, allowing full capacity discharge, should be connected to the sea chest
(5) The accessible area of the strainer plates should be no less than four (4) times the inlet pipe sectional area.

Where it is impractical to meet the requirements of *Additional ice strengthening requirements (sea inlet, cooling water systems and fire main) design)* above, two smaller sea chests may be arranged for alternating the intake and discharge of the cooling water, provided the provisions outlined in *Additional ice strengthening requirements (sea inlet, cooling water systems and fire main) design)* are complied with. Heating coils, if necessary, may be installed in the upper part of the sea chest. The use of ballast water for cooling purposes while in the ballast condition may be acceptable as an additional means but is not to be considered a permanent substitute for the above required sea inlet chest or chests.

Part IV
Crew Health, Safety and Welfare

Chapter 16
Extreme Low Temperature Safety

16.1 Introduction

Due in part to the significant environmental conditions that a vessel and its crew are operating in, particular attention must be paid to the vessel's personnel so that they remain effective in performing their duties safely and efficiently. Working in freezing weather environments has significant implications on human capabilities and unless proper precautions are made can be hazardous to a person's health. In recognition of these implications on human health and performance due to working in cold climes, this chapter is offered to provide basic information on human performance and health hazards; information for design or selection of clothing; design of equipment to be operated in frigid conditions; information that can be used to help generate freezing weather operations' safety and operating procedures; and information that can be used to preserve the health of persons working in cold environments. The information that follows is simply information and carries no weight as Rules or requirements leading to Notations or Certifications. The information is provided for those vessel designers and owners/operators that would like information of the sort provided as a reference to consider during ship design and operation. The material in this chapter is applicable to extreme freezing weather working conditions. Some of the hazards discussed here are a concern under exceptional circumstances which force personnel to stay in cold temperature for extended periods. Adequate supplies of protective clothing and thermal insulating materials are to be provided for all persons onboard at any time. For example, hand protection (such as mittens and or gloves) are to be provided; head and eye protection gear is to be provided to reduce loss of body heat and protect vision from ultraviolet rays; head and eye protection is to be compatible and usable with communications equipment; foot protection gear is to be provided. Slip-resistant, insulated safety footwear is to be provided. For heavy work, a felt-lined (or similar insulating material) rubber-bottomed, leather-topped boot with a removable felt insole is preferred. An extra pair of safety shoes for inside work is to be provided. Personnel protective equipment is to be properly maintained and stored.

A. Olsen, *Ship Operations in Extreme Low Temperature Environments*, Springer Series on Naval Architecture, Marine Engineering, Shipbuilding and Shipping 19,
https://doi.org/10.1007/978-3-031-52513-1_16

It is worth reminding ourselves of some of the key terms and concepts associated with extreme cold temperature safety before delving deeper into the subject content. *Atrium*: A body cavity or chamber, especially either of the upper chambers of the heart that receives blood from the veins and forces it into a ventricle. *Chilblains*: An inflammation followed by itchy irritation on the hands, feet, or ears, resulting from exposure to moist cold. *Clo-units*: The rate of a person's heat loss as measured in watts per square meter of skin area per kelvin of temperature difference across the clothing; the value of insulation is measured by the reciprocal of this rate, in square meter kelvins per watt (m^2-K/W). *Ergonomics*: The applied science of equipment design, as for the workplace, intended to maximise productivity by reducing operator fatigue and discomfort. *Fibrillation*: Rapid uncoordinated twitching movements that replace the normal rhythmic contraction of the heart and may cause a lack of circulation and pulse. *Hypothermia*: Abnormally low body temperature. *Melatonin*: A hormone that plays a role in sleep. *Physiologic*: Being in accord with or characteristic of the normal functioning of a living organism. *Raynaud's Sign*: A bluish discoloration of the extremities; can occur when a spasm of the blood vessels is caused by exposure to cold or by strong emotion. *Triglyceride*: A naturally occurring ester of three fatty acids and glycerol that is the chief constituent of fats and oils. *Ventricle*: The chamber on the right side of the heart that receives venous blood from the right atrium and forces it into the pulmonary artery; any of the interconnecting cavities of the brain. *Wind Chill*: A still-air temperature that would have the same cooling effect on exposed human flesh as a given combination of temperature and wind speed; called also chill factor, wind chill factor, wind chill index.

16.2 Human Response to Cold and Arctic Exposure

The core (trunk) of the human body should remain within a small temperature range for healthy function. Excessive cooling or excessive heating will result in abnormal cardiovascular and neurological function. The skin is the organ through which a person regulates body temperature. With an average skin temperature of 33 °C (91.4 °F), conductive heat loss occurs at temperatures below this value. Therefore, it is easy to see how freezing weather performance can significantly influence normal body function. As a person cools:

- Metabolism is increased to generate more body heat—as cooling continues a person will begin to "shiver"—a visible sign that body cooling has progressed beyond a comfortable level. Increased metabolism will reduce the amount of time a person can sustain work.
- Safe manual materials handling tasks require the use of tactile senses, hand dexterity, strength, and coordination. Decreases in the ability to produce force, exhibit fine control over objects, and sustain muscular workloads occur in cold working environment.

- Work in cold environments is related to an increased risk for musculoskeletal injury.
- Motor function impairments of the arms and hands will occur long before cognitive or hypothermic- related disabilities occur. Impaired cognitive performance will lead to poor decision-making and increased risk for accident.
- Persons suffering from arthritis or rheumatism will experience increased levels of pain during freezing weather operations.

16.2.1 Decreases in Cognitive/Reasoning Ability Due to Cold Exposure

According to Pilcher, Nadler and Busch's "Effects of hot and cold temperature exposure on performance: A meta-analytic review", *Ergonomics*, (2002) 45 (10), moderately cold conditions [below 10 °C (50 °F)] created an average decrease of 14% in cognitive performance (reasoning, learning and memory tasks) across 22 studies chosen for the analysis. As operational temperatures decrease, the frequency of cognitive error will increase. Tasks requiring vigilance may be hampered after prolonged exposure to cold. Decision verification procedures should be implemented. Freezing weather operations, coupled with other physical distracters, such as noise or motion environments, will influence the quality of perception, memory and reasoning and compound the risk of decision-making error.

16.2.2 Health Hazards Related to Cold Exposure

The list of potential injuries and issues for occupational work in cold environments is lengthy. Personnel should have adequate training to enhance preparation for work in cold environments. Proper planning and precaution can deter the potential risks of cold work.

Non-freezing injury. In non-freezing injuries, parts of the skin are chilled but not frozen. Frostnip is a cold injury in which the chilled areas of skin become numb, swollen and red. The only treatment needed is warming the area for a few minutes. During warming, the area may hurt or itch intensely. No permanent damage results, although sometimes the area is particularly sensitive too cold for months or years afterward. Immersion foot (trench foot) is a cold injury that develops when a foot is kept in wet, cold socks and boots for several days. The foot is pale, clammy and cold. After warming, the foot becomes red and painful to the touch. Sometimes blisters develop, which may become infected. Rarely, this type of injury occurs in the hands. Treatment consists primarily of gently warming, drying and cleaning the foot; elevating it; and keeping it dry and warm. Medical doctors (aboard or ashore) should be consulted regarding need for treatment with antibiotics to prevent infection or whether a tetanus booster is recommended if the person's tetanus vaccination is not current. Trench foot can often be prevented by changing socks and drying the feet at least daily. Chilblains (pernio) is an uncommon reaction that may occur with repeated exposure to cold. Symptoms include itching, pain and, in rare cases, discoloured areas or blisters on the affected area (usually the leg). The condition is uncomfortable and recurrent. Preventing exposure to cold is the best treatment. Treatment includes re-warming, using blankets, lukewarm baths or heating pads positioned at low heat setting, which will allow blood flow to **return** to the peripheral tissues (Figs. 16.1, 16.2, 16.3, 16.4 and 16.5).

Freezing injury. Damage to tissue during cold exposure is commonly caused by freezing of the tissue and surrounding area. Tissue freezing is more commonly referred to as frostbite and is a common occurrence in colder climates. Frostbite occurs most often at ambient temperature below – 20 °C (– 4 °F). Frostbite occurs as a result of limbs and facial areas being exposed to low ambient temperatures and

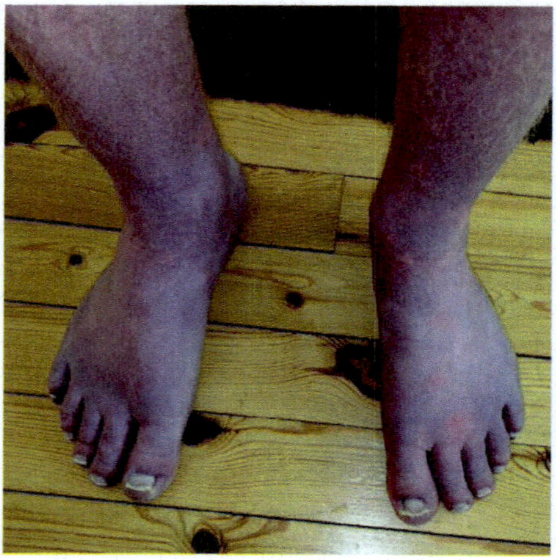

Fig. 16.1 Dependent acrocyanosis in a Norwegian 33-year-old male

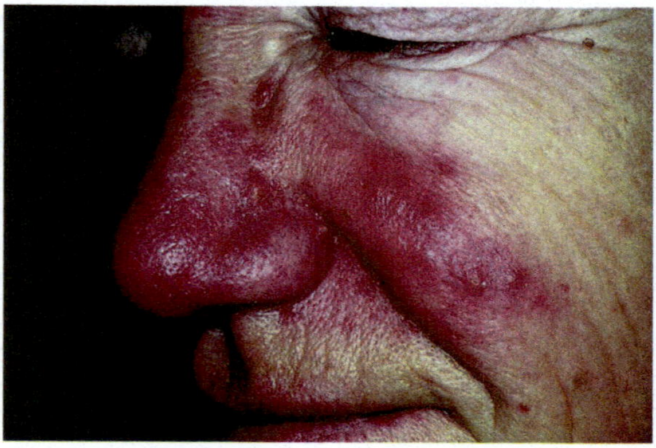

Fig. 16.2 Rosacea

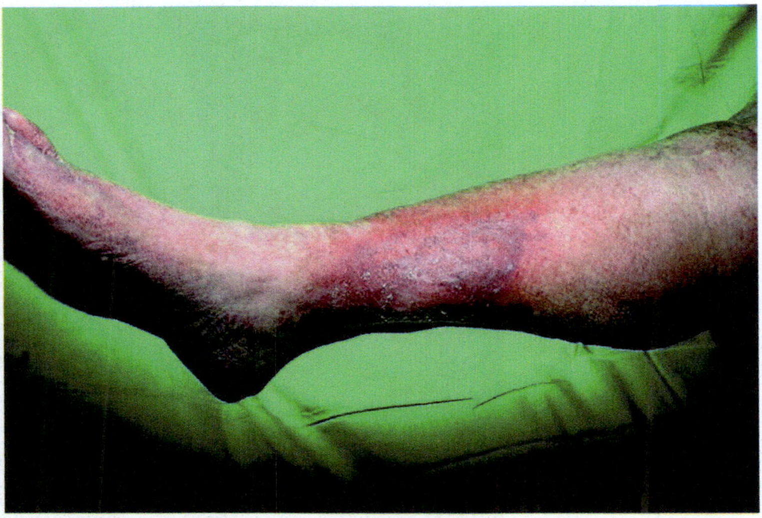

Fig. 16.3 Sclerosing panniculitis

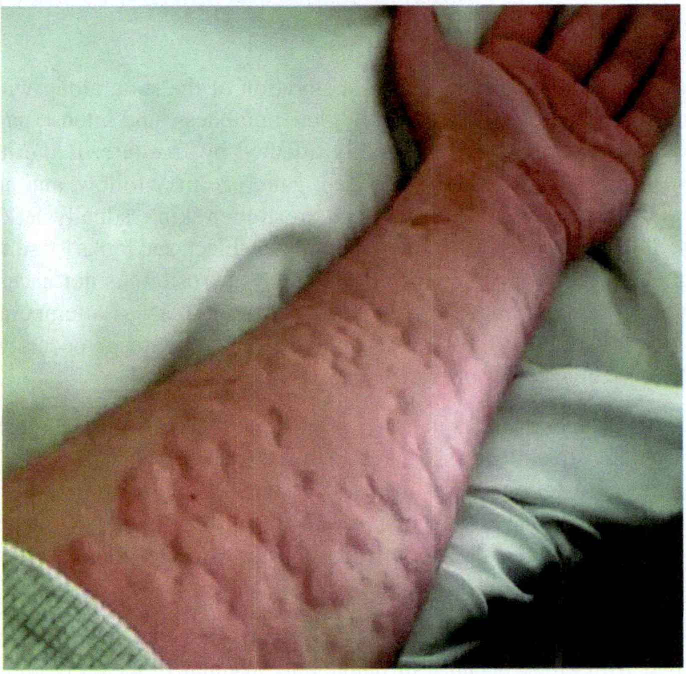

Fig. 16.4 Cold-induced urticaria

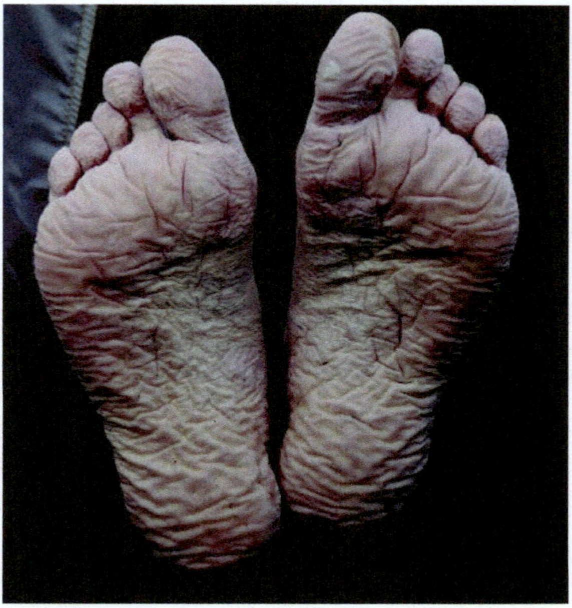

Fig. 16.5 Mild case of trench foot

wind. Frostbite is accompanied with discoloration of the skin, along with burning and/or tingling sensations, partial or complete numbness, and intense pain. Hands, feet, noses and ears are more commonly affected by frostbite. If the nerves and blood vessels have been severely damaged, gangrene may follow, and amputation may eventually be required. If left untreated, frostbitten skin gradually darkens after a few hours. Skin destroyed by frostbite is completely black and looks loose and flayed, as if burnt. The crucial factor in the development of frostbite is not necessarily the ambient temperature and wind, but whether the tissue reaches a temperature of − 4.8 °C to − 7.8 °C (23.4 °F to 18 °F). Cellular structural damage occurs when cells freeze. There are four levels of frostbite:

- First-degree frostbite results in vesicles (blisters) that damage only the outer skin
- Second-degree frostbite is associated with full thickness skin loss
- Third-degree frostbite involves the skin and underlying tissue
- Fourth-degree frostbite is the worst and results in freezing to the bone. This level of frostbite typically results in amputation of the affected tissue.

Diagnosis of whether superficial or deep tissue was damaged involves analysis of blisters; superficial injury will have clear blisters, while deeper tissue injury will have blisters filled with blood because of damage to vascular tissues. Frostbite requires emergency medical care. Procedures should be in place to provide first aid for casualties of frostbite. Means to get medical advice (from physicians either aboard or ashore) should be in place as part of these procedures. A person who has frostbite should be covered with a warm blanket and given a hot beverage because people with frostbite may also have hypothermia. Rubbing the area (particularly with snow)

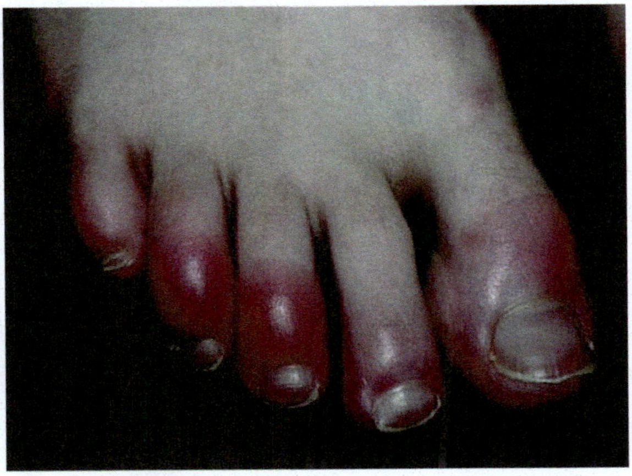

Fig. 16.6 Chillblain

leads to further tissue damage. Because the area has no sensation, it should not be warmed in front of a fire or with a heating pad or electric blanket. The frostbitten area becomes extremely painful on warming and medical attention will be urgently needed for pain management. Blisters should not be broken. If blisters break, they should be covered with antibiotic ointment. It is common for individuals who experience severe frostbite to later experience such problems as: hypersensitivity to cold, cold feet, burning/prickling/itching or tingling sensations, chronic pain, loss of sensation of touch, white fingers, excessive sweating, pain when walking, and other such problems causing a hindrance to work performance. Causes are thought to be related to permanent nerve and tissue damage, and in severe frostbite there can even be degeneration of bone and onset of arthritis (Figs. 16.6 and 16.7).

Raynaud's sign (white finger). Raynaud's sign represents a disease and relates to the body's hypersensitivity to cold, causing the arterial musculature to spasm and constrict peripheral blood flow at a much higher level than typical for a given ambient temperature. Less severe cases of Raynaud's sign, referred to as primary Raynaud's, are idiopathic (a medical condition of unknown origin), while secondary cases occur as a result of an underlying physiologic condition or environmental exposure. Secondary Raynaud's sign often occurs as a result of thickening in underlying tissue and vascular beds through mechanisms such as previous severe frostbite or exposure to vibrating tools. The term most often associated with vibration-induced impairment in hand-arm circulation is termed vibration-induced white-finger. White-finger describes the loss of colour, increased numbness, and decreased tactile sense associated with circulation impairment for individuals exposed to vibration and/or cold. At the end of a white-finger attack, blood flow returns to the area causing redness, swelling and pain. In severe cases, the symptoms can cause debilitation of performance. The condition may take a few months or several years to develop, depending on the individual physiologic tolerance and the level of cold. With the decreased ability for blood to be delivered to the peripheral areas, there are often symptoms of burning/throbbing pain, numbness, stiffness, diminished sensations

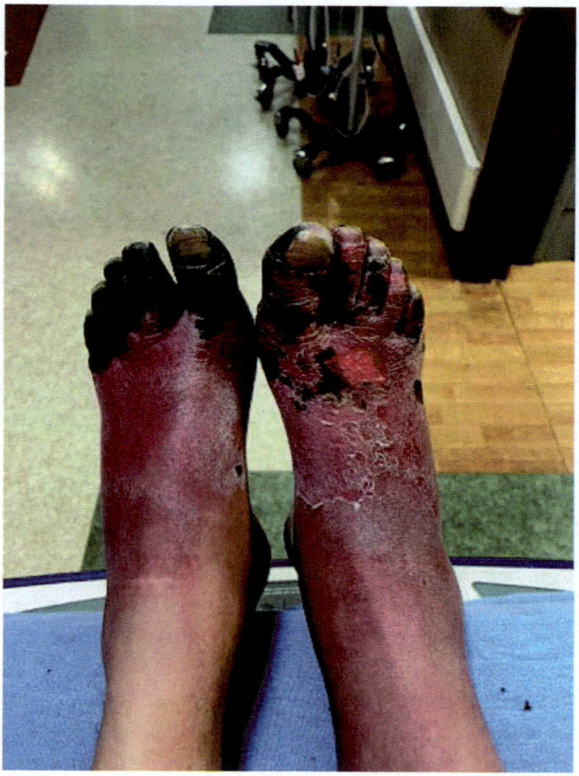

Fig. 16.7 Example of frostbite casualty

and swelling from fluid pooling. These symptoms, even at mild levels, are likely to make manual performance uncomfortable and unreliable. There is currently no known cure for Raynaud's sign. However, there are known intervention techniques to alleviate symptoms. Exercise and using warming devices to increase blood flow to affected areas has been successful (Figs. 16.8 and 16.9).

Hypothermia. Hypothermia is a rapid, progressive mental and physical collapse due to the body's warming mechanisms failing to maintain normal body temperatures. While hypothermia is often associated with immersion in freezing water, it can also occur in air when suitable freezing weather protection is not employed. Conditions of extremely low dry-ambient temperature or mildly cold ambient temperatures with wind and dampness can lead to a general cooling effect on the body. If metabolic heat production is less than the gradient of heat loss to the environment, hypothermia becomes an issue. As a result of loss of core temperature, damage to more central organs and systems presents a problem. As the temperature of the core body decreases, so does the heart and associated tissue, which results in a reduction in heart rate and therefore blood output to the body's tissues and organs. A natural protective mechanism occurs where the brain progressively shuts down cerebral (brain) areas. As core temperature drops, more areas of the brain shut down, leading

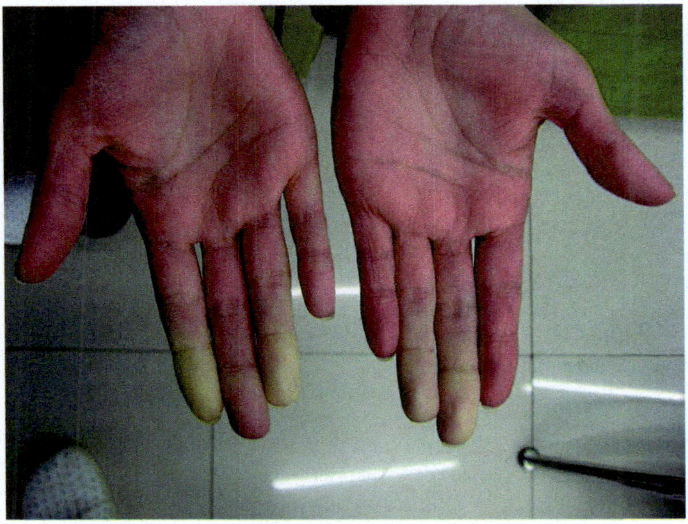

Fig. 16.8 Example of a casualty with Reynaud's Sign

Fig. 16.9 Acute photokeratitis

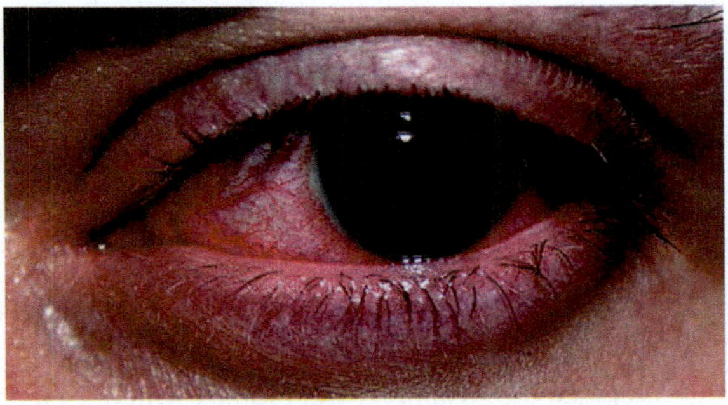

to reports of amnesia during hypothermic situations, and eventual loss of consciousness in severe hypothermia. At core temperatures less than 34 °C (93 °F), cognitive function begins to become impaired and at temperatures below 28 °C (82.4 °F), there is a risk of ventricular fibrillation ("twitching" of the heart and loss of pulse) and cardiac arrest. People with diabetes, injuries, kidney problems, epilepsy and arthritis are at a higher risk of hypothermia in comparison to healthy people. Classification of hypothermia ranges from mild to severe. Mild hypothermia involves core body temperatures between 32 and 35 °C (89.6 and 95 °F), moderate involves temperatures between 28 and 32 °C (82.4 and 89.6 °F), and severe hypothermia involves core temperatures below 28 °C (82 °F). Table 16.1 presents the symptoms of hypothermia.

Figure 16.10 presents the relationship between clothing and water survival time.

Procedures devised in consultation with physicians should be in place to guide treatment of a hypothermic person. Means to access medical advice should be in place (aboard or ashore access to physicians). The guidance contained in MSC.1/

Table 16.1 Symptoms of hypothermia

Mild hypothermia	Moderate hypothermia	Severe hypothermia
Increased heart rate resulting from an increase in venous return Shivering Excessive discharge of urine—resulting from central pooling of blood flow and resulting in rapid delivery to the kidneys Increased muscular tone Decreased nerve conduction	Impaired respiration Decreased heart rate and blood pressure Blue-gray lips, nail beds or skin colour Muscle spasms Loss of feeling in or use of arms and legs Loss of muscle function Slurred speech Blurred vision Impaired cognition Confusion Shivering stops (resulting of used energy in prior stages)	Limited or no cognitive ability Twitching of the heart muscles (atrial and ventricular fibrillation) Possible cardiac arrest Unconsciousness Death

Circ.1185/Rev.1, "Guide for Cold Water Survival" is also applicable to the treatment of hyperthermia. Treating the hypothermic casualty involves the following:

(1) Immediate treatment is much more important than any later action in reviving casualties of immersion hypothermia.
(2) Get the casualty out of the cold, wind, rain or water.
(3) Be gentle with the casualty, restricting their movements.
(4) Do not allow the casualty to walk unless necessary.
(5) Take precautions to warm the casualty up slowly.

Note: *rapidly applying heat to a hypothermia casualty can send the person into shock and cause permanent brain and organ damage.*

(1) Replace the casualty's wet clothes with dry clothes.
(2) If the casualty is conscious and alert, you can allow the casualty to drink warm liquids that do not contain alcohol or caffeine.
(3) Rewarm the casualty:

- Cover the casualty in a blanket or any other material such as gash bags that will help heat retention.
- A hot bath (the temperature no higher than the immersed hand will tolerate) is the most effective method of achieving re-warming.
- As the only non-invasive hospital treatment suitable for active core rewarming onboard ship, inhalation rewarming donates heat directly to the head, neck, and thoracic core (the critical core) through inhalation of warm, water-saturated air at 43–45 °C (107–122 °F). This method also warms the hypothalamus, the temperature regulation centre, the respiratory centre, and the cardiac centre at the base of the brainstem.

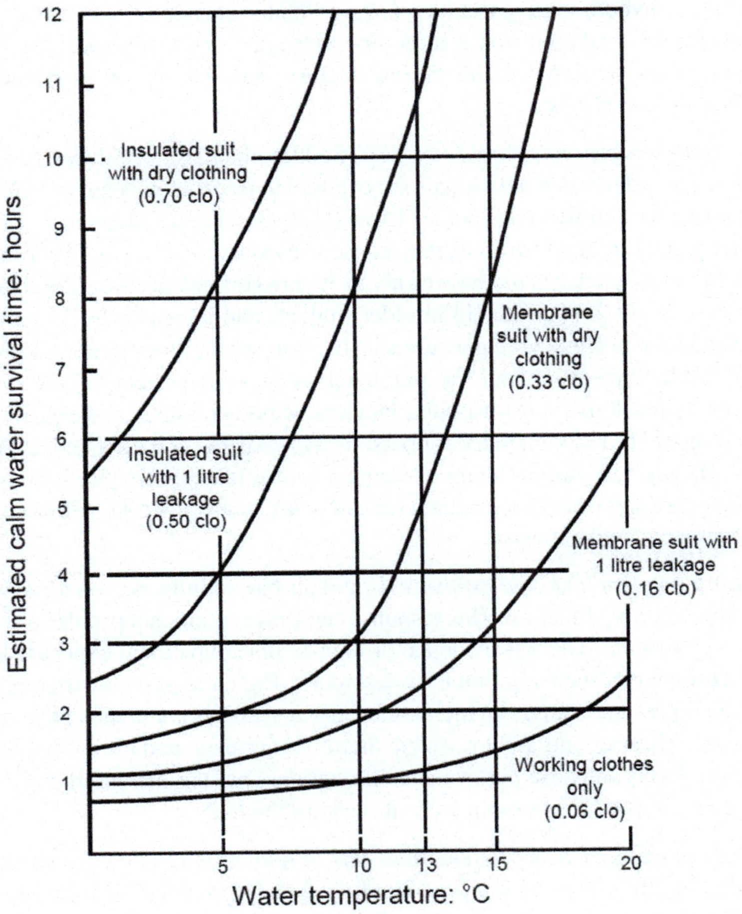

Fig. 16.10 Relationship between clothing and water survival time

- In more severe cases of hypothermia, non-clinical attempts to rewarm the body should revolve around passive rewarming, i.e., covering the body to prevent any additional heat loss and letting the body's metabolism gradually raise internal temperature. Rapid rewarming can cause immediate changes in peripheral blood distribution, cardio-pulmonary function, perfusion ratio, and nervous control of circulation; with the impaired cardiac output associated with moderate and severe hypothermia, sudden changes in peripheral blood flow can force the casualty into cardiovascular shock.

(4) If the casualty is semi-conscious, keep the casualty awake.

(5) If the casualty is unconscious and shows no signs of life, then begin CPR immediately.

(6) Severely hypothermic casualties have a greater chance of survival if greater travel time is taken to reach a facility with appropriate treatment equipment, rather than settling for the closest facility and relying on more primitive rewarming methods.

Cardiovascular function and cold exposure. Sudden changes in cardiovascular function occur every day when individuals are exposed to freezing weather after previous exposure to a thermo-neutral climate. These changes cause a sudden increase in the workload placed on the heart and may cause subsequent cardiovascular problems. Systolic blood pressure can increase by about 30 mm Hg, and diastolic blood pressure can increase about 20–22 mm Hg in older subjects and increases by 23 mm Hg in systolic and 12 mm Hg in younger subjects. In both populations there is a significant increase in blood pressure, and for individuals who are hypertensive, the strain on the heart may result in a cardiovascular incident. Blood pressure can be easily monitored and should be recommended for persons with existing coronary and respiratory disorders. Cardiovascular problems associated with cold exposure rarely occur upon the initial exposure; usually a resultant case of heart attack or stroke will occur 24 to 54 h after severe cold exposure.

Respiratory function and cold exposure. Frigid air has an influence on airway resistance and pulmonary function. This response can cause significant problems for individuals with asthma. There is evidence of long-term occupational exposure to cold causing chronic obstructive pulmonary disorders. The increased constriction of the bronchial air passages causes an increase in airway resistance and thus greater effort in breathing. These effects are not seen in thermo-neutral air, and even when present, they are not likely to cause performance limitations, but the increased constriction may further complicate symptoms of asthmatic individuals.

Physiological changes in Arctic environments unrelated to cold. Seasonal changes in circadian rhythms are common and result in changes in melatonin release. Melatonin (a hormone that plays a role in sleep) is released in response to decreased light exposure, and release is depressed with greater light. A change in melatonin release can cause disruption in the physiologic sleep–wake cycle, and this effect can reach extremes in Arctic and Antarctic environments where there are prolonged periods of light or darkness. There is also evidence of work schedules and daily routine having an influence on sleep–wake cycles; in arctic environments where light conditions persist several months followed by months of darkness, the melatonin-mediated control of circadian rhythm can clash with the social and work-mediated sleep–wake cycle. Interruptions in sleep patterns are common in arctic regions for this reason, often leading to increased fatigue and sleep deprivation. To counteract the physiologic reaction to decreased light, a strategy of matching lighting conditions to work patterns, using blinds and artificial lighting, has shown success. Releasing melatonin by natural or artificial conditions to match work schedules will allow for matching circadian rhythms to social demands and an overall improved sleep pattern. Shift work on ships is suspected to cause both physiological and psychological problems with crew, and these problems are expected to be amplified in arctic conditions.

Suntan beds have been employed in vessels operating at near-polar locations to counteract light deprivation periods of the year. Sleeping areas should be capable of being well "blacked out" to reduce problems in falling asleep due to excess light in the sleeping quarters. In snow-covered environments, the snow can function as a reflective surface for sunlight without absorbing the light in the same manner as pure-water reflective surfaces. Exposure to direct sunlight for more than 10 min requires use of protective eyewear. Using glass lenses for protective eye equipment to filter UV and IR light from entering the retina and using brimmed hats to reduce sun exposure is recommended.

16.2.3 Monitoring Environmental Conditions

Working in cold environments requires an understanding of the interaction between ambient temperature, wind speed, relative humidity, personnel protective equipment and task being performed. In order to limit the risk during operational activities due to cold stress and further prevent local cold injuries and general freezing, specific preventative measures should be evaluated and introduced during the planning and execution of the daily work activities. A plan for monitoring exposure to cold should be devised and should take account of variations in thermal conditions. Ship managers should, if practicable, eliminate the need for extended work in frigid conditions (for example, by rescheduling work to be performed in a warmer season, or by moving the work from outdoors to indoors, or separating the cold parts of a process from the workers, as far as practicable). If elimination of such work is impracticable, other measures to reduce risk from frigid conditions should be devised. Employers should ensure that workers are not positioned near very cold surfaces or, if this cannot be avoided, that the workers are protected by radiation shields. For standing tasks, the floor should, where practicable, have an insulating surface. Climatic metrics such as temperature, wind speed and humidity should be regularly monitored in the locations where outside work is to be performed. Of primary importance is a regular reporting of the wind chill or equivalent temperature. Regular communications should be maintained regarding allowable time to work outside. Indoor personnel should regularly monitor outside workers so that best work-to-rest/warming schedules are maintained. Table 16.2 presents information regarding the relationship of wind chill and exposure danger. Table 16.3 presents threshold limit values for work/warm-up schedule for four-hour shifts/watches.

16.2.4 Clothing and Personal Protective Equipment

For appropriate protection/isolation against cold climate conditions, adequate clothing should be selected and used onboard during cold periods. Such optimal clothing should be able to mitigate water and humidity during work and at the same

Table 16.2 Relationship between wind chill and exposure danger

Wind km/h	Velocity mph	Ambient temperature (°C)								
		4	− 1	− 7	− 12	− 18	− 23	− 29	−34	− 40
		Equivalent wind chill temperature (°C)								
Calm										
0	0	4	− 1	− 7	− 12	− 18	− 23	− 29	− 34	− 40
8	5	3	− 3	− 9	− 14	− 21	− 26	− 32	− 38	− 44
16	10	− 2	− 9	− 16	− 23	− 30	− 35	− 43	− 50	− 57
24	15	− 6	− 13	− 20	− 28	− 36	− 43	− 50	− 58	− 65
32	20	− 8	− 16	− 23	− 32	− 39	− 47	− 55	− 63	− 71
40	25	− 9	− 18	− 26	− 34	− 42	− 51	− 59	− 67	− 76
48	30	− 16	− 19	− 22	− 36	− 44	− 53	− 62	− 70	− 78
56	35	− 11	− 20	− 29	− 37	− 46	− 55	− 63	− 72	− 81
64	40	− 12	− 21	− 29	− 38	− 47	− 56	− 65	− 73	− 82
Little danger in less than one hour exposure of dry skin					DANGER—exposed flesh freezes within one minute		EXTREME DANGER—flesh may freeze within 30 s			

time insulate sufficiently to maintain thermal comfort during rest. The insulating effect of the clothing is influenced by several factors including temperature, wind and humidity. Clo-units determine the measure of clothing insulation. The rate of a person's heat loss is measured in watts per square meter of skin area per kelvin of temperature difference across the clothing; the value of insulation is measured by the reciprocal of this rate, in square meter kelvins per watt (m^2-K/W). One clo is equal to 0.155 m^2-K/W attire, which allows an individual to remain comfortable at an ambient temperature of 21 °C (70 °F). Garments are to be labelled with exposure protection information (for example, hours of protections at a particular temperature). Table 16.4 presents protective and functional properties for outdoor work garments.

***Note**: List of Standards names noted in the "Basis" and "Method" columns*:

- British Standard DD ENV 342: Protective Clothing—Ensembles for Protection against Cold (1998)
- British Standard DD ENV 343: Protective Clothing—Protection against Foul Weather (1998)
- British Standard EN 20,811: Textiles—Determination of Resistance to Water Penetration. Hydrostatic Pressure Test (15 November 1992)
- British Standard EN 31,092: Textiles—Determination of Physiological Properties—Measurement of Thermal and Water-Vapour Resistance Under Steady-State Conditions (Sweating Guarded-Hotplate Test) (1994)
- British Standard EN 511: Specification for Protective Gloves against Cold (30 June 2006)

Table 16.3 Threshold limit values work/warm-up schedule for four-hour shift

Air temperature Sunny sky		No noticeable wind		5 mph wind		10 mph wind		15 mph wind		20 mph wind	
°C	°F	Max work period	Nos breaks	Max work period	Nos breaks	Max work period	Nos breaks	Max work period	Nos breaks	Max work period	Nos breaks
– 26 to – 28	–15 to – 19	(Norm breaks) 1		(Norm breaks) 1		75 min	2	55 min	3	40 min	4
– 29 to -31	– 20 to – 24	(Norm breaks) 1		75 min	2	55 min	3	40 min	4	30 min	5
– 32 to – 34	– 25 to – 29	75 min	2	55 min	3	40 min	4	30 min	5	↓ Non-emergency work should cease ↓	
– 35 to – 37	– 30 to – 34	55 min	3	40 min	4	30 min	5	↓ Non-emergency work should cease ↓			
– 38 to – 39	– 35 to – 39	40 min	4	30 min	5	↓ Non-emergency work should cease ↓					
– 40 to – 42	– 40 to – 44	30 min	5	↓ Non-emergency work should cease ↓							
– 43 and below	– 45 and below	Non-emergency work should cease									

Table 16.4 Protective and functional properties for outdoor work garments

Parameter	Limit	Basis	Method
Thermal insulation (T < − 20 °C)			
• Long-term cold exposure	≥ 3 clo	ISO 11079:2007	BS DD ENV 342
• Short term cold exposure	≥ 2.5 clo	ISO 11079:2007	BS DD ENV 343
• Gloves	2.5…3.0 clo	Frostbites	BS EN 511
Air permeability			
In wind			
• Rest (100 W/m^2)	< 20 l/m^2-s	Preventing of cooling	BS EN 31092
• Heavy work (> 300 Wm^2)	20…150 l/m^2-s	Evaporation	BS EN 31092
Ventilation			
• Heavy work (> 300 Wm^2)	≥ 300 l/min		
Water vapour permeability			
• Cold weather clothing	≤ 13 m^2Pa/W	Evaporation	BS EN 31092 ISO 11092:1993
• Cold/foul weather clothing	≤ 20 m^2Pa/W	Evaporation	BS EN 31092 ISO 11092:1993
Resistance to water penetration			
• Light rain	≥ 2200 Pa	Protection against moisture	BS EN 20811 ISO 9073-16:2007
• Heavy rain	≥ 33,000 Pa	Protection against moisture	BS EN 20811 ISO 9073-16:2007

Source Adapted from a collaboration of information by Biem et al. (2003), Giesbrecht (2000), and Golden and Tipton (2002)

- ISO 11079:2007: Ergonomics of the Thermal Environment—Determination and Interpretation of Cold Stress when Using Required Clothing Insulation (IREQ) and Local Cooling Effects
- ISO 11092:1993: Textiles - Physiological Effects—Measurement of Thermal and Water-Vapour Resistance Under Steady-State Conditions (Sweating Guarded-Hotplate Test)
- ISO 9073–16:2007: Textiles—Test Methods for Nonwovens—Part 16: Determination of Resistance to Penetration By Water (Hydrostatic Pressure).

16.2.5 General Recommendations for Clothing

It is important that the clothing be as comfortable as possible. It needs to provide optimal transportation of humidity from the body to the environment, good insulation and sufficient dexterity. Each garment in the clothing ensemble is important and all pieces should be considered part of the protective system. Figure 16.5 presents heat loss versus clo factor at a comfortable skin temperature. For example, a person at

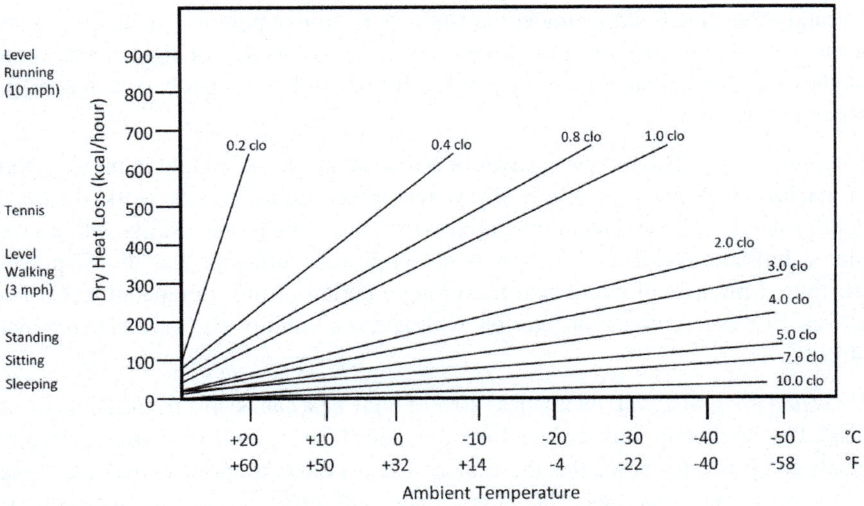

Fig. 16.11 Heat loss versus clo factor at a comfortable skin temperature

rest, generating 100 kcal per hour of metabolic heat at this activity, will require about 1.0 clo unit of insulation to be in thermal balance at 20 °C (68 °F). If the temperature were to drop to 0 °C (32 °F), he or she would require approximately 3.3 clo. With a further temperature drop to – 20 °C (–4 °F), the requirement would be approximately 4.8 clo. However, an active person generating 300 kcal/hour would only need approximately 1.8 clo at – 20 °C (– 4 °F) (Fig. 16.11).

With the addition of wind and humidity to low ambient temperature, greater clo units are required. Wind has an effect of compressing clothing, decreasing the overall amount of trapped air that creates an insulating layer. General body protection from cold should be provided. For optimal protection, the following three (3) clothing protection layers are recommended for outdoor work activities onboard (for an effective insulation of 2.6 clo):

- Inner Layer of two-layer underwear effectively absorbing and transporting perspiration: "super" underwear or comparable products against the skin Wool as second layer.
- Middle Layer which provides an isolating effect: fibre/fleece (pants and sweater/ jacket), wool sweater, woollen socks.
- Outer Layer which protects against the environment and should be waterproof.

A sufficient amount of appropriate skin protective cold cream should be available at all times onboard, located at the assigned heating huts, common rooms and sickbay. Someone should be designated to be responsible for ensuring that cold creams are available as required onboard. Dispensers should be used instead of jars containing cold cream. Clothing should have adjustable neck and sleeves to influence air exchange when internal temperature becomes unfavourable and to prevent sweating/moisture build-up under the clothing. Cotton or polypropylene long underwear is recommended for all-over warmth. Multiple layers of light, loose-fitting clothes are recommended. Use suspenders instead of belts (tight belts can constrict blood circulation).

Hand protection. Do not wear gloves or scarves that can get caught in moving parts of machinery. While glove use is always recommended for outside work, the use of cold-protective gloves impairs the ability to conduct fine motor skills, while gross motor skills are less affected. Mittens or gloves provide sufficient insulation to prevent frostbite. Mittens will retain heat for a longer period of time compared to wearing gloves. However, mittens can interfere with a person's ability to grasp and manipulate objects.

Head and eye protection. As much as 40% of body heat can be lost from an uncovered head. Use an appropriate hard hat liner to reduce heat loss when wearing protective headwear. It is often noted that the choice of safety hat takes priority over the choice of its insulating properties. This is not a desirable principal procurement criterion for freezing weather operations. Wear a warm hat with ear protection to prevent heat loss from the head. Protect vision from UV rays by wearing appropriate sunglasses while working in snow and ice on a bright day.

Foot protection. The surface contact area of the sole is important for frictional force on ice. It is suggested that shoes/boots with rounded heels increase the contact with the surface and reduce the risk of slipping. When there is a layer of water along icy surfaces, the contact with the sole of the shoe is reduced. Therefore, the addition of sharp cleats to shoes is suggested. However, wearing shoes/ boots with steel cleats on steel decks can be quite dangerous because of the loss of frictional force. Boot selection should involve procuring a boot with proper insulation for the duration of the activity, required protection and the intensity of the work. Woollen, polypropylene, or other thermal socks should be worn to protect ankles and feet. Keep snow and water out of footwear and replace damp socks with a dry pair.

Maintenance of personnel protective equipment. Dirty and oily clothing loses much of its insulation value and wearing of dirty/oily clothing should be avoided. Do not keep freezing weather clothing stored or compressed in a duffle bag for prolonged periods of time. Fluff waterfowl down garments when removing them from the duffle bag following initial boarding of vessel. Keep garments dry; brush off snow and frost before entering warm buildings or vehicles. Follow manufacturer's guidelines for cleaning clothing. Failure to do so may reduce the protective capacity of the clothing. Do not leave rips or tears in a state of disrepair. Temporary repairs can be made with electrical or duct tape.

Immersion suit protection. In work where there is a risk of accidental immersion or abandonment into cold-water, specific protective equipment is recommended:

- Dry-suits and marine anti-exposure suits are commonly worn during marine operations and have been found not to impair working ability. These provide both protection and buoyancy in the water.

- A fully insulated dry suit has a clo value of approximately 0.7, leading to survival time up to 15 h of immersion at 12 °C (53.6 °F), compared to 1–6 h without the suit.
- Protective suits should be checked regularly for wear and accidental rips. In some cases, replacement of the suit is recommended.

There are marine immersion suits available that protect the wearer from wind and cold in a dry environment and have isolative properties and buoyancy in water as well. While on land, these suits also have the ability to increase ventilation by opening collars and sleeves as needed, thereby accounting for individual variation in thermal comfort.

16.2.6 Nutrition Considerations in Cold Climates

The added weight of protective clothing and the limitations in mobility created by protective equipment will increase the mobility demands of the operator, thus increasing the metabolic needs for a given task. A mixed diet of protein, carbohydrates and fats combining for a caloric intake of 4500 kcal for an average man is recommended for demanding labour in the cold. Normal physiologic water loss is unavoidable in cold environments. There is a loss during physiologic activities such as metabolism, from sweat production inside protective equipment, and from increased urination related to cold-induced excessive discharge of urine. It is recommended that 3–6 L of daily intake of water be maintained. For warming purposes, hot non-alcoholic beverages or soup are suggested. Caffeinated drinks such as coffee should be limited because it increases urine production and contributes to dehydration. Tobacco prevents the periodic blood vessel dilation needed for preventing tissue injury. The risk of frostbite and peripheral injury may increase with tobacco use. Alcohol use should be avoided. The use of alcohol increases the flow of blood to the skin, and can lead to rapid loss of body temperature, increasing the likelihood of hypothermia. Frequent, small meals will allow a constantly high metabolism to help maintain body temperature, whereas large meals served only 2–3 times per day will result in a decreased metabolic rate after digestion and decrease the "heat production to heat loss" ratio. In the event of ship abandonment, a food stores carry-off kit should be provided, containing water and non-perishable foods. The carry-off kits should be stored in areas not subject to freezing temperatures. The carry-off kits should be stored in immediately accessible locations.

16.2.7 Workstation Design and Operational Considerations

The analysis of outdoor work situations should be performed early in design/layout development and should be updated when design changes are made that will influence personnel's exposure to cold stress. Outdoor operations analyses (an examination of the tasks to be conducted in frigid conditions) should be conducted for open work areas and semi-open work areas. The objective of these analyses is to identify and remedy task performance issues due to overall exposure to temperature, wind, icing and precipitation, including investigation of the weather protection necessary to comply with exposure limits. Environmental measurements and assessments (air temperature, humidity and wind speed and direction) should be performed on a routine basis and at regular intervals throughout the day during cold periods with temperatures below zero degrees. All personnel working in exposed locations should be monitored at regular intervals. A person is usually unable to recognise their own signs and symptoms of hypothermia. As a precaution, use of the "buddy system" to detect signs of cold injury in colleagues is recommended. Crew should be allowed sufficient time to acclimatise to an extremely hot or cold environment, including major changes in climatic conditions. Monitoring of thermal conditions should be performed and should take account of:

- All stages of work cycles and the range of temperature and humidity under which tasks are performed
- The range of clothing worn during the tasks
- Major changes in physical activity level (metabolic heat production), and
- Occasional tasks such as cleaning and maintenance of hot equipment and cold areas, and renewal of hot or cold insulation.

A log that is to be maintained by the onboard Safety Officer during operations in cold periods to keep records of the specific areas and work activities where administrative measures are introduced is recommended. Work areas where administrative measures are introduced on a regular basis should be identified by periodical review of such records. Workers in the cold will often need to urinate more frequently, and employers should ensure that suitable arrangements are available, where feasible, and that the design of protective clothing allows easy urination. Suitable protection should be given to the hands and fingers, particularly where dexterity is needed, as well as other exposed parts of the body. Employers should provide:

- Facilities for warming the hands, for example, by warm air, where appropriate
- Tools with insulated handles, especially in temperatures below the freezing point
- Measures to prevent the bare hand touching surfaces below − 7 °C (19 °F) (workplace design or protective clothing)
- Measures to prevent bare skin touching liquids below 4 °C (39 °F)
- Appropriate measures to be taken in the event of insulating clothing getting wet
- Face and eye protection, as appropriate, for outdoor work and working in snow (e.g., safety goggles against glare)
- Adequate facilities for changing
- Arrangements for cleaning such clothing and drying clothing and footwear between shifts.

Slips and falls are much more common in winter seasons.

- Walk slowly on slippery, icy surfaces to prevent danger of slipping. Stop occasionally to break the momentum
- Avoid carrying heavy loads or loads that obstruct vision during transportation
- Use approved snow removal compounds
- Use mats at entrances to prevent slippery conditions indoors caused by melting snow or ice
- Stairs, ladders and walking surfaces should be provided with non-skid surfaces
- Hand railings should be provided with non-slip surfaces or should be heated to prevent formation of ice
- Exterior stairs should be made of grated material to aid in snow and ice removal.

Post visible signage emphasising the risk for cold-related injuries at all outdoor workstation locations. Knowledge of the work area at risk for slippery conditions, augmented with proper hazard signage, is important. If individuals are aware of the freezing weather conditions, they will rely on additional muscle-based reflexes (i.e., reaction time), postural control, and muscular strength to adjust to the environment and prevent slipping and falling. Consider providing enclosed bridge wings. Consider providing a temperature transition area (such as a heat vestibule) at exit/entry points to soften the often-harsh transition from bitter cold to warmth, and vice versa). Assess possibility for temporary local shielding around working area(s) if permanent shielding is not possible. Clearly post locations for nearest re-warming or break areas. Power tools and other equipment require specific maintenance schedules for usage in freezing weather. Cold tool handles reduce grip force. Therefore, tools with larger handles may be required for freezing weather usage to accommodate the protective hand wear and reduced grip capacity. Cold exposure aggravates the effects of mechanical vibration, making manual work more difficult and painful. Effort should be made to minimise exposure to vibration at its source. Hospital/medical kits should be stocked with materials appropriate to treat freezing weather illnesses and maladies. Freezing weather illnesses training, and procedures should be provided with emphasis on both symptom identification and treatment of the cold-related illness.

16.2.8 Accommodations and Environmental Control

Personnel accommodations should be designed and arranged to protect the occupants from unfavourable environmental conditions and minimise risk of injury during normal (including ice transiting or ice breaking) operations and emergency conditions. Tables 16.5 and 16.6 provide guidance on exposure standards for vibration and vibration effects on various personnel functions. Personnel accommodations, public spaces and the equipment installed in them should be designed so that each person making proper use of them will not suffer injury during normal open water operations and emergency manoeuvring conditions. Facilities should include non-slip decking, three rigid sides, handholds and insulation from exposed hot water pipes. Personal quarters should have sufficient room to dry and store freezing weather protective equipment. Cabin heating systems should be thermostat controlled and have shielded elements, if appropriate.

Table 16.5 Exposure standards and action-limits for vibration

	Hand-transmitted vibration		Whole-body vibration			
	Exposure action value	Exposure limit value	Exposure action value		Exposure limit value	
Exposure duration	R.M.S. method		R.M.S. method	VDV method	R.M.S. method	VDV method
1 s	424.26	848.53	84.85	6.51	195.16	14.98
10 s	134.16	268.33	26.83	3.66	61.72	8.42
1 m	54.77	109.54	10.95	2.34	25.20	5.38
10 m	17.32	34.64	3.46	1.32	7.97	3.03
1 h	7.07	14.14	1.41	0.84	3.25	1.93
2 h	5.00	10.00	1.00	0.71	2.30	1.63
4 h	3.54	7.07	0.71	0.59	1.63	1.37
8 h	2.50	5.00	0.50	0.50	1.15	1.15
12 h	2.04	4.08	0.41	0.45	0.94	1.04
16 h	1.77	3.54	0.35	0.42	0.81	0.97
24 h	1.44	2.89	0.29	0.38	0.66	0.87

Source Adapted from Griffin M. J. (1990). Handbook of human vibration. London: Academic Press

Table 16.6 Vibration effects on function and performance

Vibration induced effect	Frequency (Hz) at which there is occurrence
Decreased tactile sensation	0.1–1,000,000
Motion sickness	0.1–1
Vibrating hand tools diseases	5–10
Respiration difficulties	1.5–15
Abdominal pain	5–15
Increased muscle tone	10–20
Lower back pain	7–20
Head sensations	10–20
Disturbance of speech	1–100
Urge to defecate and urinate	10–20
Speech difficulties	7–20

Source Wilson and Corlett (1990)

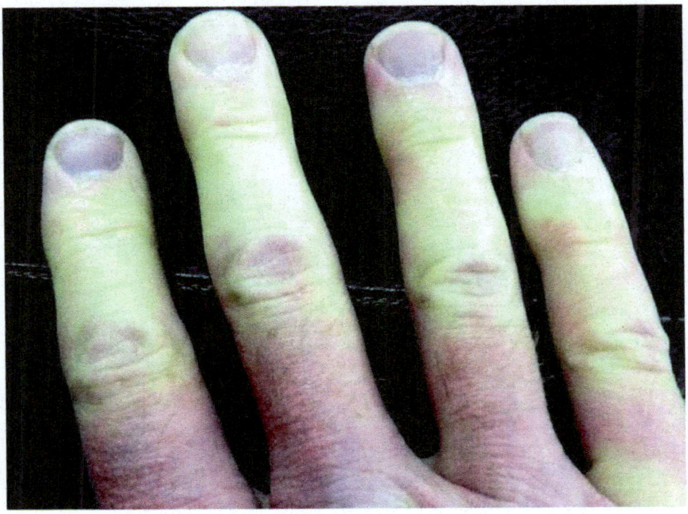

Fig. 16.12 Vibration white finger

- Protective clothing should not be dried directly upon heating system equipment or fittings, and
- Protective skin cream dispenser should be supplied in each room.

Consideration should be given for providing a sauna for crew recreational use and for rewarming cold persons (Fig. 16.12).

Chapter 17
Extreme Low Temperature Training

17.1 Introduction

Vessels operating in low temperature environments are exposed to a number of unique circumstances. Weather conditions are poor and navigation charts may be unreliable or not reflect current conditions. Shipboard polar ice charts should be updated weekly upon processing of satellite images. Local ice conditions may differ significantly from those depicted on charts. Maintaining manoeuvrability for the avoidance of locally heavy ice conditions is an important consideration when using ice charts at the route planning level, and communication systems and navigational aids present challenges to mariners. The areas that the vessels operate in are remote, making rescue and any clean-up operations difficult. Therefore, additional crew training must be undertaken, and operations manuals must be developed. Training in ice operations, navigation and winterisation are to be provided. Training is to address means to prevent and treat potential cold weather-related maladies of crew, including hypothermia and frostbite. Certifications are to be recorded, where applicable, and the records updated. The manual is to be in English and the working language of the crew. The operating section is to include, but not be limited to, the following:

- General arrangement showing location of equipment (including loose items) installed onboard to facilitate operation of the vessel in low temperature environments
- Specification of the equipment installed onboard to facilitate operation of the vessel in low temperature environments together with manufacturer's recommendations of use, operational limitations, maintenance and testing procedures, as applicable
- Relevant information related to operations in ice-covered waters, including contingency planning in the event of damage. The typical format of an operating manual is described below and normally includes sections on Normal Operation, Risk Management, and Emergency Instructions

A. Olsen, *Ship Operations in Extreme Low Temperature Environments*, Springer Series on Naval Architecture, Marine Engineering, Shipbuilding and Shipping 19,
https://doi.org/10.1007/978-3-031-52513-1_17

- Any special operating procedures specific to the vessel and its modes of operation.

The training section is to cover at least the following subject matter areas:

- Ice recognition
- Safe navigation in ice
- Conduct during escorted operations
- Instructions for drills and emergency response
- Cold weather-related maladies.

An updated listing of personnel certified (if applicable) in the use of specialised equipment carried for any type of emergency response is to be maintained. A training log is to be maintained documenting the conduct of the training and the names of the persons attending the training.

17.2 Low Temperature Training

Operations in ice require specific skills if they are to be accomplished safely and efficiently. The importance of having an experienced ice master or ice navigator has been recognised for centuries, on voyages of discovery and for commercial operations. Ice breakers and other vessels that spend much of their time in ice-congested waters may have masters and officers with sufficient expertise to provide the ice navigator function. For many other vessels, it is necessary to supplement the regular crew with supernumerary ice advisors or ice navigators, either to provide a basic capability or to comply with national regulations for local experience. Where such personnel are used, the number carried should be matched to the nature of the service and the anticipated nature of the operation. The terminology used to describe personnel with ice-related expertise is not standardised, and so its usage may lead to some confusion. Throughout this chapter, the following definitions apply:

- *Ice advisor*: an individual with expertise in the interpretation of ice information and in forecasting future ice conditions
- *Ice navigator*: a mariner (usually a master mariner) with experience in the operation of vessels in ice who can provide guidance to the vessel's master and other watchkeepers
- *Ice pilot*: a mariner individual with similar qualifications and experience to an ice navigator who may be required in some areas and authorities to undertake pilotage duties
- *Ice master*: a vessel's master with sufficient ice-going expertise to undertake the functions of ice navigator.

While these personnel require extensive specialised training, some level of training should be given to all officers and crew of vessels involved in ice operations. It should also be recognised that operations in remote regions may have extremely limited support from shore-based infrastructure, and sufficient skills should be available onboard to cover both normal and emergency situations. General guidance on training and crewing is offered in the IMO Guidelines; MSC Circular 1056, MEPC Circular 399. National administrations, including those of Canada and Russia, have specific requirements that must be respected in waters under their authority. This chapter aims to provide additional useful information that vessel owners/operators and crews can use in the planning and conduct of cold climate operations.

17.2.1 Training

Seafarers exposed to cold, and their supervisors should be trained:

- To recognise symptoms which may lead to hypothermia in themselves or others, and the steps to be taken to prevent onset and/or emergencies
- In the use of rescue and first-aid measures, ice control equipment and procedures, personal protective equipment, and other devices and procedures intended to counter the effects or working in frigid conditions
- About action to be taken in the event of increased risks of accidents because of low temperatures.
- Seafarers should also be advised of:
- The importance of physical fitness for work in cold environments
- The importance of drinking sufficient quantities of liquid and the dietary requirements providing intake of salt and potassium and other elements that are depleted due to sweating
- Effects of consuming alcohol and drugs, which can reduce their tolerance to thermal extremes.

Seafarers normally use classroom training, self-study, and operating experience to develop competencies and achieve certifications. Until recently, there have been extremely limited formal training resources for developing ice-operational knowledge. Now, several training institutes around the world can offer some level of course for ice advisors, ice navigators and/or ice masters. There are also some simulator facilities that offer training in ice recognition and other aspects of ice operations. However, these still have extremely limited fidelity in representing ship-ice interaction. On-board training remains the most valuable means of developing expertise, and minimum durations of operation and watchkeeping experience are specified by most national administrations before personnel can be qualified as ice navigators. There are no formally mandated requirements for the content of training courses in ice navigation. However, any comprehensive training program needs to address the demands of Strategic, Tactical and Close Quarters navigation. Strategic navigation is concerned with the general routing of a vessel in the geographic region in which a vessel will operate, for example, the Baltic, Northern Sea Route, Northwest Passage, etc. At the tactical level, considerations include the selection of way points according to current and forecast ice conditions. At the close quarter level, active navigation evaluates the local ice conditions and the manoeuvring of the vessel around and through ice floes.

Strategic Level (voyage planning) decision-making includes:

- Obtaining relevant ice information from ice maps, satellite imagery and medium-term forecasts
- Evaluating ice conditions and environmental forecasts to verify the feasibility of the voyage and its probable duration
- Optimising route to minimise the time that will be spent in the ice and to avoid potentially dangerous ice features
- Tactical Level (transiting) decision-making includes:
- Obtaining more local ice information from enhanced satellite imagery, airborne radar images and visual reports, reports from other vessels in the area, etc.
- Optimising the route to minimise encounters with ice that will impede vessel's progress or cause structural damage.

Close Quarters Level (manoeuvring) decision-making includes:

- Using radar and visual data to assess the properties of the ice cover
- Selecting safe speeds to reduce the risk of structural or mechanical damage
- Manoeuvring in a manner that minimises impacts against vulnerable regions of the hull, steering gear, etc.

As can be seen from the above, an essential area of expertise—and therefore training—is the ability to interpret various ice information ranging from charts and imagery to direct visual observation. Several good pictorial guides to ice recognition have been developed, examples of which are included in the reference list. Simulators with high quality graphics can also be valuable training aids. Ice advisors will have received training in these areas, but not necessarily in navigational skills. A key element of expertise for an ice navigator, pilot or master is the ability to relate the ice conditions to the capabilities of a vessel, based on its ice class, powering and other features. This can be taught in general terms by training institutes, though all vessels have their own specific capabilities that will influence performance and safety. The ice navigator and all deck officers should have some level of understanding of the relationship of the vessel's ice-strengthening class and the limiting safe ice conditions. For the IMO/IACS Polar Classes, the general definitions provided in Table 17.1, "IMO/IACS Polar Classes," offer some information (refer to Chap. 10 for additional information on ice types and conditions). Vessels operating in many Russian waters are provided with an Ice Certificate which provides more detailed information on safe speeds in different ice conditions. However, it is also necessary for operators to understand how the overall ice strengthening of any vessel can vary within the hull; and, for example, how operating in ballast conditions may expose weaker areas of the hull or appendages to ice impacts. The sides and stern of most vessels are less strong than the bow area and vessels should not be operated in ways that expose these areas to significant ice impact loads.

The manoeuvrability of most vessels is reduced dramatically by the presence of any ice, and as ice cover thickens, it may become almost impossible to alter

Table 17.1 IMO/IACS polar classes

Polar class	Ice description (based on WMO sea ice nomenclature)
PC 1	Year-round operation in all polar waters
PC 2	Year-round operation in moderate multi-year ice conditions
PC 3	Year-round operation in second-year ice which may include multi-year ice inclusions
PC 4	Year-round operation in thick first-year ice which may include old ice inclusions
PC 5	Year-round operation in medium first-year ice which may include old ice inclusions
PC 6	Summer/autumn operation in medium first-year ice which may include old ice inclusions
PC 7	Summer/autumn operation in thin first-year ice which may include old ice inclusions

course by conventional manoeuvres. This is particularly true of vessels with long parallel midbodies and no side flare (i.e., most commercial vessels). However, even ice breakers may find it necessary to change course by backing into a broken track in order to generate a change in heading. This type of manoeuvre is known as the Captain's Manoeuvre, or Star Turn, due to the appearance of the track. Another situation that may require a vessel to reverse course is the need to make repeated rams to traverse thick ice or pressure ridges. Only vessels with elevated levels of ice capability should attempt to ram, and great caution is required during the backing phase of any manoeuvre to avoid damage to propellers. It is particularly dangerous for propellers to hit ice when the shaft is stopped, which can exert extremely high bending loads on the blades. Ramming speeds should be selected to avoid beaching the vessel in an attitude that creates stability problems, or that will lead to problems of extrication. Ice navigators should be able to recognise the onset of pressure events, which can cause a vessel to become beset and may pose both direct and indirect safety hazards. This pressure can be sufficient to damage the hull, and general ice drift may carry the vessel into shallow or dangerous waters. Pressure can be managed by ice breaker intervention, or if time permits, the vessel can be oriented into rather than across the direction of compression. As pressure events are usually of reasonable short duration, stronger vessels may find it best to wait them out. The ice navigator should therefore also be sensitive to signs of reducing pressure.

A considerable number of high ice class vessels incorporate unusual design features ranging from azimuthing propulsors to bubbler, water wash and heeling systems. Ice navigators and other personnel should understand why such systems are used and how they can be operated for maximum effectiveness. Training for deck officers should address ice breaking procedures and the conduct of escorted operations. While these differ in different national waters, many general principles apply. These include ensuring that communications are established, and protocols understood. Potential problem areas for escort include manoeuvring (if the ice breaker makes too close a turn); stopping (if the ice breaker is brought to a halt by heavy ice); the risk of damage due to impact with track sides for a wide vessel escorted by a narrower ice breaker; and the risk of damage due to bow or bottom impact with large ice pieces submerged below the ice breaker. Vessel masters should understand that an escort does not take over responsibility for the safety of the escorted vessel. The points covered above are not intended to represent a complete listing of subject areas for training courses, but rather an illustration of the types of knowledge that training should aim to provide. They are also amongst the type of material that should be covered in a vessel's onboard documentation, as discussed further below.

17.3 Documentation

Under the IMO Guidelines, "all ships operating in Arctic ice-covered waters should carry onboard at all times an operating and training manual for all Ice Navigators onboard the ship." The IMO Guidelines provide a basic overview of the type of content that each manual should cover. Manuals and other documentation should conform to the intent of other requirements, such as those of the International Safety Management (ISM) Code, Contingency Planning for Shipboard Emergencies (IMO A852(20)), and other international and national standards. As such, they need to address the information needs of all crew members and procedures for making any passengers aware of specific requirements and challenges.

17.3.1 Operating Manual

This manual should contain relevant information related to operations in ice-covered waters including contingency planning in the event that the vessel suffers damage. The typical format of any operating manual should include the following sections, tailored to operations in ice and in cold temperature conditions: Normal Operation, Risk Management, and Emergency Instructions.

Normal operation. This should include measures to be taken when planning a voyage into cold or ice-infested waters, including:

- Contacting maritime administrations well in advance to confirm if specific requirements, crewing or personnel are required onboard
- Arranging for the availability of ice and weather data and forecasts
- Identifying any loading restrictions that may be required for safety or pollution prevention reasons
- Providing that the vessel is adequately outfitted and supplied for the voyage.
- As the vessel approaches the cold region, additional checks should be undertaken, such as:
- Contact the vessel traffic control services for the area
- Verify operation of VHF radio including confirming channel used by ice breakers operating in the area

- Take measures to prevent freezing damage; (e.g., piping on deck, sounding pipes and air vents, etc., are to be drained of water)
- Check drafts and trim to verify that the ice belt covers the waterline area, and that propeller and rudders are adequately submerged
- Verify the operation of searchlights and signalling devices
- Ready anchors and towlines
- Position protective covers on exposed equipment
- Configure HVAC and other systems (where possible) for freezing weather operations
- Implement any other measures required for safe navigation.

The completion of checklists should be recorded. Once the vessel is operating in cold and ice, other issues that should be covered by the Operations Manual (as applicable) may include:

- Use of specific features and systems installed on the vessel, such as water wash and bubblers, etc.
- Cold start and run-up of systems, including preheating of lubricating and hydraulic oils
- Procedures for clearing ice and snow accumulations without damage to equipment
- Recording and reporting local conditions to assist in forecasting and to comply with regulatory requirements.

Risk management. Risk Management for cold regions should cover all hazards that may result from or be accentuated by the cold temperature environment, and by operations in ice-infested waters. This includes ensuring the operability and checking the functionality of all operational and safety systems, including firefighting and lifesaving, as discussed in previous chapters. Other issues that should be addressed include monitoring of:

- Hull integrity, by sounding tanks and voids at regular intervals
- Water temperatures in tanks and systems, to provide warning of freeze-up problems
- Performance of pollution prevention systems, to prevent the discharge of wastes.

In general, risk management requires an understanding that many systems and components will be operating at or near their design limits, and their performance may degrade rapidly. Monitoring and testing should be conducted more frequently than is the case in normal operations.

Emergency instructions. Emergency instructions should address the unique aspects of certain types of incidents in cold and/or ice-infested regions. These range from communicating with support and search and rescue organisations to abandoning ship in the presence of ice. Emergency instructions cannot take the place of proper training in emergency response management but can be extremely valuable in reminding crew members of the procedures to be followed. As with all other elements of the Operating Manual, the instructions must be tailored to the specifics of the vessel and its modes of operation. They should also be integrated with the drills and exercises covered by the Training Manual and onboard training procedures.

Training manual. A training manual should be provided onboard all vessels. The contents of a manual for vessels operating in cold regions should be an informational and educational tool for personnel and a guide for the types of training exercise that should form part of risk management for cold regions operation. As such, it should cover ice recognition, safe navigation in ice, conduct of escorted operations, and instructions for drills and emergency response. Drills are covered in some detail by the *IMO Arctic Guidelines*, which addresses lifesaving, firefighting, damage control and other exercises. These should be conducted under as realistic conditions as possible consistent with maintaining the safety of those onboard. Crew members should all be given the opportunity to undertake drills while wearing freezing weather survival gear to gain a proper appreciation of how this will affect their mobility and dexterity. Where specialised equipment is carried for any type of emergency response, an adequate number of crew members should be trained in advance (and certified as necessary) in its use.

This chapter provides high level guidance relating to vessel operations in low temperature environments. These are recommendations only and cannot list all operating scenarios or equipment types that a vessel may encounter or have installed onboard. The vessel's crew will need to make regular checks of safety–critical systems and equipment so that these are still functioning or capable of functioning as intended. The vessel's Owner and Operator are responsible for developing relevant and appropriate operational guidelines and keeping these procedures up to date. Within each section is a list of operational issues that should be considered when developing the operations manual.

17.4 Deck Recommendations

17.4.1 Ice/Snow Removal

The deck and associated piping should be cleaned up periodically, particularly due to spray and inclement weather. Ice can quickly build up on the vessel, affecting safety of the crew's movement and vessel stability.

17.4.2 Crew Access

Access routes. Access ways should be checked each day when snow or under-cooled rain occurs, and any accumulation removed. De-ice chemical mix (glycol mix) should be applied to deck plating when freezing occurs. Deck drains should be opened using steam, pressure air and/or de-ice chemical mix Snow and ice should be removed by use of shovels, hardwood clubs, steam or equal. When freezing temperatures and icing are anticipated, heated walkways are activated at 5 °C (41 °F) with increasing heat as further temperature drops occur. Where necessary, additional safety lines should be run, and crew members instructed on tethering themselves.

Escape routes. Escape routes should be clearly marked, checked and kept clear of debris, snow and ice accumulation at least once a day or more often when required.

Emergency exits. Gaskets, hinges, dowels and locks should be checked and treated with approved de-ice chemicals before the temperature drops below 0 °C (32 °F). Exits should be checked and kept clear of debris, snow and ice accumulation once a day or more often when required. Snow and ice should be continuously removed using shovels, hardwood clubs, steam and/or compressed air. Pre-treatment with de-ice chemicals should be applied at least each month or when required. For drilling rigs, the Derrick man's escape (Geronimo Chutes) should be checked and treated with de-ice chemicals before the temperature drops below 0 °C (32 °F). The chute should be function assessed every day when the temperature is 0 °C (32 °F) or below.

External emergency lights. External emergency lights should be always operative and kept free from snow and ice. The lights should be checked daily when the temperature is below 0 °C (32 °F).

17.5 Vessel Systems and Machinery Recommendations

Weather forecasts should be monitored closely. In the event forecasts predict temperatures approaching the Minimum Anticipated Temperature, deck machinery should be periodically operated (barred) and any hydraulic systems' oil heaters turned on.

17.5.1 Firefighting Equipment Readiness

Door gaskets should be treated with de-ice treatments at least each month or when required. All snow and ice accumulation on equipment should be removed by using steam, compressed air or equal as necessary to always maintain system readiness. Fire water hoses that have been used should be drained and dried immediately after use or stored at a frost-free location. Fire mains should be drained until needed when the temperature is 0 °C (32 °F) or below. When the temperature drops below 0 °C (32 °F), all external fire equipment should be checked daily or more often when required for snow and ice accumulation. Portable fire extinguishers at anchor stations should be kept in operator's cabins when the temperature is – 15 °C (5 °F) or below. All the fire dampers directly exposed to the weather are to be checked and function assessed every day when the temperature is 0 °C (32 °F) or below. Exposed deck machinery should be checked frequently to verify correct operation.

17.5.2 Tanks

Any tank, its associated piping system and venting condition should be checked prior to loading or discharging operations.

17.5.3 Exterior Alarms

All external alarm equipment should be inspected daily when temperature is below 0 °C (32 °F).

17.5.4 Helicopter Deck

Foam monitors should be properly drained when the helicopter deck operations are completed. De-ice chemical mix should be applied to the helicopter deck. Deck border lights and floodlights should be treated with de-ice chemical mix. Portable fire extinguishers should be kept in a heated area close to the helicopter deck. Drains should be checked daily to verify heat tracing is working.

Procedures when operating in snow and freezing rain. In addition to the items above, the following procedures should be performed. Snow should be removed from helicopter deck and access/escape ways, when required, by using shovels, steam and/or pressure air. De-ice chemical mix should be applied to deck when required. Ice should be removed from deck border lights and floodlights. Steam and compressed air can be used. De-ice chemical mix should be applied at least every month or when required.

Procedures before landing and take-off. Helicopter deck should be checked before landing for debris, snow and ice accumulation. Snow and ice should be removed. Special attention should be paid to landing area of helicopter deck and to walkways/escape routes. Ice should be removed from deck border lights and floodlights. Helicopter deck firefighting system (foam monitors) should be activated. Personnel should be kept in protected spaces until departure to avoid cold exposure.

17.6 Safety Systems

17.6.1 Evacuation Systems

Launching/lifeboat stations. De-icing equipment should be deployed and ready for use at stations when the temperature drops below 0 °C (32 °F). Steam hoses should be rigged and ready for use at stations. Lashings on life jacket and survival suit boxes should be checked daily or more often when required for snow and ice accumulation. Ice can be removed by de-icing equipment. De-ice treatments should be applied at least each month or when required. Means to de-ice (e.g., glycol/water mix or salt) should be applied on deck plating as required to prevent icing when the temperature drops below 0 °C (32 °F). Lifeboat stations should be checked daily or more often for snow and ice accumulation when required. Ice and snow should be removed from decks, gratings, handrails, rope escape ladder, life vest and survival suit boxes using shovels, hardwood clubs, steam or compressed air.

Lifeboats. Lifeboat heaters in all lifeboats should be switched on when the temperature drops below 0 °C (32 °F). Hooks, latches and hinges should be checked daily or more often when required for snow and ice accumulation and treated with de-ice treatments. Radio equipment with batteries should be thoroughly checked every second day. Lifeboats should be checked two times a day or more often when required. Engine heaters, lifeboat heating fans and battery chargers should be checked for functionality each day. Snow and ice should be kept removed. Hooks, lashes and hinges should be checked two times a day or more often when required for snow and ice accumulation if required. Ice should be removed using de-icing equipment. De-ice treatments should be applied at least each month or when required. Lifeboat engine should be run each day until it reaches operation temperature. Lifeboat engine fuel should be checked daily for clouding or waxing. Lifeboat engine lubricating and hydraulic (starting) oils should have correct viscosity at design service temperature without the use of heaters.

Launching arrangements. Hooks, latches and hydrostatic release couplings should be checked daily or more often when required for snow and ice accumulation and treated with de-ice treatments. Brake guide wire and sheaves should be checked daily or more often when required for snow and ice accumulation and properly greased, if necessary. Lifeboat launching arrangements should be checked daily or more often when required for snow and ice accumulation or more often when required. Winches should be always maintained in a snow and ice-free condition. Ice on hooks, latches and hydrostatics release couplings should be removed using de-icing equipment. De-ice treatments should be applied at least every month or when required. Winches should be operated daily.

Life rafts. Latches/hydrostatic release couplings should be checked daily or more often when required for snow and ice accumulation and treated with de-ice treatments.

Snow and ice should be kept removed. De-ice treatments should be applied at least every month or when required.

Life raft launching arrangements. Winch should be checked and greased according to lubrication chart if required. Brake guide wire and sheaves should be checked daily or more often when required for snow and ice accumulation and properly greased if necessary. Life raft launching arrangements should be checked daily or more often when required for snow and ice accumulation. Snow and ice should be kept removed. Ice on lashes/hydrostatic release couplings should be kept removed using de-icing equipment. Ice should be removed from release hooks using de-icing equipment and should be treated with de-ice treatments at least each month or when required.

17.6.2 Rescue Boats

Radio equipment with batteries should be thoroughly checked for operability every second day. Rescue boat should be inspected two times a day or more often if required for snow and ice accumulation. Engine heaters and battery charger should be checked for operability each day. Ice and snow should be kept removed using de-icing equipment. Engine should be run daily until operating temperature is reached. Rescue boat engine fuel should be checked daily for clouding or waxing. Rescue boat engine lubricating oil should have correct viscosity at design service temperature without the use of heaters.

Rescue boat launching arrangement. Brake guide wire and sheaves should be checked and properly greased if necessary. Quick release hook should be checked for snow and ice accumulation and treated with de-ice treatments. Rescue boat launching arrangement should be test run each week or more often if required. Brake guide wire should be thoroughly checked daily or more often when required for snow and ice accumulation. Snow and ice should be kept removed from winches. Ice on quick release hook should be removed using de-icing equipment. De-ice cure should be applied at least every month or when required. Launching arrangement should be checked three times a day or more often when required for snow and ice accumulation.

17.6.3 Escape Systems

Escape chutes and life raft stations. De-icing equipment should be deployed and ready for use at stations when the temperature drops below 0 °C (32 °F). De-icing treatments should be applied to deck plating to prevent snow and ice accumulation. Life raft station should be checked two times per day or more often if required for snow and ice accumulation. Ice and snow should be removed from deck handrails, raft and launching arrangements using de-icing equipment.

17.6.4 Personal Equipment

Immersion suits and life jackets. If immersion suits are stored at stations, they should be checked for moisture and dried if necessary. Personal immersion suits should be stored in cabins to avoid moisture accumulating in the suits. *Life jackets.* Life jackets should be free from moisture when returned to their stowage boxes after use.

Correction to: Ship Operations in Extreme Low Temperature Environments

Correction to:
A. Olsen, *Ship Operations in Extreme Low Temperature Environments*, Springer Series on Naval Architecture, Marine Engineering, Shipbuilding and Shipping 19, https://doi.org/10.1007/978-3-031-52513-1

This book contains overlap in text with the previously published content [1] that was inadvertently omitted. The authors failed to attribute the reference [1]. The authors have now obtained permission to re-use this content from the American Bureau of Shipping.

References

Where
[1] is: American Bureau of Shipping (2024), *Rules and Guides* https://ww2.eagle.org/en/rules-and-resources/rules-and-guides.html

---

The updated version of this book can be found at https://doi.org/10.1007/978-3-031-52513-1

A. Olsen, *Ship Operations in Extreme Low Temperature Environments*, Springer Series on Naval Architecture, Marine Engineering, Shipbuilding and Shipping 19,
https://doi.org/10.1007/978-3-031-52513-1_18

Annex: Arctic Climate Data Sets

Introduction

The dominant factor for operations in polar and sub-polar regions is the occurrence of extremely low temperatures and the associated formation of ice. In low temperatures, any precipitation will be in the form of snow, or at closer to the freezing point as freezing rain, sleet or ice pellets. Visibility in any of these conditions can be very limited and ice build-up can produce a range of hazards, as described earlier. Ice accumulation due to spray is most likely in air temperatures below 2 °C (36 °F), and wind speeds of above 20 kn (10 m/s). It will worsen as wind speeds increase beyond this, and in higher sea states. In very low temperatures, sea ice can form quite rapidly once the water temperature itself falls below – 1.8 °C (28.8 °F). Vessels with little or no ice capability can find themselves at risk if caught in these conditions, which are most likely to occur towards the onset of winter. More generally, most vessels can be put at risk by ice movement, which can occur with considerable rapidity under conditions of high wind or currents. Conditions reported on ice charts or by remote imagery can change fast, particularly the reported positions of the ice edge and the location of leads through the pack. It is important for mariners to be able to recognise the conditions in which such changes can occur, and signs of the proximity of ice. These can include:

- "Ice blink"—a reflection of ice from the underside of cloud cover; most apparent when there is snow cover on the ice
- The onset of fog, which is often present near the ice edge
- The appearance of small ice floes, which can indicate that larger amounts are present nearby
- Rapid moderation of waves, when approaching the ice edge from downwind.

A. Olsen, *Ship Operations in Extreme Low Temperature Environments*, Springer Series on Naval Architecture, Marine Engineering, Shipbuilding and Shipping 19,
https://doi.org/10.1007/978-3-031-52513-1

Ice Types

For vessels intended to operate in ice, it is important to be able to distinguish between the different ice types that may be encountered, both for efficiency and or safety. The two most fundamental properties of ice cover are thickness and age, both of which are reported on standard ice charts using World Meteorological Organisation (WMO) terminology and symbols, as outlined in Table A.1. Other terminology is also used, particularly for the initial freezing stages, including frazil, grease, shuga, and slush ice. These are all thin forms of ice cover that will normally be reported as new ice. Ice services use a number of techniques to derive the thickness of ice cover, but all of these are approximate. The symbols on ice charts should always be treated with some caution. Old ice is ice that has survived one or more melt seasons. It encompasses both second-year and multi-year ice, but the term multi-year is frequently applied to either old ice form. Multi-year ice becomes much stronger than first year ice, due in part to its reduced salinity. Floes also tend to have much more variable thickness than younger ice, as they incorporate weathered ridges and other features. This and other features help experienced ice navigators to distinguish between first-year and multi-year ice. Ice "of land origin" is generally glacial ice, formed over thousands of years by the accumulation and recrystallisation of packed snow. Ice islands and icebergs enter the sea from glaciers and ice sheets and may in turn 'calve' smaller bergy bits and growlers as they degrade. Glacial ice is very hard and represents a major hazard for vessels with even the highest level of ice capability. Growlers and bergy bits have

Table A.1 Classification of ice stages

Description	Thickness	WMO code
New ice	< 10 cm	1
Nilas; ice rind	0–10 cm	2
Young ice	10–30 cm	3
Grey ice	10–15 cm	4
Grey-white ice	15–30 cm	5
First-year ice	30–200 cm	6
Thin first-year ice	30–70 cm	7
Thin first-year ice first stage	30–50 cm	8
Thin first-year ice second stage	50–70 cm	9
Medium first-year ice	70–120 cm	1
Thick first-year ice	120–200 cm	4
Old ice		7
Second-year ice		8
Multi-year ice		9
Ice of land origin		
Undetermined or unknown		x

small freeboards and can be very difficult to detect either when part of the general ice cover or in open water with moderate sea states. Due to their origin, they are usually found in proximity to icebergs, whose own presence is a good indicator of the potential risk of encountering these smaller fragments.Classification of ice stages

Ice cover is very rarely uniform or homogeneous. Near the coast, ice may be 'land fast', anchored in place by the shoreline and possibly by grounded pressure ridges. Land fast ice tends to have relatively consistent properties but may still include ridges and rubble piles. At the edge of the land fast ice, shear zones may occur where the free-floating pack and land fast ice collide. The shear zone can be a chaotic amalgamation of ridging and rubbling. It can be both difficult and dangerous to transit, especially if the pack is in motion at the time. Even the most powerful ice breakers have become trapped, and less capable vessels have suffered damage or been sunk by pressure events in shear zones. Shear zones should be transited, where necessary, with extreme caution. The general ice pack is typically a mix of ice types, thicknesses and floe sizes at various total ice concentrations. Patches or stretches of open water can be found even in the winter polar pack as floes move relative to each other. In some areas, more or less permanent polynyas of open water exist due to water upwelling. When ice floes move against each other, they may raft, increasing local thicknesses, form rubble, or generate ridging. All of these increase the difficulty of ice transit. Ridges may have sail and keel heights totalling in the tens of meters which can only be penetrated by repeated ramming. Ice charts consolidate all available information on ice cover using the "ice egg", which in most sea areas will be developed using WMO principles and terminology. An example of how the ice egg is developed is shown in Fig. A.1. Each of the fields is filled out using codes, of which the 'stage of development' code given in Table A.1 is one example. More complete explanations can be found in sources such as the Canadian government's MANICE, available online at.

http://www.ec.gc.ca/glaces-ice/default.asp?lang=En&n=4FF82CBD-1.

Air Temperature

Traditionally, many operations in polar regions have been conducted in the summer and autumn seasons when air temperatures are not normally as low as may be encountered in mid-winter. In sub-polar areas such as the Baltic and Gulf of St. Lawrence, all-year operations are the norm. For all of these services, air temperatures below – 20 °C (–4 °F) are relatively rare, though they can be experienced. Midwinter operations are becoming increasingly common on some Arctic routes, and here, temperatures can be considerably lower, with – 30 °C (– 22 °F) to – 35 °C (– 31 °F) not uncommon. Year-round temperature data can be obtained for coastal sites throughout the Arctic and for most sub-polar regions. Temperatures at sea (in the ice)

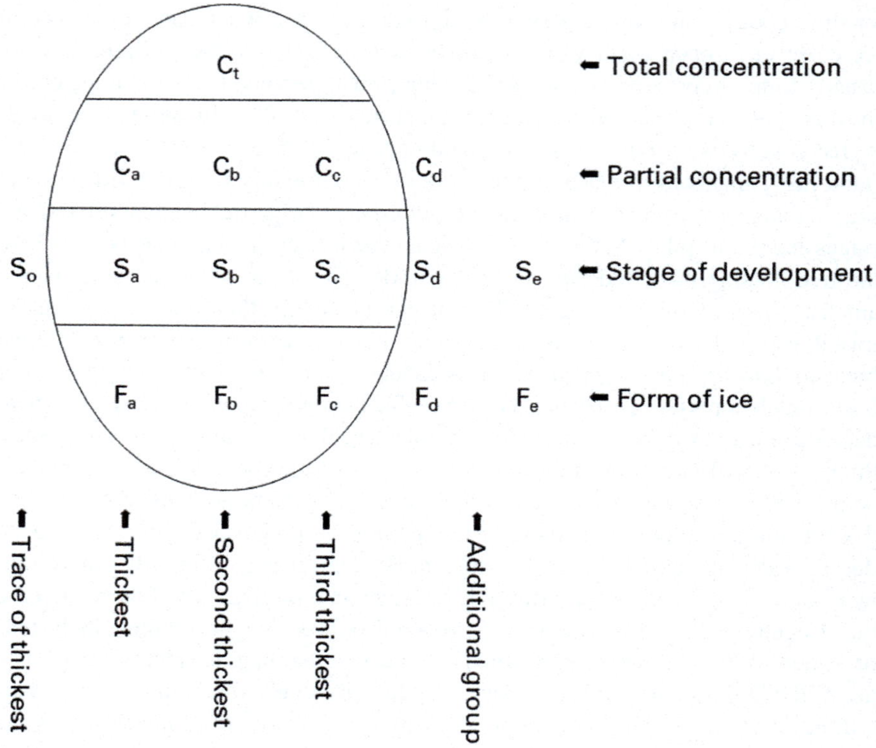

Fig. A.1 Ice egg code

are on average slightly above those on the adjacent land, therefore, port temperature data will normally represent the worst case for equipment functionality and crew comfort.

Humidity

As noted earlier, at extremely cold temperatures, the relative humidity in the air drops to well below that found typically elsewhere. Unless moderated by the vessel's HVAC system, dryness may cause crew discomfort. It may also increase the probability of static electricity build up and discharge, which can damage sensitive equipment and increase fire and explosion risk.

White Outs, Blizzards and Fog

Fog is commonly observed in areas of sea ice. Advection fog is a common occurrence in summer months when warm air masses pass over cold-water areas and sea ice, and the air mass cools below the vapor dew point. Evaporation fog or steam fog is formed when water vapour is added to air which is much colder than the vapour's source. This most commonly occurs when very cold air drifts across relatively warm water. Blowing snow has a significant impact on visibility. Depending on grain size and compaction, visibility in snow at wind speeds over 35 kn become extremely limited, to a few metres in most cases.

Hours of Darkness

Although not strictly a weather condition, the extended hours of darkness at the higher latitudes during the winter months should not be ignored. Extra lighting capacity for deck operations and additional searchlight capability for safe navigation results from this limiting visibility. Additional deck watches for areas where sea ice and glacial ice may be encountered can cause crew fatigue and should be considered when operating in any ice-covered waters. An opposite problem is the effect of almost continuous daylight during the summer months. Crew fatigue needs to be considered when operating for extended periods in ice conditions. Additional blinds for crewing quarters will provide proper resting periods. Ice blindness may be a risk for lookouts and bridge crew, who should be provided with appropriate eyewear. Navigation lights should be of sufficient brightness to be visible in arctic lighting conditions (Tables A.2, A.3, A.4, A.5, A.6, A.7, A.8, A.9, A.10, A.11, A.12, A.13 and A.14).

Table A.2 Arctic locations for land-based temperature data sets (Aasiaat, Qaasuitsup, Greenland)

Parameter	1-Jan	15-Jan	1-Feb	15-Feb	1-Mar	15-Mar	1-Apr	15-Apr
MDAT	– 8.61	– 12.67	– 13.82	– 13.62	– 14.58	– 16.59	– 11.93	– 8.33
MDLT	– 10.54	– 14.97	– 15.82	– 16.16	– 16.97	– 18.75	– 13.93	– 10.61
Rec Lo	– 21.80	– 30.00	– 33.20	– 25.73	– 32.20	– 27.20	– 22.60	– 19.10
StDev (MDLT)	5.84	6.11	7.86	5.40	9.71	6.06	5.73	4.57
Parameter	**1-May**	**15-May**	**1-Jun**	**15-Jun**	**1-Jul**	**15-Jul**	**1-Aug**	**15-Aug**
MDAT	– 2.83	– 0.64	3.10	3.89	5.68	6.80	7.30	6.93
MDLT	– 4.95	– 2.16	1.20	2.35	3.71	4.85	5.82	5.56
Rec Lo	– 12.80	– 7.60	– 2.60	– 0.80	– 0.30	1.90	1.80	2.00
StDev (MDLT)	3.23	2.63	2.29	2.06	2.53	2.62	1.88	1.83
Parameter	**1-Sep**	**15-Sep**	**1-Oct**	**15-Oct**	**1-Nov**	**15-Nov**	**1-Dec**	**15-Dec**
MDAT	5.01	3.56	0.88	– 0.92	– 4.48	– 4.57	– 5.56	– 6.98
MDLT	3.77	2.48	– 0.01	– 2.36	– 5.83	– 6.32	– 7.19	– 8.70
Rec Lo	1.60	– 1.50	– 4.50	– 7.40	– 12.80	– 13.87	– 14.07	– 14.20
StDev (MDLT)	1.17	1.74	2.19	3.19	3.14	3.49	4.65	3.79

Table A.3 Arctic locations for land-based temperature data sets (Alazeja River, Russia)

Parameter	1-Jan	15-Jan	1-Feb	15-Feb	1-Mar	15-Mar	1-Apr	15-Apr
MDAT	– 34.11	– 35.04	– 34.79	– 31.98	– 30.02	– 26.22	– 23.97	– 18.86
MDLT	– 38.79	– 38.51	– 39.33	– 36.76	– 35.25	– 32.20	– 29.15	– 24.11
Rec Lo	– 47.70	– 50.30	– 48.00	– 44.00	– 46.40	– 45.40	– 38.00	– 32.00
StDev (MDLT)	4.91	5.69	5.60	4.81	6.85	8.24	6.65	5.65
Parameter	**1-May**	**15-May**	**1-Jun**	**15-Jun**	**1-Jul**	**15-Jul**	**1-Aug**	**15-Aug**
MDAT	– 11.04	– 5.39	0.46	6.38	12.65	10.88	9.26	6.88
MDLT	– 16.52	– 9.89	– 3.54	2.11	7.11	5.99	5.13	3.49
Rec Lo	– 25.00	– 19.20	– 10.00	– 2.20	– 1.20	0.30	0.00	– 1.00
StDev (MDLT)	5.37	5.51	3.05	3.89	5.99	4.66	4.12	2.87
Parameter	**1-Sep**	**15-Sep**	**1-Oct**	**15-Oct**	**1-Nov**	**15-Nov**	**1-Dec**	**15-Dec**
MDAT	3.76	1.33	– 4.02	– 11.83	– 19.69	– 27.07	– 29.91	– 30.89
MDLT	0.96	– 0.88	– 7.09	– 14.53	– 24.59	– 31.45	– 35.67	– 37.00
Rec Lo	– 2.80	– 5.00	– 12.20	– 28.50	– 33.00	– 40.10	– 44.50	– 46.00
StDev (MDLT)	2.23	2.08	3.23	5.28	5.83	5.84	5.87	8.23

Table A.4 Arctic locations for land-based temperature data sets (Barrow, AK, USA)

Parameter	1-Jan	15-Jan	1-Feb	15-Feb	1-Mar	15-Mar	1-Apr	15-Apr
MDAT	– 23.43	– 24.81	– 27.34	– 21.91	– 26.87	– 26.66	– 20.55	– 16.81
MDLT	– 26.71	– 28.11	– 29.99	– 24.71	– 30.05	– 29.80	– 23.32	– 20.62
Rec Lo	– 34.40	– 42.20	– 46.10	– 35.00	– 42.20	– 42.80	– 33.10	– 33.30
StDev (MDLT)	4.97	6.55	8.19	7.14	6.62	5.13	5.70	5.36
Parameter	1-May	15-May	1-Jun	15-Jun	1-Jul	15-Jul	1-Aug	15-Aug
MDAT	– 10.07	– 6.19	– 1.14	2.31	4.25	6.90	5.00	5.48
MDLT	– 11.99	– 8.60	– 2.63	0.11	1.40	3.71	2.39	2.73
Rec Lo	– 19.80	– 21.00	– 8.40	– 3.00	– 1.10	– 1.70	– 1.10	– 3.00
StDev (MDLT)	5.22	4.02	2.71	2.16	3.00	3.03	2.45	3.07
Parameter	1-Sep	15-Sep	1-Oct	15-Oct	1-Nov	15-Nov	1-Dec	15-Dec
MDAT	2.80	1.83	– 2.74	– 6.36	– 9.71	– 14.79	– 18.39	– 23.00
MDLT	1.22	0.35	– 4.27	– 9.27	– 12.56	– 17.98	– 21.46	– 26.04
Rec Lo	– 2.20	– 6.70	– 12.80	– 24.20	– 23.40	– 28.90	– 35.00	– 35.00
StDev (MDLT)	2.51	2.96	2.79	4.78	5.17	4.42	5.76	6.44

Table A.5 Arctic locations for land-based temperature data sets (Clyde, Nunavut, Canada)

Parameter	1-Jan	15-Jan	1-Feb	15-Feb	1-Mar	15-Mar	1-Apr	15-Apr
MDAT	– 26.34	– 27.97	– 27.48	– 28.99	– 29.08	– 27.91	– 22.02	– 18.99
MDLT	– 29.37	– 31.17	– 30.32	– 32.51	– 31.25	– 31.90	– 25.52	– 23.33
Rec Lo	– 39.00	– 39.60	– 41.00	– 44.00	– 41.00	– 42.00	– 33.00	– 30.80
StDev (MDLT)	6.85	5.89	7.32	6.63	6.81	7.71	4.18	5.37
Parameter	1-May	15-May	1-Jun	15-Jun	1-Jul	15-Jul	1-Aug	15-Aug
MDAT	– 11.80	– 6.09	– 1.85	1.78	3.45	5.62	5.57	4.53
MDLT	– 15.13	– 9.24	– 4.00	– 0.20	1.39	2.28	2.97	2.64
Rec Lo	– 24.00	– 15.00	– 9.00	– 3.40	– 1.70	– 1.90	– 1.00	– 1.70
StDev (MDLT)	5.22	4.13	2.18	1.55	1.60	2.38	2.30	2.44
Parameter	1-Sep	15-Sep	1-Oct	15-Oct	1-Nov	15-Nov	1-Dec	15-Dec
MDAT	2.86	1.45	– 2.62	– 4.70	– 10.07	– 14.17	– 19.59	– 21.88
MDLT	41.19	– 0.42	– 5.43	– 7.53	– 13.51	– 17.15	– 22.70	– 26.05
Rec Lo	– 2.10	– 3.20	– 10.00	– 17.70	– 21.00	– 27.00	– 36.20	– 39.00
StDev (MDLT)	1.83	1.56	3.09	5.13	5.08	5.12	5.89	6.67

Table A.6 Arctic locations for land-based temperature data sets (Golomjannyj, Russia)

Parameter	1-Jan	15-Jan	1-Feb	15-Feb	1-Mar	15-Mar	1-Apr	15-Apr
MDAT	– 26.81	– 28.91	– 28.09	– 27.68	– 28.06	– 25.86	– 23.39	– 19.20
MDLT	– 29.99	– 31.42	– 31.30	– 30.27	– 30.60	– 28.70	– 26.26	– 22.50
Rec Lo	– 40.40	– 42.20	– 40.00	– 37.70	– 38.20	– 38.90	– 39.00	– 34.37
StDev (MDLT)	6.61	5.16	5.20	6.05	5.28	6.70	7.38	5.42
Parameter	**1-May**	**15-May**	**1-Jun**	**15-Jun**	**1-Jul**	**15-Jul**	**1-Aug**	**15-Aug**
MDAT	– 14.44	– 9.33	– 3.87	– 0.80	0.39	0.50	0.87	– 0.12
MDLT	– 18.07	– 11.27	– 5.49	– 1.99	– 0.72	– 0.14	0.04	– 1.01
Rec Lo	– 23.90	– 15.90	– 11.40	– 4.80	– 3.10	– 2.00	– 1.50	– 4.60
StDev (MDLT)	4.06	2.62	3.00	1.57	1.33	1.44	0.79	1.43
Parameter	**1-Sep**	**15-Sep**	**1-Oct**	**15-Oct**	**1-Nov**	**15-Nov**	**1-Dec**	**15-Dec**
MDAT	– 1.11	– 2.62	– 5.88	– 10.38	– 16.83	– 19.68	– 20.58	– 25.30
MDLT	– 2.21	– 3.89	– 7.39	– 12.10	– 19.35	– 22.78	– 23.25	– 27.82
Rec Lo	– 8.80	– 11.80	– 17.70	– 25.57	– 31.47	– 32.40	– 32.20	– 37.30
StDev (MDLT)	2.76	3.53	5.16	6.73	6.97	4.45	5.93	5.79

Table A.7 Arctic locations for land-based temperature data sets (IM. M.V. Popova, Russia)

Parameter	1-Jan	15-Jan	1-Feb	15-Feb	1-Mar	15-Mar	1-Apr	15-Apr
MDAT	– 22.70	– 22.24	– 23.40	– 24.69	– 24.54	– 24.56	– 21.09	– 16.72
MDLT	– 26.58	– 26.20	– 27.04	– 28.16	– 29.87	– 28.35	– 25.47	– 21.57
Rec Lo	– 40.43	– 38.90	– 38.07	– 41.97	– 40.70	– 41.33	– 34.33	– 34.63
StDev (MDLT)	9.31	7.88	8.86	9.09	6.83	7.51	5.34	7.73
Parameter	**1-May**	**15-May**	**1-Jun**	**15-Jun**	**1-Jul**	**15-Jul**	**1-Aug**	**15-Aug**
MDAT	– 11.05	– 7.24	– 2.61	0.22	2.84	4.78	6.61	5.39
MDLT	– 15.11	– 10.97	– 4.60	– 1.26	0.97	2.78	4.68	3.78
Rec Lo	– 26.90	– 20.20	– 8.60	– 4.20	– 3.00	– 0.50	0.00	1.00
StDev (MDLT)	6.25	4.65	3.03	1.62	2.63	2.20	2.53	1.90
Parameter	**1-Sep**	**15-Sep**	**1-Oct**	**15-Oct**	**1-Nov**	**15-Nov**	**1-Dec**	**15-Dec**
MDAT	4.34	2.80	– 0.12	– 3.93	– 11.29	– 13.03	– 14.60	– 18.62
MDLT	2.75	1.28	– 1.05	– 6.25	– 13.90	– 15.49	– 18.19	– 22.63
Rec Lo	0.30	– 2.50	– 6.10	– 14.03	– 22.53	– 27.93	– 29.00	– 32.38
StDev (MDLT)	1.86	2.17	2.32	4.56	6.01	7.47	7.11	9.79

Table A.8 Arctic locations for land-based temperature data sets (Malye Karmakuly, Russia)

Parameter	1-Jan	15-Jan	1-Feb	15-Feb	1-Mar	15-Mar	1-Apr	15-Apr
MDAT	– 14.55	– 12.81	– 16.42	– 14.10	– 12.09	– 10.03	– 12.59	– 9.45
MDLT	– 17.78	– 16.69	– 18.69	– 16.79	– 14.30	– 12.17	– 15.62	– 11.87
Rec Lo	– 28.90	– 25.10	– 27.63	– 28.90	– 26.50	– 25.50	– 26.50	– 23.43
StDev (MDLT)	7.55	5.93	7.07	6.94	7.47	6.20	6.18	6.96
Parameter	**1-May**	**15-May**	**1-Jun**	**15-Jun**	**1-Jul**	**15-Jul**	**1-Aug**	**15-Aug**
MDAT	– 5.83	– 3.87	– 0.06	2.70	6.11	8.27	7.70	7.14
MDLT	– 8.22	– 5.92	– 1.38	1.01	4.06	5.97	5.97	5.20
Rec Lo	– 15.97	– 14.50	– 4.70	– 1.90	– 0.60	0.70	2.87	2.39
StDev (MDLT)	4.92	4.68	2.26	2.46	4.00	2.69	2.19	1.72
Parameter	**1-Sep**	**15-Sep**	**1-Oct**	**15-Oct**	**1-Nov**	**15-Nov**	**1-Dec**	**15-Dec**
MDAT	4.89	4.00	1.76	– 1.39	– 5.65	– 6.86	– 7.15	– 10.68
MDLT	3.23	2.35	0.37	– 2.97	– 7.99	– 8.71	– 9.43	– 13.66
Rec Lo	– 0.67	– 3.00	– 3.07	– 9.40	– 16.83	– 21.70	– 23.00	– 24.30
StDev (MDLT)	2.09	3.25	1.95	3.70	4.91	6.30	6.68	6.77

Table A.9 Arctic locations for land-based temperature data sets (MYS Uelen, Chukotka Autonomous, Russia)

Parameter	1-Jan	15-Jan	1-Feb	15-Feb	1-Mar	15-Mar	1-Apr	15-Apr
MDAT	– 16.94	– 21.71	– 23.61	– 16.91	– 23.07	– 18.43	– 16.06	– 14.10
MDLT	– 19.55	– 24.25	– 25.76	– 19.91	– 25.93	– 20.62	– 18.49	– 16.63
Rec Lo	– 34.67	– 39.83	– 41.40	– 30.50	– 36.30	– 38.70	– 28.50	– 23.80
StDev (MDLT)	8.69	6.63	10.84	6.67	6.60	8.09	7.80	5.43
Parameter	**1-May**	**15-May**	**1-Jun**	**15-Jun**	**1-Jul**	**15-Jul**	**1-Aug**	**15-Aug**
MDAT	– 7.46	– 2.13	0.24	3.25	6.60	7.37	8.21	7.24
MDLT	– 8.84	– 3.52	– 1.03	1.56	4.49	5.84	6.57	5.72
Rec Lo	– 18.10	– 13.00	– 4.90	– 1.20	1.10	2.60	2.20	2.00
StDev (MDLT)	4.71	3.65	1.65	1.55	2.33	2.48	2.83	2.32
Parameter	**1-Sep**	**15-Sep**	**1-Oct**	**15-Oct**	**1-Nov**	**15-Nov**	**1-Dec**	**15-Dec**
MDAT	5.18	4.55	1.85	– 0.89	– 2.72	– 6.28	– 11.98	– 16.92
MDLT	4.19	3.36	0.90	– 2.42	– 4.27	– 9.21	– 14.02	– 19.67
Rec Lo	0.00	0.20	– 3.00	– 12.00	– 12.00	– 17.40	– 23.80	– 33.80
StDev (MDLT)	2.00	2.02	1.55	4.53	3.90	5.47	7.35	7.26

Table A.10 Arctic locations for land-based temperature data sets (Ostrov Kotelnyj, Russia)

Parameter	1-Jan	15-Jan	1-Feb	15-Feb	1-Mar	15-Mar	1-Apr	15-Apr
MDAT	– 27.76	– 29.52	– 27.57	– 28.25	– 30.78	– 27.30	– 24.84	– 17.68
MDLT	– 30.15	– 31.89	– 31.24	– 31.18	– 33.19	– 30.53	– 27.60	– 20.88
Rec Lo	– 38.20	– 38.10	– 42.30	– 39.80	– 38.30	– 37.50	– 34.60	– 32.50
StDev (MDLT)	5.59	4.56	5.35	5.02	3.35	4.76	4.61	5.35
Parameter	1-May	15-May	1-Jun	15-Jun	1-Jul	15-Jul	1-Aug	15-Aug
MDAT	– 12.58	– 8.18	– 3.35	0.22	3.14	2.77	3.20	2.86
MDLT	– 14.95	– 10.95	– 4.69	– 1.14	1.06	1.15	1.89	1.34
Rec Lo	– 22.90	– 18.00	– 10.10	– 3.80	– 2.40	– 2.20	– 1.80	– 1.20
StDev (MDLT)	4.60	4.34	2.55	1.86	1.77	1.77	2.12	2.38
Parameter	1-Sep	15-Sep	1-Oct	15-Oct	1-Nov	15-Nov	1-Dec	15-Dec
MDAT	0.94	– 0.33	– 3.41	– 8.56	– 18.20	– 21.17	– 24.93	– 26.40
MDLT	– 0.50	– 1.45	– 5.14	– 10.98	– 20.64	– 23.34	– 27.95	– 29.31
Rec Lo	– 6.90	– 6.90	– 13.10	– 23.20	– 30.70	– 31.00	– 34.30	– 37.30
StDev (MDLT)	2.81	2.59	3.76	5.38	4.82	4.62	5.73	5.52

Table A.11 Arctic locations for land-based temperature data sets (Pelly Island, Northwest Territories, Canada)

Parameter	1-Jan	15-Jan	1-Feb	15-Feb	1-Mar	15-Mar	1-Apr	15-Apr
MDAT	– 22.59	– 25.42	– 28.20	– 24.51	– 22.78	– 25.82	– 20.81	– 12.01
MDLT	– 25.53	– 28.42	– 31.68	– 27.07	– 25.28	– 28.44	– 23.88	– 14.92
Rec Lo	– 36.30	– 37.20	– 39.00	– 41.10	– 37.00	– 38.00	– 29.30	– 29.60
StDev (MDLT)	6.25	5.76	6.09	7.31	6.83	5.04	3.88	6.56
Parameter	1-May	15-May	1-Jun	15-Jun	1-Jul	15-Jul	1-Aug	15-Aug
MDAT	– 7.77	– 4.87	0.35	2.64	8.11	10.07	10.22	8.69
MDLT	– 9.91	– 6.86	– 0.89	1.14	5.56	7.61	7.72	6.92
Rec Lo	– 20.30	– 16.40	– 6.30	– 2.00	0.00	1.00	1.90	0.70
StDev (MDLT)	5.53	4.02	2.03	2.22	5.01	4.16	3.20	2.67
Parameter	1-Sep	15-Sep	1-Oct	15-Oct	1-Nov	15-Nov	1-Dec	15-Dec
MDAT	5.74	3.12	– 0.81	– 4.89	– 11.00	– 18.42	– 20.33	– 24.69
MDLT	3.82	1.92	– 1.80	– 5.72	– 13.72	– 20.72	– 23.16	– 26.94
Rec Lo	0.80	– 2.00	– 5.10	– 13.00	– 22.60	– 32.50	– 33.48	– 35.90
StDev (MDLT)	2.62	2.45	2.14	3.96	4.81	5.30	5.96	6.34

Table A.12 Arctic locations for land-based temperature data sets (Resolute, Nunavut, Canada)

Parameter	1-Jan	15-Jan	1-Feb	15-Feb	1-Mar	15-Mar	1-Apr	15-Apr
MDAT	– 28.30	– 31.14	– 32.66	– 32.37	– 34.03	– 30.38	– 24.86	– 20.94
MDLT	– 31.49	– 33.87	– 35.82	– 35.10	– 36.56	– 33.16	– 27.55	– 23.71
Rec Lo	– 40.80	– 42.00	– 45.00	– 40.50	– 44.00	– 44.00	– 36.00	– 33.00
StDev (MDLT)	4.74	4.47	3.79	4.60	6.52	5.43	5.98	5.75
Parameter	1-May	15-May	1-Jun	15-Jun	1-Jul	15-Jul	1-Aug	15-Aug
MDAT	– 15.77	– 9.45	– 3.89	0.60	4.31	4.35	4.33	2.24
MDLT	– 18.32	– 11.96	– 5.87	– 1.21	2.01	2.03	2.47	0.57
Rec Lo	– 25.60	– 20.00	– 14.00	– 8.00	– 2.00	– 1.00	– 1.00	– 2.00
StDev (MDLT)	4.27	5.23	4.14	2.71	1.83	2.45	3.25	1.79
Parameter	1-Sep	15-Sep	1-Oct	15-Oct	1-Nov	15-Nov	1-Dec	15-Dec
MDAT	– 0.21	– 2.66	– 7.22	– 10.91	– 17.81	– 21.58	– 25.07	– 25.79
MDLT	– 1.92	– 4.62	– 9.29	– 13.56	– 20.00	– 24.38	– 27.44	– 28.63
Rec Lo	– 9.00	– 12.20	– 13.50	– 21.00	– 33.00	– 34.00	– 36.00	– 42.30
StDev (MDLT)	2.75	3.39	2.79	5.14	6.93	6.47	4.59	7.36

Table A.13 Arctic locations for land-based temperature data sets (Reykjavik, Iceland)

Parameter	1-Jan	15-Jan	1-Feb	15-Feb	1-Mar	15-Mar	1-Apr	15-Apr
MDAT	0.25	– 0.02	0.60	1.87	– 0.58	0.85	1.55	3.82
MDLT	– 1.94	– 2.07	– 2.17	– 0.82	– 3.13	– 1.38	– 1.21	1.25
Rec Lo	– 10.00	– 9.00	– 14.00	– 6.00	– 14.00	– 8.30	– 8.30	– 3.40
StDev (MDLT)	3.69	3.90	5.24	2.99	4.73	3.88	4.04	3.13
Parameter	1-May	15-May	1-Jun	15-Jun	1-Jul	15-Jul	1-Aug	15-Aug
MDAT	4.57	6.21	8.50	9.70	11.20	11.36	12.19	11.43
MDLT	1.96	3.04	6.00	7.10	9.00	8.71	9.86	8.92
Rec Lo	– 2.70	– 2.20	1.00	4.00	6.60	7.00	7.40	5.00
StDev (MDLT)	2.85	2.87	2.14	1.71	1.51	1.27	1.06	2.45
Parameter	1-Sep	15-Sep	1-Oct	15-Oct	1-Nov	15-Nov	1-Dec	15-Dec
MDAT	10.83	8.48	6.98	5.67	3.09	1.26	0.51	1.93
MDLT	8.61	6.09	4.39	3.21	0.35	– 1.04	– 2.46	– 0.54
Rec Lo	5.00	1.00	1.00	– 7.20	– 6.20	– 7.00	– 8.70	– 6.00
StDev (MDLT)	2.22	2.96	2.67	4.31	3.82	4.88	3.98	4.14

Table A.14 Arctic locations for land-based temperature data sets (Tromsø, Langnes, Norway)

Parameter	1-May	15-May	1-Jun	15-Jun	1-Jul	15-Jul	1-Aug	15-Aug
MDAT	– 2.40	– 1.69	– 3.20	– 2.65	– 3.09	– 0.92	– 0.58	1.46
MDLT	– 5.20	– 3.97	– 6.04	– 5.35	– 5.44	– 3.27	– 3.69	– 1.49
Rec Lo	– 13.30	– 13.30	– 14.00	– 15.80	– 14.00	– 11.00	– 10.00	– 8.00
StDev (MDLT)	4.17	3.92	5.08	4.83	4.84	4.32	3.93	3.27
Parameter	1-May	15-May	1-Jun	15-Jun	1-Jul	15-Jul	1-Aug	15-Aug
MDAT	3.32	4.62	7.07	8.91	11.77	13.32	12.88	10.99
MDLT	0.50	1.98	4.45	6.46	8.10	9.69	9.78	8.13
Rec Lo	– 5.10	– 2.00	1.60	4.00	4.70	6.00	6.00	5.00
StDev (MDLT)	3.05	2.82	1.77	2.09	2.06	2.30	2.16	1.83
Parameter	1-Sep	15-Sep	1-Oct	15-Oct	1-Nov	15-Nov	1-Dec	15-Dec
MDAT	9.75	7.86	5.37	3.70	1.24	– 0.63	– 0.21	0.20
MDLT	6.77	5.38	2.37	1.17	– 0.92	– 2.56	– 2.54	– 2.19
Rec Lo	0.80	0.00	– 1.00	– 5.00	– 8.00	– 7.00	– 11.00	– 10.00
StDev (MDLT)	2.91	3.09	2.45	3.95	4.00	2.88	3.74	4.83

Glossary

Aged ridge Ridge which has undergone considerable weathering. These ridges are best described as undulations.

Anchor ice Submerged ice attached or anchored to the bottom, irrespective of the nature of its formation.

Area of weakness A satellite-observed area in which either the ice concentration or the ice thickness is significantly less than that in the surrounding areas. Because the condition is satellite observed, a precise quantitative analysis is not always possible, but navigation conditions are significantly easier than in surrounding areas.

Atrium A body cavity or chamber, especially either of the upper chambers of the heart that receives blood from the veins and forces it into a ventricle.

Auroral zones Zones of variable RF propagation in higher latitudes, due to charged particles ejected from the sun and deflected by the earth's magnetic fields.

Bare ice Ice without snow cover.

Beaufort gyre A current system centred near 78° N 140° W which moves ice in a counter-clockwise rotation before dispersing it into the Transpolar Drift or along the coasts of Ellesmere Island and Greenland.

Belt A large feature of ice arrangements, longer than it is wide; from 1 km (0.62 mi) to more than 100 km (62 mi) in width.

Bergy bit A large piece of floating glacier ice, generally showing less than 5 m (16.4 ft) above sea level but more than 1 m (3.2 ft) and normally about 100–300 m^2 (328–984 ft^2) in area. It is smaller than an ICEBERG but larger than a GROWLER. A typical bergy bit is about the size of a small house.

Bergy water An area of freely navigable water in which glacier ice is present in concentrations of less than 1/10. There may be sea ice present, although the total concentration of all ice shall not exceed 1/10.

Beset Situation of a vessel surrounded by ice and unable to move.

Big floe See Floe.

A. Olsen, *Ship Operations in Extreme Low Temperature Environments*, Springer Series on Naval Architecture, Marine Engineering, Shipbuilding and Shipping 19,
https://doi.org/10.1007/978-3-031-52513-1

Bight Extensive crescent-shaped indentation in the ice edge, formed by either wind or current.

Blue ice The oldest and hardest form of glacier ice, distinguished by a slightly bluish or greenish colour.

Brash ice Accumulations of floating ice made up of fragments not more than 2 m (6.5 ft) across, the wreckage of other forms of ice.

Bummock A downward projection from the underside of an ice field; the counterpart of a hummock.

Calving The breaking away of a mass of ice from an ice wall, ice front, or iceberg.

Chilblains An inflammation followed by itchy irritation on the hands, feet, or ears, resulting from exposure to moist cold.

Close ice Floating ice in which the concentration is 7/10 to 8/10, composed of floes mostly in contact.

Clo-units The rate of a person's heat loss as measured in watts per square meter of skin area per kelvin of temperature difference across the clothing; the value of insulation is measured by the reciprocal of this rate, in square meter kelvins per watt (m^2 K/W).

Compact ice Floating ice in which the concentration is 10/10 and no water is visible.

Compacted ice edge Close, clear-cut ice edge compacted by wind or current; usually on the windward side of an area of ice.

Compacting Pieces of floating ice are said to be compacting when they are subjected to converging motion, which increases ice concentration and/or produces stresses which may result in ice deformation.

Concentration boundary A line approximately the transition between two areas of drift-ice with distinctly different concentrations.

Concentration The ratio expressed in tenths of the sea surface actually covered by ice to the total area of sea surface, both ice-covered and ice-free, at a specific location or over a defined area.

Consolidated ice Floating ice in which the concentration is 10/10 and the floes are frozen together.

Consolidated pack ice Pack ice in which the concentration is 10/10 and the floes are frozen together.

Consolidated ridge A ridge in which the base has frozen together due to melting or other processes.

Crack Any fracture which has not parted.

Dark Nilas Nilas which is under 5 cm (1.9 in.) in thickness and is very dark in colour.

Deformed ice A general term for ice which has been squeezed together and in places forced upwards (and downwards). Subdivisions are rafted ice, ridged ice, and hummocked ice.

Dehydration A lack of sufficient fluid intake to make up for fluids lost.

Diffuse ice edge Poorly defined ice edge limiting an area of dispersed ice; usually on the leeward side of an area of pack ice.

Displacement The displacement, D, is the moulded displacement in metric tonnes (long tonnes) at the upper ice waterline.

Diverging Ice-fields or floes in an area are subjected to diverging or dispersive motion, thus reducing ice concentration and/or relieving stresses in the ice.

Dried ice Sea ice from the surface of which melt-water has disappeared after the formation of cracks and thaw holes. During the period of drying, the surface whitens.

Drift-ice Sea ice that is drifting freely.

Embrittlement The process of materials becoming brittle from extreme cold weather.

Ergonomics The applied science of equipment design, as for the workplace, intended to maximise productivity by reducing operator fatigue and discomfort.

Extension of land glaciers Limited areas may be aground. The seaward edge is termed an ice front.

Fast ice boundary The ice boundary at any given time between fast ice and drift-ice.

Fast ice edge The demarcation at any given time between fast ice and open water.

Fast ice Sea ice which forms and remains attached to the shore, to an ice wall, to an ice front, between shoals or grounded icebergs. Vertical fluctuations may be observed during changes of sea level. Fast ice may be formed in situ from the sea water or by freezing of pack ice of any age to the shore, and it may extend a few meters or several hundred kilometres from the coast. Fast ice may be more than 1 year old and may then be prefixed with the appropriate age category (old, second-year or multi-year). If it is thicker than about 2 m (3.2 ft) above sea level, it is called an ice shelf.

Fast-ice boundary The ice boundary at any given time between fast ice and pack ice.

Fast-ice edge The demarcation at any given time between fast ice and open water.

Fibrillation Rapid uncoordinated twitching movements that replace the normal rhythmic contraction of the heart and may cause a lack of circulation and pulse.

Finger-rafted ice Type of rafted ice in which floes thrust "fingers" alternately over and under one another.

Finger-rafting Type of rafting whereby interlocking thrusts are formed, each floe thrusting "fingers" alternately over and under one another. Common in Nilas and grey ice.

Firn Old snow which has recrystallised into a dense material. Unlike snow, the particles are to some extent joined together; but, unlike ice, the air spaces in it still connect with each other.

First-year ice Sea ice of not more than one winter's growth, developing from young ice, with a thickness of 30 cm (11.8 in.) to 2 m (6.5 ft). First-year ice may be subdivided into thin first year ice, white ice, medium first year ice, and thick first year ice.

Flaw lead A passageway between drift-ice and fast ice which is navigable by surface vessels.

Flaw polynya A polynya between drift-ice and fast ice.

Flaw A narrow separation zone between drift-ice and fast ice, where the pieces of ice are in chaotic state; it forms when drift - ice shears under the effect of a strong wind or current along the fast ice boundary.

Floating ice Any form of ice found floating in water. The principal kinds of floating ice are lake ice, river ice, and sea ice, which form by the freezing of water at the surface, and glacier ice (ice of land origin) formed on land or in an ice shelf. The concept includes ice that is stranded or grounded.

Floe Any relatively flat piece of sea ice 20 m (65 ft) or more across. Floes are subdivided according to horizontal extent. A giant flow is over 5.4 nm (10 km; 6.2 mi) across; a vast floe is 1.1 to 5.4 nm (2–10 km; 6.2 mi) across; a big floe is 500–2000 m (1640–6561 ft) across; a medium floe is 100–500 m (328–1640 ft) across; and a small floe is 20–100 m (65–328 ft) across.

Floeberg A massive piece of sea ice composed of a hummock, or a group of hummocks frozen together, and separated from any ice surroundings. It may float showing up to 5 m (16.4 ft) above sea level.

Floebit A relatively small piece of sea ice, normally not more than 10 m (32.8 ft) across, composed of a hummock(s) or part of a ridge(s) frozen together and separated from any surroundings. It typically protrudes up to 2 m (6.5 ft) above sea-level.

Flooded ice Sea ice which has been flooded by melt-water or river water and is heavily loaded by water and wet snow.

Fracture zone An area which has a great number of fractures.

Fracture Any break or rupture through very close ice, compact ice, consolidated ice, fast ice, or a single floe, resulting from deformation processes. Fractures may contain brash ice and/or be covered with Nilas and/or young ice. Length may vary from a few metres to many kilometres.

Fracturing Pressure whereby ice is permanently deformed, and rupture occurs. Most commonly used to describe breaking across very dose ice, compact ice, and consolidated ice.

Frazil ice Fine spicules or plates of ice suspended in water.

Free communication Occurs if the hull is ruptured so that one or more compartments are open to the sea.

Free surface Effect from ship's tank(s) or void(s) when only partially filled and the liquid contents "slosh" back and forth.

Freshwater icing When ice form on the ship's surfaces form drops of rain, damp snow or other fresh water source.

Friendly ice From the point of view of the submariner, an ice canopy containing many large skylights or other features which permit a submarine to surface. There must be more than 10 such features per 30 nm (56 km; 37 mi) along the submarine's track.

Frost smoke Fog-like clouds due to the contact of cloud air with relatively warm water, which can appear over openings in the ice, or leeward of the ice edge, and which may persist while ice is forming.

Frostbite The freezing of any part of the body. The water in the tissue cells turns to ice and disrupts the normal function of the tissue.

Frostnip Superficial frostbite, affecting only the skin or the tissue immediately beneath it.

Glacier berg An irregularly shaped iceberg.

Glacier ice Ice in or originating from a glacier, whether on land or floating on the sea as icebergs, bergy bits, or growlers.

Glacier tongue Projecting seaward extension of a glacier, usually afloat. In the Antarctic, glacier tongues may extend over many tens of kilometres.

Glacier A mass of snow and freshwater ice continuously moving from higher to lower ground or, if afloat, continuously spreading. The principal forms of glacier are: inland ice sheets, ice shelves, ice streams, icecaps, ice piedmonts, cirque glaciers, and various types of mountain (valley) glaciers.

Glaze A coating of ice, generally clear and smooth but usually containing some air pockets, formed on exposed objects by the freezing of a film of super cooled water deposited by rain, drizzle, fog, or possibly condensed from super cooled water vapor. Glaze is denser, harder and more transparent than either rime or hoarfrost Also called glaze ice, glazed frost verglas.

Grease ice Ice at that stage of freezing when the crystals have coagulated to form a soupy layer on the surface. Grease ice is at a later stage of freezing than frazil ice and reflects little light, giving the sea a matte appearance.

Green water Solid seawater (waves), not sea spray, which comes over the ship when in heavy seas. Is dangerous because it can damage equipment, break windshields and antennas and can wash people overboard.

Grey ice Young ice 10–15 cm (3.9–5.9 in.) thick. Less elastic than Nilas, it breaks on swell. Usually rafts under pressure.

Grey-white ice Young ice 15–30 cm (5.9–11.8 in.) thick. Under pressure, more likely to ridge than raft.

Grounded hummock Hummocked, grounded ice formation. There are single grounded hummocks and lines (or chains) of grounded hummocks.

Grounded ice Floating ice which is aground in shoal water.

Growler A piece of ice smaller than a bergy bit or floeberg, often transparent but appearing green or almost black in colour. It extends less than 1 m (3.2 ft) above the sea surface and its length is less than 6 m (20 ft). A growler is large enough to be a hazard to shipping but small enough that it may escape visual or radar detection.

Halo Commonly a ring of light of radius 22 or 46 with the sun or moon at the centre, caused by refraction of light by ice crystals in the atmosphere. Occasionally, a faint circle with a radius of 90 appears around the sun.

Hostile ice From the point of view of the submariner, an ice canopy containing no large skylights or other features which permit a submarine to surface.

Hummock (1) A hillock of broken ice which has been forced upwards by pressure. It may be fresh or weathered. The submerged volume of broken ice under the hummocks, forced downwards by pressure, is called a bummock; (2) a natural elevation of the earth's surface resembling a hillock, but smaller and lower.

Hummocked ice Sea ice piled haphazardly one piece over another to form an uneven surface. When weathered, hummocked ice has the appearance of smooth hillocks.

Hummocking The pressure process by which sea ice is forced into hurnmocks. When the floes rotate in the process it is termed screwing.

Hypothermia A condition that occurs when the body is unable to maintain adequate warmth and the body core temperature drops below normal.

Hypothermia Abnormally low body temperature.

Ice accretion The building up of ice which occurs when airborne moisture comes in contact with cold metal.

Ice belt The ice belt is that reinforced portion of the shell and hull appendages that overlaps the upper and lower ice waterlines and is subject to the design ice loads. The required ice belt overlap extends from 1.5 m (4.9 ft) below the lower ice waterline to 1.0 m (3.2 ft) or 1.5 m (4.9 ft) above the upper ice waterline, depending upon Polar Class. In the bow area, the overlap increases linearly to 2.0 m (6.5 ft) above the upper ice waterline at the stem.

Ice boundary The demarcation at any given time between fast ice and drift-ice, or between areas of drift-ice of different concentrations.

Ice breccia Ice of different stages of development frozen together.

Ice cake Any relatively flat piece of sea ice from 2 m to 20 m (6.5–65.6 ft) across.

Ice canopy Drift-ice from the point of view of the submariner.

Ice cover The ratio of an area of ice of any concentration to the total area of sea surface within some large geographic locale; this locale may be global, hemispheric, or prescribed by a specific oceanographic entity such as Baffin Bay or the Barents Sea.

Ice edge The demarcation at any given time between the open sea and sea ice of any kind.

Ice field Area of floating ice consisting of any size of floes, which is greater than 10 km (6.2 mi) across.

Ice foot A narrow fringe of ice attached to the coast, unmoved by tides, and remaining after the fast ice has moved away.

Ice free An area with no ice present. If ice of any kind is present, the area is not ice-free.

Ice front The vertical cliff forming the seaward face of an ice shelf or other floating glacier varying in height from 2–50 m (6.5–164 ft) or more above sea-level.

Ice island A large piece of floating ice, protruding about 5 m (16.4 ft) above sea-level, which has broken away from an ice shelf. It has a thickness of 30–50 m (98.4–164 ft) and an area of from a few thousand square metres to 500 km^2 (310 mi^2) or more and is usually characterised by a regularly undulating surface which gives it a ribbed appearance from the air.

Ice isthmus A narrow connection between two ice areas of very close or compact pack-ice. It may be difficult to pass, yet sometimes being part of a recommended route.

Ice jam An accumulation of broken river ice or sea ice caught in a narrow channel.

Ice keel From the point of view of the submariner, a downward-projecting ridge on the underside of the ice canopy; the counterpart of a ridge. Ice keels may extend as much as 50 m (164 ft) below sea-level.

Ice limit Climatological term referring to the extreme minimum or extreme maximum extent of the ice edge in any given month or period based on observations over a number of years. Term should be preceded by minimum or maximum.

Ice massif A variable accumulation of close or very close pack-ice covering hundreds of square kilometres which is found in the same regions every summer.

Ice of land origin Ice formed on land or in an ice shelf, found floating in water. The concept includes ice that is stranded or grounded.

Ice patch An area of floating ice less than 10 km (6.2 mi) across.

Ice port An embayment in an ice front, often of a temporary nature, where ships can moor alongside and unload directly onto the ice shelf.

Ice rind A brittle shiny crust of ice formed on a quiet surface by direct freezing or from grease ice, usually in water of low salinity. Thickness to about 5 cm (1.9 in.). Easily broken by wind or swell, commonly breaking into rectangular pieces. Nourished by annual snow accumulation and often also by the seaward.

Ice shelf A floating ice sheet of considerable thickness showing 2–50 m (6.5–164 ft) or more above sea- level, attached to the coast. Usually of great horizontal extent and with a level or gently undulating surface

Ice stream Part of an inland ice sheet in which the ice flows more rapidly and not necessarily in the same direction as the surrounding ice. The margins are sometimes clearly marked by a change in the direction of the surface slope but may be indistinct.

Ice under pressure Ice in which deformation processes are actively occurring and hence is a potential impediment or danger to shipping.

Ice wall An ice cliff forming on the seaward margin of a glacier which is not afloat. An ice wall is aground, the rock basement being at or below sea-level.

Iceberg tongue A major accumulation of icebergs projecting from the coast, held in place by grounding and joined together by fast ice.

Iceberg A large mass of floating ice broken away from a glacier or a large mass of ice, floating or around, that has calved from a glacier, i.e., freshwater ice.

Iceberg A massive piece of ice greatly varying in shape, showing more than 5 m (16.4 ft) above the sea surface, which has broken away from a glacier, and which may be afloat or aground. Icebergs may be described as tabular, dome shaped, pinnacled, drydock, glacier or weathered, blocky, tilted blocky, or drydock icebergs. For reports to the International Ice Patrol, they are described with respect to size as small, medium, or large icebergs.

Ice-blink A whitish glare on low clouds above an accumulation of distant ice.

Iceblink A whitish glare on the underside of low clouds caused by the sun's reflection off the surface of pack ice.

Ice-bound A harbour, inlet, etc. is said to be ice-bound when navigation by ships is prevented on account of ice, except possibly with the assistance of an ice-breaker.

Icefoot A narrow fringe of ice attached to the coast, unmoved by tides and remaining after the fast ice has moved away. It may be formed on a slight swell from grease ice, shuga, as a result of the breaking of ice rind, Nilas, or under severe conditions of swell or waves, of grey ice.

Jammed brash barrier A strip or narrow belt of new, young, or brash ice (usually 100–5000 m (328–16,404 ft wide), formed at the edge of either drift-ice or fast ice, or at the shore. It is heavily compacted mostly due to wind action and may extend 2–20 m (6.5–65.6 ft) below the surface but does not normally have appreciable topography. A jammed brash barrier may disperse with changing winds but can also consolidate to form a strip of unusually thick ice as compared to the surrounding pack-ice.

Lake ice Ice formed on a lake, regardless of observed location.

Land sky Dark streaks or patches or a greyness on the underside of extensive cloud areas, due to the absence of reflected light from bare ground. Land sky is not as dark as water sky. The clouds above ice or snow covered surfaces have a white or yellowish white glare called ice blink.

Large fracture More than 500 m (1640 ft) wide.

Large ice-field An ice-field over 20 km (12.4 mi) across.

Lead Any fracture or passageway through sea ice which is navigable by surface vessels; long open cracks.

Length The vessel's length, L, is in accordance with Rule 3(j) (General definitions) of the COLREGs, wherein the words "length" and "breadth" of a vessel mean her length overall and greatest breadth. Length overall (*LOA*) means the length of a ship measured from the foremost point of the stem to the aftermost part of the stern. Breadth, *B*, means the width of a vessel at the widest point. For Polar Class vessels, this measured on the upper ice waterline, in m (ft).

Level ice Sea ice which is unaffected by deformation.

Light Nilas Nilas which is more than 5 cm (1.9 in.) in thickness and rather lighter in colour than dark Nilas.

Looming When as a result of temperature inversion accompanied by rapid decrease in humidity, refraction becomes greater than normal, objects which are normally below the horizon become visible.

Main frame Main frames are real, or in the case of longitudinal framing, imaginary transverse frames, whose spacing corresponds to that of the vessel clear of the ice strengthening area, or of the vessel if it were not ice-strengthened.

Marginal ice zone (MIZ) The region from the ice edge (where ice is first encountered) to a point that is sufficiently far from the ocean boundary so as not to be affected by the presence of the open ocean, i.e., no evidence of wave action, swell, etc.

Mean ice edge Average position of the ice edge in any given month or period based on observations over a number of years. Other terms which may be used are mean maximum ice edge and mean minimum ice edge.

Medium first-year ice First-year ice 70–120 cm (27.5–47.2 in.) thick

Medium fracture 200 to 500 m (656–1640 ft) wide.

Medium ice field An ice-field 15–20 km (9.3–12.4 mi) across.

Melatonin A hormone that plays a role in sleep.

Metabolism The rate of burning of calories by the body.

Mirage An optical phenomenon in which objects appear distorted, displaced, raised or lowered, magnified, multiplied, or inverted due to varying atmospheric refraction which occurs when a layer of air near the earth's surface differs greatly in density from surrounding air.

Multi-year ice Sea ice which has survived at least two summer's melt. Features a light blue colour and pressure ridges that are somewhat rounded and gradual in contour. May be up to 45 m (147.6 ft) thick in ridges but is normally 2–3 m (6.5–9.8 ft) thick in level areas.

NAVSAT Navigational satellite.

Navy/NOAA Joint Ice Centre (JIC) Housed at the Naval Polar Oceanography Centre, Suitland, Maryland. Provides sea ice tracking services, including global sea analysis and forecasting.

New ice A general term for recently formed ice which includes frazil ice, grease ice, slush, and shuga. These types of ice are composed of ice crystals which are only weakly frozen together (if at all) and have a definite form only while they are afloat.

New ridge Ridge newly formed with sharp peaks and slope of sides usually 40°. Fragments are visible from the air at low altitude.

Nilas A thin elastic crust of ice, easily bending on waves and swell and under pressure, thrusting in a pattern of interlocking "fingers." Nilas has a matte surface and is up to 10 cm (3.9 in.) in thickness. It may be subdivided into dark Nilas and light Nilas. See also finger rafting.

Nip Ice is said to nip when it forcibly presses against a ship. A vessel so caught, though undamaged, is said to have been nipped.

Nipped Beset in the ice with the surrounding ice forcibly pressing against the hull.

NOAA National Oceanic and Atmospheric Administration, Washington, D.C.

NPOC Naval Polar Oceanographic Centre, located in Suitland, Maryland.

Old ice Sea ice which has survived at least one summer's melt. Most topographic features are smoother than on first-year ice. Old ice may be subdivided into second-year ice and multi year ice.

Open ice Floating ice in which the concentration is 4/10 to 6/10, with many leads and polynyas; the floes are generally not in contact with one another.

Open pack ice Pack ice in which the concentration is 4/10 to 6/10, with many leads and polynyas, and the floes generally not in contact with one another.

Open water A large area of freely navigable water in which, sea ice is present in concentrations of less than 1/10. When there is no sea ice present, the area should be termed ice-free, even though icebergs may be present.

Pack-ice Term used in a wide sense to include any area of sea ice, other than fast ice, no matter what form it takes or how it is disposed.

Pancake ice Predominantly circular pieces of ice from 30 cm (11.8 in.) to 3 m (9.8 ft) in diameter, and up to about 10 cm (3.9 in.) in thickness with raised rims due to pieces striking against one another. It may be formed on a slight swell from grease ice, shuga, or slush or as a result of the breaking of ice rind, Nilas, or under severe conditions of swell or waves, of grey ice. It also sometimes forms at some depth, at an interface between water bodies of different physical characteristics, from where it floats to the surface; its appearance may rapidly cover wide areas of water.

Physiologic Being in accord with or characteristic of the normal functioning of a living organism.

Polar regions The regions poleward of the Arctic and Antarctic Circles.

Polynya A non-linear shaped area of water enclosed by ice. Polynyas may contain brash ice and/or be covered with new ice, Nilas, or young ice; submariners refer to these as skylights. Sometimes the polynya is limited on one side by the coast and is called a shore polynya or by fast ice and is called a flaw polynya. If it recurs in the same position every year, it is called a recurring polynya.

Propulsion machinery output The propulsion machinery Output, *P*, is the maximum output in kW that the machinery can continuously deliver to the propeller(s). If the output is restricted by technical means or by any regulations applicable to the vessel, *P* is to be taken as the restricted output. If additional power sources are available for propulsion power (e.g., shaft motors), in addition to the power of the main engine(s), they shall also be included in the total propulsion machinery output. The propulsion machinery output used for the calculation of the hull scantlings shall be clearly stated on the shell expansion drawing.

Puddle An accumulation of melt-water on ice, mainly due to the melting of snow, but in the more advanced stages also to the melting ice. Initial stage consists of patches of melted snow.

Rafted ice A type of deformed ice formed by one piece of ice overriding another.

Rafting Pressure processes whereby one piece of ice overrides another. Most common in the new, and young ice.

Ram An underwater ice projection from an ice wall, ice front, iceberg, or floe. Its formation is usually due to a more intensive melting and erosion of the unsubmerged part.

Raynaud's sign A bluish discoloration of the extremities; can occur when a spasm of the blood vessels is caused by exposure to cold or by strong emotion.

Recurring Polynya A polynya which recurs in the same position every year.

Respiration Water lost when cold, dry air enters the lungs, heats up, picks up moisture and is breathed out.

Ridge A line or wall of broken ice forced up by pressure. May be fresh or weathered. The submerged volume of broken ice under a ridge, forced downwards by pressure, is termed an ice keel.

Ridged ice zone An area in which much ridged ice with similar characteristics has formed.

Ridged ice Ice piled haphazardly one piece over another in the form of ridges or walls. Usually found in first-year ice.

Ridging The pressure process by which sea ice is forced into ridges.

River ice Ice formed on a river, regardless of observed location.

Rotten ice Sea ice which has become honeycombed and is in an advanced state of disintegration.

Rubble field An area of extremely deformed sea ice of unusual thickness formed during the winter by the motion of pack-ice against, or around, a protruding rock, islet or other obstruction.

Sastrugi, (sing. sastruga) Sharp, irregular ridges formed on a snow surface by wind erosion and deposition. On mobile floating ice, the ridges are parallel to the direction of the prevailing wind at the time they were formed.

SATCOM Satellite communication.

Sea ice Any form of ice found at sea which has originated from the freezing of sea water.

Sea smoke A phenomenon which occurs when very cold air over open water produces steaming on the surface, occasionally to a height of several hundred feet.

Seasonal sea ice zone (SSIZ) Characterised by the periodic presence of ice cover and is not synonymous with the polar region.

Second-year ice Old ice which has survived only one summer's melt. Because it is thicker than first-year ice, it stands higher out of water. In contrast to multi-year ice, summer melting produces a regular pattern of numerous small puddles. Bare patches and puddles are usually greenish-blue.

Shear ridge field Many shear ridges side by side.

Shear ridge An ice ridge formation which develops when one ice feature is grinding past another. This type of ridge is more linear than those caused by pressure alone.

Shearing An area of ice is subject to shear when the ice motion varies significantly in the direction normal to the motion, subjecting the ice to rotational forces. These forces may result in phenomena similar to flaw.

Shore ice ride-up A process by which ice is pushed ashore as a slab.

Shore lead A lead between drift-ice and the shore, or between drift-ice and an ice front.

Shore melt Open water between the shore and the fast ice, formed by melting and/or due to river discharge.

Shore polynya A polynya between drift-ice and the coast, or between drift-ice and an ice front.

Shuga An accumulation of spongy white ice lumps a few centimetres across; they are formed from grease ice or slush and sometimes from anchor ice rising to the surface.

SINS Ship's inertial navigation system.

Skylight From the point of view of the submariner, thin places in the ice canopy, usually less than 1 m (3.2 ft) thick and appearing from below as relatively light, translucent patches in dark surroundings. The under-surface of a skylight is normally flat. Skylights are called large if big enough for a submarine to attempt to surface through them (120 m (393 ft), or small if not.

Slush Snow that is saturated and mixed with water on land or ice surfaces, or as a viscous floating mass in water after a heavy snowfall.

Small blindness Occurs when the retina of the eye is burned by infrared or ultraviolet rays reflected from ice and snow.

Small fracture 50 m to 200 m (164–656 ft) wide.

Small ice cake Any relatively flat piece of sea ice less than 2 m (6.5 ft) across.

Small ice-field An ice-field 10 to 15 km (6.2–9.3 mi) across.

Snow barchan A horseshoe shaped snowdrift with the ends pointing downwind.

Snow covered ice Ice covered with snow.

Snow drift An accumulation of wind-blown snow deposited in the lee of obstructions or heaped by wind eddies. A crescent-shaped snowdrift, with ends pointing downwind, is known as a snow barchan.

Soaking Long-term exposure of the ship to sub-zero temperatures and near freezing seas. The ship is soaked when it reaches equilibrium with its environment.

Spray icing Occurs at air temperature below freezing when the spray of seawater hitting the ship's surfaces freezes and creates a shell of ice.

Standing floe A separate floe standing vertically or inclined and enclosed by rather smooth ice.

Stranded ice Ice which has been floating and has been deposited on the shore by retreating high water.

Strip Long narrow area of floating ice, about 1km (0.62 mi) or less in width, usually composed of small fragments detached from the main mass of ice and run together under the influence of wind. swell. or current.

Tabular iceberg A flat-topped iceberg with length-to-height ratio greater than 5:1. Most tabular icebergs form by calving from an ice shelf and show horizontal banding.

Thaw holes Vertical holes in sea ice formed when surface puddles melt through to the underlying water.

Thick first-year ice First-year ice over 120 cm (47.2 in) thick.

Thin first-year ice/white ice First-year ice 30–70 cm (11.8–27.5 in.) thick. May sometimes be subdivided into first stage, 30–50 cm (11.8–19.6 in) thick, and second stage, 50–70 cm (19.6–27.5 in.) thick.

Tide crack Crack at the line of junction between an immovable ice-foot or ice wall and fast ice, the latter subject to the rise and fall of the tide.

Tongue A projection of the ice edge up to several kilometres in length, caused by wind or current.

Top heavy The result of topside weight increase, resulting in the ship's centre of gravity being raised.

Trans polar drift Ice flowing form the Pacific Gyre, carried from the eastern Siberian Sea across the pole and generally down the eastern coast of Greenland.

Trench foot Also known as immersion foot, is a cold injury to the feet (and possibly the hands) resulting from the prolonged exposure to dampness and temperatures below 20 °C (68 °F).

Triglyceride A naturally occurring ester of three fatty acids and glycerol that is the chief constituent of fats and oils.

UNREP Underway.

Upper and lower ice waterlines The upper and lower ice waterlines upon which the design of the vessel has been based are to be indicated on the Certificate of Classification. The upper ice waterline (UIWL) is to be defined by the maximum drafts fore, amidships and aft. The lower ice waterline (LIWL) is to be defined by the minimum drafts fore, amidships and aft. The lower ice waterline is to be determined with due regard to the vessel's ice-going capability in the ballast loading conditions. The propeller is to be fully submerged at the lower ice waterline.

Urea A white crystalline or powdery compound synthesised from ammonia and carbon dioxide used to help remove or prevent the formation of ice.

Ventricle The chamber on the right side of the heart that receives venous blood from the right atrium and forces it into the pulmonary artery; any of the interconnecting cavities of the brain.

VERTREP Vertical replenishment.

Very close ice Floating ice in which the concentration is 9/10 to less than 10/10.

Very close pack ice Pack ice in which the concentration is 9/10 to less than 10/10.

Very open ice Floating ice in which the concentration is 1/10 to 3/10 and water preponderates over ice.

Very small fracture 1–50 m (3.2–164 ft) wide.

Very weathered ridge Ridge with peaks very rounded, slope of sides usually 20° to 30°.

Viscosity Fluid flow resistance.

Water sky Dark streaks on the underside of low clouds indicating the presence of water features in the vicinity of sea ice.

Weathered ridge Ridge with peaks slightly rounded and slope of sides usually 30° to 40°. Individual fragments are not discernible.

Weathering Processes of ablation and accumulation which gradually eliminate irregularities in an ice surface.

Whiteout A phenomenon which occurs when snow obliterates surface features, and the sky is covered with a uniform layer of cirro stratus or alto stratus clouds so that there are no shadows. The horizon disappears and earth and sky blend together, forming an unbroken expanse of white without features.

Wind chill A still-air temperature that would have the same cooling effect on exposed human flesh as a given combination of temperature and wind speed; called also chill factor, wind chill factor, wind chill index.

WMO World Meterological Organisation.

Young coastal ice The initial stage of fast ice formation consisting of Nilas or young ice. Its width varying from a few metres up to 100–200 m (328–656 ft) from the shoreline.

Young ice Ice in the transition stage between Nilas and first-year ice, 10–30 cm (3.9–11.8 in.) in thickness. May be subdivided into grey ice and grey-white ice.

References

1. Arctic and Antarctic Research Institute weekly regional ice charts for Eurasian seas in vector WMO SIGRID-3 format. Digital media. http://wdc.aari.ru/datasets/d0004/
2. BASIS—A data bank for Baltic sea ice and sea surface temperatures. http://www.trafi.fi/filebank/a/1352716465/eb4c1ad7b437abb6da7338cbac5e5544/10716-No_34_BASIS_-A_data_bank_for_Baltic_sea_ice_and_sea_surf.pdf,
3. Bercha, F. G. (2005). R*ecent developments in Arctic EER (escape, evacuation and rescue)*. The Bercha Group.
4. BIMCO. (2005). *Ice handbook*. Jjorn Tomsen A/S.
5. Canadian Coast Guard. (2012). Ice navigation in Canadian waters.
6. Canadian Ice Service weekly regional ice charts in vector WMO SIGRID-3 format. Digital media. ftp://sidads.colorado.edu/DATASETS/NOAA/G02171/, http://wdc.aari.ru/datasets/d0031/
7. Finnish Maritime Administration. (2005). Winter Navigation 2004–2005.
8. Harrington, R. L. (Ed.). (1992). *Marine engineering*. Society of Naval Architects and Marine Engineers.
9. Historical Arctic and Antarctic Research Institute10-days period Arctic Ocean ice charts for1933–1992 period in WMO SIGRID format. Digital media. http://wdc.aari.ru/datasets/d0001/north/aari/
10. Historical Baltic Sea ice services 3–4 days periodicity ice charts for 1960–1979 period in WMO SIGRID format, based on BASIS data. Digital media. http://wdc.aari.ru/datasets/d0001/north/bsim/
11. Historical Canadian Ice Service weekly regional ice charts for 1968–1998 period in WMO SIGRID format. Digital media. http://wdc.aari.ru/datasets/d0001/north/cis/
12. IACS. (2011). UR I1—Polar class descriptions and application; UR I2—Structural requirements for polar class ships; UR I3—Machinery requirements for polar class ships (effective for vessels contracted for construction after 1 July 2007).
13. Ice Chart Colour Code Standard. JCOMM Technical Report Series No. 24, 2004, WMO/TD-No. 1215. http://jcomm.info/index.php?option=com_oe&task=viewDocumentRecord&docID=49
14. IMO Guidelines for Ships Operating in Polar Waters (2010) (Resolution A.1024(26)).
15. IMO Life Saving Appliances (2003).
16. IMO MEPC.163(56) Guidelines for Ballast Water Exchange in the Antarctic Treaty Area.
17. IMO MSC Circ. 1175 "Guidance on Shipboard Towing and Mooring Equipment".
18. IMO MSC Circular 1056/MEPC Circular 399 (2002). Guidelines for Ships Operating in Arctic Ice-Covered Waters.

A. Olsen, *Ship Operations in Extreme Low Temperature Environments*, Springer Series on Naval Architecture, Marine Engineering, Shipbuilding and Shipping 19,
https://doi.org/10.1007/978-3-031-52513-1

19. IMO MSC.1/Circ.1185/Rev. 1 (2012), Guide for Cold Water Survival.
20. IMO Resolution A.749(18) (1993) and updates (2002), Code on Intact Stability for All Types of Ships Covered by IMO Instruments.
21. IMO Resolutions MSC.207(81), MSC.218(82) and MSC.272(85) "Adoption of Amendments to the International Life-Saving Appliance (LSA) Code".
22. IMO. (2002). MSC.152(78), Adoption of amendments to the International Convention for the Safety of Life at Sea, 1974, as amended. 5-20-04 (Lifeboats, EPIRB's, Immersion Suits).
23. IMO. MSC. 385(94), International Code for Ships Operating in Polar Waters (Polar Code) Kitagawa, H., et al. (2000). *The northern sea route—The shortest sea route linking East Asia and Europe.* Ship & Ocean Foundation.
24. International Labour Organisation. (2001). *Ambient factors in the workplace.* Geneva.
25. Lamb, T. (Ed.). (2004). Ship design and construction. In *Ice-capable ships* (Vol. II, Chap. 40). Society of Naval Architects and Marine Engineers.
26. MacKinnon, S. (2006). *Operational and human factors in cold maritime environments.* Memorial University.
27. Mahmood, M., & Revenga, A. (2006). *Design aspects of winterized and Arctic LNG carriers—A classification perspective.* American Bureau of Shipping.
28. Ministry of Transport of the Russian Federation. (1996). Guide to navigation though the northern sea route.
29. Oil Companies International Marine Forum. (2010). The Use of large tankers in seasonal first-year ice and severe sub-zero conditions.
30. Oil Companies International Marine Forum. (2014). Offshore Vessel operations in ice and/or severe sub- zero temperatures.
31. Parent, O., & Ilinca, A. (2011). Anti-icing and de-icing for wind turbines: Critical review. *Cold Regions Science and Technology, 65*(1), 88–96.
32. Russian Maritime Register of Shipping. (2014). *Rules for the classification and construction of sea-going ships.* St. Petersburg.
33. Ryerson, C. (2009). *Assessment of superstructure ice protection as applied to offshore oil operations safety: Ice protection technologies, safety enhancements, and development need.* U.S. Army Cold Regions Research and Engineering Laboratory. ERDC-CRREL TR-09-4, 342.
34. Ryerson, C. (2011). Ice protection of offshore platforms. *Cold Regions Science and Technology, 65*(1), 97–110.
35. Ryerson, C. (2013). *Icing Management for coast guard assets.* U.S. Army Cold Regions Research and Engineering Laboratory. ERDC-CRREL TR-13-7, 288.
36. Santos-Pedro, V. M. (2005). *Ice Regime system, exporting Canada's experience.* Transport Canada.
37. Sea Ice information services in the world. WMO/TD, 2010. No. 574. http://jcomm.info/index.php?option=com_oe&task=viewDocumentRecord&docID=96073
38. SIGRID format for Gridded Sea Ice Data. WMO Commission for Marine Meteorology Abridged final report of the tenth session. Paris, February 8–17, 1989.
39. SIGRID-3: A Vector Archive Format for Sea Ice Charts. JCOMM Technical Report Series No. 24, 2004, WMO/TD-No. 1214. http://jcomm.info/index.php?option=com_oe&task=viewDocumentRecord&docID=4
40. Toomey, P. R. M., Lloyd, M., & House, D. J. (2010). The ice navigation manual Swedish Maritime Administration (2005). Winter Navigation 2004–2005.
41. Transport Canada. (2011). Winter navigation on the river and Gulf of St. Lawrence: Practical Notebook for Marine engineers and Deck Officers. TP 14335 E.
42. Tsoy, L. G. (2002). R*eport on Current regulations and practices and impact of IMO Regulations).* Central Marine Research and Design Institute.
43. Tunik, A. (1998). *Major factors in tanker design for oil transportation from Arctic regions.* American Bureau of Shipping.
44. US National Ice Center Arctic sea ice charts and climatologies for 1972–2007 period in gridded format. http://nsidc.org/data/g02172.html, http://wdc.aari.ru/datasets/d0001/north/nic/ease-grid/

45. Walsh and Chapman Northern Hemisphere Sea Ice Data Set. http://www.cgd.ucar.edu/cas/guide/Data/walsh.html
46. WMO No. 716. http://wdc.aari.ru/wmo/docs/sigrid/WMO_716_rec_11%28SIGRID%29.pdf
47. Yang, M., Khan, F., Lye, L., Sulistiyono, H., Dolny, J., & Oldford, D. (2013). Risk-based winterisation for vessels operations in Arctic environments. *Journal of Ship Production and Design, 29*(4), 199–210.

The manufacturer's authorised representative in the EU is Springer Nature Customer Service Centre GmbH, Europaplatz 3, 69115 Heidelberg, Germany. If you have any concerns regarding our products, please contact ProductSafety@springernature.com

Printed and bound by CPI Group (UK) Ltd, Croydon, CR0 4YY
01/10/2026
02225900-0019